Focus on Structural Biology

Volume 9

Series editor
Robert Kaptein, Bijvoet Center for Biomolecular Research, Utrecht University, The Netherlands

More information about this series at http://www.springer.com/series/5990

Jiapu Zhang

Molecular Structures and Structural Dynamics of Prion Proteins and Prions

Mechanism Underlying the Resistance to Prion Diseases

Springer

Jiapu Zhang
School of Sciences Information
Technology and Engineering
The Federation University
Ballarat, VIC, Australia

ISSN 1571-4853
Focus on Structural Biology
ISBN 978-94-017-7317-1 ISBN 978-94-017-7318-8 (eBook)
DOI 10.1007/978-94-017-7318-8

Library of Congress Control Number: 2015949386

Springer Dordrecht Heidelberg New York London

Printed on acid-free paper

Springer Science+Business Media B.V. Dordrecht is part of Springer Science+Business Media (www.springer.com)

To my parents Chuanting Zhang
and Xian'ai Liu

Preface

Studies conducted over the past few decades have demonstrated that rabbits are immune to prion diseases (also known as transmissible spongiform encephalopathies (TSEs)). Unlike conventional infectious diseases – which require that a microorganism bring DNA, RNA or both into the body – prion diseases can be caused by the body's own proteins only (without DNA, RNA or both). Mostly, they attack the brain and other neural tissue and are currently untreatable and universally fatal. High temperature 120∼130 °C for 4 h still cannot kill diseased prions. Creutzfeldt–Jakob disease (CJD), Gerstmann–Sträussler–Scheinker (GSS) syndrome, fatal familial insomnia (FFI) and kuru are the main prion diseases affecting humans. The widely known "mad cow" disease (bovine spongiform encephalopathy (BSE) in cattle), scrapies ("la tremblante" in French language) in sheep and goats, transmissible mink encephalopathy (TME), feline spongiform encephalopathy (FSE), chronic wasting diseases (CWDs) in elks and deers, exotic ungulate encephalopathy (EUE), etc. are also diseases of prion nature. These conditions are degenerative, with death occurring several months to several years after infection. Prion contains no nucleic acids , and it is a misshapen protein that acts like an infectious agent. There is no cure and potential treatments are highly speculative. However, a great deal of effort is being expended to find out why rabbits are immune to prion diseases. From 2005 to 2012, molecular biologists had already identified the nuclear magnetic resonance (NMR) and X-ray structures of rabbit prion protein and its mutants. In addition rabbits, horses, dogs and buffaloes are immune to the degenerative conditions too, and by the end of 2013, except for buffalo their NMR molecular structures were identified too. The author has simulated how these proteins (including those of humans, mice and elks) change shape as their temperature and pH change. This book reports the clear change of rabbit prion protein structure in higher temperature and low pH conditions and concludes that salt bridges are clearly contributing to the structural stability of rabbit prion protein so that this secret revealed could help in developing a treatment to prion diseases. Part I of this book will do the molecular dynamics (MD) simulations on all the species mentioned above.

X-ray crystallography is a powerful tool to determine the protein's three-dimensional (3D) structure. However, it is time-consuming and expensive, and not all proteins can be successfully crystallised, particularly membrane proteins. Although NMR spectroscopy is indeed a very powerful tool in determining 3D structures of membrane proteins, it is also time-consuming and costly. Due to the unstable, noncrystalline and insoluble nature of some proteins, the technologies such as X-ray crystallography, NMR spectroscopy, dual polarisation interferometry, etc. cannot produce the 3D structures of these proteins. In this condition, molecular modelling (MM) might have been the only strategy to get the 3D structures. Neurodegenerative diseases are amyloid fibril diseases. Amyloid fibrils are unstable, noncrystalline and insoluble. This book will also do MM for prion amyloid fibrils from the point of view of global optimisation (GO) algorithm computations. The MM works will be done in Part II of this book.

The book is ideal for researchers of normal prion proteins and diseased prions as well as for teachers in biochemistry, biophysics, bioinformatics, material sciences and engineering, operations research, applied mathematics, theoretical physics, etc.

Acknowledgments: All the MD computational results of this book were calculated on the Tango machines of VPAC (Victorian Partnership for Advanced Computing, www.vpac.org) and the Tambo and Merri machines of VLSCI (Victorian Life Sciences Computation Initiative, www.vlsci.org.au) of Victoria, Australia. I am very grateful to the staff of VPAC, VLSCI, Melbourne, Ballarat and Swinburne Universities for all their great help and support to my projects numbered pBall0004, pMelb0062 and VR0063. I gratefully acknowledge the help on molecular dynamics skills from colleagues Drs. Titmuss SJ and Zhang JF and on global/local optimisation skills from Professor Bagirov AM et al. I wish to thank all those who have helped me in one way or another. This book systematically united my research results of recent years but without any research fund support; thus, I am in debt to my family (especially to my wife Vanessa) and acknowledge them deeply.

Melbourne, Australia Jiapu Zhang
June 2015

Contents

About the Author

Dr Jiapu Zhang was born in 1971 in Fei County, Shandong, P.R. China, in a family inclined to the field of medicine. *Firstly, Dr Zhang received formal education from universities and research organisations.* He got his bachelor's degree in 1993 and 1st master's degree in research in 1996, both from Qufu Normal University, P.R. China; 2nd master's degree in research in 2000 from the National University of Singapore, Republic of Singapore; and a PhD degree in 2005 from Federation University, Australia. During 1996–1998, he lectured in Shandong University (South Campus) as an assistant professor. During 2005–2008, he finished his formal postdoctoral fellowship training from CSIRO (Commonwealth Scientific and Industrial Research Organisation), Australia. *Secondly, Dr Zhang also received great honours from the world and Australia on his prion research.* In 2011 the report of MIT (Massachusetts

Institute of Technology) Technology Review on his prion results caused worldwide impact so that formal organisations such as Physics Society of Iran, Moscow Technological Institute, etc. reported his prion results in seven languages. In recent years, the NHMRC (National Health and Medical Research Council) of Australia has assigned him a review task of prion projects every year. In 2006 he once won one CSIRO Award (together with Drs. Varghese JN and Epa VC). All these offered Dr Zhang great honours in prion research field.

Acronyms

3D	Three-dimensional
$\alpha 1$	α-helix 1
$\alpha 2$	α-helix 2
$\alpha 3$	α-helix 3
$\beta 1$	β-strand 1
$\beta 2$	β-strand 2
BufPrP	Buffalo prion protein
BSE	Bovine spongiform encephalopathy
CDT	Canonical dual theory
CG	Conjugate gradient
CJD	Creutzfeldt–Jakob disease
COM	Centre-of-mass
CWD	Chronic wasting disease
DG	Discrete gradient
DGP	Distance geometry problem
DogPrP	Dog prion protein
DSSP	Database/Dictionary of Secondary Structures of a Protein
DWP	Double well potential
ElkPrP	Elk prion protein
EM	Energy minimisation
ESP	Electrostatic potential
EUE	Exotic ungulate encephalopathy
fCJD	Familial Creutzfeldt–Jakob disease
FFI	Fatal familial insomnia
FSE	Feline spongiform encephalopathy
GA	Genetic algorithm
GO	Global optimisation
GSS	Gerstmann–Straussler–Scheinker syndrome
H1	α-helix 1
H2	α-helix 2
H3	α-helix 3

HB	Hydrogen bond
HoPrP	Horse prion protein
HYD	Hydrophobic interaction
iCJD	Iatrogenic Creutzfeldt–Jakob disease
LJ	Lennard–Jones
MC	Monte Carlo
MCMC	Markov chain Monte Carlo
MDGP	Molecular distance geometry problem
MD	Molecular dynamics
MM	Molecular modelling
MoPrP	Mouse prion protein
HuPrP	Human prion protein
LBFGS	Limited-memory Broyden–Fletcher–Goldfarb–Shanno
NMR	Nuclear magnetic resonance
NPT	Constant number of particles, pressure and temperature
NVT	Constant number of particles, volume and temperature
PDB	Protein data bank
PMEMD	Particle mesh Ewald molecular dynamics
PRNP	Prion protein gene
PrP	Prion protein
PrP^{C}	Normal cellular prion protein
PrP^{Sc}	Misshapen diseased infectious prion
QM/MM	Quantum mechanics and molecular mechanics
PSO	Particle swarm optimisation
RaPrP	Rabbit prion protein
RE	Replica exchange
REMD	Replica exchange molecular dynamics
RMS	Root mean square
RMSD	Root mean square deviation
S1	β-strand 1
S2	β-strand 2
SA	Simulated annealing
SANDER	Simulated annealing with Nmr-derived energy restraints
saPMCA	Serial automated protein misfolding cyclic amplification
SASA	Solvent-accessible surface area
SB	Salt bridge
sCJD	Sporadic Creutzfeldt–Jakob disease
SD	Steepest descent
SDCG	Steepest descent conjugate gradient
SE	Spongiform encephalopathy
SEC	Surface electrostatic charge
SPRN	Shadoo gene
TME	Transmissible mink encephalopathy
TS	Tabu search
TSE	Transmissible spongiform encephalopathy

vdW	van der Waals
vCJD	Variant Creutzfeldt–Jakob diseases
VLSCI	Victorian Life Sciences Computation Initiative
VMD	Visual molecular dynamics
VPAC	Victorian Partnership for Advanced Computing

Chapter 1
Basic Knowledge

1.1 Protein Structures and Functions

A protein is a macromolecule consisted of about at least 20 or 30 amino acids. The definition of protein is very different from the definition of peptide (even polypeptide): protein usually has function and structure, but peptide or polypeptide has no a stable three-dimensional structure.

In the world in all there are about 20 standard amino acids: (named in 3-letter) Ala, Asp, ..., (or named in one-letter A, D, ...). Each amino acid contains a central carbon atom ($C\alpha$) with a hydrogen atom (H), an amine group (H_3N), a carboxylic acid group (COOH) and a side chain (R) that varies between different amino acids. The key elements of an amino acid are carbon, hydrogen, oxygen, and nitrogen, each amino acid has four backbone atoms C, CA, O, N (www.johnkyrk.com/aminoacid.html). Corresponding to each amino acid there is/are its/their DNA or RNA codes, for example, Ala has GCT, GCC, GCA, GCG in all four DNA codons or in all four RNA codons GCU, GCC, GCA, GCG. The side chain determines the property of each amino acid, e.g. strictly hydrophobic: Ala (A), Val (V), Leu (L), Ile (I), Phe (F), Pro (P), Met (M); charged: Asp (D), Glu (E), Lys (K), Arg (R); polar: Ser (S), Thr (T), Cys (C), Asn (N), Gln (Q), His (H), Tyr (Y), Trp (W); Gly (G) has one hydrogen atom as its side chain (R) with higher degree of freedom thus Gly (G) is commonly associated with turns in a protein secondary structure and sometimes acts as a helix breaker.

The amino acids in a protein are joined together by the (so-called) peptide bonds between the carboxyl and amino groups of adjacent amino acid residues, when the carboxyl group of one amino acid reacts with the amino group of another amino acid, releasing a molecule of water (H_2O), and (at last) each protein has a N-terminal and a C-terminal.

Each protein has primary structure (i.e. amino acid sequence), secondary structure (i.e. α-Helix, π-Helix, 3_{10}-Helix, β-Sheet, Turn, Bend), and 3-dimensional tertiary and quaternary structures. The primary structure of a protein is held together

J. Zhang, *Molecular Structures and Structural Dynamics of Prion Proteins and Prions*, Focus on Structural Biology 9, DOI 10.1007/978-94-017-7318-8_1

by polypeptide covalent bonds. The tertiary structure is the 3D conformation in space of a protein. A quaternary structure is formed when proteins behave in a collective manner. Here we introduce secondary structure in details as in [662]. The protein backbone is composed of repeated atoms –N–Cα–C–. The ϕ, ψ and ω torsion angles govern the conformation of the protein backbone. The ϕ angel is the rotation about the N–Cα bond (angle subtended by the two planes formed by C–N–Cα–C). The ψ angle is the rotation about the Cα–C bond (angle subtended by the two planes formed by N–Cα–C–N). The ω angle is the rotation about the C–N bond (angle subtended by the two planes formed by Cα–C–N–Cα). Usually the peptide plane is formed by ω with an angle close to 180°, therefore the two torsion angles ϕ and ψ are usually used and we denote (ϕ, ψ) as the dihedral angles. The secondary structure also refers to the ordered local structural features observed in proteins. These can be classified into several main categories: helices, sheets, turns and loops. *Helices* are formed when a sequence of amino acids adopts a winding spiral motif which is stabilized by the formation of hydrogen bonds (HBs) between atoms of the protein backbone. Due to the helix dipole moment created by the slight negativity of the carboxyl oxygen of each residue, helices are typically quite strong. 3_{10}-helices, α-helices and π-helices are three commonly observed types of helices: 3_{10}-helices have HBs between residues i and $i + 3$ with (ϕ, ψ) dihedral angles of (−74.0, −4.0), α-helices have HBs between residues i and $i + 4$ with (ϕ, ψ) dihedral angles of (−57.8, −47.0), and π-helices have HBs between residues i and $i + 5$ with (ϕ, ψ) dihedral angles of (−57.1, −69.7). *β-sheets* are composed of a sequence of elongated peptide chains which are stabilized by the formation of HBs between backbones, perpendicular to the plane of the strands. The HBing in β-sheets is weaker than the HBs in helices. *Turns and loops* enable a protein to adopt a wider range of conformation which can be often related to protein activity; e.g. they allow the protein to fold into compact tertiary structures that can minimize the solvent accessible surface area (SASA).

Protein is a macromolecule. A molecule is consisted of a network of atoms by the peptide bonds, covalent bonds (e.g. disulfide bonds), and noncovalent bonds such as HBs, SBs (salt bridges), van der Waals (vdW) contacts, and hydrophobic interactions (HYDs). A HB usually has an angle of 150∼180° between the donor-hydrogen-acceptor atoms and a donor-acceptor atoms' distance no further than 3 or 3.2 or 3.5 Å. A SB is considered to be formed if the distance between any of the O atoms of acidic residues (Asp, Glu) and the N atoms of basic residues (Arg, His, Lys, Hsp) are within the cutoff distance (usually less than 4 Å). A HYD, for example, can be seen between aromatic side chains. Each atom consists of negatively charged electron clouds, positively charged protons and electrically neutral neutrons. Protein 3D structure is usually determined by X-ray, NMR, etc.

A protein usually has some functions. The way a protein folds is absolutely essential to determine its function. There are many such molecules which resemble each other and differ only in how they bend, but their functions are totally different. The S2-H2 loop has been reported to play an important role to stabilize the structural stability of rabbit and horse prion proteins [56, 166, 206, 329, 335, 336, 390, 463,

557, 558, 625, 626]. In this loop, prion proteins differ from each other only in one or two residues, but how these proteins bend at this loop might produce different functions for different species. The author recommends books such as [45, 65] to readers to know more knowledge about protein structures and functions.

1.1.1 Prion Proteins and Prions

A *prion protein* (called PrP) is a normal, healthy protein. PrP is everywhere in the brain keeping nerves happy. PrPs are encoded by a chromosomal gene and the product of the PrP gene in normal animals is the cellular prion protein PrP^C [543] found on the membranes of cells. PrP^C is a molecular weight of 35–36 kDa and a mainly α-helical structure. PrP^C(1–123) is unstructured, but PrP^C(124–228) is well structured, with three α-helices and a two short antiparallel β-strands (denoted as H1, H2, H3, S1 and S2). PrP^C has a NH2-terminal tail (containing a signal peptide at 1–23 and five or six amino-terminal octapeptide repeats with sequence PHGGGWGQ), a short COOH-terminal tail, a glycosylphosphatidylinositol (GPI) (at sites Asn180 and Asn196) membrane anchor at the COOH-terminal tethers PrP to cell membranes, a disulfide bond between Cys178 and Cys213 linking H2 and H3, and a highly conserved (central) hydrophobic region in 109–136, etc.

A *prion* (the name is derived from the words protein and infection) is an infectious agent composed of protein in a misfolded form [506], which does not contain nucleic acids (either DNA, RNA, or both). Prions are misshapen proteins (called PrP^{Sc}) with a mainly β-sheet structure.

Some interesting structural characters of PrPs are listed as follows.

- *Sheep and goat PrP*. Sheep PrP polymorphisms of the PrP-VRQ form and PrP-ARQ form are associated with increased susceptibility, whereas PrP-ARR is associated with resistance (however PrP-ARR polymorphisms are susceptible to atypical scrapie), PrP27-30 is the necessary part causing scrapies. Q171 in sheep PrP is R171, and this position is a easy infection point (where 177-Glu-Glu will be easier than 177-Glu-Arg or 177-Arg-Arg) and this residue will affect the species barrier between sheep and others (at 136(A/V), 154(R/H) and 171(R/H/Q) there are 5 prion diversities ARQ, ARR, AHQ, ARH and VRQ [40, 197, 492]). At 108 and 189 there are associated polymorphisms [433]. Let142Met of goat PrP will immunize against scrapies. Polymorphisms at sites 136, 154, and 171 are associated with varying susceptibility to scrapie. The polymorphisms of goat PrP happen at residues 21, 23, 49, 142, 143, 154, 168, 220, 240 [706].
- *Human PrP*. In human PrP, D178N, E200K will cause CJD. P102L, A117V, F198S will cause GSS. M129V, M129 is easy to be infected by PrP^{Sc}. 19 % mutations are in structured region. Special positions for species barrier from PrP^C to PrP^{Sc}: Q186, I184, I203, I205 are associated with the transference efficiency N155, M139, M138, R148, W145, N143 are associated with different species,

Q172, Q168, T215, Q219 are the binding positions with protein X [487]. N181 and N197 are two points of N-linked glycosylation sites. Asn171Ser, Glu219Lys are associated with GSS diseases.

- *Mouse PrP*. F108 in MousePrP is L108. Muramoto et al. [437] found the knockout of residues 23–88 and 144–147 does not affect the transformation from PrP^C to PrP^{Sc}, but the deletion of 95–107, 108–121, or 122–140 will completely inhibit the transformation [437]. Kuwata et al. [357] found the peptide PrP(106–126) will form into amyloid fibrils [357] and PrP(106–126) sometimes is replaced as a model of PrP^{Sc} studied in vitro. The studies of transgenic (Tg) mice showed PrP(90–140) may be the interaction region of PrP^C and PrP^{Sc}. Pro101Leu, Leu108Met, Val111Met will change the incubation time of mouse PrP. At position 138, Ile of mouse PrP will help the transformation from PrP^C to PrP^{Sc}, but Met of Syrian hamster PrP will block the transformation from PrP^C to PrP^{Sc}. At position 155, mouse is Tyr but Syrian hamster is Asn thus residue 155 can affect the PrP species barrier of mouse and Syrian hamster. Asn99Gly, Leu108Met, Asn173Ser and Val214Ile will clearly disturb the transformation from PrP^C to PrP^{Sc} for mice. In mouse, Syrian hamster PrPs the residue at 145 is Trp.
- *Primate animals*. Monkeys can be infected by BSEs and scrapies, and bovines, monkeys, sheep and goats all have same residues at 101–143, especially at 108, 109, and 112 [706].
- *Special segments*. In all the PrP sequences, 113–128 is the most conserved region. 144–152 is highly conserved in humans, bovines, sheep and goats, rabbits, minks and primate animals etc. 132–156 and 144–152 are acknowledged to play important roles in the PrP conformational change. Paramithiotis et al. [459] reported that the side-chains of Tyr-Tyr-Arg are buried inside the structure of PrP^C but exposed on the molecular structure surface in PrP^{Sc} [459].
- *Ions binding*. Regarding ions binding with PrPs, Turi et al. [575] studied HuPrP(91–115), HuPrP(76–114)H85A and HuPrP(84–114)H96A and found that His96 refers Ni^{2+} and His111 prefers Cu^{2+} [323, 575].

1.2 Protein Structural Dynamics and Molecular Dynamics

To really understand the stability of molecules and their action mechanism, we should consider not only the static X-ray or NMR structures concerned but also the dynamical information obtained by simulating their internal motions or dynamic process. Many marvelous biological functions in proteins and DNA and their profound dynamic mechanisms, such as switch between active and inactive states, cooperative effects, allosteric transition, intercalation of drugs into DNA, and assembly of microtubules, can be revealed by studying their internal motions. Recently, MD simulations have been used to study the switch mechanism of human Rab5a [600], the inhibition mechanism of PTP1B [602], the gating and inhibition mechanism of the M2 proton channel from influenza A viruses [610] based on the NMR structure [470, 519], the personalized drug design [608, 612], the enzyme-ligand binding interaction [611, 613], the binding mechanism of H5N1 influenza virus neuraminidase with ligands [250], the metabolic mechanism [615], and the binding mechanism of calmodulin with chrysin [384], etc. Thus, basic MD techniques are very necessary to be simply introduced in this chapter.

"The MD computational method calculates the time dependent behavior of a molecular system. The MD simulation method is based on Newton's second law or the equation of motion. From a knowledge of the force on each atom, it is possible to determine the acceleration of each atom in the system. Integration of the equations of motion then yields a trajectory that describes the positions, velocities and accelerations of the particles as they vary with time. From this trajectory, the average values of properties can be determined. The method is deterministic; once the positions and velocities of each atom are known, the state of the system can be predicted at any time in the future or the past" (www.ch.embnet.org/MD_tutorial). Setting up and running MD simulations basically consists of Initialization (including Optimization), Heating the system, Equilibration and Production phase, and the analysis of MD simulations basically focuses on root mean square deviation (RMSD) and Radius of Gyration. RMSD and Radius of Gyration are two indicators for structural changes in a protein. The radius of gyration is the mass weighted scalar length of each atom from the center-of-mass (COM). The RMSD is used to measure the scalar distance between atoms of the same type for two structures. For instance, you used the RMSD between Cα atoms to measure the fit between protein homologs. In this calculation, you will use the RMSD of heavy-atoms to compare the spatial deviation between structures in time and the original structure (at time = 0 ps). The author recommends books such as [214] to readers to know more knowledge about MD.

1.3 Molecular Modeling

Basically, MM considers the global optimization of total potential energy (1.1) of a molecule [397]

$$
\begin{aligned}
E_{\text{total}} = & \sum\nolimits_{\text{bonds}} K_r (r - r_{eq})^2 \\
& + \sum\nolimits_{\text{angles}} K_\theta (\theta - \theta_{eq})^2 \\
& + \sum\nolimits_{\text{dihedrals}} \frac{V_n}{2} [1 + \cos(n\phi - \gamma)] \\
& + \sum\nolimits_{i<j}^{\text{vdW}} \left[\frac{A_{ij}}{R_{ij}^{12}} - \frac{B_{ij}}{R_{ij}^{6}} \right]
\end{aligned}
$$

$$+\sum_{i<j}^{\text{electrostatic}}\left[\frac{q_i q_j}{\epsilon R_{ij}}\right]$$

$$+\sum_{\text{HBs}}\left[\frac{C_{ij}}{R_{ij}^{12}}-\frac{D_{ij}}{R_{ij}^{10}}\right]+\ldots \tag{1.1}$$

(where r is the bond length, θ is the bond angle, ϕ is the dihedral angle, γ is the dihedral phase, subscript eq denotes equilibrium of their respective properties, q is the charge of the particle, R_{ij} is the distance between the centers of particles i and j, and all others are constants under a force field and each force field usually defines different constants) and the structural dynamical behavior (so-called MD [214]) of the molecule with propagation of time. In mathematical formula (1.1), "covalent bonds (bonds, angles, dihedrals) are considered to be too strong to be broken or modified at physiological temperatures, so the only degrees of freedom of complex molecules can be considered as those associated with non-bonded pairs (vdW, electrostatic, HBs). This is the reason why in many applications the vdW and electrostatic (and HBs) items are the only ones taken into account in (global) optimization, as all the other terms are assumed to contribute a constant term to the total energy" [397]. The greatest and most successful MM example should be the discovery of double-helical 3D structure of DNA [479]. The author has done MM works for the unstructured region PrP(1–123) of prions, and studied the covalent bonds, angles, dihedrals, HBs, etc. for two copper binding sites (regions PrP(57–89) and PrP(92–96) respectively) of prions in the use of Quantum Mechanics and Molecular Mechanics (QM/MM). In this book the author will do the noncovalent-bond MM works on the AGAAAAGA region (113–120) of prions (other protein aggregation diseases such as Alzheimer's Disease, Huntington's Disease, Parkinson's Disease and frontotemporal dementia can also be better understood by constructing (and reconstructing) amyloid molecular structure models). First we will give brief reviews on the algorithms to minimize the total potential energy (1.1) or the vdW part of it (i.e. the Lennard-Jone (LJ) potential energy).

1.3.1 Algorithms to Minimize the Total Potential Energy and the Lennard-Jones Potential Energy

In molecular mechanics, current generation potential energy functions provide a reasonably good compromise between accuracy and effectiveness. Here we firstly review several most commonly used classical potential energy functions and their optimization methods used for the energy minimization. To minimize a potential energy function, about 95 % efforts are spent on the LJ potential of vdW interactions; we also give a detailed review on some effective computational optimization methods listed in the Cambridge Cluster Database to solve the problem of LJ clusters.

In molecular mechanics, current potential energy functions provide a reasonably good accuracy to structural data obtained from X-ray crystallography and NMR spectroscopy, and dynamic data obtained from spectroscopy and inelastic neutron scattering and thermodynamic data. Currently, AMBER, CHARMM, GROMOS and OPLS/AMBER are among the most commonly used classical potential energy functions [6, 87, 397, 458, 548]. The energy, E, is a function of the atomic positions, R, of all the atoms in the system these are usually expressed in term of Cartesian coordinates. The value of the energy is calculated as a sum of bonded (or internal) terms E_{bonded}, which describe the bonds, angles and bond rotations in a macromolecule, and a sum of non-bonded (or external) long-range terms $E_{non-bonded}$ [6, 87, 548]:

$$\begin{aligned} E_{potential} &= E_{bonded} + E_{non-bonded} \\ &= (E_{bond-stretch} + E_{angle-bend} + E_{rotate-along-bond} \\ &\quad (+E_{Urey-Bradley} + E_{improper} + U_{CMAP})) \\ &\quad +(E_{van-der-Waals} + E_{electrostatic} + E_{hydrogen-bonds}). \end{aligned} \tag{1.2}$$

For example, for AMBER and CHARMM force fields [87, 548] respectively, the potential energy functions are [87]:

$$\begin{aligned} E_{AMBER} = & \sum_{bonds} k_b(b-b_0)^2 + \sum_{angles} k_\theta(\theta-\theta_{eq})^2 + \sum_{dihedrals} (V_n/2)[1+\cos(n\phi-\gamma)] \\ & + \sum_{i=1}^{N-1}\sum_{j=i+1}^{N}\left[\frac{A_{ij}}{R_{ij}^{12}} - \frac{B_{ij}}{R_{ij}^{6}} + \frac{q_i q_j}{\varepsilon R_{ij}}\right], \end{aligned} \tag{1.3}$$

$$\begin{aligned} E_{CHARMM} = & \sum_{bonds} k_b(b-b_0)^2 + \sum_{angles} k_\theta(\theta-\theta_0)^2 + \sum_{dihedrals} k_\phi[1+\cos(n\phi-\delta)] \\ & + \sum_{Urey-Bradley} k_u(u-u_0)^2 + \sum_{impropers} k(\omega-\omega_0)^2 + \sum_{\phi,\psi} V_{CMAP}(\phi,\psi) \\ & + \sum_{nonbonded}\left(\varepsilon\left[\left(\frac{R_{min_{ij}}}{R_{ij}}\right)^{12} - \left(\frac{R_{min_{ij}}}{R_{ij}}\right)^{6}\right] + \frac{q_i q_j}{\varepsilon R_{ij}}\right), \end{aligned} \tag{1.4}$$

where $b, \theta, \phi, R_{ij}, u, \omega, \psi$ are basic variables (b is the bond length of two atoms, θ is the angle of three atoms, ϕ is the dihedral angle of four atoms, R_{ij} is the distance between atoms i and j), and all other mathematical symbols are constant parameters specified in various force fields respectively. This paper will discuss how to effectively and efficiently use computational optimization methods to solve the minimization problem of the potential energy in Eq. (1.2), i.e.

$$min \quad E_{potential}. \tag{1.5}$$

Firstly, for Eq. (1.5), we consider why we should perform energy minimization (EM). There are a number of reasons:

(i) To remove nonphysical (or bad) contacts/interactions. For example, when a structure that has been solved via X-ray crystallography, in the X-ray crystallization process, the protein has to be crystallized so that the position of its constituent atoms may be distorted from their natural position and contacts with neighbors in the crystal can cause changes from the in vitro structure; consequently, bond lengths and angles may be distorted and steric clashes between atoms may occur. Missing coordinates obtained from the internal coordinate facility may be far from optimal. Additionally, when two sets of coordinates are merged (e.g., when a protein is put inside a water box) it is possible that there are steric clashes/overlap presented in the resulting coordinate set (www.charmmtutorial.org).
(ii) In MD simulations, if a starting configuration is very far from equilibrium, the forces may be excessively large and the MD simulation may fail [6].
(iii) To remove all kinetic energy from the system and to reduce the thermal noise in the structures and potential energies [6].
(iv) Re-minimize is needed if the system is under different conditions. For example, in QM/MM one part of the system is modeled in QM while the rest is modeled in MD, re-minimize the system with a new condition is needed (www.charmmtutorial.org).

To perform EM is to make the system reaching to a equilibration state. EM of Eq. (1.5) can be challenging, as there are many local minima that optimization algorithms might get stuck in without finding the global minima – in most cases, this is what will actually happen. Thus, how much and how far we should minimize should be well considered. Over-minimization can lead to unphysical "freezing" of the structure and move too much from its original conformation; if not minimized enough and exactly, for example, the normal mode calculation cannot arrive at the bottom of its harmonic well. However, in MD, because the output of minimization is to be used for dynamics, it is not necessary for the optimization to be fully converged but a few hundreds or tens of local optimization search are good and kind enough. To make enough local optimization, usually, after we put the protein into a solvent (e.g. waters), first we restrain the protein by holding the solute fixed with strong force and only optimize the solvent, next holding the solute heavy atoms only, and then holding the CA atoms only, and lastly remove all restraints and optimize the whole system.

Secondly, for Eq. (1.5), we consider what optimization algorithms we should use. In packages of [6, 51, 87] etc., the following three local search optimization methods have been used.

(i) Steepest descent (SD) method is based on the observation that if the real-valued function $E(x)$ is defined and differentiable in a neighborhood of a point x_0 then $E(x)$ decreases fastest if one goes from x_0 in the direction of the negative gradient of $E(x)$ at x_0. SD method is the simplest algorithm, it simply moves the coordinates in the negative direction of the gradient (hence in the direction of the force – the force is the (negative) derivative of the potential), without consideration of build ups in previous steps – this is the fastest direction making the potential energy decrease. SD is robust and easy to implement. But SD is not the most efficient especially when closer to minimum and in the vicinity of the local minimum. This is to say, SD does not generally converge to a local minimum, but it can rapidly improve the conformation when system is far from a minimum – quickly remove bad contacts and clashes.

(ii) Conjugate gradient (CG) method is a method adds an orthogonal vector to the current direction of optimization search and then moves them in another direction nearly perpendicular to this vector. CG method is fast-converging and uses gradient information from previous steps. CG brings you very close to the local minimum, but performs worse far away from the minimum. CG is slower than SD in the early stages of minimization, but becomes more efficient closer to the energy minimum. In GROMACS CG cannot be used with constraints and in this case SD is efficient enough. When the forces are truncated according to the tangent direction, making it impossible to define a Lagrangian, CG method cannot be used to find the EM path.

(iii) L-BFGS method is a Quasi-Newton method that approximates the reverse of Hessian matrix $[\nabla^2 E(x)]^{-1}$ of $E(x)$ for the Newton method search direction $-[\nabla^2 E(x)]^{-1}\nabla E(x)$. L-BFGS method is mostly comparable to CG method, but in some cases converges 2$\sim$3 times faster with super-linear convergent rate (because it requires significantly fewer line search steps than Polak-Ribiere CG). L-BFGS of Nocedal approximates the inverse Hessian by a fixed number of corrections from previous steps. In practice L-BFGS converges faster than CG.

(iv) The combination of CG and LBFGS, so-called lbfgs-TNCG-BFGS method is a preconditioned truncated Newton CG method, it requires fewer minimization steps than Polak-Ribiere CG method and L-BFGS method, but L-BFGS can sometimes be faster in the terms of total CPU times.

If a global optimization is required, approaches such as simulated annealing (SA), parallel tempering method (super SA, also called replica exchange [688]), Metropolis algorithms and other Monte Carlo methods, Simplex method, Nudged Elastic Band method, different deterministic methods of discrete or continuous optimization etc. may be utilized. The main idea of SA refinement is to heat up the system such that the molecule of interest has enough energy to explore a wide range of configurational space and get over local optimal energy barriers. Relatively large structural rearrangements are permitted at these high temperatures. As the temperature is cooled gradually, the structural changes proceed in smaller steps, continuing to descend toward the global energy minimum.

For solving Eq. (1.5), without considering

$$E_{hydrogen-bonds} = \sum_{i=1}^{N-1}\sum_{j=i+1}^{N}\left[\frac{C_{ij}}{R_{ij}^{12}} - \frac{D_{ij}}{R_{ij}^{10}}\right],$$

about 95 % of the CPU time of calculations is spent at

$$min \quad E_{van-der-Waals} = \sum_{i=1}^{N-1} \sum_{j=i+1}^{N} \left[\frac{A_{ij}}{R_{ij}^{12}} - \frac{B_{ij}}{R_{ij}^{6}} \right], \tag{1.6}$$

where C_{ij}, D_{ij} are constants. In [397], this problem is also called Lennard-Jones (LJ) Atomic Cluster Optimization problem (where within the field of atomic clusters only nonbonded interactions are accounted for and particles are considered to be charge-free; e.g. real clusters of metals like gold, silver, and nickel). It is very necessary to up to date review some effective and efficient computational methods for solving Eq. (1.6):

- Hoare and Pal's work [284–286] may be the early most successful results on LJ problem. The idea is using build-up techniques to construct the initial solutions which are expected to represent low energy states, and using those initial solutions as starting points for a local search method to relax to the optimal solution [285]. The starting seed is the regular unit tetrahedron with atoms at the vertexes, the obvious global optimal solution for $N = 4$. Beginning with this tetrahedron, Hoare and Pal [285, 286] added one atom at a time to construct a sequence of polytetrahedral structures and at last got good results up to $N = 66$ [284–286]. For example, for $N = 5$ its globally optimal trigonal bi-pyramid (bi-tetrahedron) structure is gotten by adding an atom at the tetrahedral capping position over a triangular face; following the bi-tetrahedron structure, the optimal structure of $N = 6$ is tri-tetrahedron (another known optimal structure for $N = 6$ is octahedron (using tetrahedral capping over triangular faces and half-octahedral capping over square faces), which is not a polytetrahedron); for $N = 7$ its best structure constructed is the pentagonal bi-pyramid, a structure with a five-fold axis of symmetry. Many computer science data structure procedures such as greedy forward growth operator and reverse greedy operator can make the build-up technique work well. The application of methods of studying noncrystalline clusters to the study of "spherical" face centred cubic (fcc) microcrystallites was described in [286]. In [284] the chief geometrical features of the clustering of small numbers of interacting particles were described.
- The data structure of Northby [436] in finding the good starting solution is the lattice based structure. The lattice structures consist of an icosahedral core and particular combinations of surface lattice points. A class of icosahedral packings was by constructed in [409] adding successively larger icosahedral shells in layers around a core central atom; this icosahedral lattice can be described as 20 slightly flattened tetrahedrally shaped fcc units with 12 vertices on a sphere centered at the core atom. Atoms within each triangular face are placed in staggered rows in a two dimensional hexagonal close-packed arrangement. Each atom in the interior of a face in a given shell is a tetrahedral capping position relative to three atoms in the underlying shell. Northby [436] relaxed the structure of [409] to get his IC and FC multilayer icosahedral lattice structures [436]. The

IC lattice can be referred to the FORTRAN code in [653]; it consists of all those sites which will comprise the outer shell of the next complete Mackay [409] icosahedron. FC lattice is a slight modification of IC lattice in that its outer shell maintains icosahedral symmetry and consists of points at the icosahedral vertices and the stacking fault positions of the outer IC shell. Basing on the IC and FC lattices, Northby [436] gave his algorithm first finding a set of lattice local minimizers and then relaxing those lattice minimizers by performing continuous minimization starting with those lattice minimizers [436]. The algorithm was summarized as Algorithm 1 and Algorithm 2 of [653].

- The great majority of the best known solutions of Northby [436] are icosahedral in character. The hybridization of global search and local search methods, usually, is more effective to solve the large scale problem than the global search method or local search method working alone. Catching those two ideas, Romero et al. [500] combined a genetic algorithm with a stochastic search procedure on icosahedrally derived lattices [32, 500]. The structures of the optimal solutions gotten in [500] are either icosahedral or decahedral in character. The best results of [632] for N = 82, 84, 86, 88, 92, 93, 94, 95, 96, 99, 100 were gotten by using a genetic algorithm alone. Deaven et al. [172] also using the genetic algorithm got the optimal value known for the magic number $N = 88$ [172].
- The successful works to improve Northby's results in [436] were mainly done by Xue [654, 654], Leary [370], and Doye et al. [190, 191].

  - Xue [654] introduced a modified version of the Northby algorithm. He showed that in some cases the relaxation of the outer shell lattice local minimizer with a worse potential function value may lead to a local minimizer with a better value. In Northby's algorithm [436] the lattice search part is a discrete optimization local search procedure, which makes a lattice move to its neighboring lattice with $O(N^{\frac{5}{3}})$ time complexity. In [654] Xue introduced a simple storage data structure to reduce the time complexity to $O(N^{\frac{2}{3}})$ per move, and then used a two-level simulated annealing algorithm within the supercomputer CM-5 to be able to solve fastly the LJ problem with sizes as large as 100,000 atoms. In [653] by employing AVL trees [290] data structure Xue furthermore reduced the time complexity to $O(\log N)$ if NN (nearest neighbor) potential function is used. Xue [653] relaxed every lattice local minimizer found instead of relaxing only those lattice local minimizers with best known potential function value by a powerful Truncated Newton local search method, and at last got the best results known for $N = 65, 66, 134, 200, 300$.
  - Leary [370] gave a successful Big Bang Algorithm for getting the best values known of $N = 69, 78, 88, 107, 113, 115$. In [370], the FCC lattice structure is discussed and its connections are made with the macrocluster problem. It is also concluded in [370] that almost all known exceptions to global optimality of the well-known Northby multilayer icosahedral conformations for microclusters are shown to be minor variants of that geometry. The Big Bang Algorithm contains three steps: Step 1 is an initial solution generating

procedure which randomly generates each coordinate of the initial solution with the independently normal distribution; Step 2 is to generate the new neighborhood solution by discrete-typed fixed step steepest descent method, which is repeated until no further progress is made; Step 3 is to relax the best solution gotten in Step 2 by a continuous optimization method – conjugate gradient method.

  - Doye et al. [191] investigated the structures of clusters by mapping the structure of the global minimum as a function of both cluster size and the range of the pair potential which is appropriate to the clusters of diatomic molecule, C_{60} molecule, and the ones between them both. For the larger clusters the structure of the global minimum changes from icosahedral to decahedral to fcc as the range is decreased [190]. In [191], Doye et al. predicted the growth sequences for small decahedral and fcc clusters by maximisation of the number of NN contacts.

- Calvo et al. [84] gave some results on quantum LJ Clusters in the use of Monte Carlo methods [84].
- Xiang et al. [639] presented an efficient method based on lattice construction and the genetic algorithm and got global minima for $N = 310 \sim 561$. In 2004, Xiang et al. [638] continued to present global minima for $N = 562 \sim 1000$.
- Barron-Romero [35] found the best solutions for $N = 542 \sim 543, 546 \sim 548$ in the use of a modified peeling greedy search method [35].
- Takeuchi [565] found best solutions for $N = 506$, 521, $537 \sim 538$ and 541 by a clever and efficient method "using two operators: one modifies a cluster configuration by moving atoms to the most stable positions on the surface of a cluster and the other gives a perturbation on a cluster configuration by moving atoms near the center of mass of a cluster" [565].
- Lai et al. [362] found best solutions for $N = 533$ and 536 using the dynamic lattice searching method with two-phase local search and interior operation [362, 363, 661].
- Algorithms to get the structures at the magic numbers $N = 17$, 23, 24, 72, 88 (the exceptions to [500]):

  - Freeman and Doll [213] presented the best value for $N = 17$ when the thermodynamic properties of argon clusters were studied by a combination of classical and quantum Monte Carlo methods. The poly-icosahedral growth of Farges et al. [200] starts from a 13-atom primitive icosahedron containing a central atom and 12 surface atoms [200]. On each one of the five tetrahedral sites, surrounding a particular vertex, a new atom is added and finally a sixth atom is placed on top to create a pentagonal cap. In this way a 19-atom structure being made of double interpenetrating icosahedra, which is a 13-atom icosahedra sharing 9 atoms, is obtained; i.e., for three pentagonal bipyramids each one shares an apex with its nearest neighbour. In this way a 23-atom model consisting of three interpenetrating icosahedra is gotten for the best value known.

- Wille [630] used the SA method yielding low-lying energy states whose distribution depends on the cooling rate to find the best solution known for $N = 24$ [630].
- Coleman et al. [154] proposed a build-up process to construct the optimal solution structures. The HOC (half octahedral cap) structure of the optimal solution for $N = 72$ is found by a prototype algorithm designed using the anisotropic effective energy simulated annealing method at each build-up stage [154].
- Wales and Doye [596] gave the lowest values known for $N = 192, 201$ [596]. Their method is so-called basin-hopping method, in which first the transformed function $\tilde{f}(x) = \min\{f(x)\}$ was defined and performed starting from x by the PR conjugate gradient method and then the energy landscape for the function $\tilde{f}(x)$ was explored using a canonical Monte Carlo simulation.
- Pintér [472] has developed techniques for moving along sequences of local minima with decreasing energies to arrive at good candidates for global optima and got the best value known on $N = 185$.

Thus, we finished the above outline of some successful optimization algorithms used to solve Eqs. (1.5)–(1.6).

Part I
Molecular Dynamics of Prion Proteins

Introduction

Prion diseases are invariably fatal and highly infectious neurodegenerative diseases affecting humans and animals. The neurodegenerative diseases such as CJD, vCJD, iCJD, fCJD, sCJD, GSS, FFI, KURU in humans, Scrapie in sheep, BSE or 'mad-cow' disease in cattle, CWD in white-tailed deer, elk, mule deer, moose, TME in mink, FSE in cat, EUE in nyala, oryx, greater kudu, and SE in ostrich etc belong to prion diseases. By now there have not been some effective therapeutic approaches or medications to treat all these prion diseases [8, 486, 624].

Rabbits are one of the few mammalian species reported to be resistant to infection from prion diseases isolated from other species [46, 205, 331, 398, 432, 441, 505, 542, 591]. Thus, the book studies rabbit prion protein in order to get some clues for the treatment of prion diseases.

Prion diseases are "protein structural conformational" diseases. The normal cellular prion protein (PrP^C) is rich in α-helices but the infectious prions (PrP^{Sc}) are rich in β-sheets amyloid fibrils. The conversion of PrP^C to PrP^{Sc} is believed to involve a conformational change from a predominantly α-helical protein (about 42 % α-helix and 3 % β-sheet) to a protein rich in β-sheets (about 30 % α-helix and 43 % β-sheet) [85, 169, 257, 447, 457, 491]. Fortunately, the X-ray and NMR protein structures of rabbit prion protein and its S173N and I214V mutants were released into protein data bank (www.rcsb.org) (PDB ID codes 3O79, 2FJ3, 2JOH, 2JOM) recently. Hence, the conformational changes may be amenable to study in this book by MD techniques of these X-ray and NMR structures.

A prion (the name is derived from the words protein and infection) is an infectious agent composed of protein in a misfolded form [506], which does not contain nucleic acids (either DNA, RNA, or both). Prions are misshapen proteins (called PrP^{Sc}).

A prion protein (called PrP) is a normal, healthy protein. PrP is everywhere in the brain keeping nerves happy. PrPs are encoded by a chromosomal gene and the product of the PrP gene in normal animals is the cellular prion protein PrP^C [543]

found on the membranes of cells. PrP^C has 228 amino acids (in rabbits), a molecular weight of 35–36 kDa and a mainly α-helical structure. PrP^C(1–123) is unstructured, but PrP^C(124–228) is well structured, with three α-helices and two short antiparallel β-strands (denoted as H1, H2, H3, S1 and S2), and a disulfide bond linking H2 and H3 (residues numbered 178 and 213). This book mainly reports the MD results on the structured region PrP^C(124–228) of rabbits (compared with humans, mice, elks, dogs, horses, and buffaloes).

The rest of Part I is arranged as follows. Chapter 2 reports the MD works done on the rabbit homology structure PrP^C(120–229). Chapter 3 will reports the MD works done for the NMR structures of the wild-type and mutants of rabbit PrP^C(124–228) (denoted as $RaPrP^C$(124–228)). Chapter 4 does the MD works of $RaPrP^C$(124–228) compared with NMR human PrP^C(125–228) and NMR mouse PrP^C(124–226). Chapter 5 does the MD works of $RaPrP^C$(124–228) compared with NMR dog PrP^C(121–231) and NMR horse PrP^C(119–231). Chapter 6 does the MD works compared with homology buffalo PrP^C(124–227). Chapter 7 does the MD works compared with NMR elk PrP^C(121–231). Chapter 8 does the MD works on the rabbit X-ray structure PrP^C(126–230). Chapters 2, 3, 4, 5, 6, 7, and 8 concludes a clue on SBs which reveal a secret of rabbits' immunity to prion diseases. Chapter 9 discusses other clues such as surface electrostatic charge (SEC) distributions, copper bindings (e.g. the weak HGGGW octarepeat copper binding region PrP^C(57–89) and the strong GGGTH copper binding region PrP^C(92–96)), and π-interactions covering the β2-α2 loop and H1 that might also contribute to reveal the secrets of prion diseases. Chapter10 does MD on the helix HuPrP(109–136) and introduces detailed bioinformatics on PrP central hydrophobic region. Each chapter of Part I can be read independently.

Chapter 2
The Homology Structure and Dynamics

2.1 Molecular Modeling

X-ray crystallography and NMR spectroscopy are two powerful tools to determine the protein 3D structure. However, not all proteins can be successfully crystallized, particularly for membrane proteins. Although NMR spectroscopy is indeed very powerful in determining the 3D structures of membrane proteins, same as X-ray crystallography, it is still very time-consuming and expensive. Under many circumstances, due to the noncrystalline and insoluble nature of some proteins, X-ray and NMR cannot be used at all. Theoretical methods and computational approaches, however, allow us to obtain a description of the protein 3D structure at a submicroscopic level.

Prion diseases belong to neurodegenerative diseases which include Parkinson's, Alzheimer's, Huntington's diseases, etc. All these neurodegenerative diseases are amyloid fibril diseases. Now we know that the atomic structures of all amyloid fibrils revealed steric zippers, with strong vdW interactions between β-sheets and HBs to maintain the β-strands [517]. Thus, keeping the vdW and HB contacts we can do modeling on molecular structures of some amyloid fibril regions of neurodegenerative diseases, which are hard to get any structural information because of the unstable, noncrystalline and insoluble nature of amyloid fibrils.

The RaPrP^C(120–229) homology model used in this chapter was constructed by mutations using the NMR structure of HuPrP^C(125–228) (1QLX.pdb) as the template.

J. Zhang, *Molecular Structures and Structural Dynamics of Prion Proteins and Prions*, Focus on Structural Biology 9, DOI 10.1007/978-94-017-7318-8_2

Table 2.1 Simulation conditions for the homology model [700]

pH value	Specie	Truncated octahedral box (Å)	Total atoms	Ions added	TIP3P waters added
Neutral pH	HuPrP	75.868	19,484	3Na+	5929
	MoPrP	67.447	13,422	2Na+	3918
	RaPrP	78.130	21,469	2Na+	6572
Low pH	HuPrP	74.834	18,530	16Cl−	5599
	MoPrP	67.335	13,208	14Cl−	3836
	RaPrP	80.896	23,847	14Cl−	7354

2.2 The Homology Rabbit Prion Protein at 500 K

The author with his colleagues studied the MD of RaPrP^C(120–229) homology structure (6EPA.pdb) [700]. The MD simulations used Amber 8 [90] PMEMD program, with explicit water at different temperatures and pH values. The simulation conditions are listed in Table 2.1.

The RMSD and radius of gyration results are shown in Figs. 2–3 of [700], where we may see that rabbit prion protein has more stable structural dynamical behavior compared to the human and mouse prion proteins at 500 K under neutral pH environment. This is also shown in Fig. 4 of [700] of snapshots for human, mouse and rabbit prion proteins at 5, 10, 15, 20, 25, and 30 ns respectively. Figure 4 of [700] shows that the helices of HuPrP and MoPrP were unfolded but RaPrP still keeps the helical structures at 500 K under neutral pH environment. The performance of RaPrP also shows that 500 K is a not a very bad temperature chosen for theoretical research for RaPrP. Under low pH environment at 500 K, these helical structures of RaPrP were unfolded. One of the reasons of the rabbit prion protein unfolding is due to the remove of the SBs such as N177-R163 (Fig. 5 of [700]). We may see in Fig. 5 of [700] that the SB/HB between Arginine 163 and Aspartic acid 177 is conserved through a large part of the simulations and contributes to the protein stability of rabbit prion protein structure. Simulations at low pH value, where this SB is absent, show RMSD and radius of gyration values for the rabbit prion protein to be of the same magnitude as the human and mouse prion proteins. In this rest of this chapter we check the MD results at 450 K.

2.3 Molecular Dynamics Methods at 300 and 450 K

MD simulations used the ff03 force field of the AMBER 8 package, in neutral and low pH environments. The systems were surrounded with a 12 Å layer of TIP3PBOX water molecules and neutralized by sodium ions using XLEaP module of AMBER 8. The solvated proteins with their counterions were minimized mainly by the steepest descent method and then a small number of conjugate gradient steps were performed on the data, in order to remove bad HB contacts. Then the solvated proteins were heated from 0 to 300 K during 300 ps and then kept at 300 K for 700 ps, both in constant NVT ensembles using Langevin thermostat algorithm. The SHAKE algorithm and PMEMD algorithm with nonbonded cutoffs of 12 Å were used during the heating. The solvated proteins were also heated to 450 K in the same way. Enough equilibrations had been done in constant NPT ensembles under Berendsen thermostat during 5 ns until the RMSD, PRESS, VOLUME and DENSITY were sufficiently stable, where the RMSD values did not fluctuate very much within a few picoseconds. After equilibrations, production MD phase was carried out at 300 or 450 K for 30 ns using constant pressure and temperature ensemble and the PMEMD algorithm with nonbonded cutoffs of 12 Å during simulations. Step size for equilibration is 0.5 fs, and 1 fs is for the production runs. All simulations were performed on the Tango facilities of VPAC. The structures were saved to file every 1000 steps.

2.4 Results at 450 K and Discussion

The MD simulations done at room temperature 300 K displayed very little fluctuation and no variation among rabbit, human and mouse prion proteins. In what follows we only show and analyze the results at 450 K. Radii of gyrations of rabbit, mouse and human prion proteins are shown in Fig. 2.1a. In Fig. 2.1a, we see that under neutral pH environment the values of radii of gyrations of rabbit prion protein are almost leveling off around 15 Å during the whole simulation but for mouse and human prion proteins the values have steadily increased. The RMSDs from the minimized structures of homology rabbit prion protein and human and mouse prion proteins under neutral pH environment are also shown in Fig. 2.1. In Fig. 2.1b

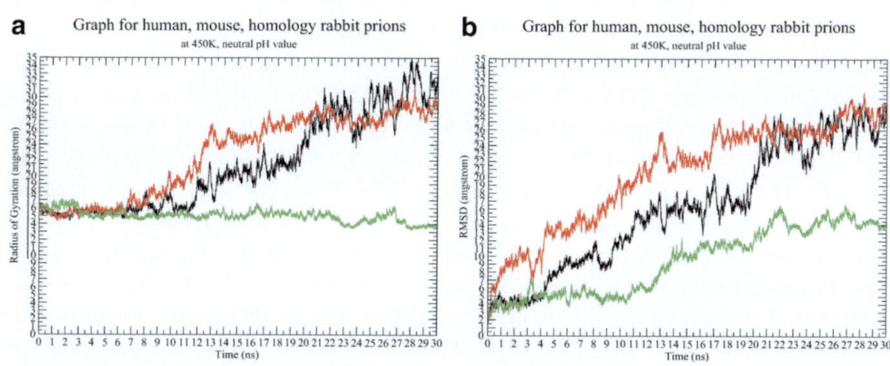

Fig. 2.1 Radius of gyration (**a**) and RMSD (**b**) graphs for human, mouse, and homology rabbit prion proteins (*black*: human, *red*: mouse, *green*: homology rabbit) at 450 K in neutral pH environment

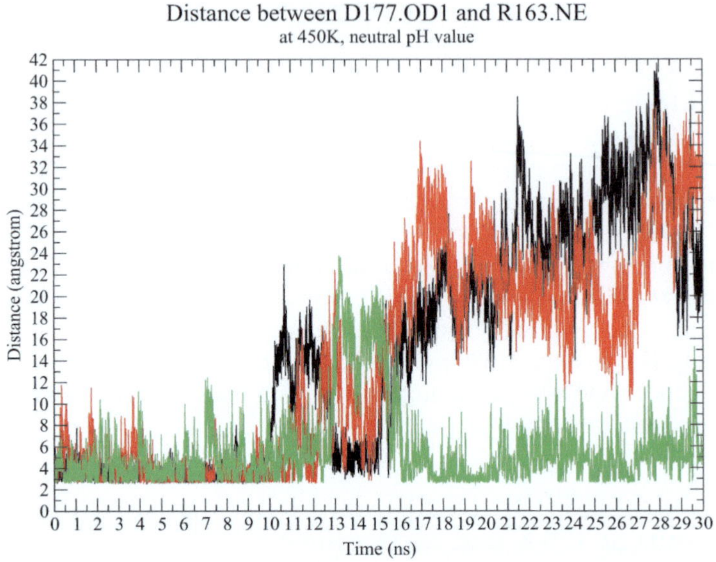

Fig. 2.2 SB D177-R163 graph for human, mouse, homology rabbit prion proteins (*black*: human, *red*: mouse, *green*: homology rabbit)

we see that: for rabbit prion protein the RMSD values have generally been leveling off; for human and mouse prion proteins the RMSDs have steadily increased during the whole simulations. Thus, we may conclude that rabbit prion protein has a more stable structure.

SBs and HBs have been investigated during the whole simulations. We find that SB/HB (i.e. polar contact) D177-R163 is almost conserved throughout the simulation (except for 3 ns' broken in the middle of simulation) (see Fig. 2.2) and contributes to the stability of rabbit prion protein structure; the occupied rate of SB D177-R163 is nearly 90 % for rabbit prion protein. Simulations under low pH environment, where this SB is absent, show RMSD and radius of gyration values for the rabbit prion protein to be of the same magnitude as the human and mouse prion proteins. Experimental results showed a reduced stability if D178 was mutated into N178 for human prion protein [388]. Recently, on 3rd March 2009, the NMR structure of human prion protein variant D178N (PDB ID: 2K1D) was released [427]. In order to perform an analysis where D177 is mutated into N177 in the rabbit prion protein, completely same procedures of the minimizations, enough equilibrations (5 ns) and 12 ns' productions were done at 450 K for the wild-type rabbit prion protein (PDB ID: 2FJ3), the D178N mutant of human prion protein, and human and mouse prion proteins. These simulations are independent to the simulations of Sect. 2.2. The secondary structural performance of the D178N mutant under neutral pH environment is same as that of the wild-type rabbit prion protein under low pH environment (Fig. 2.3); in Fig. 2.3 we may see the difference between rabbit prion protein and human and mouse prion proteins for the percentage changes

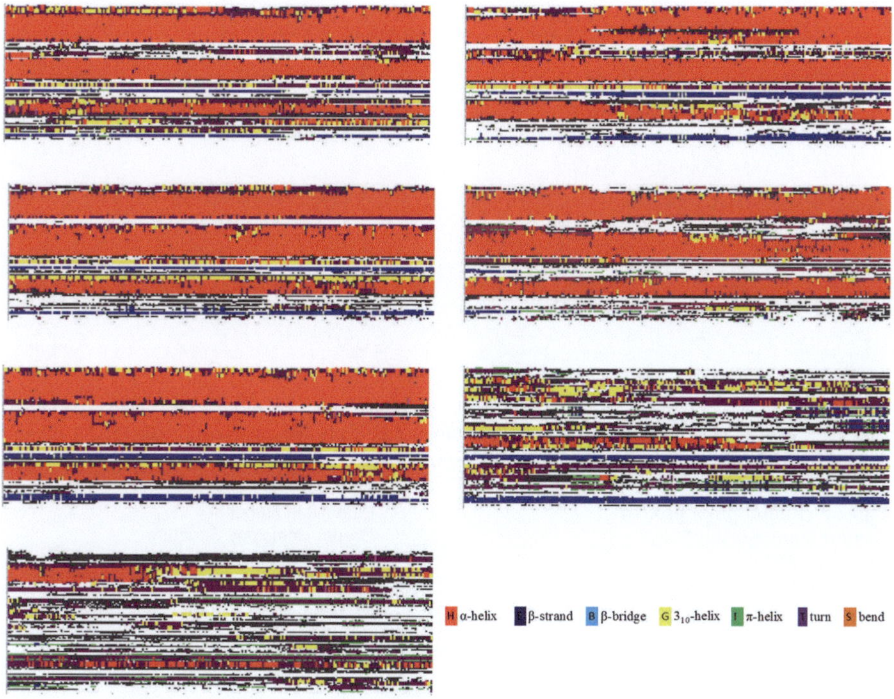

Fig. 2.3 Secondary Structure graph for wild-type human, mouse, rabbit prion proteins and the D178N mutant (from up to down) (*red*: α-helix, *blue*: β sheet; x axis: time, y axis: residue number; *left column*: neutral pH, *right column*: low pH) of the new simulations

of α-helices from neutral pH environment to low pH environment. Thus, D177 might be one of the residues that inhibit formation of abnormal isoform of rabbit prion protein. The SB R163-D177 is conserved in rabbit prion protein but not always conserved in human and mouse prion proteins throughout the whole simulations (Fig. 2.2); the reason should be due to the stable structural behavior of rabbit prion protein. Moreover, though the SB R163-D177 in the structure of prion proteins spans only 14 residues and it should have a rather impact on the local structure, human and mouse prion proteins do not have the stable local structure kept by the SB R164-D178, which should link the middle of H2 and the coil between the S2 and the H2. Rabbit prion protein has the stable local structure behavior. Occupied rates of all the trajectories of 30 ns of the SB R163-D166 are 3.3 % for rabbit prion protein, 10 % for mouse prion protein and 13.3 % for human prion protein. SB D201-R155 is also found important to the structural stability of rabbit prion protein; at 300 K, the occupied rates of all the trajectories of 30 ns are 70.82 % for human prion protein, 100 % for mouse and rabbit prion proteins; at 450 K, the occupied rate for this SB are 0 % for human and mouse prion proteins, 28.59 % for rabbit prion protein. But during 15–30 ns, the SB D201-R155 does not exist for rabbit prion protein. The HB between the backbone atoms of the first residue L124 and the last residue A228 also contributes to the structural stability of rabbit prion protein, but its occupied rate is very low. In Fig. 2.3, under neutral pH environment the short 12 ns' simulations

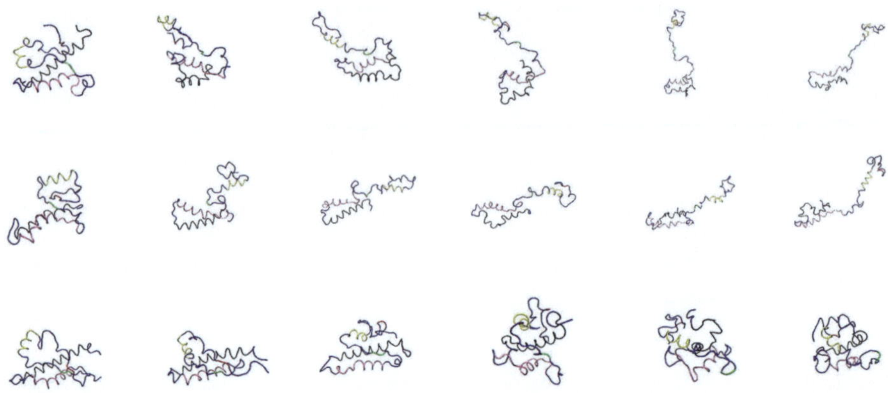

Fig. 2.4 Snapshots of human, mouse and homology rabbit prion proteins at 5, 10, 15, 20, 25 and 30 ns separately (*1st row*: human, *2nd row*: mouse, *3rd row*: homology rabbit)

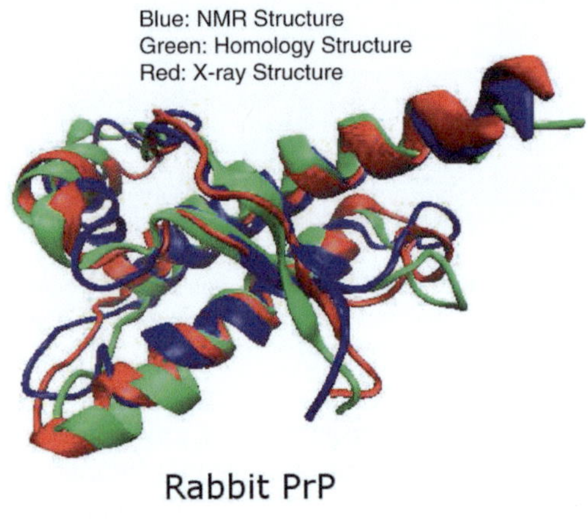

Fig. 2.5 Rabbit prion protein NMR, homology and X-ray structures (2FJ3.pdb, 6EPA.pdb, and 3O79.pdb)

cannot show much difference of the secondary structure among wild-type rabbit, human and mouse prion proteins (where longer simulations should be done).

More stable dynamical behaviour from the rabbit prion protein structure compared to the human and mouse prion proteins is also shown in the snapshots of human, mouse and rabbit prion proteins at 5, 10, 15, 20, 25 and 30 ns separately: Fig. 2.4.

2.5 Concluding Remarks

RMSDs and Radii of Gyrations are two important indexes to evaluate the structural stability of a protein. Studies of the RMSDs and Radii of Gyrations of human, mouse and homology rabbit prion proteins show that rabbit prion protein has a more stable structure than human and mouse prion proteins. The polar contact (i.e. SB & HB) between D177 and R163 is found to play an important role in keeping rabbit prion protein structure much more stable.

Lastly, we illuminate the figure (Fig. 2.5) of rabbit prion protein, including the homology, NMR and X-ray structures (6EPA.pdb, 2FJ3.pdb, and 3O79.pdb respectively). We superpose the homology structure onto the NMR structure and find the RMSD value is 3.2031669 Å. Similarly, we superpose the X-ray structure onto the NMR structure and we get their RMSD value is 2.7918559 Å. This implies to us that the homology structure 6EPA.pdb made in 2004 by Epa [700] is as effective as the X-ray structure 3O79.pdb released into PDB Bank on date 2010-11-24 (last modified on 2011-02-02) [336]; this also confirms that the results of this chapter are correct.

Chapter 3
The NMR Structure and Dynamics of the Wild-Type and Mutants

3.1 Overview

Prion diseases are invariably fatal neurodegenerative diseases that affect humans and animals. Unlike most other neurodegenerative diseases, these can be highly infectious. They include CJD, GSS, FFI, Kuru in humans, scrapie in sheep, and BSE or 'mad-cow' disease in cattle, etc. Transmission across the species barrier to humans, especially in the case of BSE in Europe and CWD in North America, is a major public health concern. Since 1996, vCJD has been found even in young people in UK. However, there is no effective therapeutic approach for treating all these diseases.

Rabbits are the only mammalian species reported to be resistant to infection from prion diseases isolated from other species [591]. At the end of 2007, the NMR structure of rabbit prion protein (124–228) was deposited into PDB with PDB ID code 2FJ3 [381]; the NMR structure of rabbit prion protein mutation S173N (PDB ID code 2JOH) and the NMR structure of rabbit prion protein mutation I214V (PDB ID code 2JOM) were released in early 2008; this gives us a golden opportunity to study the inhibition mechanism of rabbit prion protein and to find an effective therapeutic approach to prion diseases. This chapter studies the inhibition mechanism at a molecular structural level.

To really understand the stability of rabbit prion molecules and their action mechanism, we should consider not only the static structures concerned but also the dynamical information obtained by simulating their internal motions or dynamic process. Prion protein MD simulations usually have been done on the C-terminal structured region, some C-terminal mutants, copper binding segments (e.g. HGGGW, GGGTH), and amyloid fibril segments (e.g. AGAAAAGA). Recently Sawaya et al. [517] revealed the common structural features shared by all amyloid fibrils [517]. Some prion MD works focus on prion amyloid fibrils. Inhibition mechanisms are also studied through MD simulations. This chapter will study the inhibition mechanism of rabbit prion protein RaPrP^{C}(124–228) through MD

J. Zhang, *Molecular Structures and Structural Dynamics of Prion Proteins and Prions*, Focus on Structural Biology 9, DOI 10.1007/978-94-017-7318-8_3

simulations compared with its mutants. Simulation results confirm the structural stability of wild-type rabbit prion protein, and show that the SB between D177 and R163 greatly contributes to the structural stability of rabbit prion protein.

3.2 Materials and Methods

Simulation initial structures for the rabbit prion protein, its S173N mutant and I214V mutant were separately built on RaPrP^C(124–228) (PDB entries 2FJ3, 2JOH, 2JOM). Simulations were done under low pH and normal pH environments separately. All the simulations were done starting from the same initial velocity, and all the simulations were performed with the AMBER 9 package [88], with analysis carried out using functionalities in AMBER 9 and AMBER 7 CARNAL [92]. Graphs were drawn by XMGRACE of Grace 5.1.21, DSSP [326].

All simulations used the ff03 force field of the AMBER 9 package, in neutral and low pH environments (where residues HIS, ASP, GLU were changed into HIP, ASH, GLH separately by the XLEaP module of AMBER 9 in order to get the low pH environment). The systems were surrounded with a 12 Å layer of TIP3PBOX water molecules and neutralized by sodium ions using XLEaP module of AMBER 9 (14 Cl–, 14 Cl–, 13 Cl– and 5909 waters, 4185 waters, 4729 waters were added for the rabbit prion protein wild-type, S173N mutant and I214V mutant separately for the low pH environment). The solvated proteins with their counterions were minimized mainly by the steepest descent method and then a small number of conjugate gradient steps were performed on the data, in order to remove bad HB contacts. Then the solvated proteins were heated from 100 to 450 K step by step during 3 ns. The thermostat algorithm used is the Langevin thermostat algorithm in constant NVT ensembles. The SHAKE algorithm and PMEMD algorithm with nonbonded cutoffs of 12 Å were used during the heating. Equilibrations were done in constant NPT ensembles under Langevin thermostat for 5 ns. After equilibrations, production MD phase was carried out at 450 K for 15 ns using constant pressure and temperature ensemble and the PMEMD algorithm with nonbonded cutoffs of 12 Å during simulations. Step size for equilibration was 0.5 and 1 fs for the production runs. All simulations were performed on the Tango facilities of VPAC, Australia. The structures were saved to file every 1000 steps. For prion proteins some MD works have done at 500 K (e.g. [198, 523, 525, 659]); simulation results on protein structures and their dynamics of this chapter showed that the force field parameters are suitable to allow simulations at 450 K.

3.3 Results and Discussion

Radii of gyrations of rabbit prion protein and its I214V mutant and S173N mutant have generally been leveling off around 15 Å whether under neutral pH environment or under low pH environment, without very large differences. But the RMSDs

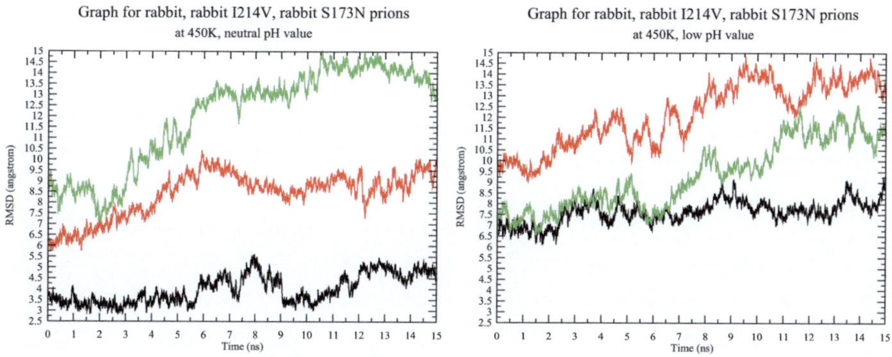

Fig. 3.1 RMSD graphs for rabbit prion protein, its I214V mutant and S173N mutant (*black*: rabbit prion protein, *red*: rabbit prion protein I214V mutant, *green*: rabbit prion protein S173N mutant)

calculated from the minimized structures of rabbit prion protein and its I214V and S173N mutants differ by a large amount (Fig. 3.1). In Fig. 3.1 we see that the RMSDs of rabbit prion protein level off across the whole simulation at around 4 Å under the neutral pH environment and around 7 Å under the low pH environment. But we see that in Fig. 3.1 the RMSDs of rabbit I214V and S173N mutants increase steadily across all the simulations of 15 ns. For rabbit prion protein under neutral pH environment, the RMS Fluctuations and B-factors are at constant values with respect to residue numbers; the mutants do not have this property. In what follows the HB and SB analyses will reveal some bioinformatics which contributes to the structural stability of rabbit prion protein.

The mutations at I214 and S173 make several HBs and SBs at I214 and S173 broken. Thus, the mutants become much more unstable than the wild-type. If we only show HBs that are resident for more than 5 % of the simulation, there are HBs that have occupied at these points whether for the wild-type or for the mutants. There is not a HB at V214 of the I214V mutant whether under the neutral or low pH environments. The HB occupied rates of all the trajectories of 15 ns are: for the I214V mutant at S173 10.97 % for the low pH environment and 21.9 % for the normal pH environment, for the wild-type at S173 5.47 % for the low pH environment and 6.77 % for the normal pH environment and at I214 13.8 % for the normal pH environment, and for the S173N mutant at I214 12.73 % for the low pH environment and 17.67 % for the normal pH environment and at N173 5.07 % for the normal pH environment. There always exists a SB between D201 and R155 for the wild-type and the I214V and S173N mutants, to make H3 and H1 salt-linked. The disulfide bond links helices H2 and H3 and keeps several strong SBs linked. One strong SB is the one between D177 and R163 for the wild-type, which keeps the linking of the middle of helix H2 and the coil at S2-H2. However, the mutations make this SB broken so that we cannot see the SB between D177 and R163 for the I214V and S173N mutants. Under low pH environment the SB between D177 and R163 is removed and leads to the collapse of the stable helical structure of

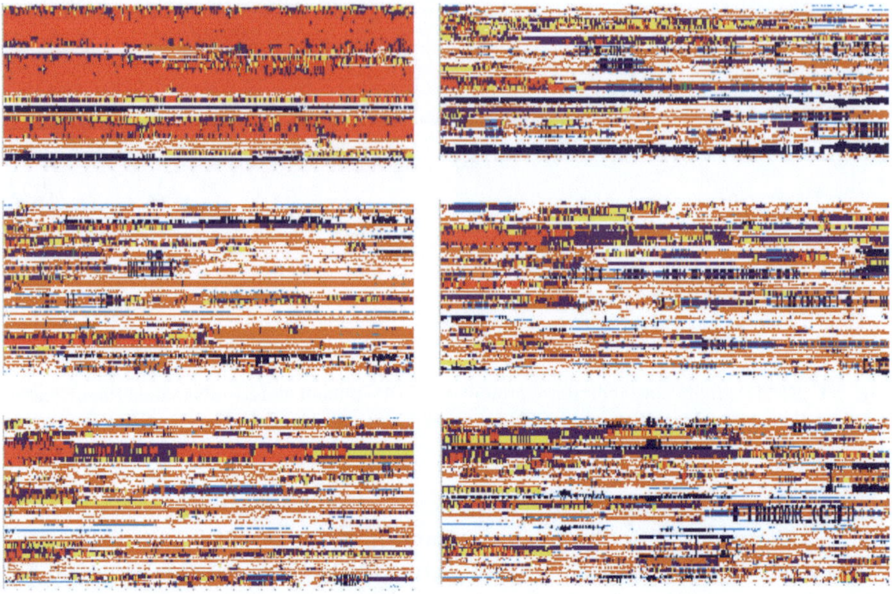

Fig. 3.2 Secondary Structure graph for rabbit prion protein, its S173N mutant and I214V mutant (from up to down) (*red*: α-helix, *blue*: β-sheet; x-axis: time (0–15 ns), y-axis: residue number; *left column*: neutral pH, *right column*: low pH)

wild-type rabbit prion protein (Fig. 3.2). Thus, the SB between D177 and R163 greatly contributes to the structural stability of rabbit prion protein. Because many experimental researches have indicated that a stable monomer PrP^C protein structure should be above pH 4.4 [289, 372, 412, 623], the low pH environment in this chapter should be less than pH 4.4. In Fig. 3.2 we see for the wild-type rabbit prion protein the unwinding and aggregation of helices H1, H3, the aggregation of the two β-strands, and partial unwinding and aggregation of helix H2.

3.4 A Concluding Remark

The above analyses confirm the structural stability of wild-type rabbit prion protein and the SB between D177 and R163 greatly contributes to the structural stability of rabbit prion protein.

Chapter 4
Compared with the NMR Structure and Dynamics of Humans and Mice

4.1 Overview

Rabbits are one of the few mammalian species reported to be resistant to infection from prion diseases isolated from other species [591]. Recently, the NMR molecular structures of wild-type, mutant D173N and mutant I214V rabbit prion proteins (124–228) were released into the Protein Data Bank with PDB ID codes 2FJ3, 2JOH, 2JOM respectively. Chapter 3 studied these NMR structures by MD simulations and simulation results at 450 K under low and neutral pH environments confirmed the structural stability of wild-type rabbit prion protein. Chapter 3 did (1) the simulations under low pH environments, (2) the simulations at 450 K, (3) the simulations by Amber 9, (4) the mutants of rabbit prion protein, and (5) the equilibrations in constant NPT ensembles under Langevin thermostat for 5 ns and the productions under Langevin thermostat at 450 K. But, in Chap. 3, the author did not carry on the MD simulations at 450 K under low and neutral pH environments for human prion protein (125–228) (PDB ID code 1QLX) and mouse prion protein (124–226) (PDB ID code 1AG2); this chapter will do this work. Findings of this chapter agree with the findings of Chap. 3. In order to further confirm the findings, this chapter will do longer simulations than Chap. 3. Simulation results of this chapter show that the SBs D177-R163, D201-R155 greatly contribute to the structural stability of rabbit prion protein. We also find that the HB H186-R155 partially contributes to the structural stability of rabbit prion protein.

4.2 Materials and Methods for 300 and 450 K

Simulation initial structure for the rabbit prion protein was built on RaPrP^C(124–228) (PDB entry 2FJ3). Identical simulations were also done for human prion protein (HuPrP^C(125–228)) and mouse prion protein (MoPrP^C(124–226)). The initial

J. Zhang, *Molecular Structures and Structural Dynamics of Prion Proteins and Prions*, Focus on Structural Biology 9, DOI 10.1007/978-94-017-7318-8_4

simulation structures of human and mouse prion proteins were built on PDB entries 1QLX [673] and 1AG2 [494] respectively. Simulation methods are completely same as the ones of Sect. 3.2. 16 Cl–, 14 Cl–, 14 Cl–, and 5599 waters, 3836 waters, 5909 waters were separately added for the human, mouse, rabbit prion proteins under low pH environment. The step size of 2 fs is typical for the SHAKE algorithm at 300 K. At 450 K, 1 fs was used as the step size. For prion proteins some MD works have been done at 500 K (e.g. [198, 523, 525, 659]); this chapter does the MD simulations at 450 K and these simulations are not physically relevant.

Simulations were done under low pH and normal pH environments respectively. All the simulations were performed with the AMBER 9 package [88], with analysis carried out using functionalities in AMBER 9 and AMBER 7 CARNAL [92]. Graphs were drawn by XMGRACE of Grace 5.1.21, DSSP [326]. The AMBER ff03 force field was used. The vdW and electrostatic interactions were treated by SHAKE algorithm and PMEMD algorithm with nonbonded cutoffs of 12 Å. The systems were surrounded with TIP3PBOX water molecules and neutralized by sodium ions, and optimized to remove bad hydrogen contacts. Then the systems were heated from 100 to 450 K step by step during 3 ns. The thermostat algorithm used is the Langevin thermostat algorithm in constant NVT ensembles. Equilibrations were done in constant NPT ensembles under Langevin thermostat for 5 ns. After equilibrations, production MD phase was carried out at 450 K for 20 ns using constant pressure and temperature ensemble.

4.3 Results and Discussion for 300 and 450 K

The MD simulations done at room temperature 300 K whether under neutral or low pH environment display very little fluctuation and no variation among rabbit, human and mouse prion proteins. At 450 K there are fluctuation and variation among rabbit, human and mouse prion proteins, but their backbone atom RMSDs respectively calculated from their minimized structures and their radii of gyrations do not have great differencc even under low pH environment (Figs. 4.1 and 4.2). Their secondary

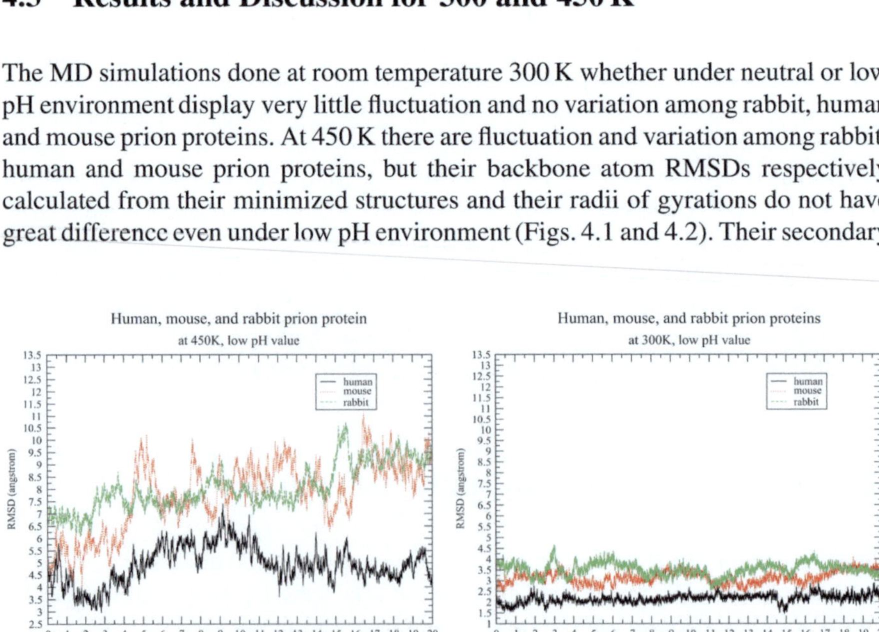

Fig. 4.1 Backbone atom RMSD graphs for human, mouse, rabbit prion proteins

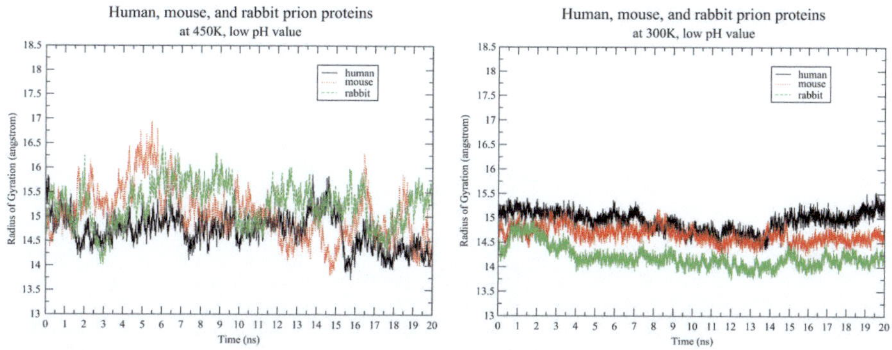

Fig. 4.2 Radius of gyration graphs for human, mouse, rabbit prion proteins

structures under neutral pH environment at 450 K, under low (Fig. 4.3) and neutral pH environments at 300 K, do not change very much either. However, the secondary structures under low pH environment at 450 K have great differences between rabbit prion protein and human and mouse prion proteins (Figs. 4.4 and 4.5): the α-helices of rabbit prion protein were completely unfolded and began to turn into β-sheets but those of human and mouse prion proteins were not changed very much. These results indicate the C-terminal region of $RaPrP^C$ has lower thermostability than that of $HuPrP^C$ and $MoPrP^C$. Under the low pH environment, the SBs D177-R163, D201-R155 were removed (thus the free energies of the SBs changed the thermostability) so that the structure nearby the central helices 1–3 was changed for rabbit prion protein.

There always exist SBs (where the oxygen-nitrogen distance cut-off calculated for the SBs is 3.2 Å) between D202-R156, D178-R164 for human and mouse prion proteins, between D201-R155, D177-R163 for rabbit prion protein. The disulfide bond between C178-C213 links the H2 and H3 and keeps several strong SBs linked. The SB between D177-R163 keeps the linkage of the middle of H2 and the coil at S2 to H2. The SB between D201-R155 makes the 3rd and 1st α-helices linked. The salt link distances of D177-R163 and D201-R155 are illustrated in Figs. 4.6, 4.7 and 4.8. Except for the short period of break of the SB N202-R156 of human prion protein during 8.5–10.5 ns, the SBs are always occupying human, mouse, and rabbit prion proteins. Under low pH environment, all these SBs were removed. They did not change the secondary structures of human and mouse prion proteins very much. However, rabbit prion protein is very sensitive to the low pH environment: the remove of these SBs led to the collapse of the stable helical structure of wild-type rabbit prion protein. Structural stability of a protein is determined by factors of HB, vdW force,HYD, and SB; change from neutral to low pH environments will break the SBs only. Thus, we might be able to say that SBs D177-R163, D201-R155 greatly contribute to the structural stability of rabbit prion protein. $RaPrP^C$(124–228) apparently resists the formation of the scrapie form, and the enhanced stability of the C-terminal ordered region especially H2 through the D177-R163 salt-bridge

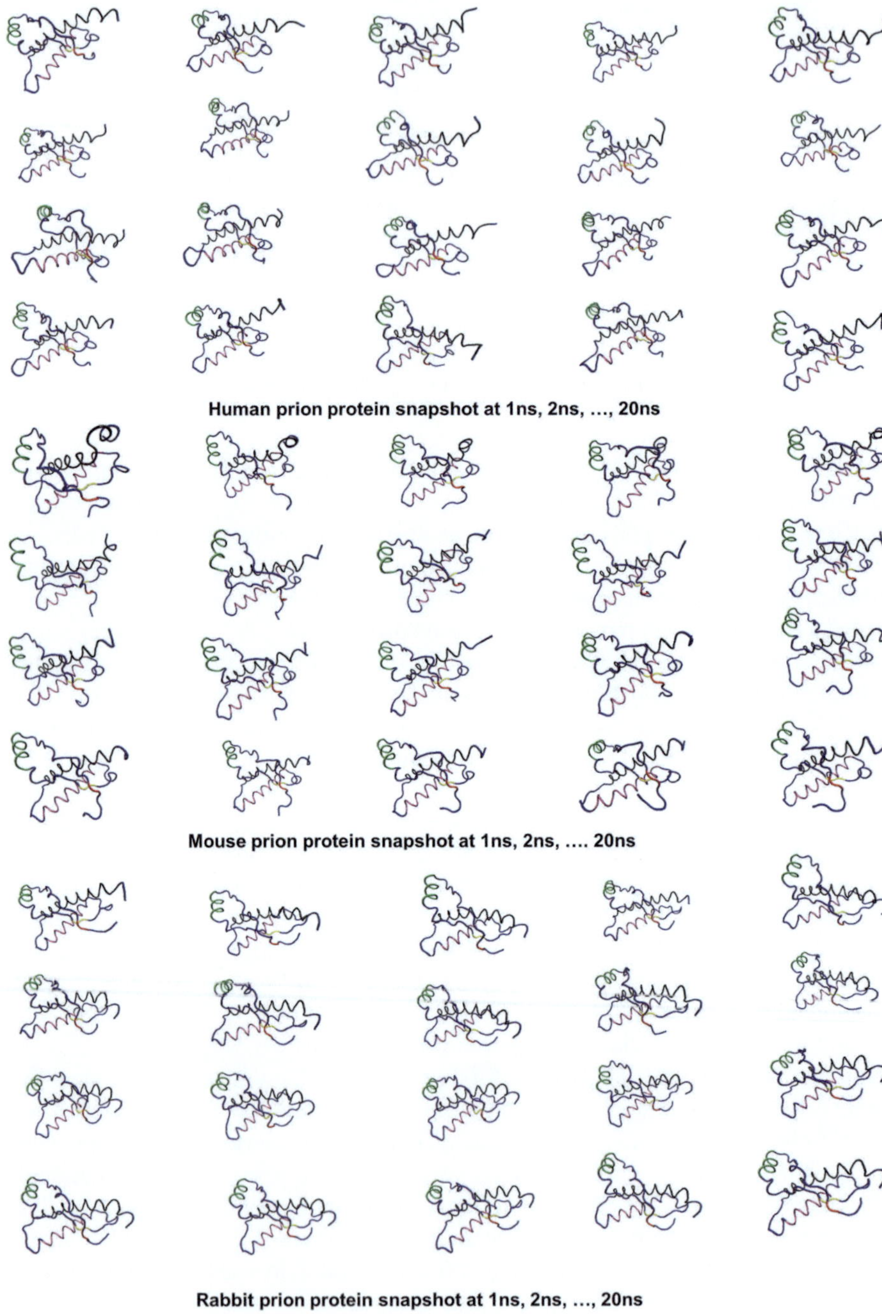

Fig. 4.3 Snapshots of human, mouse, and rabbit prion proteins at 300 K under low pH environment

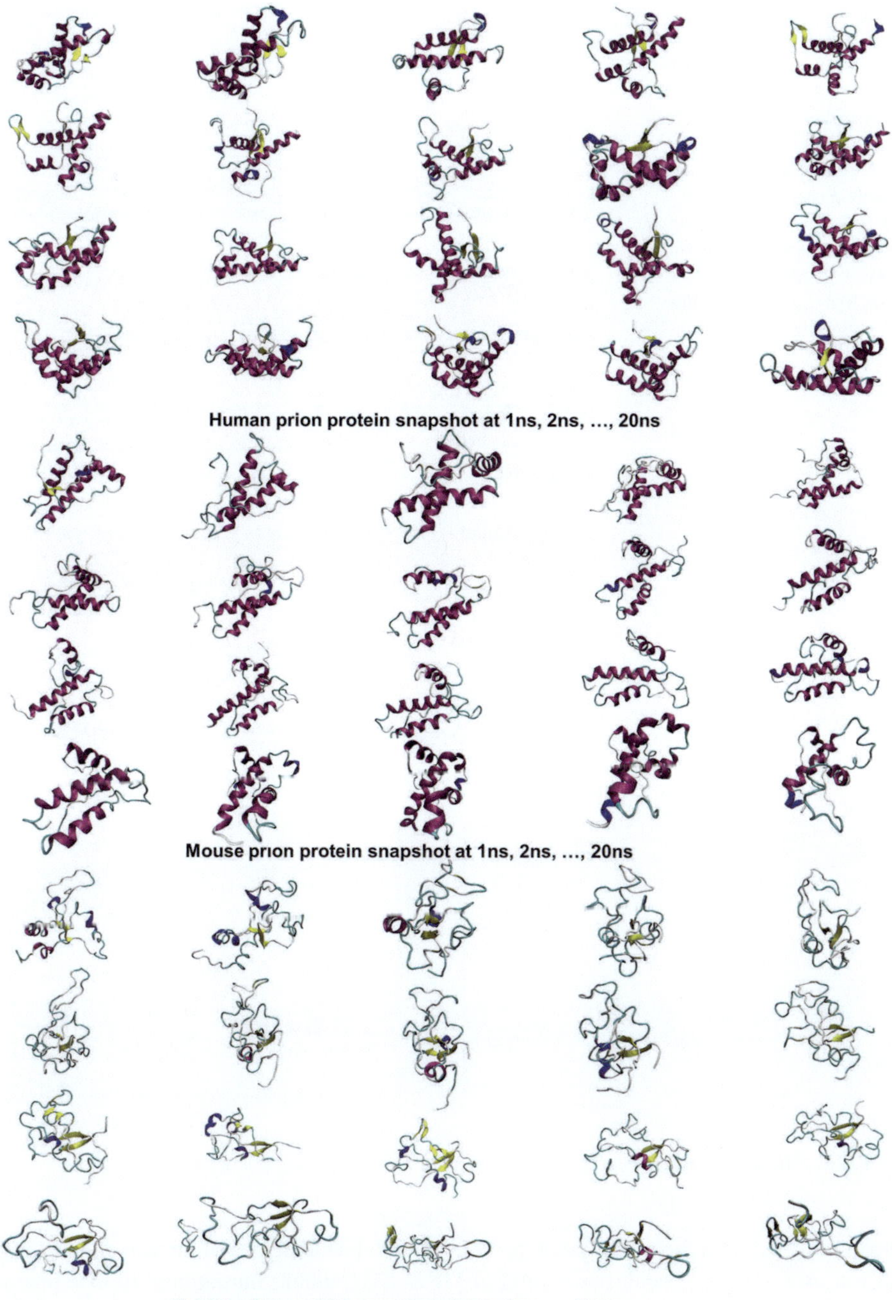

Fig. 4.4 Snapshots of human, mouse, and rabbit prion proteins at 450 K under low pH environment

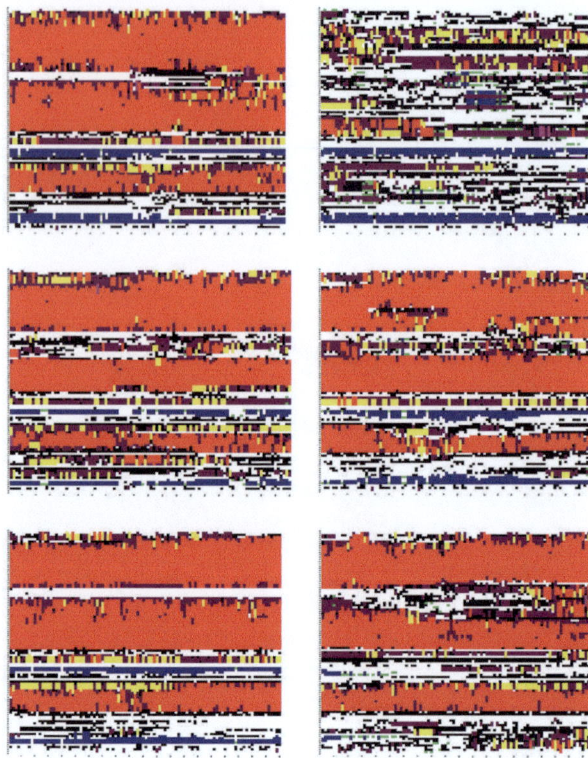

Fig. 4.5 Secondary structures of rabbit, human and mouse prion proteins (from up to down) at 450 K under neutral and low (from left to right) pH environments. (*red*: α-Helix, *pink*: π-helix, *yellow*: 3/10-helix, *green*: β-bridge, *blue*: β-sheet, *purple*: turn, *Black*: bend; x-axis: time (0–20 ns), y-axis: residue numbers)

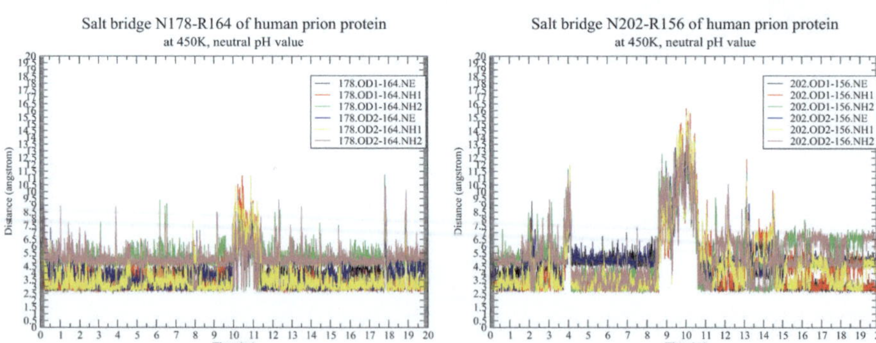

Fig. 4.6 SBs of human prion protein at 450 K under neutral pH environment

formation renders the rabbit prion protein stable. The analyses of human and mouse prion proteins agree with the results of [182, 183], where human and mouse prion proteins were showed lack of stability and correlated mutations that reduce the frustration in the second half of H2 in mammalian prion proteins could inhibit the formation of PrP^{Sc}.

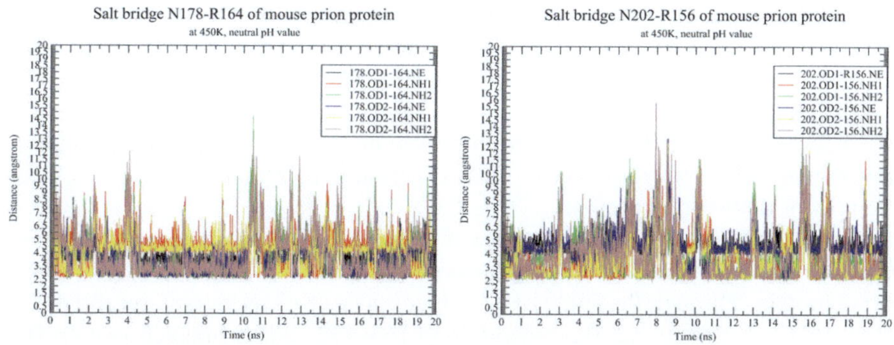

Fig. 4.7 SBs of mouse prion protein at 450 K under neutral pH environment

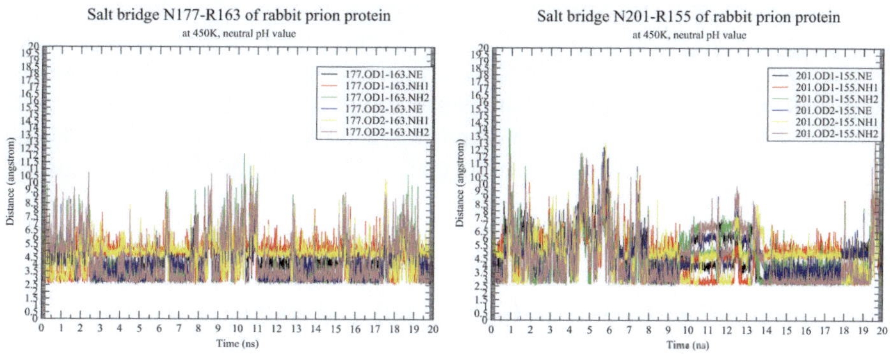

Fig. 4.8 SBs of rabbit prion protein at 450 K under neutral pH environment

The MD simulation experience of the NMR structure of human prion protein variant D178N (PDB ID: 2K1D, released on 03-MAR-2009) [427] of the author confirmed that the secondary structural performance of the D178N mutant under neutral pH environment at 450 K is same as that of the wild-type rabbit prion protein under low pH environment at 450 K. Though the SB R163-D177 in the structure of prion proteins spans only 14 residues and it should have a rather impact on the local structure, human and mouse prion proteins do not have the stable local structure kept by the SB R164-D178. This agrees with experimental results of [388], which showed a reduced stability if D178 was mutated into N178 for human prion protein. The MD simulation experience of the homology structure of rabbit prion protein of Zhang et al. [700] also confirmed that rabbit prion protein has a more stable structure than human and mouse prion proteins, and the SB and HB between D177 and R163 plays a key important role to this stability [700].

Recently Zhong [712] reported that for human prion protein (PDB entry 1QM0) the mutation of H187 into R187 makes the HB H187-R156 broken, and the strong electrostatic repulsion between R187 and R156 drives both positively charged side chains away from their original positions leaving their hydrophobic core to be solvent accessible [712]. For rabbit prion protein at 450 K under neutral pH environment, there is the HB H186-R155 during more than 26.44 % time of the 20 ns production trajectories; however, for mouse (PDB entry 1AG2) and human (PDB entry 1QLX) prion proteins at 450 K under the neutral pH environment, the HB between H187 and R156 has been broken during the whole 20 ns production trajectories. The HB H186-R155 clearly partially contributes to the structural stability of rabbit prion protein.

Besides pH deviations and the choice of temperatures, point mutations from wild-type can identify more contributing factors for the extraordinary stability of rabbit prion protein. Through the study of wild-type and its S173N variant, recently Wen et al. [626] suggested that the ordered loop 165–172 and its interaction with H3, together with the unique distribution of surface electrostatic potential, significantly contribute to the unique structural characteristics of $RaPrP^C$.

The choice of temperatures is important and the choosing 450 K for this chapter is very significant. At 300 K, the MD results display little difference among rabbit, human and mouse prion protein structures whether under neutral or low pH environment. At 450 K, this present chapter found that the structure of rabbit is much more stable than these of human and mouse; this agrees with the experimental results of rabbit prion protein and shows that the choosing 450 K of this chapter is significant.

4.4 Concluding Remarks for the 300 and 450 K

The chapter is a straight forward MD simulation study of rabbit prion protein (monomer cellular form) which apparently resists the formation of the scrapie form. The analyses of MD simulation results confirmed the structural stability of rabbit prion protein under neutral pH environment. The main point is that the enhanced stability of the C-terminal ordered region especially H2 through the D177-R163 salt-bridge formation renders the rabbit prion protein stable. The SB D201-R155 linking H3 and H1 also contributes to the structural stability of rabbit prion protein.

We also find that the HB H186-R155 partially contributes to the structural stability of rabbit prion protein.

4.5 Molecular Dynamics at 350 K

350 K might be a practical temperature for some experimental laboratory works. Here the MD simulations at 350 K are done for wild-type rabbit, human and mouse prion proteins in the use of the following MD methods.

All simulations used the ff03 force field of the AMBER 11 [91] package, in neutral and low pH environments (where residues HIS, ASP, GLU were changed into HIP, ASH, GLH respectively by the XLEaP module of AMBER 11 in

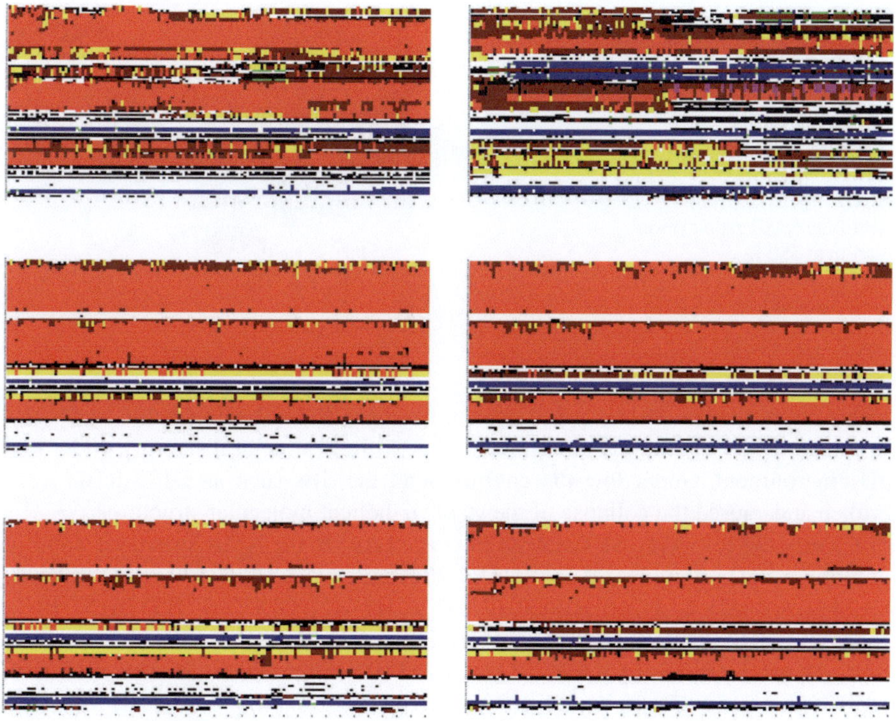

Fig. 4.9 Secondary structures of *rabbit, human* and *mouse* prion proteins (from up to down) at 350 K under *neutral* to *low* pH environments (from left to right) (X-axis: 0–30 ns (from left to right), Y-axis: residue numbers 124–228/125–228/124–226 (from up to down))

order to get the low pH environment). The systems were surrounded with a 12 angstrom layer of TIP3PBOX water molecules and neutralized by sodium ions using XLEaP module of AMBER 11. The solvated proteins with their counterions were minimized mainly by the steepest descent method and then a small number of conjugate gradient steps were performed on the data, in order to remove bad HB contacts. Then the solvated proteins were heated from 100 to 300 K during 1 ns (with step size 1 fs) and from 300 to 350 K during 1 ns (with step size 2 fs). The thermostat algorithm used is the Langevin thermostat algorithm in constant NVT ensembles. The SHAKE algorithm and PMEMD algorithm with nonbonded cutoffs of 12 Å were used during the heating. Equilibrations were done in constant NPT ensembles under Langevin thermostat for 2 ns. After equilibrations, production MD phase was carried out at 350 K for 30 ns using constant pressure and temperature ensemble and the PMEMD algorithm with nonbonded cutoffs of 12 Å during simulations. Step size for the production runs is 2 fs. The structures were saved to file every 1000 steps.

Seeing Fig. 4.9, we know that, at 350 K, human and mouse prion proteins have stable molecular structures whether under neutral or low pH environments; but for

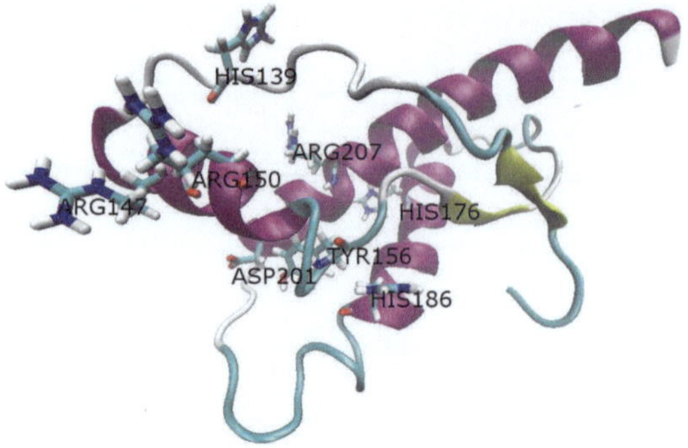

Fig. 4.10 Some special SBs ARG207-HIS176, TYR156-HIS186, HIS139-ARG150, ASP201-ARG147, ASP201-ARG150, ASP201-HIS186, ARG155-GLU151 of wild-type NMR rabbit prion protein at 350 K

rabbit prion protein. Rabbit prion protein has been found having stable molecular structures under neutral pH environment, but without structural stability under low pH environment. Under low pH environment, the SBs such as D177-R163 were broken and caused the collapse of the stable α-helical molecular structures.

Clearly the following SBs play an important role to the NMR structural stability of rabbit prion protein: (1) GLU210-ARG207-GLU206-LYS203 (99.78 %, 88.85 %, 82.74 %, H3-H3), GLU210-HIS176 (74.31 %, H3-H2), GLU206-HIS176 (57.10 %, H3-H2), ARG207-HIS176 (0.52 %, H3-H2), ASP177-ARG163 (19.54 %, H2-S2); (2) ARG150-ASP146-ARG147-ASP143 (91.38 %, 100 %, 86.43 %, H1-H1), HIS139-ARG150 (50.96 %), HIS139-ASP146 (92.62 %); (3) ASP201-ARG155 (10.07 %, H3-H1), ASP201-ARG150 (2.61 %, H3-H1), ASP201-ARG147 (0.01 %, H3-H1), ASP201-HIS186 (0.50 %, H3-H2); and (4) ARG155-ASP201 (10.07 %, H1-H3), TYR156-HIS186 (H1-H2, 71.69 %), ARG155-GLU151 (20.70 %, H1-H1), ARG155-GLU195 (0.06 %), where '%' denotes the percentage during the whole simulation of 30 ns. Compared with human and mouse NMR prion proteins, rabbit NMR prion protein has some special SBs which contribute to its structural stability at 350 K during the simulation of 30 ns (Fig. 4.10) (human, mouse, dog and horse NMR prion proteins have not these SBs).

Under low pH environment at many levels of temperatures with different starting MD velocities, rabbit prion protein always unfolds its α-helical structures into β-sheet structures. Prion diseases are just caused by the conversion from predominant α-helices of PrP^C into rich β-sheets of PrP^{Sc}. This implies to us we might have found some secrets being revealed prion diseases from rabbit prion protein. In the next chapter, we will furthermore confirm this finding compared with dog and horse prion proteins.

Chapter 5
Compared with the NMR Structure and Dynamics of Dogs and Horses

In 2008, canine mammals including dogs (canis familials) were the first time academically reported to be resistant to prion diseases [474]; and in 2010 horses were reported to be resistant to prion diseases too [336]. By now all the NMR structures of rabbit, dog and horse prion proteins had been released into protein data bank [381, 406, 463]. Thus, at this moment it is very worth studying the NMR molecular structures of horse, dog and rabbit prion proteins to obtain insights into their immunity prion diseases.

5.1 Rabbits Compared with Dogs

This section studies the molecular structural dynamics of wild-type dog prion protein compared with wild-type rabbit prion protein. The comparison analyses with rabbit prion protein show that the dog prion protein has stable molecular structures whether in neutral or low pH environments. The SBs such as D177-R163 contribute to the structural stability of wild-type rabbit prion protein in neutral pH environment was found by this section.

5.1.1 Overview

In Chap. 4 the author studied the NMR structure of wild-type rabbit prion protein by MD simulations, and simulation results at 450 K in low and neutral pH environments confirmed the structural stability of wild-type rabbit prion protein in neutral pH environment, but in low pH environment the wild-type rabbit protein protein is without structural stability. In this section another set of MD simulation starting from different initial velocities (so-called *seed2*) will be repeated for the wild-type

J. Zhang, *Molecular Structures and Structural Dynamics of Prion Proteins and Prions*, Focus on Structural Biology 9, DOI 10.1007/978-94-017-7318-8_5

rabbit prion protein. We call the MD simulation initial velocity of Chap. 4 *seed1*. For both *seed1* and *seed2*, the dog prion protein (PDB ID: 1XYK) is studied in this section.

All the MD simulations in this section confirmed the structural stability of wild-type dog prion protein in both neutral and low pH environments. The analyses of SBs, HBs and HYDs for dog prion protein will be done in order to seek reasons of the stability.

5.1.2 Materials and Methods at 300 and 450 K

The MD simulation materials and methods for dog and rabbit prion proteins are completely same as the ones of Chap. 4. Simulation initial structures for the dog and rabbit prion proteins were built on DogPrP^C(121–231) (PDB entry 1XYK) and RaPrP^C(124–228) (PDB entry 2FJ3), respectively. Simulations were done in low pH and normal pH environments respectively. The simulations of dog prion protein were done starting from two sets of initial velocities (*seed1* and *seed2*). The simulations of rabbit prion protein were done starting from the same *seed1* and *seed2*, i.e. the same two sets of initial velocities. MD simulation experience of the author showed that other additional seeds could not make much difference for the MD simulations. All the simulations were performed with the AMBER 9 package [88], with analysis carried out using functionalities in AMBER 9 [88] and AMBER 7 [92] CARNAL program. Graphs were drawn by XMGRACE of Grace 5.1.21, DSSP [326].

All simulations used the ff03 force field of the AMBER 9 package, in neutral and low pH environments (where residues HIS, ASP, GLU were changed into HIP, ASH, GLH respectively by the XLEaP module of AMBER 9 in order to get the low pH environment). The systems were surrounded with a 12 Å layer of TIP3PBOX water molecules and neutralized by sodium ions using XLEaP module of AMBER 9. 15 Cl−, 14 Cl− and 4337 waters, 5909 waters were added for the dog and rabbit prion wild-type proteins respectively for the low pH environment. The solvated proteins with their counterions were minimized mainly by the steepest descent method and then a small number of conjugate gradient steps were performed on the data, in order to remove bad HB contacts. Then the solvated proteins were heated from 100 to 450 K step by step during 3 ns. The thermostat algorithm used is the Langevin thermostat algorithm in constant NVT ensembles. The SHAKE algorithm and PMEMD algorithm with nonbonded cutoffs of 12 Å were used during the heating. Equilibrations were done in constant NPT ensembles under Langevin thermostat for 5 ns. After equilibrations, production MD phase was carried out at 450 K for 30 ns using constant pressure and temperature ensemble and the PMEMD algorithm with nonbonded cutoffs of 12 Å during simulations. Step size for equilibration was 0.5 fs, and 1 fs for the production runs. The structures were saved to file every 1000 steps.

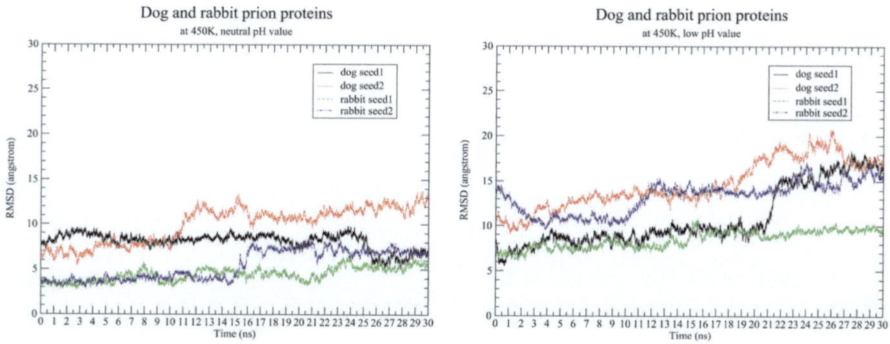

Fig. 5.1 Backbone atom RMSD graphs for rabbit and dog prion proteins

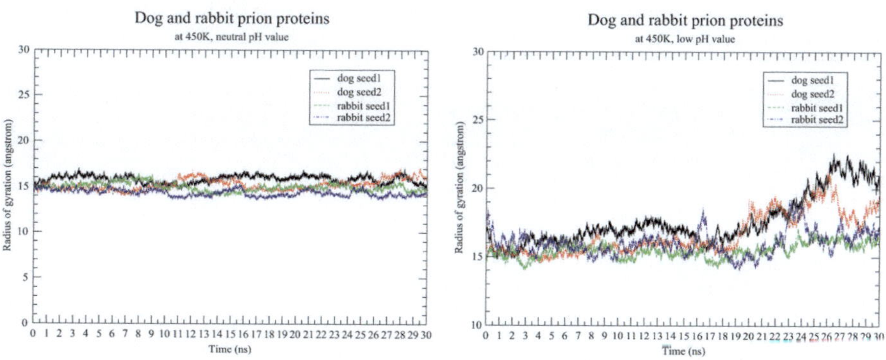

Fig. 5.2 Radius of gyration graphs for rabbit and dog prion proteins

5.1.3 Results and Discussions for 300 and 450 K

For both the *seed1* and *seed2*, the MD simulations done at room temperature 300 K whether in neutral or low pH environment display very little fluctuation and no variation between dog and rabbit prion proteins. At 450 K there are fluctuation and variation among dog and rabbit prion proteins, but we cannot find their real difference between their backbone atom RMSDs (which were calculated respectively from their minimized structures) and their radii of gyrations (Figs. 5.1–5.2).

However, a clear difference is found between dog and rabbit prion proteins. It is the secondary structures of dog and rabbit prion proteins in low pH environment at 450 K: the α-helices of rabbit prion protein were completely unfolded and began to turn into β-sheets but those of dog prion protein were not changed very much (Figs. 5.3–5.4).

These results indicate the C-terminal region of RaPrP^C has lower thermostability than that of DogPrP^C. We found that, for both the *seed1* and *seed2*, rabbit has

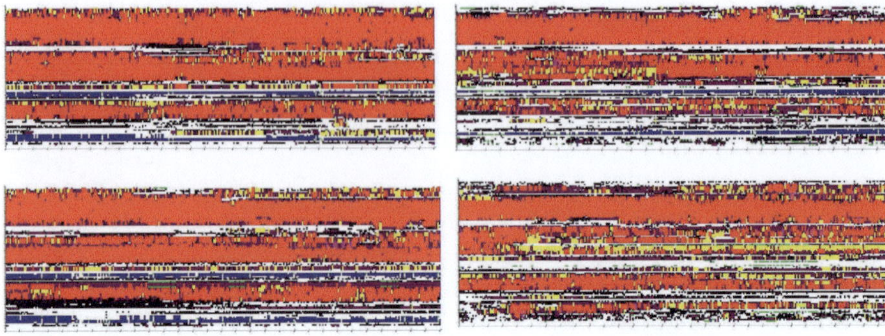

Fig. 5.3 Secondary structures of rabbit (*left*) and dog (*right*) prion proteins at 450 K in *neutral* pH environment: up – *seed1*, down – *seed2* (*red*: α-Helix, *pink*: π-helix, *yellow*: 3_{10}-helix, *green*: β-bridge, *blue*: β-sheet, *purple*: turn, *black*: bend; horizontal axis: time (0–30 ns from *left* to *right*), vertical axis: residue numbers (from down to up, 124–228 for rabbit and 121–231 for dog))

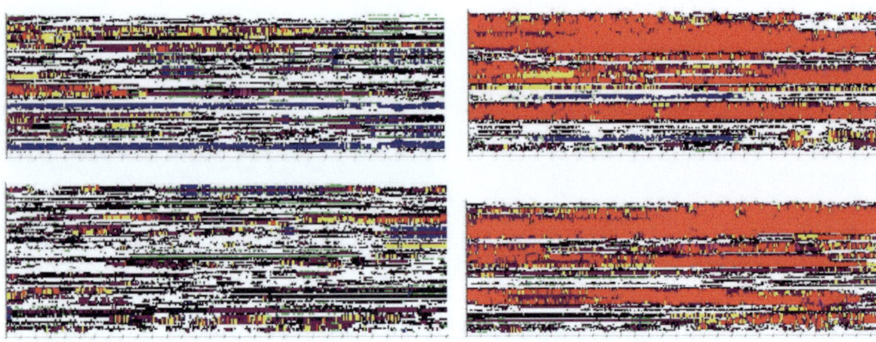

Fig. 5.4 Secondary structures of rabbit (*left*) and dog (*right*) prion proteins at 450 K in *low* pH environment: up – *seed1*, down – *seed2* (*red*: α-Helix, *pink*: π-helix, *yellow*: 3_{10}-helix, *green*: β-bridge, *blue*: β-sheet, *purple*: turn, *black*: bend; horizontal axis: time (0–30 ns from *left* to *right*), vertical axis: residue numbers (from down to up, 124–228 for rabbit and 121–231 for dog))

two important SBs D177-R163, D201-R155 for the whole 30 ns in neutral pH environment (the remove of these SBs led to the collapse of its stable helical structure), but for dog without the SB D202-R156. This means the SB D177-R163 contributes to the structural stability of rabbit prion protein (Figs. 5.5–5.7).

As reported in Chap. 4, recently the helix-capping motif between residues 166 and 174 (i.e. the S2-H2 loop) was also reported to contribute to the structural stability of rabbit prion protein [336, 517], and recently again Wen et al. (2010) reported that the ordered loop 165–172 contributes to the unique structural stability of RaPrP^C [626]. Sweeting et al. (2009) reported that the rabbit prion mutants of S170N, S174N, and S170N/S174N break the HB network of 170–174 residues [557]. We may see that the motif, the loop 165–172, and the 170–174 amino acids are just between R163 and D177. If we only show the SBs that are resident for more than 5 % of the whole 30 ns simulations, the following SBs contribute to the stable

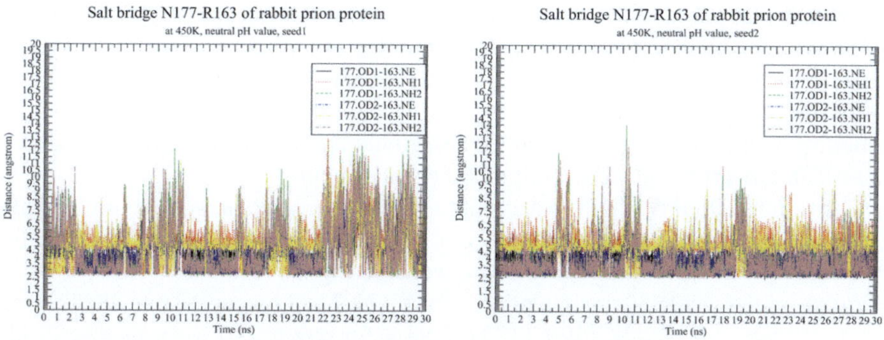

Fig. 5.5 The SB D177-R163 contributes to structural stability of $RaPrP^C$

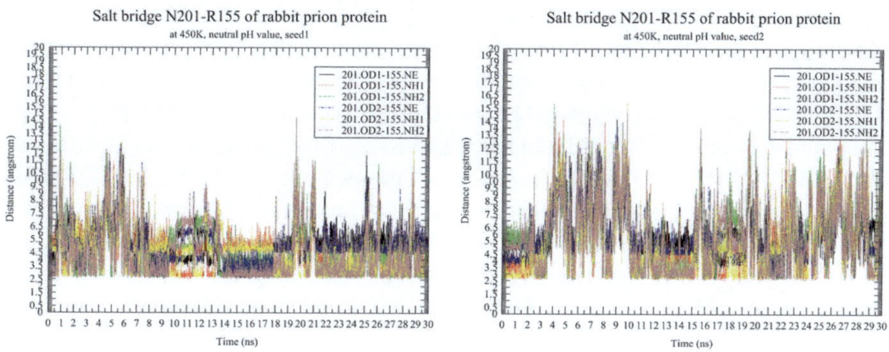

Fig. 5.6 The SB D201-R155 contributes to structural stability of $RaPrP^C$

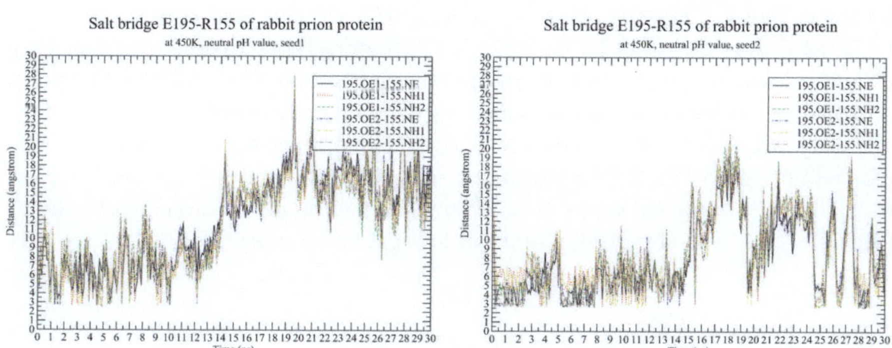

Fig. 5.7 The SB E195-R155 contributes to structural stability of $RaPrP^C$

structures of rabbit prion protein in neutral pH environment: D177-R163 (occupied rate 38.33 % for *seed1*, 47.67 % for *seed2*), H186-R155 (38.33 % for *seed1*, 10 % for *seed2*), R155-E195 (7 % for *seed1*, 21.33 % for *seed2*), E145-R135 (35.33 % for *seed2*), D146-H139 (72.67 % for *seed1*), D201-R155 (6.67 % for *seed1*, 5.67 %

for *seed2*) and D201-H186 (5.33 % for *seed1*). The HB occupied rates for H186-R155 and E195-R155 respectively are: H186-R155 (17.70 % for *seed1*), E195-R155 (5.53 % for *seed1*, 27.83 % for *seed2*). Thus, we may say that SBs D177-R163 and D201-R155 greatly and H186-R155 partially contribute to structural stability of rabbit prion protein. In this section we also found that the SB E195-R155 partially contributes to structural stability of rabbit prion protein.

Why does DogPrP^C still have stable molecular structures even in low pH environment? in low pH environment, dog prion protein has 22 and 27 strong HBs respectively for *seed1* and *seed2*, if only considering the resident of HBs for more than 5 % of the whole 30 ns simulation. HBs C179-T183 (occupied rate 28.08 % for *seed1*, 47.50 % for *seed2*), T190-T193 (12.93 % for *seed1*, 6.07 % for *seed2*), T188-T192 (24.85 % for *seed2*), H187-T191 (11.17 % for *seed2*), and N173-R177 well maintain H2. H3 is well maintained by HBs Q212-T216 (29.68 % for *seed1*, 61.33 % for *seed2*), T218-S222 (21.07 % for *seed1*, 21.22 % for *seed2*), M213-T216, and M213-Q217. The HBs T193-T199 (17.60 % for *seed1*) and N197-T201 (14.77 % for *seed1*) well connect H2, H3 and the loop between the two helices. Moreover, Q160-H187 (9.45 % for *seed1*) connects H2 with the loop between H1 and H2; N153-T199 maintains the connection of H1 and H3; and S132-T163, S132-N153 (11.85 % for *seed2*), T128-T163 (11.60 % for *seed2*) links the two antiparallel S1 and S2. Thus, we may see that the helical structure of dog prion protein has been well maintained by these HBs during the long simulations of 30 ns. Numerical computational results also show that HYDs contribute greatly to the structural stability of dog prion protein in low pH environment. During the whole 30 ns of simulations, for both *seed1* and *seed2*, the following strong HYDs are more than 90 % resident in the core of the protein: V210-V209, V209-M206, M206-I205, M206-M203, I205-M203, V209-I205, M213-V210, V215-M213, M213-V209, M206-V210 (H3), V184-V180, V176-F175 (H2), V210-V180 (linking H3 and H2), and I139-L138, L138-P137, M134-A133, M129-L130, V121-V122 (before H1) , V166-P165 (after H1). Compared with dog prion protein, except for one between A224-A223, rabbit prion protein does not have other HYDs which resides with more than 90 % occupied rate.

Lastly, we show some snapshots at certain important time points in the MD time evolution. In Fig. 5.1 we may see a large variation at 10–12 ns and 15–17 ns respectively for dog and rabbit prion proteins in neutral pH environment for *seed2* (Figs. 5.8–5.9). For dog prion protein in low pH environment, in Fig. 5.2 we may

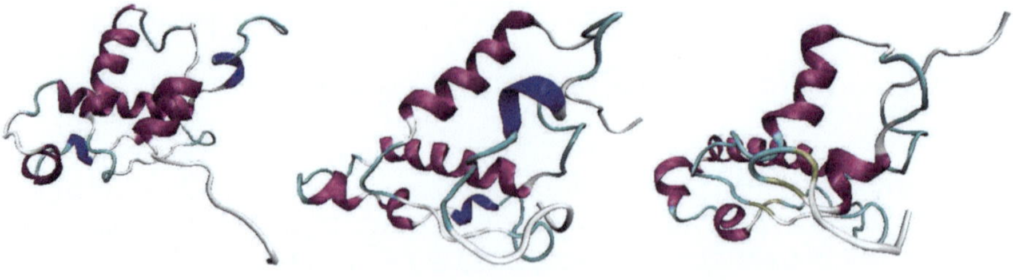

Fig. 5.8 Dog prion protein at 10 11, 12 ns, in neutral pH environment for *seed2*

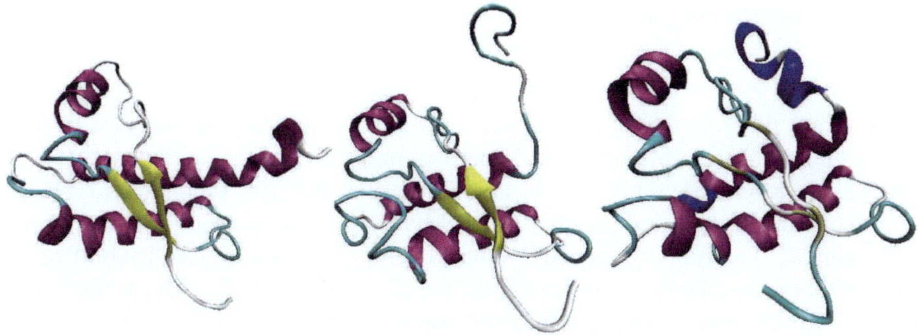

Fig. 5.9 Rabbit prion protein at 15 16, 17 ns, in neutral pH environment for *seed2*

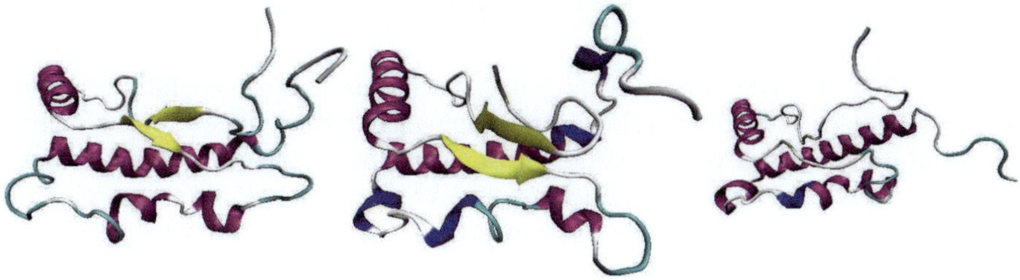

Fig. 5.10 Dog prion protein at 15 17, 20 ns, in low pH environment for *seed1*

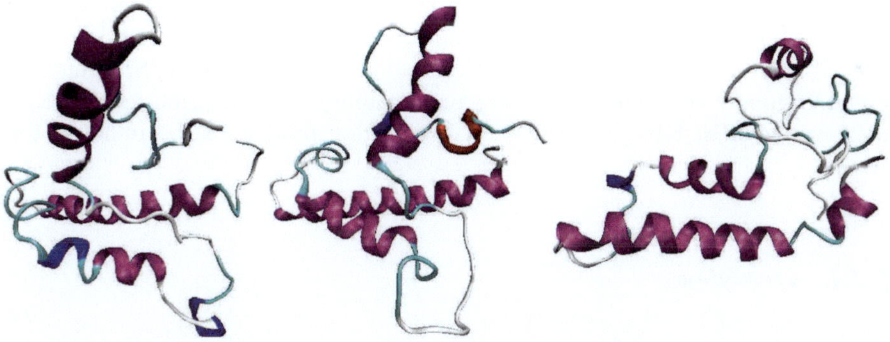

Fig. 5.11 Dog prion protein at 15 17, 20 ns, in low pH environment for *seed2*

see a large variation at 17 ns for both the *seed1* and *seed2*, thus we illuminate the structures before and after 17 ns: at 15 ns and at 20 ns (Figs. 5.10–5.11).

By Figs. 5.9–5.10 we may know that, the large variation at point 17 ns is due to the unfolding of β-sheets because of the break of HBs between the two β-sheets.

5.1.4 Concluding Remarks

This section carried out MD simulations of the structured regions of dog and rabbit prion proteins from different sets of initial velocities. The analyses of MD simulation results confirmed the structural stability of dog prion protein (even in low pH environment). This section also found that the SBs such as D177-R163 contribute to the structural stability of wild-type rabbit prion protein in neutral pH environment. The identity rate and similarity rate between dog and rabbit prion proteins are respectively 92.16 % and 98.04 %. Except two residues at positions 159 and 174 are different, all other different residues are in H3.

5.2 Rabbits Compared with Horses

Prion diseases are invariably fatal and highly infectious neurodegenerative diseases affecting humans and animals. However, by now there have not been some effective therapeutic approaches or medications to treat all these prion diseases. Rabbits, dogs, and horses are the only mammalian species reported to be resistant to infection from prion diseases isolated from other species. Recently, the S2-H2 loop has been reported to contribute to their protein structural stabilities. The author has found that rabbit prion protein has a strong SB ASP177-ARG163 (like a taut bow string) keeping this loop linked. This section confirms that this SB also contributes to the structural stability of horse prion protein. Thus, the region of S2-H2 loop might be a potential drug target region. Besides this very important SB, other four important SBs GLU196-ARG156-HIS187, ARG156-ASP202 and GLU211-HIS177 are also found to greatly contribute to the structural stability of horse prion protein. Rich databases of SBs, HBs and HYDs for horse prion protein can be found in this section.

5.2.1 Overview

Prion diseases such as CJD, vCJD, GSS, FFI, Kuru in humans, scrapie in sheep, BSE or 'mad-cow' disease and CWD [532] in cattle are invariably fatal and highly infectious neurodegenerative diseases affecting humans and animals. However, by now there have not been some effective therapeutic approaches or medications to treat all these diseases [8, 486, 624]. In 2008, canine mammals including dogs (canis familials) were the first time academically reported to be resistant to prion diseases [474] (where, before 2008, dogs and horses were reported to resist prion diseases in media). Rabbits are the mammalian species known to be resistant to infection from prion diseases from other species [591]. Horses were academically reported to be resistant to prion diseases too [463]. Thus, it is very worth studying the molecular structures of dog, rabbit and horse prion proteins to obtain insights into the immunity of dogs, rabbits and horses to prion diseases. MD simulations provide an excellent method to understand the structural stability as well as ligand interactions of biological systems. It should be noted that recently there have been a devoted effort by MD to understand the human prion proteins [414, 712, 713]. This author has already investigated by MD simulations the dog and rabbit prion proteins in the above section. This section is focusing on the MD simulation studies on the horse prion protein C-terminal structured region, as well as a comparison of rabbit, dog and horse prion proteins.

Rabbits, dogs and horses are resistant to prion diseases. The infectious prion (PrP^{Sc}) is an abnormally folded form of the normal cellular prion (PrP^{C}) and the conversion of PrP^{C} to PrP^{Sc} is believed to involve conformational change from a predominantly α-helical protein (42 % α-helix, 3 % β-sheet) to a protein rich in β-sheets (30 % α-helix, 43 % β-sheet). For a protein structure, its stability is maintained by its HBs, HYDs, SBs, and vdW contacts. Thus, when we study their NMR structure and structural dynamics, we have to consider these factors. The secondary structure of rabbit prion protein turned into β-sheet structures from α-helical structures under low pH environment. The change of pH environments causes the change of secondary structure from α-helices into β-sheets, because the SBs disappear under low pH environment. Thus, we might say that the SB provides the stability and it might be a drug target.

All the MD simulations in this paper confirmed the structural stability of wild-type horse prion protein under both neutral and low pH environments. The analyses of SBs, HBs and HYDs for horse prion protein will be done in order to seek reasons of the stability (where the SBs, HBs and HYDs will be presented in this section at the residue-residue level).

5.2.2 *Materials and Methods for 350 K*

The MD simulation materials and methods for horse prion protein are completely same as the ones of Sect. 5.1.2. Simulation initial structure for the horse prion protein was built on HoPrP^{C}(119–231) (PDB entry 2KU4). The low pH in the simulations is achieved by the change of residues HIS, ASP, GLU into HIP, ASH, GLH respectively and the Cl− ions added by the XLEaP module of AMBER 11. The neutral pH in the simulations is achieved by the change of residues HIS into HID and the Na+ ions added by the XLEaP module of AMBER 11. 16 Cl− and 6281 waters were added for the horse prion protein under low pH environment, and 2 Na+ and 6679 waters were added under neutral pH environment. After equilibrations, 30 ns' production MD simulations were done using constant pressure and 350 K temperature ensemble for the horse prion protein for both the *seed1* and *seed2* defined in Sect. 5.1.1. The two seeds are two different initial velocities. Different initial conditions should produce the same thermodynamic quantities after equilibration. This will firmly ensure that our research findings in this section are correct.

The studies on the rabbit prion protein at 300, 450, 500 K in Chaps. 2–4 and the dog prion protein at 300, 450 K (Sect. 5.1 of this chapter) have confirmed the research findings on the SB ASP178-ARG164 and the secondary structure change of rabbit prion protein from neutral to low pH environments. 350 K is a practical experimental laboratory temperature for prion proteins reported. Thus, 350 K is set for $HoPrP^C$ in this section.

5.2.3 *Results and Discussions for 350 K*

For both the *seed1* and *seed2*, the MD simulations at room temperature 300 K whether under neutral or low pH environments display very little fluctuation. At 350 K there is fluctuation and variation for different pH values, but we cannot find their real difference between (i) their backbone atom RMSDs (Fig. 5.12), which were calculated respectively from the minimized structure, (ii) their radii of gyrations (Fig. 5.12), and (iii) their secondary structures (Fig. 5.13).

Why does $HoPrP^C$ still have stable molecular structures even under low pH environment? Under low pH environment, horse prion protein has 45 and 38 strong HBs respectively for *seed1* and *seed2*, if only considering the resident of HBs for more than 5 % of the whole 30 ns simulations. By analysis, during the long simulations of 30 ns we can see that all these HBs well maintain the three α-helices and the two β-sheets, and especially their interactions linked by the following HBs (the first percentage is for *seed1* and the second percentage is for *seed2*):

- SER170-TYR218 (linking the S2-H2 loop with H2, 78.00 %, 65.52 %),
- ASP202-TYR157 (linking H3 with H2, 66.05 %, 65.44 %),
- TYR157-ARG136 (linking the S1-H1 loop with H2, 68.27 %, 71.04 %),
- MET134-ASN159 (linking the S1-H1 loop with the H1-S2 loop, 29.83 %, 21.47 %),

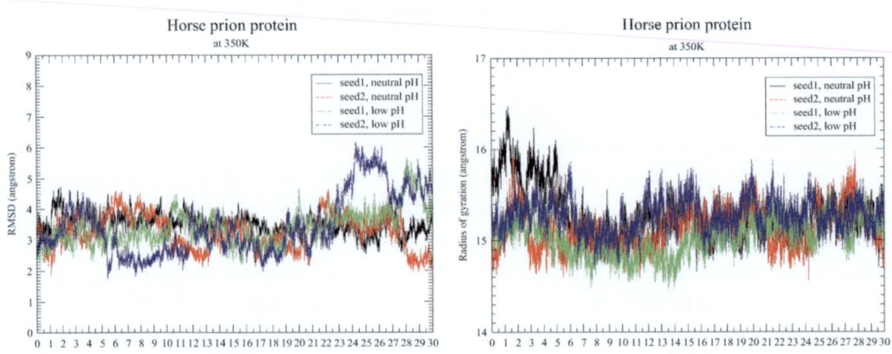

Fig. 5.12 The RMSD and radius of gyration graphs for horse prion protein at 350 K under neutral pH environment

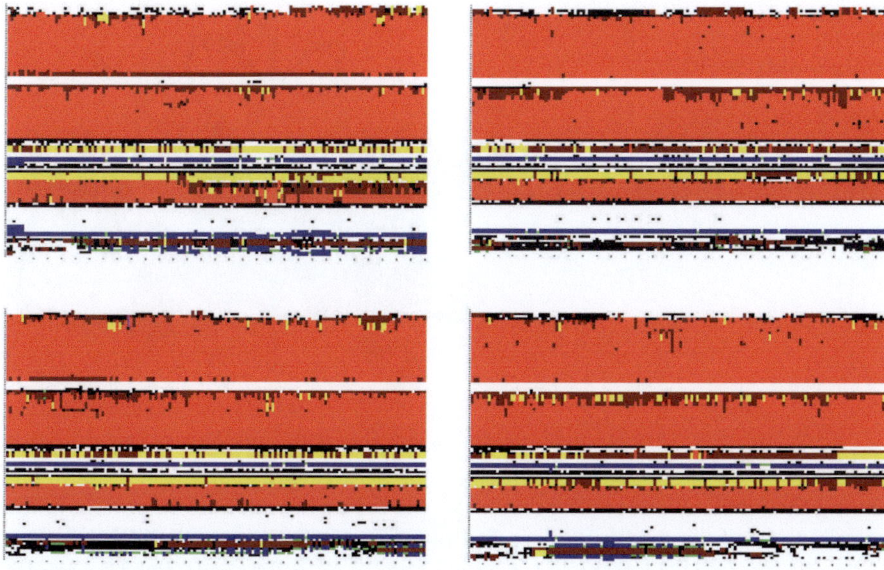

Fig. 5.13 Secondary structures of horse prion protein at 350 K in *neutral* (*left*) and *low* (*right*) pH environments: up – *seed1*, down – *seed2* (*red*: α-helix, *pink*: π-helix, *yellow*: 3_{10}-helix, *green*: β-bridge, *blue*: β-sheet, *purple*: turn, *black*: bend; horizontal axis: time (0–30 ns from *left* to *right*), vertical axis: residue numbers (from down to up, 119–231))

- GLY131-GLN160 (linking the S1-H1 loop with the H1-S2 loop, 26.67 %, 37.03 %),
- HIS140-ARG208 (linking the S1-H1 loop with H3, 43.85 %, 10.80 %),
- SER132-GLN217 (linking the S1-H1 loop with H3, 29.70 %, 12.90 %),
- PRO137-TYR149 (linking the S1-H1 loop with H1, 17.18 %, 23.39 %),
- PRO158-ARG136 (linking the S1-H1 loop with S2, 9.53 %, 6.01 %),
- ARG156-HIS187 (linking H1 and H2, 6.64 %, 33.70 %), and
- GLU221-TYR163 (linking H3 and H2, 11.40 %, 6.22 %).

HYDs also contribute greatly to the structural stability of horse prion protein in low pH environment. In all there are 1760 and 2006 HYDs respectively for *seed1* and *seed2* during the whole 30 ns of simulations, and for *seed1* and *seed2* respectively, there are 383 and 363 strong HYDs are100 % resident in the core of the protein. The three HYDs CYS214-CYS179, CYS214-VAL176, TYR162-LEU130 should be well noticed, where there is a disulfide bond between CYS179 and CYS214 linking H2 and H3, VAL176 is in H2, and TYR162-LEU130 (but 99.06 % for *seed2*) are just respectively in the S1 and S2 of the antiparallel β-sheet (we noticed that there is not a HB between TYR162 and LEU130).

Under the neutral pH environment, there are 33 SBs respectively for *seed1* and *seed2*, which contribute to the structural stability of horse prion protein. These SBs well keep the structural stability of the three helices and their interactions.

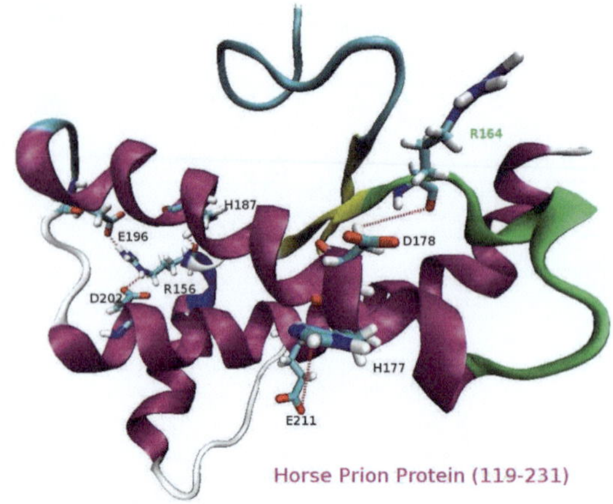

Fig. 5.14 SBs GLU211-HIS177, GLU196-ARG156-HIS187, ARG156-ASP202 and ASP178-ARG164 of horse prion protein (*dash line*: a SB between the two residues, *green color*: the S2-H2 loop)

During the whole 30 ns, among these SBs the following ones are important in the contributions of the three α-helices and two β-sheets' structural stability of horse prion protein: GLU211-HIS177 (occupied rate 27.48 % for *seed1* and 17.77 % for *seed2*), GLU196-ARG156 (occupied rate 17.17 % for *seed1* and 43.47 % for *seed2*), ARG156-HIS187 (occupied rate 9.14 % for *seed1* and 75.60 % for *seed2*), ARG156-ASP202 (occupied rate 0.25 % for *seed1* and 6.69 % for *seed2*), ASP178-ARG164 (occupied rate 10.45 % for *seed1* and 4.30 % for *seed2*). The positions of all these important SBs can be seen in Fig. 5.14. We can see that the residue ARG156 of H1 forms a network of three SBs (and HBs) linking H2 and H3.

Recently, some researchers [206, 336, 390, 463, 557, 625, 626] reported that the S2-H2 loop plays an important role to stabilize the structural stability of rabbit and horse prion proteins. This conclusion can be confirmed from the HB, HYD, and SB databases on horse prion protein presented by this paper. All their reports are not addressing the SB problem of this loop. Like Sect. 5.1 and Chaps. 2–4, this section also reports the important SB ASP178-ARG164 (like a taut bow string) keeping this loop linked. The important SBs reported in this section, together with the highly ordered S2-H2 loop and its interactions with H3, maintain then structural stability of horse prion protein in a perfect way.

Lastly, we make some comparison of horse prion protein with dog and rabbit prion proteins. Their sequences alignment is shown in Fig. 5.15. In Fig. 5.15, "*" means that the residues in that column are identical in all sequences in the alignment, ":" means that conserved substitutions have been observed, "." means that semi-conserved substitutions are observed, the RED color takes place at small (small + hydrophobic (incl.aromatic-Y)) residues, the BLUE color takes place at acidic residues, the MAGENTA color takes place at Basic-H residues,GREEN color takes place at Hydroxyl + sulfhydryl + amine + G residues, and Grey color takes place at unusual amino/imino acids etc.

We may see in Fig. 5.15 that there are more identities and similarities between horse and dog prion proteins, compared with the identities and similarities between

```
CLUSTAL 2.1 multiple sequence alignment

1XYK_6_DoPrP.pdb      --VVGGLGGYMLGSAMSRPLIHFGNDYEDRYYRENMYRYPDQVYYRPVDQYSNQNNFVRD 58
2FJ3_14_RaPrP.pdb     ------LGGYMLGSAMSRPLIHFGNDYEDRYYRENMYRYPNQVYYRPVDQYSNQNSFVHD 54
2KU4_3_HoPrP.pdb      GSVVGGLGGYMLGSAMSRPLIHFGNDYEDRYYRENMYRYPNQVYYRPVSEYSNQKNFVHD 60
                            **********************************:*******.:****:.**:*

1XYK_6_DoPrP.pdb      CVNITVKQHTVTTTTKGENFTETDMKIMERVVEQMCVTQYQKESEAYYQRGAS 111
2FJ3_14_RaPrP.pdb     CVNITVKQHTVTTTTKGENFTETDIKIMERVVEQMCITQYQQESQAAYQRA-- 105
2KU4_3_HoPrP.pdb      CVNITVKQHTVTTTTKGENFTETDVKIMERVVEQMCITQYQKEYEAFQQRGAS 113
                      ************************:***********:****:* :*  **.
```

Fig. 5.15 Multiple sequence alignment of dog, rabbit, and horse prion proteins

horse and rabbit prion proteins, between dog and rabbit prion proteins. This point can be furthermore confirmed by the following data of the Needleman-Wunsch or Smith-Waterman Pairwise Sequence Alignment and the jCE algorithm or jCE Circular Permutation Pairwise Structure Alignment. For horses and dogs, the Identities are 89.38 % (query) and 90.99 % (subject) and the Similars are 93.81 % (query) and 95.50 % (subject) for the Sequence Alignment, and the Identity is 89.11 % and the Similarity is 93.07 % for the Structure Alignment. For horses and rabbits, the Identities are 87.61 % (query) and 71.74 % (subject) and the Similars are 93.81 % (query) and 76.81 % (subject) for the Sequence Alignment, and the Identity is 71.72 % and the Similarity is 77.78 % for the Structure Alignment. For dogs and rabbits, the Identities are 72.46 % (query) and 90.09 % (subject) and the Similars are 76.81 % (query) and 95.50 % (subject) for the Sequence Alignment, and the Identity is 92.93 % and the Similarity is 98.99 % for the Structure Alignment. The 3D structural Identity and Similarity of horse and dog prion proteins (89.11 %, 93.07 %) are clearly very larger than those of horse and rabbit prion proteins (71.72 %, 77.81 %). This explains the reasons why rabbit prion protein differs very much from dog and horse prion proteins in secondary structures under low pH environment. The rabbit/ horse and dog prion proteins have a strong SB ASP177-ARG163/ASP178-ARG164 (like a taut bow string) keeping the S2-H2 loop linked. But the SB ASP178-ARG164 does not exist for human and mouse prion proteins (Chap. 4). For horses and humans (1QLX.pdb), the Identities are 86.73 % (query) and 46.67 % (subject) and the Similars are 51.43 % (subject) for the Sequence Alignment. For horses and mouses (1AG2.pdb), the Identities are 80.53 % (query) and the Similars are 87.61 % (query) for the Sequence Alignment. Thus, compared with the sequence alignments of horses-dogs (89.38 %, 90.99 %, 93.81 %, 95.50 %) and horses-rabbits (87.61 %, 71.74 %, 93.81 %, 76.81 %), the percentages of Identities and Similars of the sequence alignments of horses-humans (86.73 %, 46.67 %, 51.4 %) and horses-mouses (80.53 %, 87.61 %) are clearly very less. This is due to human and mouse prion proteins are non-resistive prion proteins, but dog, horse and rabbit prion proteins are prion disease resistive proteins. Rabbit prion protein differs from dog prion protein in that, except two residues at positions 158 and 173 are different, all other different residues are in H3 (Sect. 5.1.4). However, compared with horse prion protein, rabbit prion protein has different residues at positions 158, 159, 172, and 173. In the C-terminal end of H3, at positions 219, 222 and 228 rabbit prion protein differs from dog and horse prion proteins. We also find that horse prion protein has different residues from dog and rabbit prion proteins at positions 167, 168, 173, 212, 216. The residues at all these special positions should specially contribute to the structural stability of these resistive prion proteins.

5.2.4 *Concluding Remarks*

Rabbits, dogs, and horses are the only mammalian species reported to be resistant to infection from prion diseases isolated from other species. Recently, the S2-H2 loop has been reported to contribute to their protein structural stabilities. The author has found that rabbit prion protein has a strong SB ASP177-ARG163 (like a taut bow string) keeping this loop linked. This section confirms that this SB also contributes to the structural stability of horse prion protein. Thus, the region of S2-H2 loop should be a potential drug target region. Besides this very important SB, other four important SBs GLU196-ARG156-HIS187, ARG156-ASP202 and GLU211-HIS177 are also found to greatly contribute to the structural stability of horse prion protein.

5.2.4.1 Salt Bridges: A Clue of Rabbits Resisting Prion Diseases

The change of pH values from neutral one to low one will destroy SBs of a molecule. The author has found that, for prion proteins, at 350 K during long (30 ns) MD simulations rabbits are the unique mammalian species unfolding its stable α-helices into β-sheets, compared with dogs, horses, humans and mice (Fig. 5.16). This clearly implies to us that SBs might be a clue of rabbits which resist prion diseases.

Clearly the following SBs play an important role to the NMR structural stability of rabbit prion protein:

(1) *GLU210-ARG207-GLU206-LYS203, GLU210-HIS176, GLU206-HIS176, ARG207-HIS176, ASP177-ARG163;*
(2) *ARG150-ASP146-ARG147-ASP143, HIS139-ARG150, HIS139-ASP146;*
(3) *ASP201-ARG155, ASP201-ARG150, ASP201-ARG147, ASP201-HIS186 and*
(4) *ARG155-ASP201, TYR156-HIS186, ARG155-GLU151, ARG155-GLU195.*

Compared with human and mouse NMR prion proteins, rabbit NMR prion protein has some special SBs (Fig. 5.17) which contribute to its structural stability at 350 K

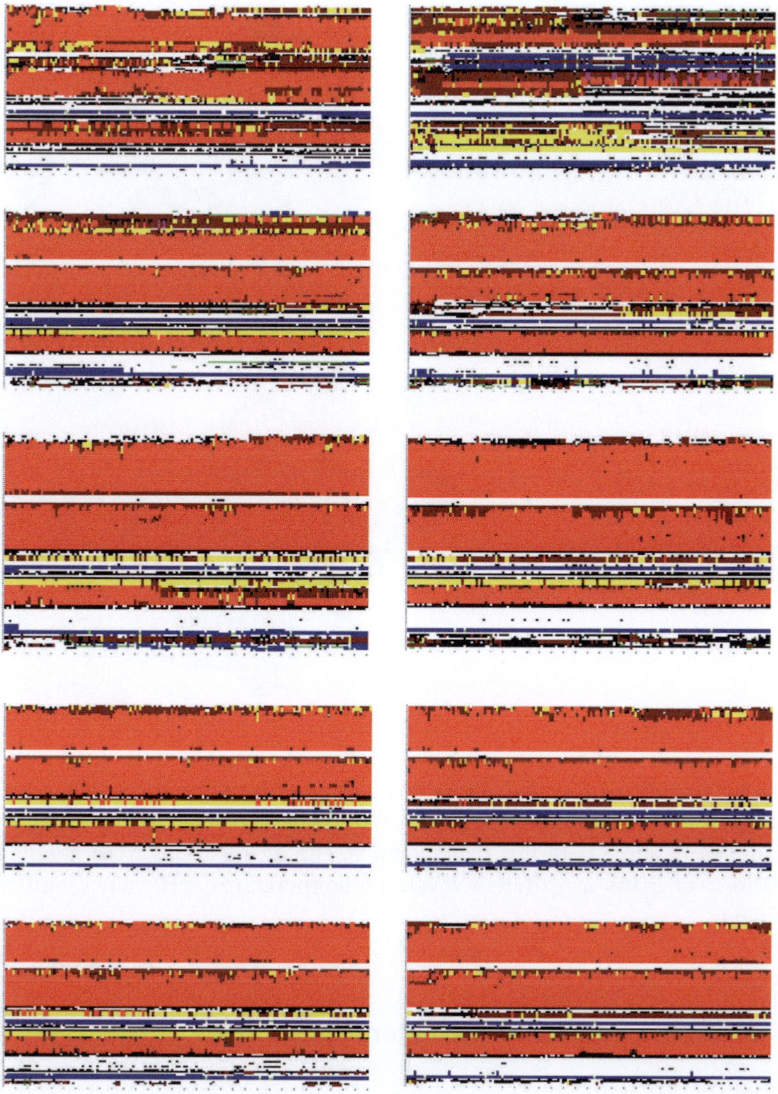

Fig. 5.16 Secondary structures of rabbit, dog, horse, human and mouse prion proteins: from up to down – rabbit, dog, horse, human, mouse; from *left* to *right* – neutral pH, low pH (*red*: α-helix, *pink*: π-helix, *yellow*: 3_{10}-helix, *green*: β-bridge, *blue*: β-sheet, *purple*: turn, *black*: bend; horizontal axis: time (0–30 ns from *left* to *right*), vertical axis: residue numbers). We can see the clear conversion of α-helices into β-sheets of rabbit prion protein from neutral pH to low pH environments

during the long MD simulation of 30 ns but human, mouse, dog and horse NMR prion proteins do not have these SBs. Thus, we might be able to say that SB is a clue of rabbits resisting prion diseases.

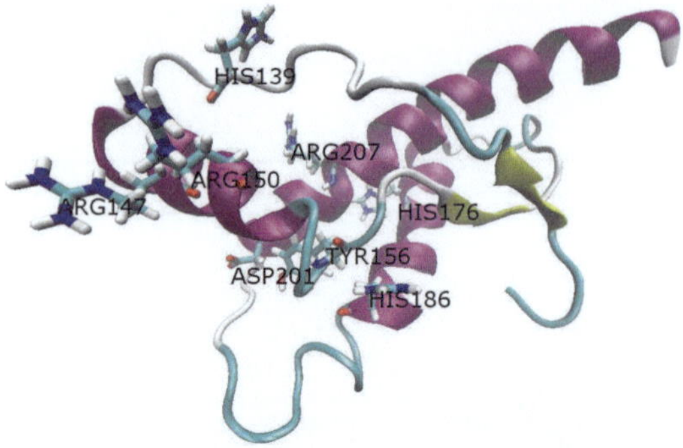

Fig. 5.17 Some special salt bridges *ARG207-HIS176, TYR156-HIS186, HIS139-ARG150, ASP201-ARG147, ASP201-ARG150, ASP201-HIS186, ARG155-GLU151* of wild-type NMR rabbit prion protein at 350 K

Fig. 5.18 Distributions of the surface charges of rabbit prion protein NMR structured 124–228 region under neutral pH environment with Na+ ions added (*Blue* is for positive charge whereas *red* is for negative charge)

The electrostatics in formula (1.1) was said to play an important role to stabilize rabbit prion protein structure [390, 625, 626] (see Fig. 5.18) (but the author does not completely agree with this conclusion by the observations from his long MD simulations of 30 ns at different levels of temperatures, pH values, initial MD velocities).

5.3 Rabbits Compared with Horses and Dogs at 350 K

Sections 4.5 and 5.2.2 did MD simulations at 350 K for rabbits and horses respectively, but Sect. 5.1 has not done MD simulations at 350 K for dogs yet. Thus, in this section, the MD simulations at 350 K for dogs will be done and the comparison analysis of rabbits with horses and dogs will be done. Some common point of all these three immunity species of prion diseases will be seeking.

5.3.1 Materials and Methods

The MD simulation materials and methods for horse, dog and rabbit prion proteins are completely same as the ones of Sects. 4.5 and 5.2.2. Simulation initial structure for the horse, dog and rabbit prion proteins were built on HoPrP^C(119–231) (PDB entry 2KU4), DogPrP^C(121–231) (PDB entry 1XYK) and RaPrP^C(124–228) (PDB entry 2FJ3), respectively. Simulations were done under low pH and normal pH environments respectively. All the simulations were performed with the AMBER 11 package [91], with analysis carried out using functionalities in AMBER 9 [88] and AMBER 7 CARNAL [92]. Graphs were drawn by XMGRACE of Grace 5.1.21, DSSP [326].

All simulations used the ff03 force field of the AMBER 11 package, in neutral and low pH environments (where residues HIS, ASP, GLU were changed into HIP, ASH, GLH respectively by the XLEaP module of AMBER 11 in order to get the low pH environment). The systems were surrounded with a 12 Ålayer of TIP3PBOX water molecules and neutralized by sodium ions using XLEaP module of AMBER 11. The solvated proteins with their counterions were minimized mainly by the steepest descent method and then a small number of conjugate gradient steps were performed on the data, in order to remove bad HB contacts. Then the solvated proteins were heated from 100 to 300 K during 1 ns (with step size 1 fs) and from 300 to 350 K during 1 ns (with step size 2 fs). The thermostat algorithm used is the Langevin thermostat algorithm in constant NVT ensembles. The SHAKE algorithm and PMEMD algorithm with nonbonded cutoffs of 12 Å were used during the heating. Equilibrations were done in constant NPT ensembles under Langevin thermostat for 2 ns. After equilibrations, production MD phase was carried out at 350 K for 30 ns using constant pressure and temperature ensemble and the PMEMD algorithm with nonbonded cutoffs of 12 Å during simulations. Step size for the production runs is 2 fs. The structures were saved to file every 1000 steps.

5.3.2 Results and Discussion

At 350 K there are fluctuation and variation among rabbit, dog and horse prion proteins, but their backbone atom RMSDs respectively calculated from their minimized structures and their radii of gyrations do not have great difference even under low pH environment. Their secondary structures under neutral pH environment at 350 K do not change very much either. However, the secondary structures under low pH environment at 350 K have great differences between rabbit prion protein and dog and horse prion proteins (Fig. 5.19): the α-helices of rabbit prion protein were completely unfolded and began to turn into β-sheets but those of dog and horse prion proteins were not changed very much. These results indicate the C-terminal region of RaPrP^C has lower thermostability than that of DogPrP^C and HoPrP^C. Under the low pH environment, the important SBs such as D177-R163, D201-R155

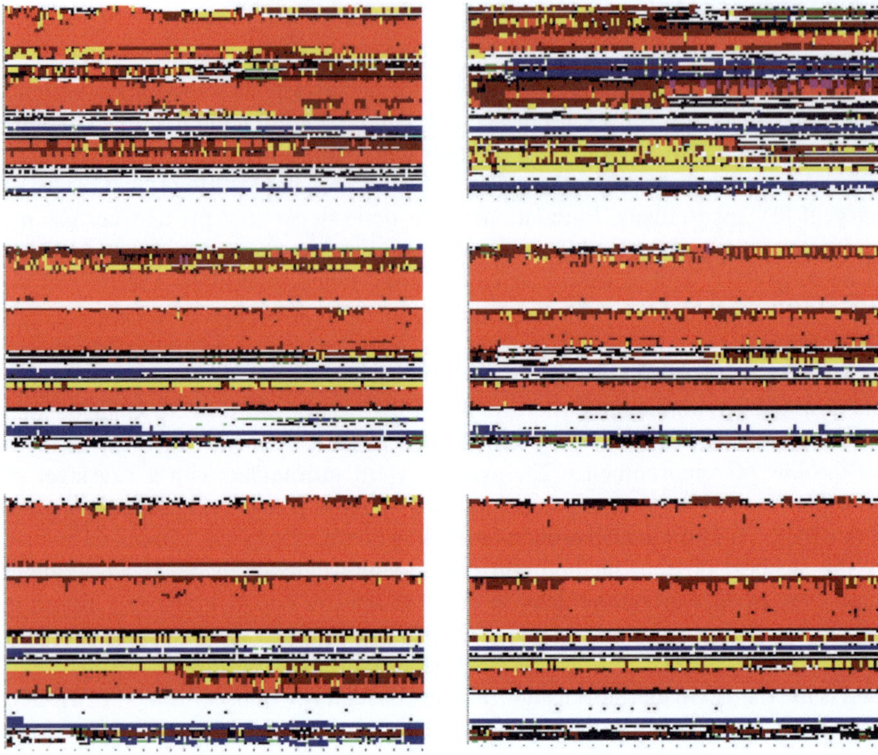

Fig. 5.19 Secondary structures of *rabbit, dog* and *horse* prion proteins (from up to down) at 350 K under *neutral* to *low* pH environments (from *left* to *right*) (X-axis: 0–30 ns (from *left* to *right*), Y-axis: residue numbers 124–228 / 121–231 / 119–231 (from up to down))

were removed (thus the free energies of the SBs changed the thermostability) so that the structure nearby the central helices 1–3 was changed for rabbit prion protein.

There always exists a strong SB (where the oxygen-nitrogen distance cut-off calculated for the SBs is 3.2 Å) between D177-R163 for rabbit prion protein, and D178-R164 for dog and horse prion proteins (Fig. 5.20) under neutral pH environment at 300, 350, and 450 K. We can see in Fig. 5.20 that this SB is just like a taut bow string linking S2 and H2. It contributes to the structural stability of all these prion proteins. Thus, the region of S2-H2 loop should be a potential drug target region according to recent reports on the S2-H2 loop of rabbit, horse and dog prion proteins [206, 336, 390, 463, 557, 625, 626].

Lastly, we compare rabbit with dog and horse prion proteins through the pairwise sequence and structure alignments. We find that there are more identities and similarities between horse and dog prion proteins, compared with the identities and similarities between dog and rabbit prion proteins, and between horse and rabbit prion proteins. This point can be confirmed by the following data of the Needleman-Wunsch or Smith-Waterman Pairwise Sequence Alignment and the jCE algorithm or jCE Circular Permutation Pairwise Structure Alignment. (1) For dogs and horses,

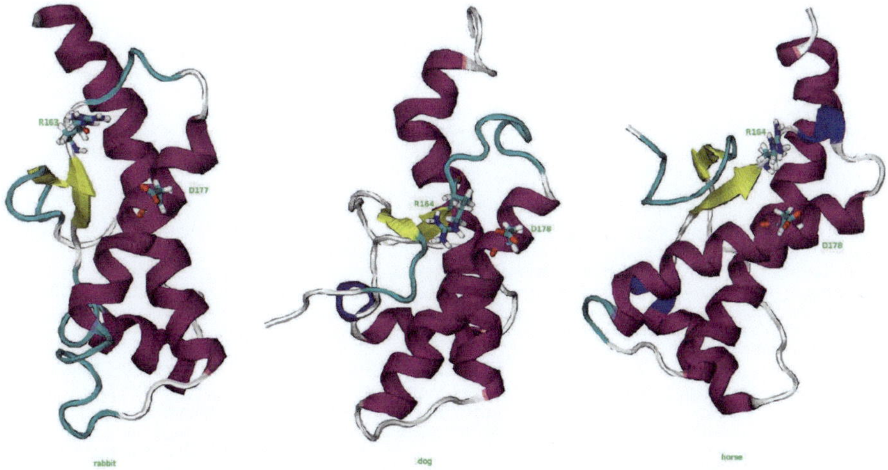

Fig. 5.20 The SB D177-R163 of rabbit prion protein, D178-R164 of dog and horse prion proteins. It is like a taut bow string of the S2-H2 loop

the Identities are 89.38 % (query) and 90.99 % (subject) and the Similars are 93.81 % (query) and 95.50 % (subject) for the Sequence Alignment, and the Identity is 89.11 % and the Similarity is 93.07 % for the Structure Alignment. (2) For dogs and rabbits, the Identities are 90.09 % (query) and 72.46 % (subject) and the Similars are 95.95 % (query) and 77.18 % (subject) for the Sequence Alignment, and the Identity is 92.93 % and the Similarity is 98.99 % for the Structure Alignment. (3) For horses and rabbits, the Identities are 87.61 % (query) and 71.74 % (subject) and the Similars are 93.81 % (query) and 76.81 % (subject) for the Sequence Alignment, and the Identity is 71.72 % and the Similarity is 77.78 % for the Structure Alignment. This might explain the reasons why rabbit prion protein differs very much from dog and horse prion proteins in secondary structures in low pH environment.

5.3.3 A Concluding Remark

The section is a straight forward MD simulation study of rabbit, dog and horse prion proteins which apparently resist the formation of the scrapie form. The analyses of MD results confirmed the structural stability of rabbit prion protein under neutral pH environment, and of dog and horse prion proteins under neutral and low pH environments. The main point is that the enhanced stability of the C-terminal ordered region especially helix 2 through the D177-R163 (D178-R164) SB formation renders the rabbit, dog and horse prion proteins stable. This SB might be a potential drug target for prion diseases.

```
CLUSTAL 2.1 multiple sequence alignment

                90123456789012345678901234567890 123456789012345678901234567
HorsePrP        MVKSHVGGWILVLFVATWSDVGLCKKRPKPGG-WNTGGSRYPGQGSPGGNRYPPQGGGGW 59
DogPrP          MVKSHIGSWILVLFVAMWSDVGLCKKRPKPGGGWNTGGSRYPGQGSPGGNRYPPQGGGGW 60
RabbitPrP       --MAHLGYWMLLLFVATWSDVGLCKKRPKPGGGWNTGGSRYPGQSSPGGNRYPPQGGG-W 57
HumanPrP        --MANLGCWMLVLFVATWSDLGLCKKRPKPGG-WNTGGSRYPGQGSPGGNRYPPQGGGGW 57
MousePrP        --MANLGYWLLALFVTMWTDVGLCKKRPKPGG-WNTGGSRYPGQGSPGGNRYPPQGG-TW 56
                   :::* *:* ***: *:*:*********** ***********.************  *

                8901234567890123456789012345678901234 5678901234567890123456
HorsePrP        GQPHGGGWGQPHGGGWGQPHGGGWGQPHGGGGWGQGG-SHGQWNKPSKPKTNMKHVAGAA 118
DogPrP          GQPHGGGWGQPHGGGWGQPHGGGWGQPHGGGGWGQGG-THSQWNKPSKPKTNMKHVAGAA 119
RabbitPrP       GQPHGGGWGQPHGGGWGQPHGGGWGQPHGGG-WGQGG-THNQWGKPSKPKTSMKHVAGAA 115
HumanPrP        GQPHGGGWGQPHGGGWGQPHGGGWGQPHGGG-WGQGGGTHSQWNKPSKPKTNMKHMAGAA 116
MousePrP        GQPHGGGWGQPHGGSWGQPHGGSWGQPHGGG-WGQGGGTHNQWNKPSKPKTNLKHVAGAA 115
                **************.*******.******** ***** :*.**.*******.:**:****

                119        130       140       150       160       170
                789012345678901234567890123456789012345678901234567890123456
HorsePrP        AAGAVVGGLGGYMLGSAMSRPLIHFGNDYEDRYYRENMYRYPNQVYYRPVSEYSNQNNFV 178
DogPrP          AAGAVVGGLGGYLLGSAMSRPLIHFGNDCEDRYYRENMYRYPNQVYYRSVDQYNNQSTFV 179
RabbitPrP       AAGAVVGGLGGYMLGSAMSRPLIHFGNDYEDRYYRENMYRYPNQVYYRPVDQYSNQNSFV 175 -1
HumanPrP        AAGAVVGGLGGYMLGSAMSRPIIHFGSDYEDRYYRENMHRYPNQVYYRPMDEYSNQNNFV 176
MousePrP        AAGAVVGGLGGYMLGSAMSRPMIHFGNDWEDRYYRENMYRYPNQVYYRPVDQYSNQNNFV 175
                ************:********:****.* *********:*********.:.:*.**..**

                 180       190       200       210       220          231
                789012345678901234567890123456789012345678901234567  8901234
HorsePrP        HDCVNITVKQHTVTTTTKGENFTETDVKIMERVVEQMCITQYQKEYEAFQQ--RGASVVL 236
DogPrP          HDCVNITVKQHTVTTT-KGENFTETDIKMMERVVEQMCITQYQRESEAYYQ--RGASVIL 236
RabbitPrP       HDCVNITVKQHTVTTTTKGENFTETDIKIMERVVEQMCITQYQQESQAAYQ--RAAGVLL 233 -1
HumanPrP        HDCVNITIKQHTVTTTTKGENFTETDVKMMERVVEQMCITQYERESQAYYQ--RGSSMVL 234
MousePrP        HDCVNITIKQHTVTTTTKGENFTETDVKMMERVVEQMCVTQYQKESQAYYDGRRSSSTVL 235
                *******:******** *********:*:*********:***::* :*  :  *.:. :*

                5678901234567890123
HorsePrP        FSSPPVVLLIFFLIFLIVG 255
DogPrP          FSSPPVILLVSFLIFLIVG 255
RabbitPrP       FSSPPVILLISFLIFLIVG 252
HumanPrP        FSSPPVILLISFLIFLIVG 253
MousePrP        FSSPPVILLISFLIFLIVG 254
                ******:**: ********
```

Fig. 5.21 Horse, dog, rabbit, human and mouse prion protein sequence alignment

5.4 Alignment Analyses

Lastly, we make the sequence alignment of PrP from horse, dog, rabbit, human and mouse protein (Fig. 5.21). In Fig. 5.21, "*" means that the residues in that column are identical in all sequences in the alignment, ":" means that conserved substitutions have been observed, "." means that semi-conserved substitutions are observed, the RED color takes place at small (small+ hydrophobic (incl.aromatic-Y)) residues, the BLUE color takes place at acidic residues, the MAGENTA color takes place at Basic-H residues,GREEN color takes place at Hydroxyl + sulfhydryl + amine + G residues, and Grey color takes place at unusual amino/imino acids etc. For the structural domain, in Fig. 5.21 we can see some special residues listed in Table 5.1 for horse, dog, human and mouse prion proteins, which might contribute to characters of each structure respectively. Rabbits differ from horses, dogs, humans and mice at: S173 (N174 for horse, T174 for dogs, N174 for humans and mice), Q219 (K220

Table 5.1 Alignment analysis of special residues for HoPrP, DogPrP, HuPrP, and MoPrP

Horse	S167 (others are D), Y222 (others are S), Q226 (others are Y), V241 (others are I), F245 (others are S)
Dog	L129 (others are M), S165 (others are P), N170 (others are S), S173 (others are N), V244 (others are I)
Human	I138 (immunities are L), S143 (others are N), H155 (others are Y), M166 (others are V), I183 (immunities are V), E219 (others are Q), S230 (immunities are A)
Mouse	I183 (immunities are V), V215 (others are I), D217 (others are Q), S230 (immunities are A)

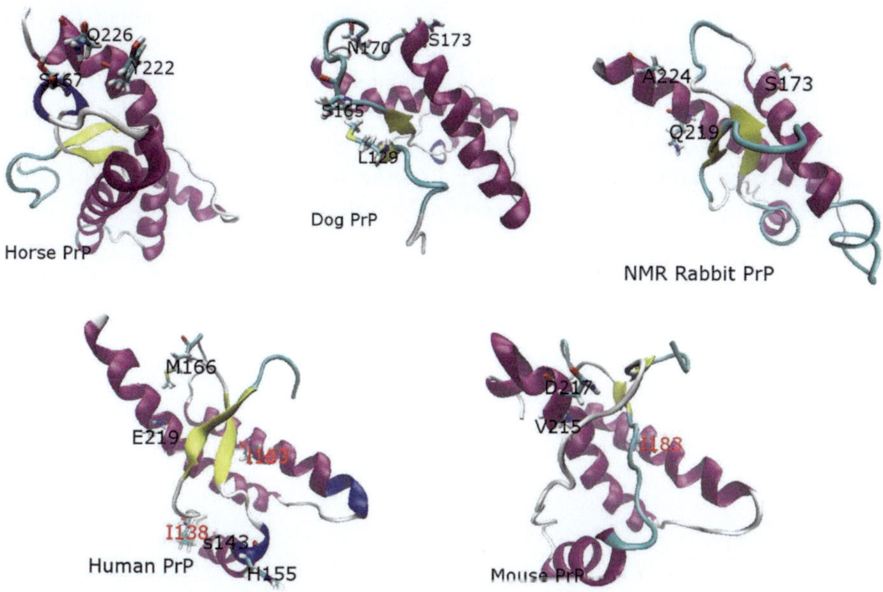

Fig. 5.22 Special residues owned only by HoPrP, DogPrP, RaPrP, HuPrP, and MoPrP respectively

for horses and humans, R220 for dogs and mice), A224 (F225 for horses, Y225 for dogs, humans and mice), L232 (I233 for dogs, V233 for horse, humans and mice), and G228 (others are S229). For rabbits, at positions 89 and 97 the residues are special from all others (G89 (others are N90), S97 (others are N98)). These special residues are illuminated in Fig. 5.22. Some recent researches are focusing on the loop between S2 and H2, i.e. PrP(164–171) [18, 206, 336, 463, 625, 626]; we may see in Fig. 5.22 that the immune animals horses, dogs and rabbits have some residues in this loop different from humans and mice.

5.5 Conclusion

To really reveal the secrets of prion diseases is very hard. Prion proteins have two regions: unstructured region and structured region. Rabbits, horses, and dogs were reported having immunity to prion diseases. Fortunately, by the end of 2010 all the NMR molecular structures of rabbit, horse, and dog prion proteins had been released into PDB bank already; for rabbit prion protein, its X-ray structure was also released into PDB bank in the end of 2010. Prion diseases are 'structural conformational' diseases. This book timely presents a clue to reveal some secrets in the view of the dynamics of prion molecular structures. MD results of the author nearly in the passing 10 years show to us a common conclusion: under low pH environment at many levels of temperatures with different starting MD velocities, rabbit prion protein always unfolds its α-helical structures into β-sheet structures. Prion diseases are just caused by the conversion from predominant α-helices of PrP^C into rich β-sheets of PrP^{Sc}. Hence, we should furthermore study rabbits, horses and dogs, compared with humans and mice in order to reveal some secrets of prion diseases; for us, it is a long shot but certainly worth pursuing.

Chapter 6
Compared with a Homology Structure and Dynamics of Buffaloes

6.1 Introduction

Prion diseases are a class of fatal neurodegenerative diseases including human CJD, cattle BSE (or called 'mad cow' disease), sheep scrapie and others. Among them, the cattle BSE is highly contagious and lethal and can infect humans through the food chain – this is a major public health concern. Only in UK (United Kingdom), in 2000 it was reported there are more than 180,000 cattle infected with 'mad cow' disease [69]. Bovines and buffalo both belong to bovids, and there is only 1 different residue at position 143 in their structural region PrP(124–227) by the alignment of amino acid sequences in GenBank. However, by now, not a single case of TSE-infected buffalo has been reported [298, 302, 456, 488, 576, 708]. This article is to summarize our recent work of BufPrP and to report recent findings of BufPrP from molecular structure and structural dynamics points of view.

First we briefly review the research results on 'buffalo prion' protein listed in the PubMed (http://www.ncbi.nlm.nih.gov/pubmed/?term=buffalo+prion). In 1998, Iannuzzi et al. (1998) found the high degree of gene and chromosome banding conservation among bovids. "The assignment of PRNP (prion protein gene) to river buffalo and goat chromosomes allows us to indirectly assign the bovine syntenic group U11 to specific chromosomes, since it is the first in situ localization on buffalo 14 and goat 13" [298]. In 2009, Oztabak et al. (2009) reported "unlike domestic cattle and bison, no indel polymorphisms of the PRNP promoter and intron 1 were examined in any population of the water buffalo (Bubalus bubalis)". The same authors found that "frequencies of allele, genotype, and haplotype of the indel polymorphisms in PRNP of the Anatolian water buffalo are significantly different those from cattle and bison PRNP indel polymorphisms" [456]. In 2012, Imran et al. (2012) reported that "the bovine PRNP E211K polymorphism is absent from domesticated Pakistani bovids", "there were significant differences between Pakistani and worldwide cattle in terms of allele, genotype and haplotype frequencies". As a result, it is concluded that "Pakistani cattle are relatively more (genetically)

J. Zhang, *Molecular Structures and Structural Dynamics of Prion Proteins and Prions*, Focus on Structural Biology 9, DOI 10.1007/978-94-017-7318-8_6

resistant to classical BSE than European cattle" [302]. In 2012, Zhao et al. (2012) did comparative analysis of the Shadoo gene (SPRN) between cattle and buffalo and found out the following three results: (i) a 12-bp insertion/deletion polymorphism is not revealed in buffalo gene, (ii) mutations 102Ser→Gly, 119Thr→Ala, 92Pro>Thr/Met, 122Thr>Ile and 139Arg>Trp present different genotypic and allelic frequency distributions between cattle and buffalo, (iii) the activity of SPRN promoter in buffalo is significantly higher than cattle and there are higher relative expression levels of Shadoo protein in cerebrum from buffalo than from cattle [488, 708]. In 2014, Uchida et al. (2014) investigated the frequencies of 23-bp insert/deletion (indel) polymorphism in the promoter region (23indel) and 12-bp indel polymorphism in intron 1 region (12indel), octapeptide repeat polymorphisms and single nucleotide polymorphisms (SNPs) in the bovine PRNP of cattle and water buffalo in Vietnam, Indonesia and Thailand. It was found that: (i) "the frequency of the deletion allele in the 23indel site (of these water buffalo) was significantly low in cattle of Indonesia and Thailand and water buffalo", (ii) "the deletion allele frequency in the 12indel site was significantly low in all of the cattle and buffalo", (iii) "in some Indonesian local cattle breeds, the frequency of the allele with 5 octapeptide repeats was significantly high despite the fact that the allele with 6 octapeptide repeats has been reported to be most frequent in many breeds of cattle", and (iv) "four SNPs observed in Indonesian local cattle have not been reported for domestic cattle (before)" [576].

From previous studies it is clear that buffalo is a low susceptibility species resisting to prion diseases, and the study of the protein structure or its structural dynamics of BufPrP^C becomes very important to understand the structure and function. In this chapter, we will study BufPrP^C from the protein structure or structural dynamics points of view, in order to reveal why BufPrP^C is so stable and resistant to prion diseases.

To date, there is no structural data available on BufPrP, although many experimental studies have shown that BufPrP is very stable so that it resists to the infection of diseased prions [298, 302, 456, 488, 576, 708]. Thus, in Sect. 6.2 we will establish a homology structure for BufPrP^C. As we all know, prion diseases are caused by the conversion from PrP^C to PrP^{Sc}; in structure the conversion is mainly from α-helices to β-sheets (generally PrP^C has 42 % α-helix and 3 % β-sheet, but PrP^{Sc} has 30 % α-helix and 43 % β-sheet [169, 257, 321, 447, 457, 491]) (where the structural region of a PrP^C consists of β-strand 1 (β1), α-helix 1 (H1), β-strand 2 (β2), α-helix 2 (H2), α-helix 3 (H3), and the loops are linked each other). The conformational changes may be amenable to study by MD techniques. Hence, in Sect. 6.2 we will use MD to study the homology structure BufPrP^C(124–227). It is revealed that how BufPrP^C is resistant to become into PrP^{Sc} under neutral or low pH environments. These observations and their analyses will be done in Sect. 6.3 Results and Discussion. At last, this chapter will summarize the findings of proposed reasons why buffalo is resistant to prion diseases in the Concluding Remarks section.

6.2 Materials and Methods

6.2.1 *Homology Structure for BufPrP^C(124–227)*

The multiple sequence alignment analysis tool used here is ClustalW2 [365]. By alignments of the whole (unstructured region + structured region) sequence of 32 PrPs:

Human PrP (NM_000311.3),
Mouse PrP (NM_011170.3),
Rat PrP (NM_012631.2),
Rabbit PrP (NM_001082021.1),
Horse PrP (NM_001143798.1),
Dog PrP (NM_001013423.1),
Hamster PrP (M14055.1),
Gold Hamster PrP (XM_005068660.1),
Cat PrP (EU341499.1),
Cat2 PrP (AF003087.1),
Elk PrP (EU082291.1),
Bovine PrP (NM_001271626.1),
Sheep PrP (NM_001009481.1),
Goat PrP (JF729302.1),
Pig PrP (NM_001008687.1),
Turtle PrP (AJ245488.1),
Chicken PrP (NM_205465.2),
Frog PrP (NM_001088711.1),
Red Deer PrP (EU032287.1),
Donkey PrP (AY968590.1),
Ermine PrP (EU341505.1),
Atalantic Salmon PrP (EU163438.1),
Common Carp PrP (DQ507237.1),
Giant Panda PrP (NM_001304886.1),
Black Buck PrP (AY720706.1),
Chamois PrP (AY735496.1),
Chimpanzee PrP (NM_001009093.3),
Wapiti PrP (AF016227.2),
Rhesus PrP (NM_001047152.1),
Deer PrP (AY330343.1), Cattle PrP (NM_181015.2),
Buffalo PrP (KC.1137634),

where the codes in the brackets are nucleotide codes in GenBank of http://www.ncbi.nlm.nih.gov/nuccore/, we found Bovine, Cattle and Black Buck have (the highest) 97.35 % similarity with Buffalo (Fig. 6.1). In the structured region, Black Buck has 2 residues different from Buffalo, while Bovine and Cattle only have 1 difference from Buffalo (Fig. 6.1). Cattle and Black Buck have no NMR or

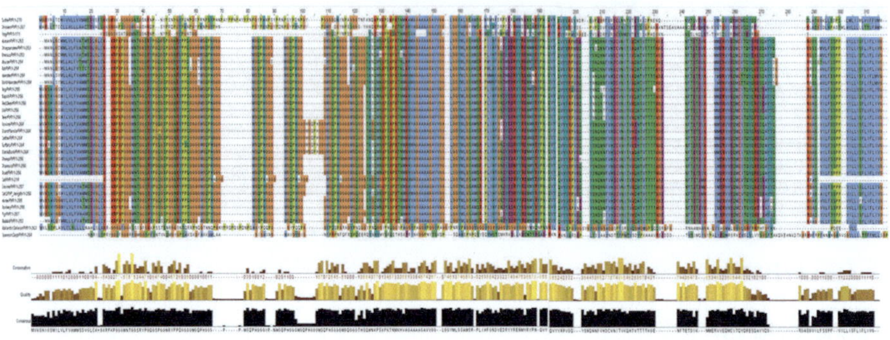

Fig. 6.1 This colored map shows the conserved and the non-conserved region between the buffalo and other 31 PrP sequences

X-ray structure. Thus we choose Bovine. In PDB Bank (www.rcsb.org), bovine PrP has the following PDB entries: 1DWY.pdb, 1DWZ.pdb, 1DX0.pdb, 1DX1.pdb,and 1SKH.pdb, etc. In the PubMed on 'bovine prion protein molecular dynamics', we found 1DWY.pdb [9, 101], 1DWZ.pbb [282], 1DX0.pdb [354] were used. 1DWZ.pdb has 20 structures and, by clustering the 20 structures, we picked up the Number 9 from these 20 structures and we superposed it to 1DWY.pdb and found their RMSD (Root Mean Square Deviation) value is 0 Å (however if we superposed it to 1DX0.pdb, the RMSD is 1.22117 Å). Thus, we also chose 1DWY.pdb as [101] (at the same time in order to conveniently make comparisons with the results of bovine PrP in [101]).

The BufPrP(124–227) homology model used in this chapter was constructed by one mutation S143N at position 143 using the NMR structure 1DWY.pdb of Bovine PrP(124–227) (where the experimental temperature is 293 K, pH value is 4.5, and pressure is 1 ATM, etc.). The homology structure constructed is reasonable and sound correct.

6.2.2 Molecular Dynamics (MD) Techniques

The MD methods employed are the same as the previous studies [679, 703, 704]. Briefly, all simulations used the ff03 force field of the AMBER 11 package [91]. The systems were surrounded with a 12 Å layer of TIP3PBOX water molecules and neutralized by sodium ions using the XLEaP module of AMBER 11. To remove the unwanted bad contacts, the systems of the solvated proteins with their counter ions had been minimized mainly by the steepest descent method and followed by a small number of conjugate gradient steps on the data, until without any amino acid clash checked by the Swiss-Pdb Viewer 4.1.0 (http://spdbv.vital-it.ch/). Next, the solvated proteins were heated from 100 to 300 K in 1 ns duration. Three sets of initial velocities denoted as seed1, seed2 and seed3 are performed in parallel for stability (this will make each set of MD starting from different MD initial velocity, implemented in Amber package we choose three different odd-real-number values for “ig”) – but for the NMR structure and the X-ray structure of RaPrP^C, each set of the three has the same “ig” value in order to be able to make comparisons. The thermostat algorithm used is the Langevin thermostat algorithm in constant NVT ensembles. The SHAKE algorithm (only on bonds involving hydrogen) and PMEMD (Particle Mesh Ewald Molecular Dynamics) algorithm with non-bonded cutoff of 12 Å were used during heating. Equilibrations were reached in constant NPT ensembles under Langevin thermostat for 5 ns. After equilibrations, production MD phase was carried out at 300 K for 25 ns using constant pressure and temperature ensemble and the PMEMD algorithm with the same non-bonded cutoff of 12 Å during simulations. The step size for equilibration was 1 and 2 fs in the MD production runs. The structures were saved to file every 1000 steps. During the constant NVT and then NPT ensembles of PMEMD, periodic boundary conditions have been applied.

In order to obtain the low pH (acidic) environment, the residues HIS, ASP, GLU were changed into their zwitterion forms of HIP, ASH, GLH, respectively, and Cl- ions were added by the XLEaP module of the AMBER package. Thus, the SBs of the system (residues HIS, ASP, GLU) under the neutral pH environment were broken in the low pH environment (zwitterion forms of HIP, ASH, GLH).

6.3 Results and Discussion

6.3.1 BufPrP Homology Structure Has 5 Hydrogen Bonds at Asn143

In whole sequences of 264 residues we found bovine and cattle have 97.35 % similarity with buffalo. In the structured region PrP(124–227), bovine and cattle only have 1 difference from buffalo at residue 143. Asn143 plays an important role in β1-H1-β2 [572]. At Asn143 of BufPrP(124–227), we found there are 5 HBs (Fig. 6.2): ASN143.O–GLU146.N (3.11 Å) (forming a “pincette” motif [395]), ASN143.O–ASP147.N (3.34 Å), ASN143.OD1–ASP144.N (2.38 Å), ASN143.OD1–TYR145.N (2.66 Å), ASN143.N–GLU146.OE1 (2.97 Å).

6.3.2 BufPrP Is Stable Under Neutral or Low pH Environments at Room Temperature

We show our 25 ns’ MD results of secondary structures in Fig. 6.3 and Table 6.1. Generally, we may see that, whether under neutral or low pH environments, there are about 51 % α-helix and 3.85 % β-sheet – almost same as the α-and-β percentages of

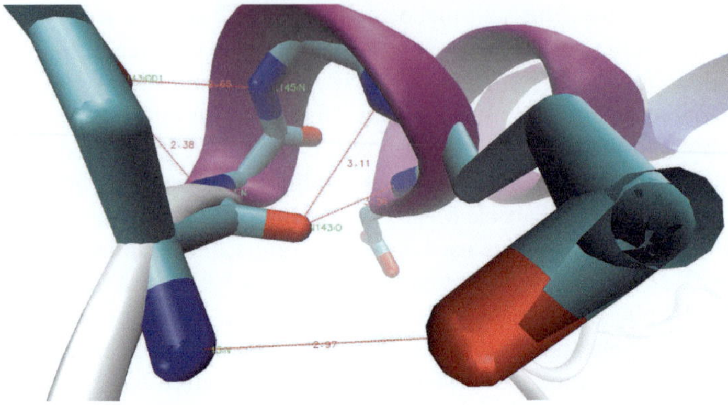

Fig. 6.2 At residue Asn143 of Homology BufPrP(124–227), there are 5 HBs: ASN143. O–GLU146.N (3.11 Å), ASN143.O–ASP147.N (3.34 Å), ASN143.OD1–ASP144.N (2.38 Å), ASN143.OD1–TYR145.N (2.66 Å), ASN143.N–GLU146.OE1 (2.97 Å)

a normal cellular PrP^C (42 % α-helix and 3 % β-sheet). However, we should notice that under neutral pH environment there are 51.42 % α-helix and 3.84 % β-sheet, but under low pH environment there are 50.97 % α-helix and 3.887 % β-sheet; for seed1 the percentage of E (i.e. the extended strand (participates in β-ladder)) is increasing and the percentage of H (i.e. α-helix) is decreasing from neutral pH environment to low pH environment (Table 6.3), and for seed2 the percentage of E (i.e. the extended strand (participates in β-ladder)) is increasing from neutral pH environment to low pH environment. The reason is that the low pH environment can greatly weaken some SBs of the neutral pH environment and BufPrP is not very sensitive to the low pH environment. All in all, BufPrP is stable under neutral or low pH environments at room temperature. The above observations can also be found from Figs. 6.4∼6.7.

Seeing Figs. 6.4∼6.5, from the RMSDs and Radii Of Gyrations values of the 25 ns' MD we may say 25 ns is short but good enough for a small protein like PrP, the systems reached enough equilibrations (the variation of RMSDs and Radius Of Gyrations values is less than 1.5 Å and in the normal interval), and the three seeds are valid so that we cannot find great differences among the three seeds.

Seeing Figs. 6.6∼6.7, we know that the variations of B-factor and RMSF values are in the loops β1-α1, β2-α2, α2-α3 but not clearly in the short loop α1-β2. These loops are the most solvent-accessible surface areas. We also cannot clearly see the greatly differences between neutral and low pH environments, among the three seeds. The three α-helices and the two anti-parallel β-strands are not variable very much during the 25 ns' MD simulations.

As we all know, the stability of a protein is maintained by its SBs, HBs, HYDs, vdWs and disulfide bonds (for PrP monomer there exists a disulfide bond (S-S) between CYS179 and CYS214) etc., to drive to be able to perform the biological function of the protein.

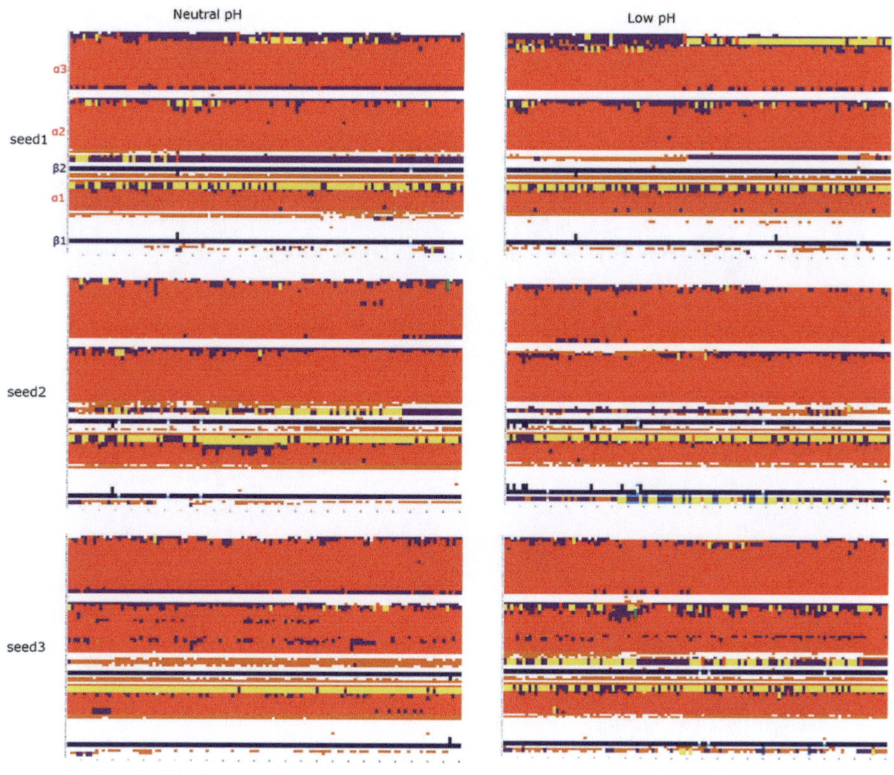

Fig. 6.3 Secondary Structure graphs homology BufPrP^C at 300 K (x-axis: time (0–25 ns), y-axis: residue number (124–227); *left column*: neutral pH, *right column*: low pH; up to down: seed1–seed3). H is the α-helix, I is the π-helix, G is the 3-helix or 3_{10} helix, B is the residue in isolated β-bridge, E is the extended strand (participates in β-ladder), T is the HBed turn, and S is the bend

Table 6.1 Percentages (%) of elements of secondary structure under neutral and low pH environments for BufPrP at 300 K during 25 ns' MD

		B	G	I	H	E	T	S
Neutral pH	Seed1	1.54e-4	2.86	4.62e-4	50.58	3.85	10.39	6.66
	Seed2	4.81e-4	3.49	4.01e-4	53.04	3.79	6.45	7.29
	Seed3		2.98		50.64	3.88	6.7	8.74
Low pH	Seed1	1.54e-4	3.62		48.22	3.89	9.55	7.05
	Seed2	0.35	3.12		53.7	4.02	6.59	5.94
	Seed3	7.69e-4	4.12	3.85e-4	50.98	3.75	8.56	6.95

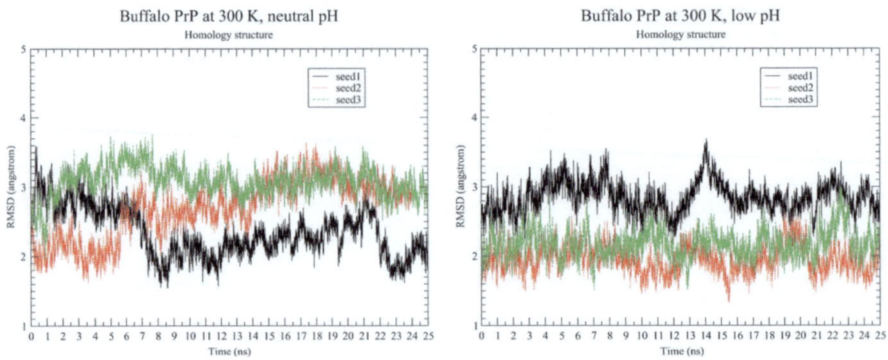

Fig. 6.4 RMSD of BufPrP at 300 K, neutral and low pH values (*left*: neutral pH, *right*: low pH) during 25 ns' MD

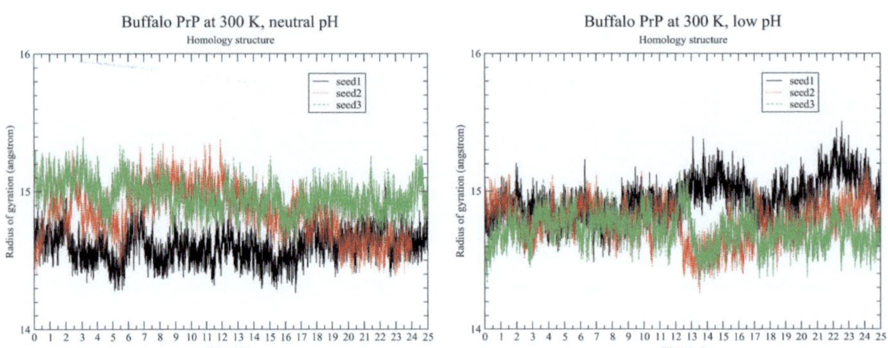

Fig. 6.5 Radius of gyration of BufPrP at 300 K, neutral and low pH values (*left*: neutral pH, *right*: low pH) during 25 ns' MD

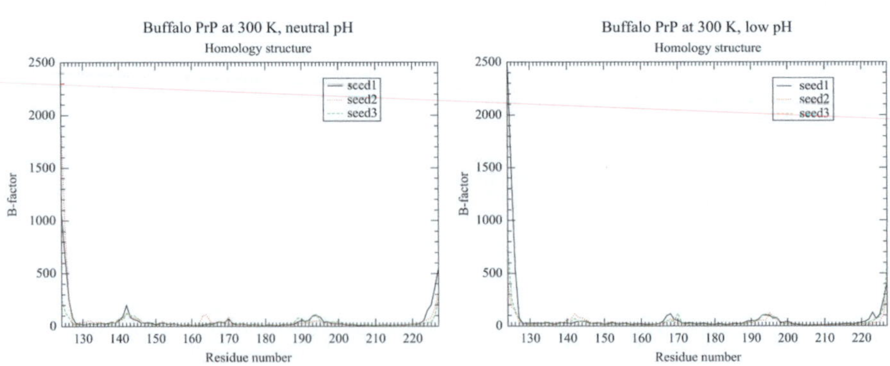

Fig. 6.6 B-factor of BufPrP at 300 K, neutral and low pH values (*left*: neutral pH, *right*: low pH) during 25 ns' MD

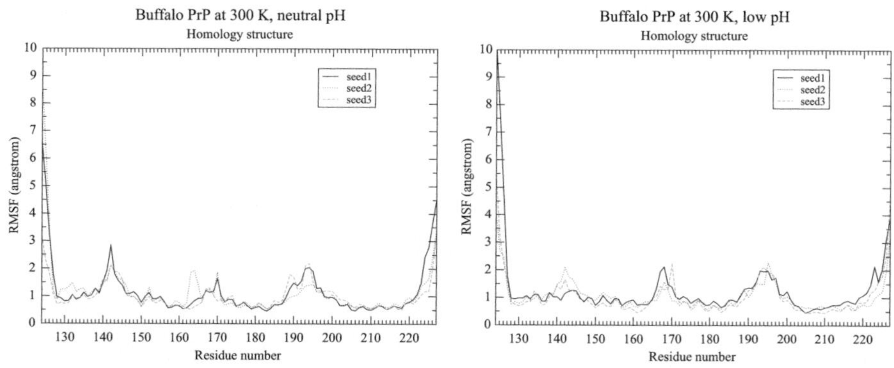

Fig. 6.7 RMSF of BufPrP at 300 K, neutral and low pH values (*left*: neutral pH, *right*: low pH) during 25 ns' MD

The following SBs (with percentages in the brackets for seed1∼seed3 respectively) contribute to the structural stability of BufPrP(124–227):

- special SBs:
  - ASP178–ARG164 (88.98 % (seed1), 11.69 % (seed2), 0.04 % (seed3)) – linking the $\beta 2$-$\alpha 2$ loop,
  - ASP202–ARG156 (1.9 % (seed1), 1.15 % (seed2), 23.87 % (seed3)),
  - GLU196–ARG156 (96 % (seed1), 9.35 % (seed2), 16.46 % (seed3)),
  - GLU211–HIS177 (86 % (seed1), 2.45 % (seed2), 94.57 % (seed3)) – linking H3 and H2,
  - HIS187–ARG156 (82 % (seed1), 52.40 % (seed2), 64.73 % (seed3)) – linking H2 and the 3_{10}-helix after H1.
- SBs in H1:
  - ASP147–ARG148 (100 % (seed1), 100 % (seed2), 100 % (seed3)),
  - HIS155–ARG156 (99.74 % (seed1), 100 % (seed2), 100 % (seed3)),
  - ASP147–HIS140 (45.48 % (seed1), 19.49 % (seed2), 12.78 % (seed3)),
  - GLU152–ARG148 (40.67 % (seed1), 21 % (seed2), 31.58 % (seed3)),
  - GLU152–ARG151 (37.31 % (seed1), 42.57 % (seed2), 25.84 % (seed3)),
  - ASP144–ARG148 (29.12 % (seed1), 88.38 % (seed2), 74.26 % (seed3)),
  - ASP147–ARG151 (19.83 % (seed1), 54.03 % (seed2), 28.46 % (seed3)),
- SBs in H2:
  - ASP178–HIS177 (8.95 % (seed1), 26.51 % (seed2), 8.17 % (seed3)),
  - GLU186–LYS185 (93.97 % (seed1), 92.51 % (seed2), 97.56 % (seed3)),
- SBs in H3:
  - GLU211–ARG208 (99.46 % (seed1), 99.64 % (seed2), 92.30 % (seed3)),
  - GLU207–LYS204 (98.24 % (seed1), 99.90 % (seed2), 99.93 % (seed3)),

Table 6.2 Percentages (%) of some HBs (between two residues) under neutral and low pH environments for BufPrP at 300 K during 25 ns' MD

		G131–Q160	P137–Y150	S170–Y218	Y157–R136
Neutral pH	Seed1	36.36	59.92	72.48	100
	Seed2	62.99	63.83	13.91	100
	Seed3	40.21	79.71	91.49	100
Low pH	Seed1	60.08	18.53	48.87	100
	Seed2	56.37	47.50	89.78	100
	Seed3	21.78	20.40	53.96	100

		H187–T191	C179–T183	T188–T192	Y149–N153	E186–T190
Neutral pH	Seed1	92.65	80.41	89.87	48.80	30.77
	Seed2	85.48	83.57	60.93	44.50	63.57
	Seed3	77.01	85.75	81.55	62.91	19.46
Low pH	Seed1	87.7	87.66	73.12	46.38	46
	Seed2	83.86	93.47	65.03	41.49	18
	Seed3	72.94	94.69	60.51	54.38	43

- GLU221–ARG220 (96.14 % (seed1), 56.49 % (seed2), 44.08 % (seed3)),
- GLU207–ARG208 (57.44 % (seed1), 32.68 % (seed2), 78.63 % (seed3)),

• SBs in the H2-H3 loop:

- GLU196–LYS194 (64.74 % (seed1), 55.05 % (seed2), 23.03 % (seed3)).

The HBs contributed to the structural stability of BufPrP(124–227) are listed as follows: GLY131–GLN160 (linking β1-α1 loop and α1-β2 loop), PRO137–TYR150 (linking β1-α1 loop and H1), SER170–TYR218 (linking β2-α2 loop and H3), TYR157–ARG136 (linking α1-β2 loop and β1-α1 loop), HIS187–THR191 (in H2), CYS179–THR183 (in H2), THR188–THR192 (in H2), TYR149–ASN153 (in H1), GLU186–THR190 (in H2) (the percentages of their occupied rates can be seen in the Table 6.2). Focusing on the β2-α2 loop BufPrP^C(164–172), besides HB SER170–TYR218 (occupied rates in Table 6.2, see Fig. 6.9) and HB ARG164–ASP178 (with occupied rates 89.65 % (seed1, neutral pH), 47.17 % (seed2, neutral pH), 10.58 % (seed3, neutral pH), 20.59 % (seed3, low pH), this HB is also a SB (Fig. 6.8)), we found some HBs in this loop listed in Table 6.3, where we can find the residues in the β2-α2 loop are not only HBed contacting with the C-terminal H3 residues but also the N-terminal residues GLY124$\sim$TYR128 of BufPrP^C(164–172).

To maintain the structural stability of BufPrP^C(124–227), there are the following HYDs with 100 % occupied rates:

- ILE139–LEU138–PRO137, MET134–ALA133 (in β1-α1 loop),
- LEU130–MET129 (in β1),
- VAL166–PRO165 (in β2-α2 loop),
- ILE215–MET213–VAL210–VAL209–MET206–MET205–ILE203, VAL209–MET205, ILE203–MET206 (in H3),
- VAL210–VAL180–VAL184, MET213–VAL180 (linking H3 and H2).

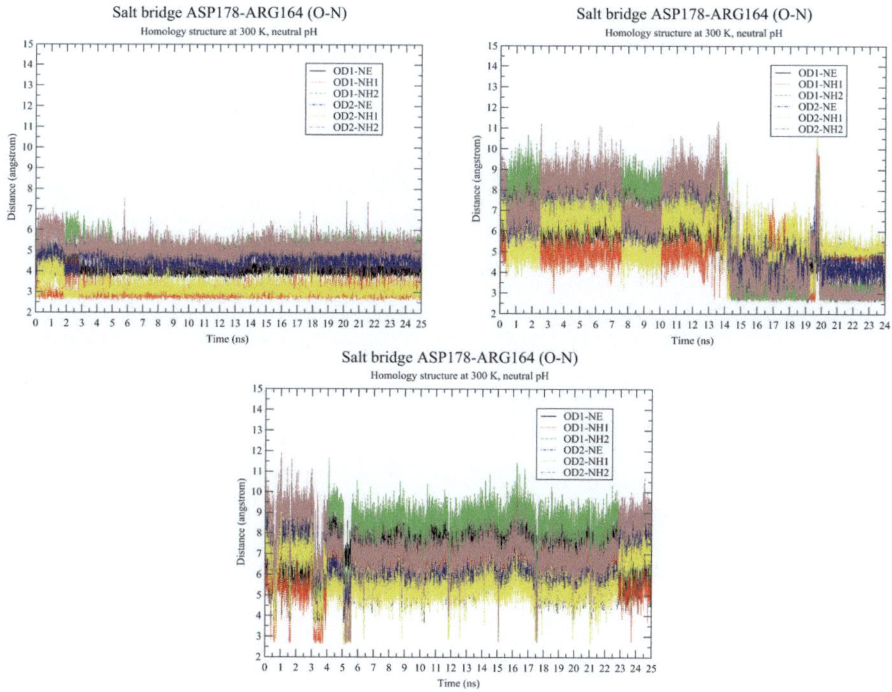

Fig. 6.8 SB ASP178–ARG164 (O–N) of BufPrP at 300 K during 25 ns' MD (*left∼right*: seed1∼seed3). The occupied rates are 88.98 %, 11.69 %, 0.04 % for seed1∼seed3 respectively. The occupied rates for the HB between these two residues are 89.65 %, 47.17 %, 10.58 % for for seed1∼seed3 respectively under the neutral pH environment, and 20.59 % for seed3 under the low pH environment

However, under low pH environment, the following HYDs VAL176–ILE215, MET213–VAL180, MET213–VAL161–VAL210, ILE203–PHE198 of neutral pH become weak or lost because of the disturbance of some SBs removed in low pH environment (especially weaken the HYDs VAL176–ILE215, VAL180–MET213, VAL161–VAL210, VAL161–MET213 linking $\beta 2$-$\alpha 2$ loop and the C-terminal of H3), although the HYD PRO165–VAL166 in $\beta 2$-$\alpha 2$ loop is always conserved whether under neutral or low pH environments.

For BufPrP^C(124–227), in the $\beta 2$-$\alpha 2$ loop, by Tables 6.3∼6.2 we cannot find all the HBs, HYDs and polar contacts constructing the helix-cap motif as wild-type rabbit PrP^C [336, 556]. Between PRO165 and GLN168, there is no direct HB, but there are HBs PRO165–ARG164 and ARG164–GLN168 making PRO165 and GLN168 linking indirectly. For residue VAL166, there is no HB VAL166–TYR169, instead, there are HBs VAL166–SER170, VAL166–TYR216 (linking the loop and H3); strong HYD exists between VAL166 and PRO165. For residue GLN168, there exist HBs GLN168–ARG164, GLN168–ASP167, GLN168–LEU125,

Table 6.3 Percentages (%) of some HBs linking one/two residue(s) of the $\beta 2$-$\alpha 2$ loop (BufPrP(164–172)) under neutral and low pH environments for BufPrP at 300 K during 25 ns' MD

		R164–D167	R164–Q168	R164–Y169	R164–N174	P165–R164
Neutral pH	Seed1					
	Seed2		23.01	5.21		
	Seed3				10.58	
Low pH	Seed1					
	Seed2	6.33				7.82
	Seed3					

		V166–S170	D167–Q168	D167–S170	Y169–N174	Y169–D178
Neutral pH	Seed1	16.98	5.89			13.20
	Seed2		5.53			64.4
	Seed3		18.65			
Low pH	Seed1			40.25		
	Seed2				7.00	
	Seed3					

		V166–Y226	D167–Y225	S170–S222	Q172–Q219	R164–L125
Neutral pH	Seed1					33.78
	seed2	9.26	21.84		13.82	
	Seed3		13.50			
Low pH	Seed1			6.17		
	Seed2					
	Seed3				5.58	

		R164–G126	R164–G127	Q168–L125	Q168–G126	Y169–G124	Y169–Y128
Neutral	Seed1	26.22	7.24		7.94	11.02	
	Seed2			5.44			
	Seed3	72.47					
Low	Seed1		5.68				
	Seed2	20.65	6.27				29.20
	Seed3		7.46				

GLN168–GLY126 in the $\beta 2$-$\alpha 2$ loop and linking N-terminal residues of BufPrP^C(124–227). For residue TYR169, there exist HBs TYR169–ARG164, TYR169–ASN174, TYR169–ASP178, TYR169–GLY124, TYR169–TYR128. However, we think structure of the $\beta 2$-$\alpha 2$ loop of BufPrP^C should have a 3_{10}-helix (Fig. 6.11, seed2 in neutral pH environment, seed3 in low pH environment) as that of rabbits, horses, elks, tammar wallaby, bank voles [463].

In [188], cation-π interaction (a non-covalent binding force that plays a significant role in protein stability) is well studied for human PrP and its mutants, in the use of C-α-distance between two residues involved in cation-π interaction.

For BufPrP, we found the π-π stackings Y162–Y128–Y163, Y169–F175, Y225–Y226, Y145–Y149 and the π-cations Y128–R164–Y169, Y145–R148–Y149 have C-α-distances less than 6.0 Å during the whole 25 ns of MD. At the same time we found other important π-π stackings such as F175–Y218 and π-cations such as F141–R208, F198–R156, H155–R136 (Table 6.4). We can see around the β2-α2 loop there is a π-chain/circle Y128–F175–Y218–Y163–F175–Y169–R164–Y128(–Y162) as reported in [692]. By the way, we found the "C-α-distance" to calculate π-cations of [188] is not a perfect way for calculations (Table 6.4).

Table 6.4 Some π-π stackings and π-cations under neutral and low pH environments for BufPrP at 300 K at some snapshots for seed1∼seed3 (where "low" and "neutral" stand for the low and neutral pH environments respectively)

π-π stack	Seed1, neutral	Seed1, low	Seed2, neutral	Seed2, low	Seed3, neutral	Seed3, low
F175–Y218	at 1,2,3,4, 5,9,10,13, 15,16,18,19, 20,21,23, 24,25 ns		9,13,16,18, 19,21,23 ns	1,3,4,5,7,8, 10,11,12,13, 14,16,17,18, 20,23,23, 24 ns		8,10,13, 17,23 ns
Y128–Y169				14,18,19, 21 ns		
Y128–F175				5,6,7,8,14, 15,1820,22, 23,24 ns		
Y218–Y225		12,13,15,16, 20,22,24 ns				
F141–Y150		12,13 ns	1,23,24 ns	9,21,22, 24 ns		7 ns
Y163–F175		8,14,18, 19 ns				
H187–F198		3,6,7,13, 15,17				
Y218–Y226	11 ns					
Y162–Y128	2,8,10 ns	3,4,7,8,10, 11,18,20,22, 24 ns				
Y128–Y163						
Y169–F175	10 ns		1,2,3,4,6,7, 8,11,12,13, 14,18,20 ns			3,7,9 ns
Y225–Y226	6,16 ns	3,8 ns	9,10,13,14, 16,18,19,21, 22 ns	12,13,15, 19 ns	4,5,10,11, 12,13,16, 23,24 ns	1,5,6,14, 15,18,20, 21,22,23, 24,25 ns
Y145–Y149						16 ns

(continued)

Table 6.4 (continued)

π-cation	Seed1, neutral	Seed1, low	Seed2, neutral	Seed2, low	Seed3, neutral	Seed3, low
F141–R208	14,15,18, 23 ns		3,4,6,14 ns	4,5,12,19, 20,23	6,8,9,10, 16,18,19, 20 ns	7,8,9,10, 11,12,13, 14,17,18, 19,20,21, 22,24 ns
F198–R156	1,2,3,7,11, 12,13,14,15, 16,17,18,19, 20,22,23,24, 25 ns	10,11,12, 17 ns	1,2,3,4,5, 12,13,14, 17 ns	23,24 ns	3,7,8,10, 12,14,17, 18,19,20, 21,22,23, 24,25 ns	5,12,14, 16 ns
H155–R136	4,5 ns	1,3,6,7,8, 14,15,17, 18,19,20,23, 25 ns	5 ns	3,5,7, 11,12,14, 20 ns	8,10,13, 19,21, 23 ns	2,3,4, 8,10,12, 15,24 ns
Y163–R220			4 ns	1,2,3,4, 5,7,8, 10,11 ns		
F141–R204	1,2,3,4,5, 16,17,20, 25 ns					
H187–F198		6,7,8, 9 ns				
H155–K194					6,7, 14 ns	
H140–R151	8 ns					
Y128–R164	6		1,4,5,7, 10,14			7,12, 25 ns
R164–Y169				1,2,4,6, 7,8,9,10, 12,16,17, 25 ns	11,12,20, 22 ns	2 ns
Y145–R148	21 ns	6,9 ns	3,6,8,9, 10,13,14, 15,16,17, 18,19 ns	10 ns		9,22 ns
R148–Y149						16 ns

6.3.3 Some Special Contributions to the Stable BufPrP

At last, we will list four special contributions to the stability of BufPrP^C(124–227) as follows.

- We have found one focus of prion protein structures is at the β2-α2 loop and its interacted C-terminal of H3 [54–56, 83, 145, 146, 166, 252, 293, 301, 349, 355, 356, 371, 463, 466, 556, 625, 626, 672, 699]. This article found there is a SB ASP178-ARG164 (O–N) in BufPrP^C (Fig. 6.8, with 88.98 % occupied rate for seed1), which just keeps this loop being linked, and there is a HB SER170–TYR218 (O–H) of BufPrP^C (Fig. 6.9, Table 6.2), which just keeps this loop and C-terminal of H3 being linked.
- It was said that if the mutation H187R is made at position 187 then the hydrophobic core of PrP^C will be exposed [712]. We found there is a very strong SB HIS187–ARG156 (N–O) (Fig. 6.10) linking H2 and the 3_{10}-helix after H1. The mutation H187R will make the SB HIS187–ARG156 (N–O) broken.

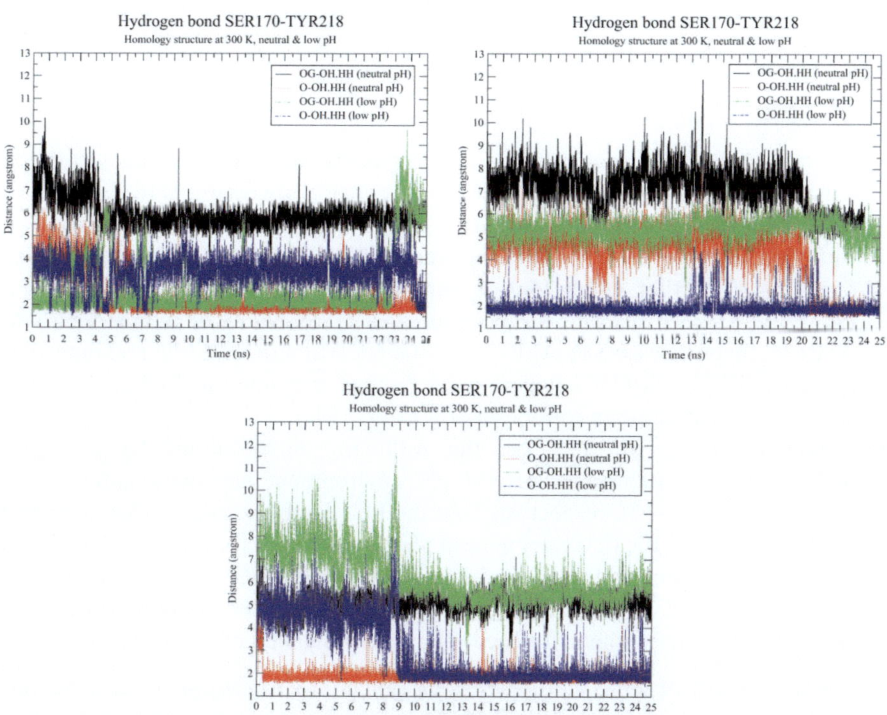

Fig. 6.9 HB SER170-TYR218 (O–H) of BufPrP at 300 K during 25 ns' MD (*left*~*right*: seed1~seed3). The occupied rates for this HB are 72.48 %, 13.91 %, 91.49 % for seed1~seed3 respectively under the neutral pH environment, and 48.87 %, 89.78 %, 53.96 % for seed1~seed3 respectively under the low pH environment

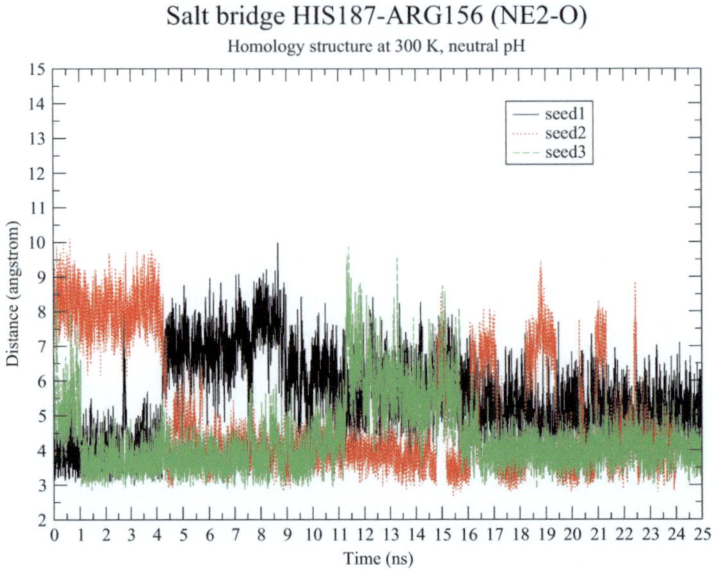

Fig. 6.10 SB HIS187–ARG156 (NE2–O) of BufPrP at 300 K during 25 ns' MD. The occupied rates are 82 %, 52.40 %, 64.73 % for seed1∼seed3 respectively

- For bovine PrP^C, at low pH, hydrophobic contacts with M129 nucleated the nonnative β-strand, and at mid-pH, polar contacts involving Q168 and D178 facilitated the formation of a hairpin at the flexible N-terminus [101]. For $BufPrP^C$, we found there is a HYD between MET129 and LEU130 with occupied rate 100 % whether under low or neutral pH environments, and there is a HB Y169–D178 instead of the polar contact Q168–D178 of bovine PrP^C. At ASP178, there are two SBs ASP178–ARG164, ASP178–HIS177, and one polar contact ASP178–ARG164 (where polar contact is defined including both HBs and SBs formed between residues/atoms as in [101]).
- Seeing Fig. 6.11, we may get the following observations: (i) the optimal/minimized structure in neutral pH environment has two 3_{10}-helices at 125–127 and 152–156 respectively but the first 3_{10}-helix had quickly unfolded since the start of MD; (ii) under neutral pH environment, for seed1, at 10, 15 and 20 ns the C-terminal end of H3 unfolded – this agrees with the observation from Fig. 6.3, at 25 and 30 ns the 3_{10}-helix at the C-terminal end of H1 unfolded; (iii) under low pH environment, for seed1 the C-terminal end of H3 unfolded.

Lastly, we will observe the electrostatic potential surface charge distributions of $BufPrP^C$(124–227). For rabbit PrP^C, it carries a continuous area of positive charges on the surface [626] (mainly constructed by residues HIS139, HIS176, ARG150, LYS203, ARG147 in Fig. 6.12), which is distinguished from other PrP^C. Observing Fig. 6.12, we may see around the β2-α2 loop buffalo does not have a large land of continuous area of positive charges as rabbit. For rabbit, there are two positively charged residues ARG163 and ARG227 linking the β2-α2 loop and

Fig. 6.11 Snapshots at minimized/optimal structure, 5, 10, 15, 20, 25 ns (columns from *left* to *right*) for MD of homology structure of BufPrP^{C}(124–227) at 300 K. The first four rows are for neutral pH environment and the last four rows are for low pH environment. The second and sixth rows are for seed1, the third and seventh rows are for seed2, and the fourth and eighth rows are for seed3

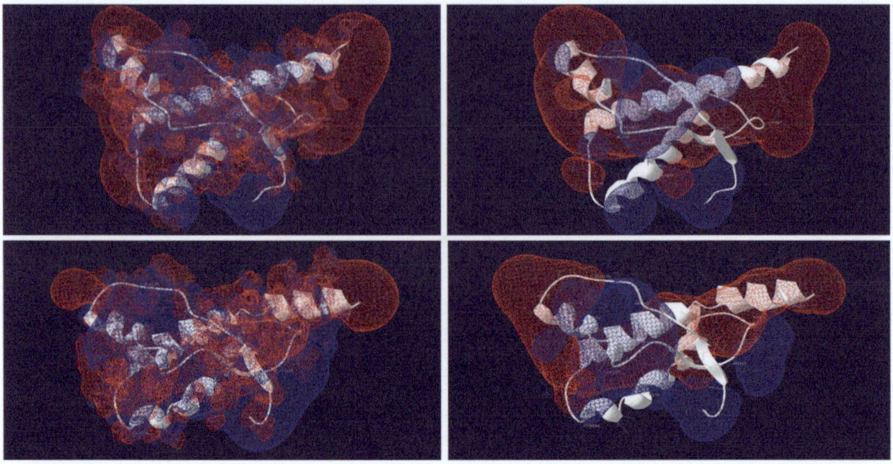

Fig. 6.12 Surface electrostatic charge distributions of the energy minimized/optimal structures of BufPrP^C(124–227) and rabbit PrP^C(124–228) (2FJ3.pdb). The first row is for buffalo and the second row is for rabbits. The first column is for all the atoms partial charged and the second column is for all the residues charged. Poisson-Boltzmann method is used. *Blue*: positive charge, *red*: negative charge. This figure was drawn by Swiss-PdbViewer 4.1.0

the C-terminal end of H3. FirstGlance in Jmol (bioinformatics.org/firstglance/fgij/) was used to detect all the charges of BufPrP^C(124–227): 10 + (7 Arg, 3 Lys) (4 His) and 14–(5 Asp, 8 Glu, 1 C-termini), in defining the following SBs under the neutral pH environment: ASP178.OD1–ARG164.NH1 (4.6319 Å), GLU186.OE2–LYS62.NZ (3.3250 Å), GLU200.OE2–LYS204.NZ (3.5305 Å), ASP202.OD2/1–ARG156.NH2/NE (3.4993 Å), GLU186.OE1/2–ARG156.NH1/2 (4.4191 Å). We used Maestro 10.1 2015–1 (Academic use only) free package to draw the Poisson-Boltzmann electrostatic potential surface charges of our energy minimized/optimal structure (also confirmed by Swiss-PdbViewer 4.1.0) and the average structures of 25 ns' MD of homology BufPrP^C at 300 K in neutral pH environment (Fig. 6.13). In Fig. 6.13 (also in Fig. 6.12), we can observe that the positively charged surface (blue colored) at ARG164 in the $\beta 2$-$\alpha 2$ loop, and a large continuous positively charged surface constructed by residues ARG136, ARG151, HIS140, ARG208, LYS204 together. For BufPrP^C(124–227), LYS194, ARG156, or ARG148 constructs discrete area of the positively charged electrostatic surface respectively.

6.4 Compared with Rabbits

We cannot see great structural difference between buffaloes and rabbits. First we do the sequence alignment of the four species with low susceptibility to prion diseases (Fig. 6.14).

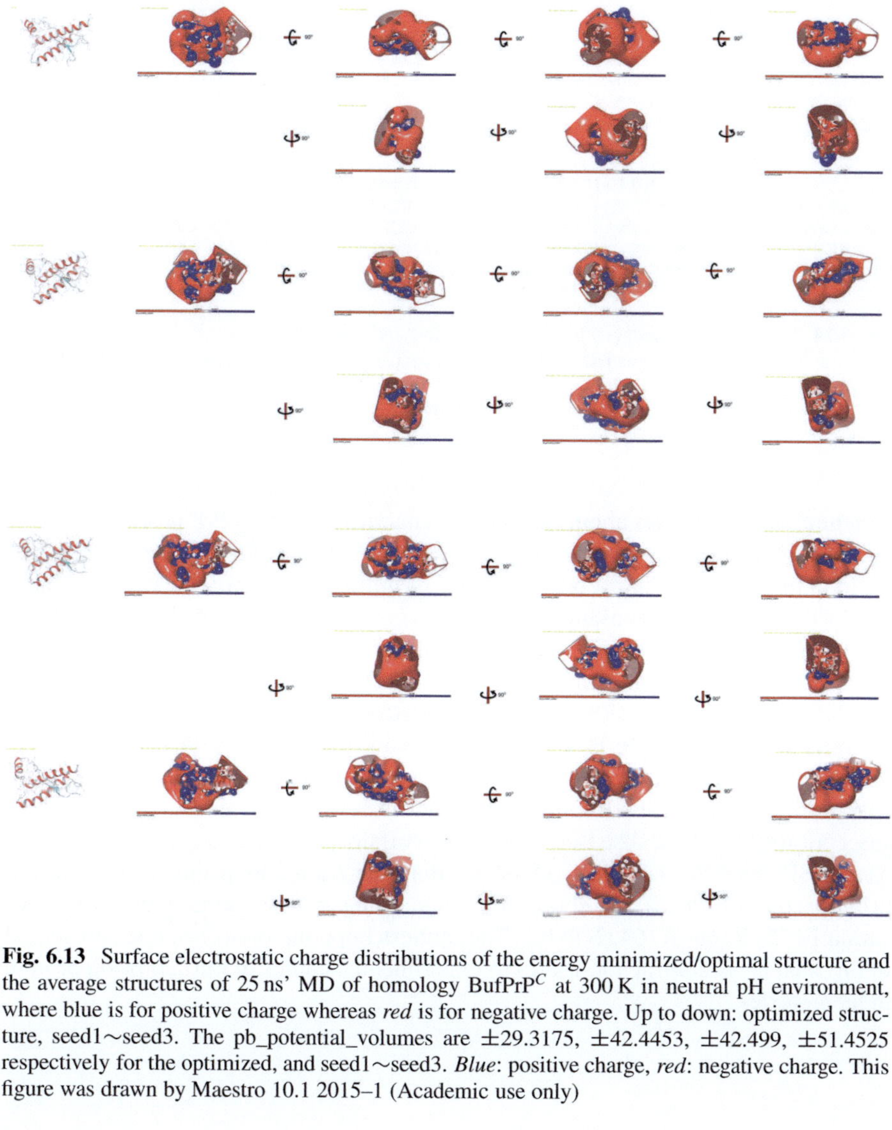

Fig. 6.13 Surface electrostatic charge distributions of the energy minimized/optimal structure and the average structures of 25 ns' MD of homology BufPrP^C at 300 K in neutral pH environment, where blue is for positive charge whereas *red* is for negative charge. Up to down: optimized structure, seed1~seed3. The pb_potential_volumes are ±29.3175, ±42.4453, ±42.499, ±51.4525 respectively for the optimized, and seed1~seed3. *Blue*: positive charge, *red*: negative charge. This figure was drawn by Maestro 10.1 2015–1 (Academic use only)

6.5 Concluding Remarks

This chapter constructed a molecular structure of buffalo prion protein and then did MD study on this molecular structure. Clearly, this homology structure is useful as a reference for biochemical laboratories and later for NMR or X-ray structural laboratories. Buffalo is a low susceptibility species resisting to prion diseases and buffalo prion protein is very stable. To date there is no structural data available. Protein structure of buffalo PrP was constructed by this chapter and we also present

```
                 124  130       140       150       160       170       180
                    456789012345678901234567890123456789012345678901234567890123
BuffaloPrP          GLGGYMLGSAMSRPLIHFGNDYEDRYYRENMHRYPNQVYYRPVDQYSNQNNFVHDCVNIT 60
RabbitPrP           GLGGYMLGSAMSRPLIHFGNDYEDRYYRENMYRYPNQVYYRPVDQYSNQNSFVHDCVNIT 60
HorsePrP            GLGGYMLGSAMSRPLIHFGNDYEDRYYRENMYRYPNQVYYRPVSEYSNQNNFVHDCVNIT 60
DogPrP              GLGGYLLGSAMSRPLIHFGNDCEDRYYRENMYRYPNQVYYRSVDQYNNQSTFVHDCVNIT 60
                    *****:*************** *********:*********.*.:*.**..*********
                        190       200       210       220    227
                    45678901234567890123456789012345678901234567
BuffaloPrP          VKEHTVTTTTKGENFTETDIKMMERVVEQMCITQYQRESQAYYQ- 104
RabbitPrP           VKQHTVTTTTKGENFTETDIKIMERVVEQMCITQYQQESQAAYQ- 104
HorsePrP            VKQHTVTTTTKGENFTETDVKIMERVVEQMCITQYQKEYEAFQQ- 104
DogPrP              VKQHTVTTT-KGENFTETDIKMMERVVEQMCITQYQRESEAYYQR 104
                    **:****** *********:*:**************:* :*  *
```

Fig. 6.14 Sequence alignment of buffalo, dog, rabbit and horse PrPs.
At 155 and 186, buffaloes have HIS and GLU residues respectively, but others have TYR and GLN residues respectively. In 3D structures, the RMSD values with dogs, horses and rabbits are 8.35, 18.17, and 19.84 Å respectively and imply to us buffaloes differ from dogs, horses and rabbits very much

structural bioinformatics about molecular dynamics of buffalo PrP protein. Same as that of rabbits, dogs or horses, the SB ASP178–ARG164 (O–N) (keeping the β2-α2 loop linked) contributes to the stability of buffalo prion protein. We also found HB SER170–TYR218 (linking the β2-α2 loop with the C-terminal end of α-helix H3), and SB HIS187–ARG156 (N–O) (linking α-helices H2 and H1) contribute to the stability of buffalo prion protein. At D178, there is a HB Y169–D178 and a polar contact R164–D178 for BufPrP^C instead of a polar contact Q168–D178 for bovine PrP^C. Buffalo is a species with low susceptibility to prion diseases, thus the bioinformatics of this chapter might be useful to the structure-based drug design of prion diseases. Rabbits, dogs and horses are also the species with low susceptibility to prion diseases; we found buffaloes, rabbits, dogs and horses all have a SB ASP178–ARG164 (O–N) keeping the β2-α2 loop linked. For buffalo prion protein, in the β2-α2 loop, there is a strong π-π stacking and a strong π-cation F175–Y169–R164.(N)NH2. The authors hope the bioinformatics presented in this chapter is helpful and useful to experimental studies of buffalo prion protein in laboratories.

Chapter 7
Compared with the NMR Structure and Dynamics of Elks

We used the NMR structure of ElkPrP (1XYW.pdb) to do MD simulations, and the MD methods are completely same as those in Chaps. 3∼6.

7.1 Results of Molecular Dynamics at 300 K

At 300 K, we may see from Fig. 7.1 that the two β-strands are changed into β-bridges from neutral to low pH environments. For seed2, in neutral pH environment during 10∼25 ns, the two β-strands are changed into β-bridges, and in low pH environment during 18∼30 ns, the two β-strands are changed into π-helices. Form Fig. 7.1, we also know that from neutral to low pH environments the N-terminal ends of H1 and H2 are unfolded into β-strands. This implies to us that elks are a high susceptibility species to prion diseases and from neutral to low pH environments the structures of ElkPrP will be easily changed from α-helices into β-sheet amyloids, even at room temperatures.

However, at 300 K, the three α-helices of ElkPrP are still conserved during the 30 ns of MD simulations whether under neutral pH or low pH environments thus make ElkPrP structures not changing greatly (Figs. 7.1, 7.2 and 7.3).

Seeing Fig. 7.2, for seed1, neutral pH environment, ElkPrP seems to have special snapshots during 25∼30 ns. This might be due to the C-terminal of H2 being changed into 3_{10}-helix structures (Fig. 7.1).

7.2 Results of Molecular Dynamics at 450 K

The structures of ElkPrP changed a lot at 450 K. H1 and H2 have unfolded under low pH environment for seed3 (Fig. 7.4) and the two β-strands disappeared, under neutral pH environment, for seed1 during the whole 30 ns, for seed2 during

J. Zhang, *Molecular Structures and Structural Dynamics of Prion Proteins and Prions*, Focus on Structural Biology 9, DOI 10.1007/978-94-017-7318-8_7

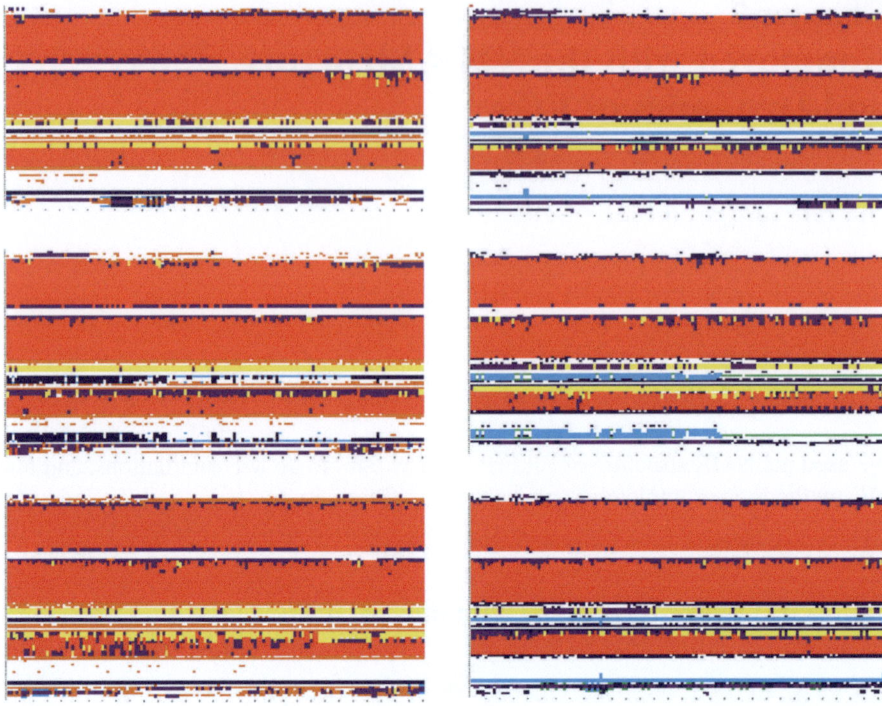

Fig. 7.1 Secondary Structure graphs of ElkPrP^C at 300 K (x-axis: time (0–30 ns), y-axis: residue number (121–231); *left column*: neutral pH, *right column*: low pH; up to down: seed1–seed3; $\alpha 1$ is α-helix 1, $\alpha 2$ is α-helix 2, $\alpha 3$ is α-helix 3, $\beta 1$ is β-strand 1, and $\beta 2$ is β-strand 2). H is the α-helix, I is the π-helix, G is the 3-helix or 3_{10} helix, B is the residue in isolated β-bridge, E is the extended strand (participates in β-ladder), T is the HBed turn, and S is the bend

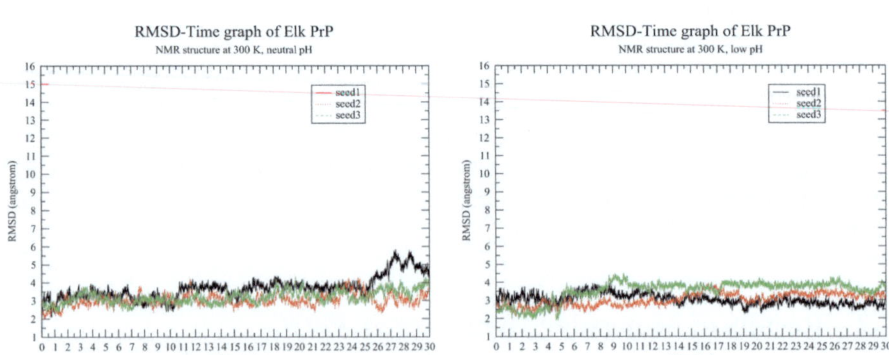

Fig. 7.2 RMSD of ElkPrP at 300 K, neutral and low pH values (*left*: neutral pH, *right*: low pH) during 30 ns' MD

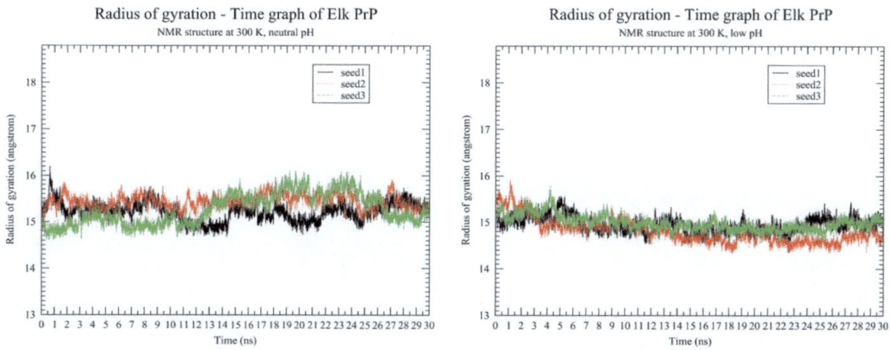

Fig. 7.3 Radius of gyration of ElkPrP at 300 K, neutral and low pH values (*left*: neutral pH, *right*: low pH) during 30 ns' MD

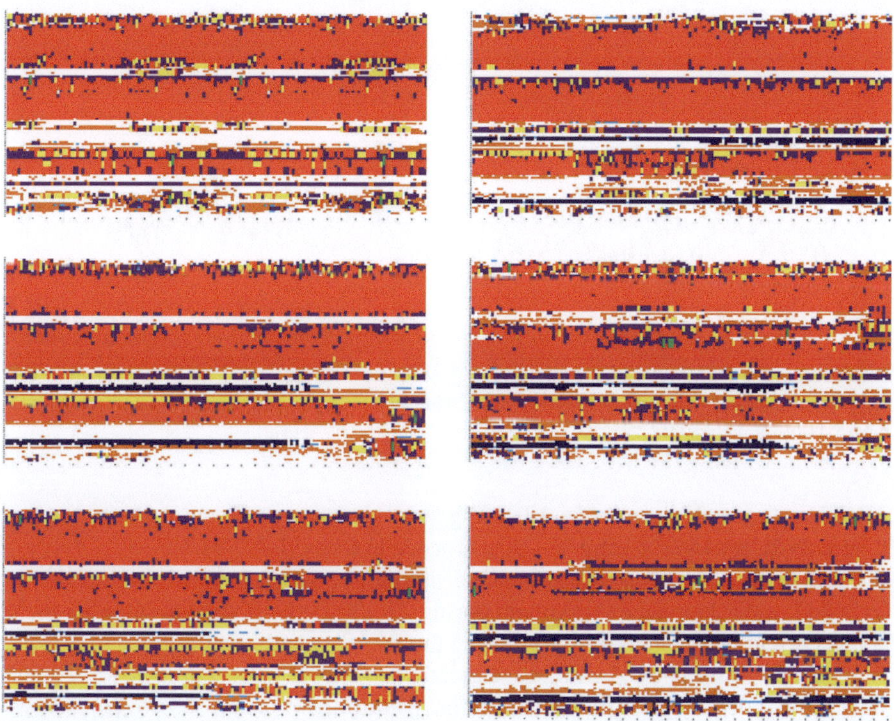

Fig. 7.4 Secondary structure graphs of ElkPrP^C at 450 K (x-axis: time (0–30 ns), y-axis: residue number (121–231); *left column*: neutral pH, *right column*: low pH; up to down: seed1–seed3). H is the α-helix, I is the π-helix, G is the 3-helix or 3_{10} helix, B is the residue in isolated β-bridge, E is the extended strand (participates in β-ladder), T is the HBed turn, and S is the bend

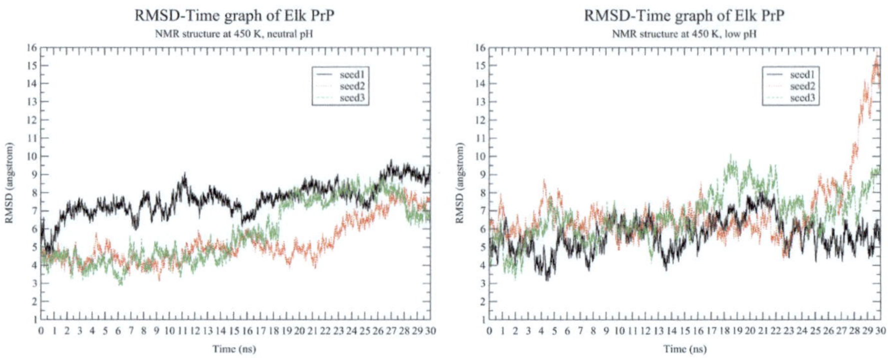

Fig. 7.5 RMSD of ElkPrP at 450 K, neutral and low pH values (*left*: neutral pH, *right*: low pH) during 30 ns' MD

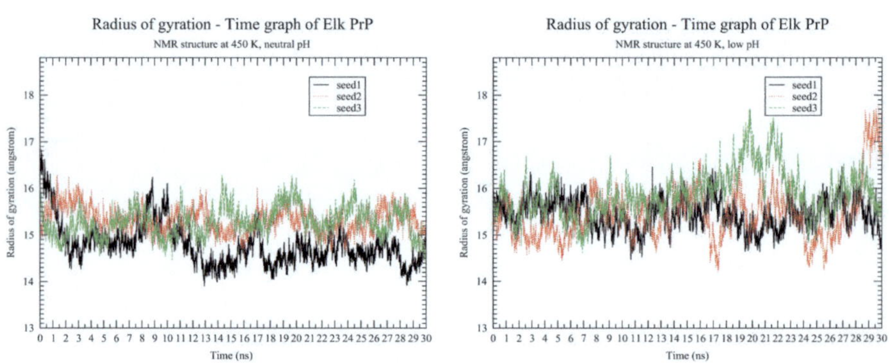

Fig. 7.6 Radius of gyration of ElkPrP at 450 K, neutral and low pH values (*left*: neutral pH, *right*: low pH) during 30 ns' MD

22∼30 ns, for seed3 during 15∼30 ns (Fig. 7.4). Under low pH environment, the two β-strands disappeared during 23∼30 ns for seed2 (Figs. 7.4–7.6).

Seeing Figs. 7.5–7.6, we know that the variations of RMSD values are in 7 Å, and for seed2 under low pH environment during 23∼30 ns the RMSD is jumping to 16 Å– this might be due to the HBs between the two β-strands are completely broken. Seeing Fig. 7.6 we know the radius of gyrations of seed2 and seed3 vary very much and imply to us the structures of ElkPrP have unfolded.

7.3 A Concluding Remark Compared with Rabbit PrP

This is a common character that ElkPrP and RaPrP both have a highly and clearly ordered $\beta 2$-$\alpha 2$ loop. The SB between ASP178 and ARG164 (Figs. 7.7∼7.8) clearly contributes to this character.

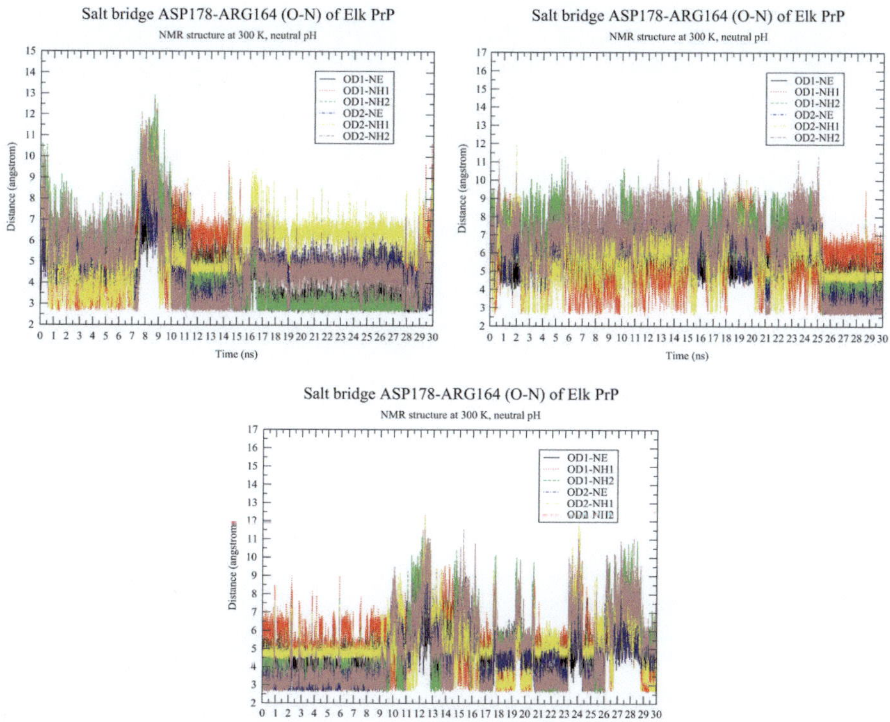

Fig. 7.7 SB ASP178–ARG164 (O–N) of ElkPrP at 300 K during 30 ns' MD (*left∼right*: seed1∼seed3)

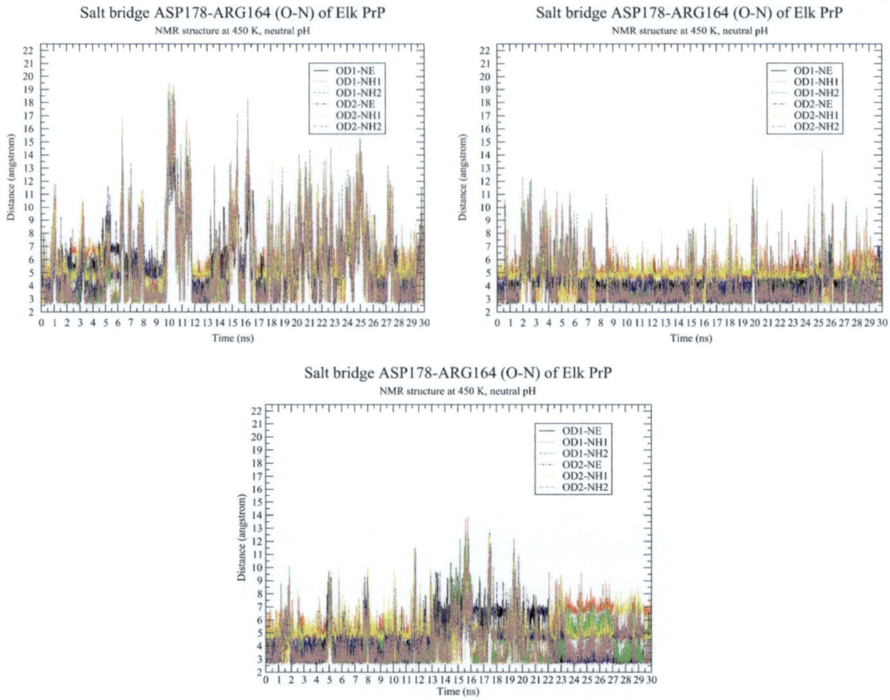

Fig. 7.8 SB ASP178–ARG164 (O–N) of ElkPrP at 450 K during 30 ns' MD (*left*∼*right*: seed1∼seed3)

Chapter 8
Compared with the X-Ray Structure and Dynamics of Rabbits

8.1 Introduction

BSE ("mad-cow" disease) belongs to a contagious type of TSEs. Scientists believe it is caused by prions (the misfolding prion proteins) but they may have not yet solved the riddle of "mad cow disease". This is due to a prion is neither a virus, a bacteria nor any microorganism so the disease cannot be caused by the vigilance of the organism immune system and it can freely spread from one species to another species. The humans exists the susceptibility of TSEs. For example, the human version of "mad cow" diseases named CJD and vCJD just happen randomly through infections of transplanted tissue or blood transfusions or consumption of infected beef products. Cat, mink, deer, elk, moose, sheep, goat, nyala, oryx, greater kudu, ostrich and many other animals are also susceptible to TSEs. However, rabbits, horses and dogs seem to be unaffected by Prions [336, 474, 591, 682, 697]. Scientists do not know the reason.

The prion protein is a naturally occurring protein in vivo. Its lesions in brain are not caused by the vigilance of the immune system. Recent studies have found that the lesions led astray as long as by contact with other normal prion proteins. The cells are arranged in accordance with the instruction of the gene and formed into proteins with different shapes and functions. But, like cardboard boxes, proteins need to be properly "folded" in order to ensure their normal work. When proteins are folded into the wrong shape, they do not work. Under normal circumstances, the cells will supervise these misfolded proteins and automatically decompose them. However, the supervision mechanism is not with 100 % insurance. Scientists found that the rate of decomposition of prions is not quick enough and these prions accumulate and change the cellular metabolism and eventually kill the cell. This leads to the death of neurons in the brain. The dead neurons decompose and release more prion proteins into the biological mechanism to cause prion diseases. The infectious diseased prion is thought to be an abnormally folded isoform (PrP^{Sc}) of a host protein known as the prion protein (PrP^{C}). The conversion of PrP^{C} to PrP^{Sc} occurs

J. Zhang, *Molecular Structures and Structural Dynamics of Prion Proteins and Prions*, Focus on Structural Biology 9, DOI 10.1007/978-94-017-7318-8_8

post-translationally and involves conformational change from a predominantly α-helical protein to one rich in β-sheet amyloid fibrils. Much remains to be understood about how the normal cellular isoform of the prion protein PrP^C undergoes structural changes to become the disease associated amyloid fibril form PrP^{Sc}. The "structural conformational" changes of PrP^{Sc} from PrP^C are just very proper to be studied by MD techniques. Classical studies have showed that rabbits have low susceptibility to be infected by PrP^{Sc} [29, 160, 204, 336, 350, 407, 441, 587, 591, 625, 626, 717]. However, recently there is one exception saying that (i) rabbit prion can be produced through saPMCA (*s*erial *a*utomated Protein Misfolding Cyclic Amplification) in vitro [106, 203] (though not by challenging rabbits directly in vivo with other known prion strains), and (ii) the rabbit prion generated is infectious and transmissible [106]. RaPrP had already had NMR [381] and X-ray [336] structures in Protein Data Bank (www.rcsb.org) with PDB entries 2FJ3, 3O79 respectively. In 2010 and 2013, the NMR and X-ray structures of some RaPrP mutants (with PDB entries 2JOH, 2JOM, 4HLS, 4HMR, 4HMM) were also already released into PDB bank [556, 626]. This chapter will study the NMR and X-ray structures of RaPrP and its I214V and S173N mutants by MD techniques, in order to explain the specific mechanism about rabbit PrP^C ($RaPrP^C$) and the conversion of $PrP^C \rightarrow PrP^{Sc}$ of rabbits. Surely, the studies should provide valuable knowledge about the rules governing the $PrP^C \rightarrow PrP^{Sc}$ conversion, which can provide some ideas for designing novel therapeutic approaches that block the conversion and disease propagation.

In the MD studies, we are not only doing simulations under neutral pH environment, but also under low pH environment, because recent studies indicate that the low pH solution system is an ideal trigger of PrP^C to PrP^{Sc} conversion [59, 240, 716].

Through some preliminary analyses of the MD simulation results, we have found that SB is a clear factor of the structural stability of RaPrP. This agrees with a recent finding on SBs about PrP^C [716]. The SBs of the native prion protein (PrP^C) have been calculated and analyzed by quantum chemical calculations by Ishikawa and Kuwata [310], and the detailed biophysical characteristics and NMR studies of $PrP^C \rightarrow PrP^{Sc}$ conversion process have also been reported by Bjorndahl et al. [59]. Recently, incorporating Chou's wenxiang diagrams [716] further summarized their research based on the NMR, CD (circular diagram) spectra and DLS (Dynamic light scattering) data at the low pH environment, and found that some SBs and the HYDs in the three helices of the prion proteins can affect the helical structural stability. These studies have provided the insight into the prion misfolding mechanism.

The rest of this chapter is organized as follows. Firstly, we will review previous research results on rabbit prion protein listed in the PubMed of NCBI. Secondly, we will present the MD simulation materials and methods for NMR and X-ray structures of $RaPrP^C$ wild-type and mutants. Thirdly, we will give analyses of MD simulation results and discussions. Lastly, concluding remarks on RaPrP are summarized.

8.2 A Detailed Review on Rabbit PrP

The symptoms of TSEs were first described for sheep in 1730 and called "scrapie" in England, "vertige" in France and "Traberkrankheit" in Germany [581]. Now we know that many species such as sheep, goats, mice, humans, chimpanzees, hamsters, cattle, elks, deers, minks, cats, chicken, pigs, turtles, etc are susceptible to TSEs. But many laboratory experiments show that rabbits, horses and dogs seem to be the resistant (or at least the low-susceptibility-rate) species to TSEs. However, recently Chianini et al. (2012) reported "rabbits are not resistant to prion infection" [106]. Thus, at this moment it is very necessary for us to give a detailed review some laboratory works (from Year 1976 to Year 2014) on RaPrP:

- In 1976, Barlow and Rennie (1976) made many attempts to infect rabbits with the ME7 scrapie (where scrapie is a prion disease in sheep and goats) and other known prion strains but all failed at last [29].
- In 1984∼1985, some antibodies to the scrapie protein were reported by Prusiner's research group [37, 44]. The antibodies to the scrapie agent were produced after immunization of rabbits with either scrapie prions or the prion protein PrP(27–30). The monospecificity of the rabbit antiserum raised against PrP(27–30) was established by its reactivity after affinity purification, and the rabbit antiserum to PrP(27–30) was successfully produced. In 1985, the characterization of antisera raised in rabbits, against scrapie-associated prion diseases and the prion human CJD, was studied in [61].
- In 1986, Takahashi et al. (1986) reported, rabbits immunized with the fraction P4 containing scrapie infectivity prepared from mouse brains raised antibodies against three major polypeptides of [562]. Cho (1986) reported that the antibody to scrapie-associated fibril protein roduced in a rabbit identifies a cellular antigen [107]. Barry et al. (1986) reported that rabbit antisera to a synthetic peptide PrP-P1 constructed based on PrP(27–30) were found by immunoblotting to react with PrP(27–30) and its precursor PrP^{Sc}(33–35), as well as with a related protease-sensitive cellular homologue PrP^{C}(33–35), this means scrapie (PrP^{Sc}) and cellular (PrP^{C}) prion proteins share polypeptide epitopes [36]. An enzyme-linked immunosorbent assay showed that rabbit antiserum to PrP(27–30) was more reactive with PrP(27–30) than with PrP-P1; conversely, antiserum to PrP-P1 was more reactive with the peptide than with the prion proteins [36]. Shinagawa et al. (1986) reported "immunization of a rabbit with the (synthetic) peptide conjugated with ovalbumin induced specific antibodies" corresponding to the N-terminal region of the scrapie prion protein (PrP^{Sc}) [528]. "Rabbit antisera were raised to SAFs (scrapie-associated fibrils) isolated from mice infected with the ME7 scrapie strain and to SAFs isolated from hamsters infected with the 263K scrapie strain" [332, 423]. Robakis et al. (1986) clearly pointed out that rabbit brain is resistant to scrapie infection [496] in 1986.
- In 1987, there were several reports on rabbits. Bockman et al. (1987) identified by immunoblotting human & mouse CJD prion proteins ($HuPrP^{Sc}$ and $MoPrP^{Sc}$) using rabbit antisera raised against hamster scrapie prion proteins ($HaPrP^{Sc}$)

[60]. Wade et al. (1987) found a 45 kD protein in scrapie-infected hamster brain has a signal to inoculate rabbits [592]. Kascsak et al. (1987) reported "MAb (monoclonal antibody) 263K 3F4 (that was derived from a mouse immunized with hamster 263K PrP^{Sc} reacted with hamster but not mouse PrP^{Sc}) recognized normal host protein of 33 to 35 kDa in brain tissue from hamsters and humans but not from bovine, mouse, rat, sheep, or rabbit brains" [333]. Wiley et al. (1987) used rabbit monospecific antisera raised against synthetic peptides corresponding to the N-terminal 13 or 15 amino acids of PrP(27–30) and rabbit antisera raised against infectious prions or PrP(27–30) purified from scrapie-infected hamster brains to immunostaine glutaraldehyde-perfused hamster brains [629]. Hay et al. (1987) found the evidence for a secretory form of the cellular prion protein (PrP^{C}) "cell-free translation studies in rabbit reticulocyte lysates supplemented with microsomal membranes gave results: while one form of HaPrP (hamster brain prion protein) was found as an integral membrane protein spanning the membrane at least twice, another form of HaPrP was found to be completely translocated to the microsomal membrane vesicle lumen" [277].

- In 1988, Caughey et al. (1988) detected the immunoprecipitation of PrP synthesis using a rabbit antibody specific for a 15 amino acid PrP peptide and concluded that "either PrP is not the transmissible agent of scrapie or the PrP is not processed appropriately in this cell system to create the infectious agent" [93]. Barry et al. (1988) undertook ELISA (enzyme-linked immunosorbent assay) and immunoblotting studies with rabbit antisera raised against three synthetic PrP peptides of PrP(27–30), PrP^{Sc}, and PrP^{C} and concluded that the three proteins are encoded by the same chromosomal gene [38]. Baron et al. (1988) found "polyclonal rabbit antiserum to SAF protein was reacted with brain sections from scrapie-infected mice, two familial cases of transmissible dementia, and three cases of Alzheimer's disease (AD)" and "evidence of the similarity of SAF protein to PrP(27–30)" [31]. Gabizon et al. (1988) found "polyclonal RaPrP antiserum raised against NaDodSO4/PAGE-purified scrapie prion protein of 27–30 kDa reduced scrapie infectivity dispersed into detergent-lipid-protein complexes" [216]. Roberts et al. (1998), "using monoclonal antibodies to a synthetic peptide corresponding to a portion of beta-protein and rabbit antiserum to hamster scrapie PrP(27–30), examined in situ amyloid plaques on sections from cases of neurodegenerative diseases" and their "results emphasize the need for classification of CNS (central nervous system) amyloids based on the macromolecular components comprising these pathologic polymers" [497].
- In 1989, Gabizon et al. (1989) reported that "polyclonal rabbit PrP antiserum raised against sodium dodecyl sulfate-polyacrylamide gel electrophoresis (SDS-PAGE)-purified PrP(27–30) reduced scrapie infectivity dispersed into DLPC (detergent-lipid-protein complexes)" [217]. "Kuru plaque is a pathognomonic feature in the brain of patients with CJD and in the brain of CJD-infected mice" [341], Kitamoto et al. (1989) reported "kuru plaques from CJD-infected mice were immunolabeled with rabbit anti-murine prion protein (PrP) absorbed with human PrP, but not so with mouse anti-human PrP" [341]. Farquhar et al. (1989)

"Two polyclonal antisera were raised in rabbits against the scrapie-associated fibril protein (PrP) prepared from sheep and mice" [201].

- In 1990, Yost et al. (1990) reported that in the rabbit reticulocyte lysate system, an unusual topogenic sequence in the prion protein fails to cause stop transfer (the polypeptide chain across the membrane of the endoplasmic reticulum) of most nascent chains [666]. Lopez et al. (1990) reported a completely translocated (secretory) topology form of the major product synthesized in rabbit reticulocyte lysates (RRL) [399].
- In 1991, Di Martino et al. (1991) reported the characterization of two polyclonal antibodies which were raised by immunizing rabbits with two non carrier-linked synthetic peptides whose amino acid sequences corresponded to codons 89–107 (peptide P1) and 219–233 (peptide P2) of the translated cDNA sequence of murine PrP protein [184]. Ikegami et al. (1991) detected the scrapie-associated fibrillar protein in the lymphoreticular organs of sheep by means of a rabbit-anti-sheep PrP (the scrapie-associated fibrillar protein) polyclonal antibody by Western blot analysis [300].
- In 1992, Hashimoto et al. (1992) did immunohistochemical study of kuru plaques using antibodies against synthetic prion protein peptides and used two synthetic peptides to immunize rabbits and produce antisera (anti-N and anti-M) [275]. Kirkwood et al. (1992) using rabbit antiserum raised against mouse PrP detected an abnormal PrP (prion protein) from the brains of domestic cattle with spongiform encephalopathy (SE) [340].
- In 1993, the Western blot analysis was performed with rabbit serum against the sheep SAF [450]. To determine if amyloid deposits be visualized by immunocytochemical techniques, Guiroy et al. (1993) used a rabbit antiserum directed against scrapie amyloid PrP(27–30) to stain formalin-fixed, formic acid-treated brain tissue sections from several animal species with natural and experimental transmissible mink encephalopathy (TME) [266]. Groschup and Pfaff (1993) reported that "rabbit antisera to synthetic peptides representing amino acid sequence 108 to 123 of PrP of cattle, sheep and mice reacted strongly with modified PrP of the homologous host but not, or only poorly, with PrP of heterogeneous origin" [261]. Miller et al. (1993) did the immunohistochemical detection of prion protein in sheep with scrapie using a primary antibody obtained from a rabbit immunized to PrP^{Sc} extracted from brains of mice with experimentally induced scrapie [426].
- In 1994, Groschup et al. (1994) investigated with eight different anti-peptide sera raised in rabbits against various synthetic peptides representing segments of the amino acid (aa) sequence 101–122 of ovine, bovine, murine and hamster PrP, and found that "the region close to the actual or putative proteinase K cleavage sites of PrP seems to exhibit high structural variability among mammalian species" [260]. Xi et al. (1994) detected the proteinase-resistant protein (PrP) in small brain tissue samples from CJD patients using rabbit polyclonal antibody against hamster PrP(27–30) [637]. Schmerr et al. (1994) used a fluorescein-labeled goat anti-rabbit immunoglobulin as an antibody and used rabbit antiserum

for immunoblot analysis, and PrP^{Sc} was solubilized and reacted with a rabbit antiserum specific for a peptide of the prion protein [518].

- In 1995∼1996, Yokoyama et al. did some works. In 1995, Yokoyama et al. (1995) used antisera raised in rabbits against three peptides PrP 150–159, PrP 165–174, and PrP 213–226 of mouse prion and concluded that rabbit antiserum against the MAP (multiple antigenic peptide) representing amino acid sequence 213–226 of mouse PrP is useful as a diagnostic tool for prion disease of animals [664]. In 1996, Yokoyama et al. (1996) detected species specific epitopes of mouse and hamster prion proteins by anti-peptide antibodies, where the antisera were produced in rabbits [663].
- In 1997, Madec et al. (1997) undertook Western blot analyses using rabbit antiserum that recognized both normal and pathologic sheep prion proteins to study the biochemical properties of PrP^{Sc} in natural sheep scrapie [411]. Loftus and Rogers (1997) cloned RaPrP open reading frame (ORF) and characterised rabbits as a species with apparent resistance to infection by prions [398]. Groschup et al. (1997) raised antisera in rabbits and chicken against 16 synthetic peptides which represent the complete amino acid sequence of ovine PrP to generate antibodies to further regions of PrP, in order to immunochemical diagnosis and pathogenetic studies on prion diseases [259]. In [348],"the Ure2p yeast prion-like protein was translated in vitro in the presence of labeled [35S]methionine in either rabbit reticulocyte lysate (RRL) or wheat germ extract (WGE) cell-free systems". In 1997, Korth et al. (1997) found that RaPrP was not recognized by a conformational antibody specific for PrP^{Sc}-like structures [350].
- In 1999, Takahashi et al. (1999) immunized rabbits with four synthetic peptides and compared the immunoreactivity of antibodies to bovine prion proteins (bovine-PrPs) from various species by immunoblotting and immunohistochemistry [564] and they identified two regions in bovine-PrP which appear suitable for raising antibodies that detect various kinds of PrPs, and one region (Ab103–121) which appears suitable for raising antibodies that detect several species of PrP [564].
- In 2000, Garssen et al. (2000) did applicability of three anti-PrP peptide sera including staining of tonsils and brainstem of sheep with scrapie [236]. "The three rabbit antibodies (R521, R505, R524) were produced, and raised to synthetic peptides corresponding to residues 94–105, 100–111, and 223–234, respectively, of the sheep prion protein" [236]. "The usefulness of all three anti-peptide sera in the immunohistochemical detection of PrP^{Sc} in brain stem and tonsils of scrapie-affected sheep was demonstrated and compared with an established rabbit anti-PrP serum" [236]. Zhao et al. (2000) using prokaryotic expressed GST-PrP fusion protein as antigen, found that "rabbits were immunized subcutaneously" [707].
- In 2001, Kelker et al. (2001) showed that "combination of authentic rabbit muscle GAPDH (glyceraldehyde-3-phosphate dehydrogenase) with tNOX (a cell surface NADH oxidase of cancer cells) renders the GAPDH resistant to proteinase K digestion" [334]. Vol'pina et al. (2001) reported "rabbits were immunized with

either free peptides or peptide-protein conjugates to result in sera with a high level of antipeptide antibodies" to the BSE prion disease [590]. Bencsik et al. (2001) identified prion protein PrP "using either RB1 rabbit antiserum or 4F2 monoclonal antibody directed against AA 108–123 portion of the bovine and AA 79–92 of human prion protein respectively" and "showed the close vicinity of these PrP expressing cells with noradrenergic fibers" [43]. In [383], "the rabbits were immuned with bovine prion protein (BoPrP^{C}) which was expressed in E. coli and anti-PrP^{C} antibody (T1) was obtained", and Li et al. (2001) could detect BSE and scrapie with T1 antibody [383].

- In 2002, Laude et al. (2002) reported "In one otherwise refractory rabbit epithelial cell line, a regulable expression of ovine PrP was achieved and found to enable an efficient replication of the scrapie agent in inoculated cultures" [367]. Takekida et al. (2002) established a competitive ELISA to detect prion protein in food products using rabbit polyclonal antibodies that were raised against bovine prion peptides [563].
- In 2003, Vorberg et al. (2003) found multiple amino acid residues (such as GLY99, MET108, SER173, ILE214) within the RaPrP inhibit formation of its abnormal isoform [591]. The authors made some substitutions of mouse PrP amino acid sequence by rabbit PrP amino acid sequence and found (i) at the N-terminal region (residues 1–111) the PrP^{Sc} formation is totally prevented, (ii) at the central region (residues 112–177), the constructed PrP failed to be converted to protease resistance, (iii) at the C-terminal region (residues 178–254) the formation of PrP^{Sc} is drastically decreased but is not abolished completely [591]. Thus, rabbit cells are negatively affected by the formation of PrP^{Sc}. Jackman and Schmerr (2003) synthesized fluorescent peptides from the prion protein and produced the corresponding antibodies in rabbits against these peptides, and at last detected abnormal prion protein in a tissue sample [312]. Gilch et al. (2003) reported "treatment of prion-infected mouse cells with polyclonal anti-PrP antibodies generated in rabbit or auto-antibodies produced in mice significantly inhibited endogenous PrP^{Sc} synthesis" and found "immune responses against different epitopes when comparing antibodies induced in rabbits and PrP wild-type mice" [244].
- In 2004, Brun et al. (2004) reported the development and further characterisation of a novel PrP-specific monoclonal antibody 2A11, which reacts with PrP^{C} from a variety of species including rabbit [71]. Sachsamanoglou et al. (2004) described "the quality of a rabbit polyclonal antiserum (Sal1) that was raised against mature human recombinant prion protein (rHuPrP)" [507]. Senator et al. (2004) investigated "the effects of cellular prion protein (PrP^{C}) overexpression on paraquat-induced toxicity by using an established model system, rabbit kidney epithelial A74 cells, which express a doxycycline-inducible murine PrP^{C} gene" [524].
- In 2005, Golanska et al. (2005) used 2 different anti-14-3-3 antibodies: rabbit polyclonal and mouse monoclonal antibodies to analyze the 14-3-3 protein in the cerebrospinal fluid in CJD [248].

- In 2006, Dupiereux et al. (2006) investigated the effect of PrP(106–126) peptide on an established non neuronal model, rabbit kidney epithelial A74 cells that express a doxycycline-inducible murine PrP^C gene [195]. Biswas et al. (2006) reported "a rabbit polyclonal anti-serum raised against dimeric MuPrP (murine prion protein) cross-reacted with p46 (a 46 kDa species) and localized the signal within the Golgi apparatus" [57]. Gao et al. (2006) reported recombinant neural protein PrP can bind with both recombinant and native apolipoprotein E (ApoE) in vitro, where the "ApoE-specific antiserum was prepared by immunizing rabbits with the purified ApoE3" [231]. Kocisko and Caughey (2006) reported "rabbit epithelial cells that produce sheep prion protein in the presence of doxycycline (Rov9) have been infected with sheep scrapie" [346]. Xiao et al. (2006) used the method "two male rabbits were immunized for 4 times with the purified protein, and the antiserum against neuron-specific enolase (NSE) protein was collected and evaluated by ELISA, Western blotting and immunohistochemistry" and concluded "high expression of HuNSE (human neuron-specific enolase) is obtained in E. coli and the prepared antiserum against HuNSE can be used potentially for diagnosis of prion-associated diseases and other nervous degeneration diseases" [640].
- In 2007, Oboznaya et al. (2007) reported "antibodies to a nonconjugated prion protein peptide 95–123 interfere with PrP^{Sc} propagation in prion-infected cells", where "rabbits were immunized with free nonconjugated peptides" [446]. Handisurya et al. (2007) reported "Immunization with PrP-virus-like particles induced high-titer antibodies to PrP in rabbit and in rat, without inducing overt adverse effects. As determined by peptide-specific ELISA, rabbit immune sera recognized the inserted murine/rat epitope and also cross-reacted with the homologous rabbit/human epitope differing in one amino acid residue. Rabbit anti-PrP serum contained high-affinity antibody that inhibited de novo synthesis of PrP^{Sc} in prion-infected cells" [270]. Bastian et al. (2007) did an experiment "spiroplasma mirum, a rabbit tick isolate that had previously been shown to experimentally induce spongiform encephalopathy in rodents, was inoculated intracranially (IC) into ruminants" and at last concluded "Spiroplasma spp. from TSE brains or ticks induce spongiform encephalopathy in ruminants" [39]. Dong et al. (2007) did the interaction analysis between various PrP fusion proteins and the tubulin in vitro, where the native tubulin was extracted from rabbit brain tissues [187].
- In 2008, it was said that "ovine prion protein renders rabbit epithelial RK13 cells permissive to the multiplication of ovine prions, thus providing evidence that species barriers can be crossed in cultured cells through the expression of a relevant ovine PrP^C" [160]. Sakudo et al. (2008) "developed a mammalian expression system for a truncated soluble form of human prion protein with the native signal peptide but without a glycosylphosphatidylinositol (GPI)-anchor site, driven by the peptide chain elongation factor 1alpha promoter in stably transfected rabbit-kidney epithelial RK13 cells, to investigate the SOD

(superoxide dismutase) activity of mammalian prion protein" and concluded "GPI-anchorless human prion protein is secreted and glycosylated but lacks superoxide dismutase activity" [512]. Shin et al. (2008) cloned a prion protein (PrP) Glu218Lys gene from Korean bovine (Bos taurus coreanae) and raised the production of rabbit anti-bovine PrP antibody [527]. Lawson et al. (2008) reported "rabbit kidney epithelial cells (RK13) are permissive to infection with prions from a variety of species upon expression of cognate PrP transgenes" [369].

- In 2009, Hanoux et al. (2009) reported "when injected into rabbits, (a synthetic peptide) CDR3L generated anti-SAF61 anti-Id polyclonal antibodies that exclusively recognized SAF61 mAb but were unable to compete with hPrP for antibody binding" [271]. Tang et al. (2009) reported fibrinogen, one of the most abundant extracellular proteins, has chaperone-like activity: it maintains thermal-denatured luciferase in a refolding competent state allowing luciferase to be refolded in cooperation with rabbit reticulocyte lysate, and it also inhibits fibril formation of yeast prion protein Sup35 (NM) [566]. Differed from the reaction with N-terminal proline/glycine-rich repeats recognizing rabbit polyclonal antibody, seven monoclonal antibodies (mAbs) against chicken cellular prion protein ($ChPrP^C$) were obtained [309]. Fernandez-Funez et al. (2009) found RaPrP does not induce neurodegeneration in the brains of transgenic flies [204].
- In 2010, Nisbet et al. (2010) created a mutant mouse PrP model containing RaPrP specific amino acids at the GPI anchor site and found that the GPI anchor attachment site (ω site) controls the ability of $PrP^C \rightarrow PrP^{Sc}$ and the residues at ω and ω+1 of PrP are important modulators of this pathogenic process [441]. Nisbet et al. (2010) recognized that "rabbits are one of a small number of mammalian species reported to be resistant to prion infection" [441]. Wen et al. (2010) using multidimensional heteronuclear NMR techniques reported that the I214V and S173N substitutions result in distinct structural changes for $RaPrP^C$ [625, 626] and concluded that the highly ordered $\beta 2$ $\alpha 2$ loop may contribute to the local as well as global stability of the RaPrP protein [625, 626]. Wen et al. (2010) also recognized "rabbits are one of the few mammalian species that appear to be resistant to TSEs due to the structural characteristics of $RaPrP^C$ itself" [626]. Fernandez-Funez et al. (2010) showed that "RaPrP does not induce spongiform degeneration and does not convert into scrapie-like conformers" [205]. Bitel et al. (2010) examined "changes in muscle tissue in a classic model of diabetes and hyperglycemia in rabbits to determine if similar dysregulation of Alzheimer $A\beta$ peptides, the prion protein (PrP), and superoxide dismutase 1 (SOD1), as well as nitric oxide synthases is produced in muscle in diabetic animals" [58]. Khan et al. (2010) found the propensity to form β-state (the β-sheet-rich structure) is greatest for hamster PrP, less for mouse PrP, but least for the PrP of rabbits, horses and dogs under different conditions and using two-wavelength CD (Circular Dichroism) method they also found a key hydrophobic staple-like helix-capping motif keeping the stability of RaPrP's X-ray crystallographic molecular structure [336].

- In 2011, Zocche et al. (2011) used the methods "rabbit aortic smooth muscle cells were challenged for 4, 8 and 18 h, with angiotensin-II, tunicamycin and 7-ketocholesterol, and rabbit aortic arteries were subjected to injury by balloon catheter", and got the results "the PrP^C mRNA expression in rabbit aortic artery fragments, subjected to balloon catheter injury, showed a pronounced increase immediately after overdistension" [720]. Mays et al. (2011) reported "PrP^{Sc} was efficiently amplified with lysate of rabbit kidney epithelial RK13 cells stably transfected with the mouse or Syrian hamster PrP gene" [420]. Julien et al. (2011) reported the different overall sensitivities toward NMR urea denaturation with stabilities in the order hamster = mouse < rabbit < bovine protein, and they also investigated the effect of the S174N mutation in rabbit PrP^C [324]. Zhou et al. (2011) found that the crowded physiological agents Ficoll 70 and dextran 70 have effects significantly inhibiting fibrillation of RaPrP [407, 717]. Fernandez-Funez et al. (2011) also acknowledged that "classic studies showing the different susceptibility to prion disease in mammals have recently found support in structural and transgenic studies with PrP from susceptible (mouse, hamster) and resistant (rabbit, horse, dog) animals" [206].
- In 2012, Chianini et al. (2012) generated rabbit PrP^{Sc} in vitro subjecting unseeded normal rabbit brain homogenate to saPMCA and found the rabbit PrP^{Sc} generated in vitro is infectious and transmissible [106] and they declared "rabbits are not resistant to prion infection" [106]. Kim et al. (2012) reported "elk prion protein ($ElkPrP^C$) has been confirmed to be capable of rendering rabbit epithelial RK13 cells permissive to temporal infection by chronic wasting disease (CWD) prions." [337]. Fernández-Borges et al. (2012) reported the results of [106] and pointed out it is not reasonable to attribute species-specific prion disease resistance based purely on the absence of natural cases and incomplete in vivo challenges; the concept of species resistance to prion disease should be re-evaluated using the new powerful tools available in modern prion laboratories, whether any other species could be at risk [203].
- In 2013, Vidal et al. (2013) studied the saPMCA and reported that rabbits are an apparently resistant species to the original classical cattle BSE prion [583]. Wang et al. (2013) reported rabbits are "insensitivity to prion diseases" [618]. Wang et al. (2013) aimed to investigate "potential mechanisms contributing to prion resistance/susceptibility by using the rabbit, a species unsusceptible to prion infection, as a model" [616] and investigated "the expression level and distribution of LRP/LR (laminin receptor precursor/laminin receptor) in rabbit tissues by real-time polymerase chain reaction and by immunochemical analysis with a monoclonal anti-67 kDa LR antibody" [616] and at last their findings confirmed the prion resistance in rabbits [616]. Sweeting et al. (2013) produced X-ray structures of mutants in the $\beta 2$-$\alpha 2$ loop and reported that the helix-capping motif in the $\beta 2$-$\alpha 2$ loop modulates β-state misfolding in RaPrP, and still acknowledged "rabbit PrP, a resistant species" [556]. Friedman-Levi et al. (2013) reported "pAb RTC and EP1802Y (EP), a rabbit α PrP mAb directed against the CITQYER ESQAYYQRGS sequence present at the C-terminal part of human

PrP, just before the PrP GPI anchor" [215]. Timmes et al. (2013) used R20 as the rabbit polyclonal anti-PrP antibody [569].

- In 2014, Yan et al. (2014) reported "the rabbit PrP does not readily form fibrils and is unlikely to cause prion diseases" and they "demonstrated "the domains beyond PrP–H2H3 (β-strand 1, α-helix 1, and β-strand 2) have a remarkable effect on fibrillization of the rabbit PrP but almost no effect on the human PrP" [658].

Throughout the above review on the research advances in RaPrP, we noticed that the rabbit prion in [106, 203] was just produced through saPMCA in vitro not by challenging rabbits directly in vivo with other known prion strains, and the saPMCA result of [106, 203] was refused by the test of cattle BSE [583]. All other RaPrP research results generally agree with each other to look rabbits as a resistant species to prion diseases (though, recently, artificial transgenic techniques could produce rabbit prions; however, "what exactly makes the rabbit species comparatively resistant to prion disease remains to be clarified." – www.sciencedaily.com/releases/2015/08/150806144421.htm)

8.3 Materials and Methods for MD of RaPrP X-Ray Structure

Many marvelous biological functions in proteins and DNA and their profound dynamic mechanisms (such as cooperative effects, allosteric transition, DNA internal motion, intercalation of drugs into DNA, and assembly of microtubules) can be revealed by studying their internal motions. Likewise, to really understand the action mechanism of prion protein, we should consider not only the static structures concerned but also the dynamical information obtained by simulating their internal motions or dynamic process. To realize this, the MD simulation is one of the feasible tools.

The MD simulations of this chapter are the continuation and extension of the ones of [679]. Zhang (2010) carried out (i) 15 ns of production phase of MD simulations and (ii) the heating phase of MD simulations is starting from 100 K with one set of initial velocity (denoted as seed2) [679]. This chapter continued to finish another 15 ns of production phase of MD simulations for seed2, and carried out other two sets of initial velocities for heating from 100 K (denoted as seed1 and seed3) of MD simulations with 30 ns of production phases.

We furthermore extended the MD simulation of [679]. Under the same simulation conditions, we also did the MD simulations with the three seeds at room temperature (300 K) for X-ray structures (PDB entries 3O79 [336], 4HMM [556]) and the NMR structures (PDB entries 2FJ3, 2JOH, 2JOM), in order to further confirm the MD results of NMR structures. For the three seeds at room temperature and at 350 K, the MD simulations are also done for human and mouse prion proteins (PDB entries 1QLX [673], 1AG2 [494] respectively), in order to make a comparison with HuPrP and MoPrP to further confirm the stability of RaPrP.

In order to get the low pH environment, the residues HIS, ASP, GLU were changed into HIP, ASH, GLH respectively and Cl- ions were added by the XLEaP module of AMBER package. Thus, the SBs of the neutral pH environment were broken in low pH environment.

8.4 Results and Discussions

For seed1∼seed3, at 450 K, during the whole 30 ns, radii of gyrations of the wild-type and the I214V and S173N mutants have been always level off around 15 Å whether in neutral pH environment or in low pH environment.

In neutral pH environment, compared with the I214V and S173N mutants (whose three α-helices have been unfolded into β-sheets), the wild-type is always at the lowest level of RMSD (root mean square deviation), RMSF (root mean square fluctuation) and B-factor values. This show to us the wild-type RaPrP is very stable under neutral pH environment. However, under low pH environment (both the wild-type and mutants have been unfolded their structures from α-helices into β-sheets), we cannot see some large differences among the wild-type, the I214V mutant and the S173N mutant.

8.4.1 450 K

The variations of the molecular structures of the wild-type and during the 30 ns for the three sets of MD simulations can be seen in Fig. 8.1, where the graphs were produced in the use of DSSP program [326], H is the α-helix, I is the π-helix, G is the 3-helix or 3_{10} helix, B is the residue in isolated β-bridge, E is the extended strand (participates in β-ladder), T is the HBed turn, and S is the bend; H1, H2, and H3 respectively denote the α-helix 1 (α1), α-helix 2 (α2) and α-helix 3 (α3) of a prion protein (we denote the β-strand before H1 as β1 and the β-strand beween H1 and H2 as β2). We can see that for the wild-type the three α-helices have been still kept during the whole 30 ns of each set under the neutral pH environment, but under low pH environment, the three α-helices of the wild-type are unfolded as of the I214V and S173N mutants under neutral or low pH environment and at last should be unfolded into rich β-sheet structures.

Under low pH environment, the SBs of the wild-type (Table 8.1) are broken thus leads to the unfolding of the stable helical structures of RaPrP. The following SBs should contribute to the stability of RaPrP: *ASP146-ARG147, GLU210-ARG207, GLU206-LYS203, GLU206-ARG207, ASP146-HIS139, GLU151-ARG150, GLU151-ARG147, GLU151-ARG155, ASP177-ARG163, GLU145-ARG135, ASP143-HIS139, GLU145-HIS139, GLU195-ARG155, ASP146-ARG150, ASP177-HIS176, ASP143-ARG147, GLU195-LYS193, HIS186-ARG155, HIS186-LYS184, ASP166-ARG163, GLU210-HIS176, HIS139-ARG135, GLU199-LYS193, GLU206-*

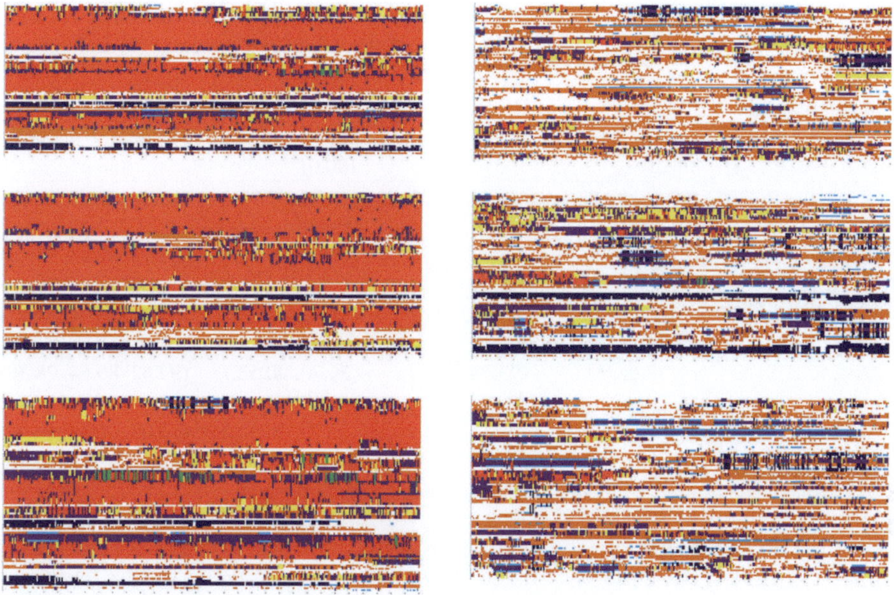

Fig. 8.1 450 K – Secondary Structure graphs for the NMR wild-type RaPrP (seed1 to seed3 from up to down) *[x-axis: time (0–30 ns), y-axis: residue number (124–228); left column: neutral pH, right column: low pH]*

HIS176. Among these SBs, compared with the SBs of I214V and S173N mutants, we find that GLU-195-ARG155 and ASP166-ARG163 are important to contribute to the structural stability of wild-type RaPrP.

Zhang (2011) found that there always exist SBs between ASP202-ARG156, ASP178-ARG164 in human and mouse prion proteins, between ASP201-ARG155, ASP177-ARG163 in RaPrP [680]. The author broke SBs of human, mouse and rabbit prion proteins by doing MD simulations from neutral to low pH environment; consequently, the secondary structures of human and mouse prion proteins were not changed very much, while the stable helical structure of wild-type RaPrP collapsed [680]. This is to say that the human and mouse prion protein structures are not affected by removing these SBs but the structure of RaPrP is affected very much by these SBs. Therefore, compared with human and mouse prion proteins, the SBs such as ASP201–ARG155, ASP177–ARG163 contribute greatly to the structural stability of RaPrP.

The SB ASP177-ARG163 is just like a taut bow-string keeping the $\beta 2$-$\alpha 2$ loop linked. This loop has been a focus on the studies of RaPrP molecular structure [50, 104, 105, 144–147, 156, 166, 252, 338, 402, 421, 463, 504, 531, 533, 534, 544, 555, 556, 626].

Recently, Garrec et al. (2013) reported that the salt linkage HIS187-ARG136 of mouse PrP causes "two misfolding routes for the prion protein" [235]. For RaPrP^C, we found that at HIS187 the SB HIS186-ARG155 contributes to the structural stability (Table 8.1).

Table 8.1 SBs of the RaPrP wild-type under neutral pH environment for the MD simulations with occupied rates for seed1, seed2, seed3 (where * denotes the occupied rate is less than 10 %)

SBs Donor-Acceptor	450 K (NMR)	300 K (NMR)	300 K (X-ray)	350 K (NMR)
ASP146-ARG147	(100.0, 100.0, 100.0)	(100.0, 100.0, 100.0)	(100.0, 100.0, 100.0)	(100.0, 100.0, 100.0)
GLU210-ARG207	(99.15, 97.58, 98.43)	(99.84, 98.01, 47.02)	(99.95, 99.98, 99.74)	(99.78, 99.81, 99.84)
GLU206-LYS203	(97.67, 92.85, 87.35)	(78.03, 98.85, 91.59)	(97.67, 99.12, 96.23)	(82.74, 82.98, 98.35)
GLU206-ARG207	(84.25, 84.80, 75.50)	(86.90, 65.61, 62.79)	(81.49, 83.23, 68.71)	(88.85, 98.84, 89.27)
ASP146-HIS139	(*, 73.65, *)	(93.74, 34.67, 44.44)	(56.70, 90.53, 61.20)	(92.62, 80.64, 28.38)
GLU151-ARG150	(72.13, 47.10, 84.25)	(94.15, 44.07, 29.83)	(47.79, 26.72, 40.83)	(76.05, 30.84, 50.65)
GLU151-ARG147	(63.13, 36.28, 64.22)	(98.55, 23.69, *)	(46.27, 42.27, 50.39)	(*, 31.43, 23.11)
GLU151-ARG155	(60.78, *, 60.07)	(56.19, *, 57.29)		(20.70, *, 57.71)
ASP177-ARG163	(47.80, 40.38, 21.92)	(23.93, 31.62, 82.38)	(70.59, 26.27, 61.89)	(19.54, *, 38.69)
GLU145-ARG135	(33.52, *, 47.13)			
ASP143-HIS139	(27.95, *, 10.95)	(*, *, 22.43)		
GLU145-HIS139	(25.00, *, 18.60)			
GLU195-ARG155	(21.88, 7.97, *)		(13.39, 55.78, 44.30)	(*, *, 32.25)
ASP146-ARG150	(21.23, 66.38, 34.25)	(92.55, 34.63, 86.09)	(15.15, *, 10.20)	(91.38, 62.11, 94.13)
ASP177-HIS176	(21.17, 16.55, 10.15)	(15.47, 11.66, *)	(21.09, 24.35, 19.69)	(*, 25.63, *)
ASP143-ARG147	(19.73, 50.30, 28.38)	(93.43, 42.51, 79.52)	(53.97, 32.05, 38.03)	(86.43, 78.38, 73.68)
GLU195-LYS193	(19.68, 21.98, 10.27)	(*, *, 25.35)	(52.20, 61.13, 20.80)	(*, 15.92, 46.75)
HIS186-ARG155	(14.01, 96.35, 12.47)	(100.0, 73.69, *)	(81.19, 35.80, 78.98)	(71.69, 100.0, *)
HIS186-LYS184	(*, 37.95, 12.95)			
ASP166-ARG163	(*, *, 54.08)			
GLU210-HIS176	(*, *, 33.37)			(74.31, *, *)
HIS139-ARG135	(*, *, 32.02)			
GLU199-LYS193	(*, *, 11.95)			
GLU199-LYS203		(11.73, *, *)		
GLU206-HIS176	(*, *, 11.83)			(57.10, *, 75.83)
ASP201-ARG155	(6.62, 6.08, *)	(3.95, 37.81, *)	(*, *, 10.05)	(10.07, *, *)
GLU199-LYS184				(46.57, *, *)
HIS139-ARG150				(50.96, *, *)
ASP166-ARG227				(*, 11.07, *)

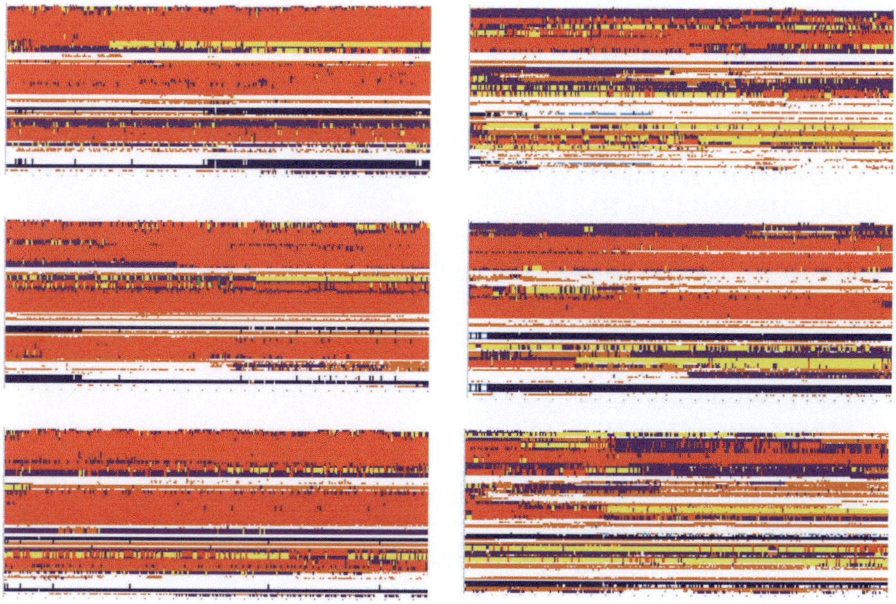

Fig. 8.2 300 K – Secondary Structure graphs for the NMR wild-type RaPrP (seed1 to seed3 from up to down) *[x-axis: time (0–30 ns), y-axis: residue number (124–228); left column: neutral pH, right column: low pH]*

8.4.2 At 300 K (the Room Temperature)

We found at the room temperature 300 K, RaPrP has a clear difference from HuPrP and MoPrP: under low pH environment, the three α-helices of RaPrP are unfolded for all the seeds (Fig. 8.2) but for HuPrP and MoPrP the three α-helices have not unfolded during the long 30 ns of MD simulations. This implies to us the broken of SBs under low pH environment has not affected the structure of HuPrP and MoPrP very much, and indicates to us the C-terminal region of RaPrP^C has lower thermostability than that of HuPrP^C and MoPrP^C.

We also compared with NMR structures with X-ray structures of RaPrP. Generally, for the wild-type RaPrP, under neutral pH environment at 300 K the three α-helices have not much unfolded during the whole 30 ns of MD simulations. For the S173N mutant of RaPrP, the secondary structure of X-ray is very similar as that of NMR during the 30 ns of MD simulations at 300 K. The RMSD value between 2FJ3.pdb (NMR, wild-type) and 3O79.pdb (X-ray, wild-type) is 2.791856 Å, and between 2JOH.pdb (NMR, S173N mutant) and 4HMM.pdb (X-ray, S173N mutant) is 2.996173 Å. Thus, in this chapter we used the NMR structures to replace the X-ray structures in our MD result analyses.

At 300 K, seeing Table 8.1 we know that the following SBs should contribute to the structural stability of RaPrP:

- ASP146-ARG147 (in H1),
- GLU210-ARG207 (in H3),
- GLU206-LYS203 (in H3),
- GLU206-ARG207 (in H3),
- ASP146-HIS139 (linking H1 $\sim$ loop $\beta 1$-$\alpha 1$),
- GLU151-ARG150 (in H1),
- GLU151-ARG147 (in H1),
- GLU151-ARG155 (in H1),
- ASP177-ARG163 (linking H2 $\sim$ loop $\beta 2$-$\alpha 2$),
- ASP143-HIS139 (in loop $\beta 1$-$\alpha 1$),
- GLU195-ARG155 (linking loop $\alpha 2$-$\alpha 3$ $\sim$ bend $\alpha 1$-$\beta 2$),
- ASP146-ARG150 (in H1),
- ASP177-HIS176 (in H2),
- ASP143-ARG147 (in H1),
- GLU195-LYS193 (in loop $\alpha 2$-$\alpha 3$),
- HIS186-ARG155 (linking C-terminals of H2 and H1),
- GLU199-LYS203 (in H3),
- ASP201-ARG155 (linking H3 with the C-terminal of H1).

All these SB stabilizing interactions added up can make an important contribution to the overall stability of RaPrP^C. Same as at 450 K, we have also found that at 300 K during the long 30 ns the SB ASP177-ARG163 is very strong and always keeps the linking of H2 and the loop $\beta 2$-$\alpha 2$ [50, 144–147, 156, 166, 252, 338, 402, 421, 463, 504, 531, 533, 534, 544, 555, 556, 626]. Observing Fig. 8.2, we can clearly see that under low pH environment the lose of the SBs in H1 made H1 unfolded completely. In Fig. 8.2, for seed2 under low pH environment, H2 has not unfolded; this almost agrees with the observation from Fig. 3 of [681].

8.4.3 At 350 K

Similarly, we found at temperature 350 K, RaPrP has a clear difference from HuPrP and MoPrP: under low pH environment, the three α-helices of RaPrP are unfolded for all the seeds (Fig. 8.3) but for HuPrP and MoPrP the three α-helices have not unfolded during the long 30 ns of MD simulations.

At 350 K, all the SBs listed in Table 8.1 added up can make an important contribution to the overall stability of RaPrP^C. Same as at 450 K, we have also found that at 350 K during the long 30 ns the SB ASP177-ARG163 is very strong and always keeps the linking of H2 and the loop $\beta 2$-$\alpha 2$ [50, 144–147, 156, 166, 252, 338, 402, 421, 463, 504, 531, 533, 534, 544, 555, 556, 626]. Observing Fig. 8.3, we can clearly see that under low pH environment the lose of the SBs in H1 made H1 unfolded completely, and H2 is almost unfolded. In Fig. 8.3, for seed3 under low pH environment, H3 is unfolded.

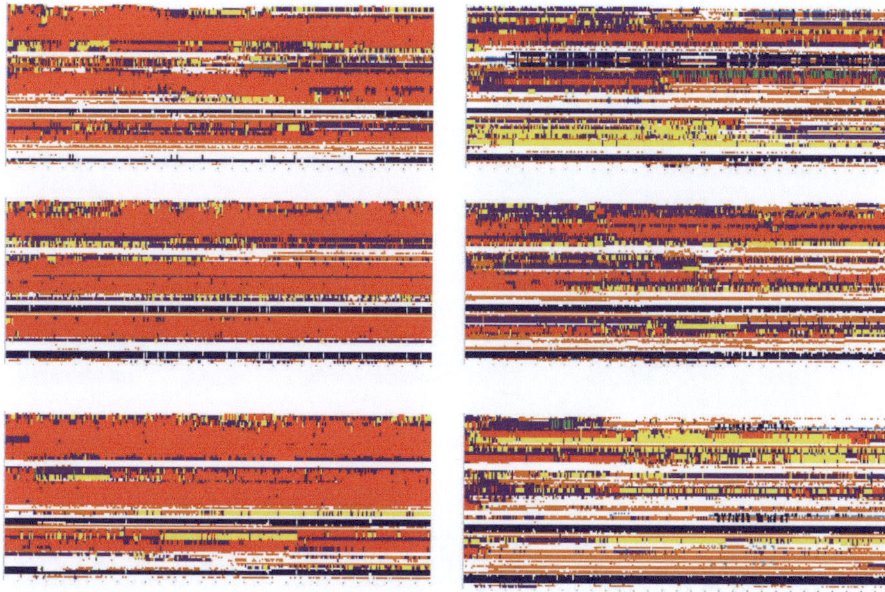

Fig. 8.3 350 K – Secondary Structure graphs for the NMR wild-type RaPrP (seed1 to seed3 from up to down) *[x-axis: time (0–30 ns), y-axis: residue number (124–228); left column: neutral pH, right column: low pH]*

8.4.4 Hydrogen Bonds

A SB is actually a combination of two noncovalent interactions: HBing and electrostatic interactions. From neutral pH to low pH, the electrostatic interactions were lost, and it also made the lost of some (weak noncovalent) HBing interactions at the same time which made the unfolding of H1$\sim$H3. This is because an α-helix is mainly maintained by HBing interactions and in an ideal α-helix there are 3.6 residues per complete rotating so a rotation of 100° per residue. We found under neutral pH environment there are HBs ASP177-ARG163, ASP201-ARG155, which have highly occupied during the long 30 ns of MD simulations for seed1$\sim$seed3 at 450, 300 and 350 K (Figs. 8.4$\sim$8.3).

8.5 Conclusion

Like the controversy on "prion" theory, there is also a big controversy over "whether rabbits are resistant to prion infection or not?". This chapter briefly reviewed the research results on RaPrP and rabbit prions. This chapter also did MD simulation studies on RaPrP and its I214V and S173N mutants with some preliminary analyses

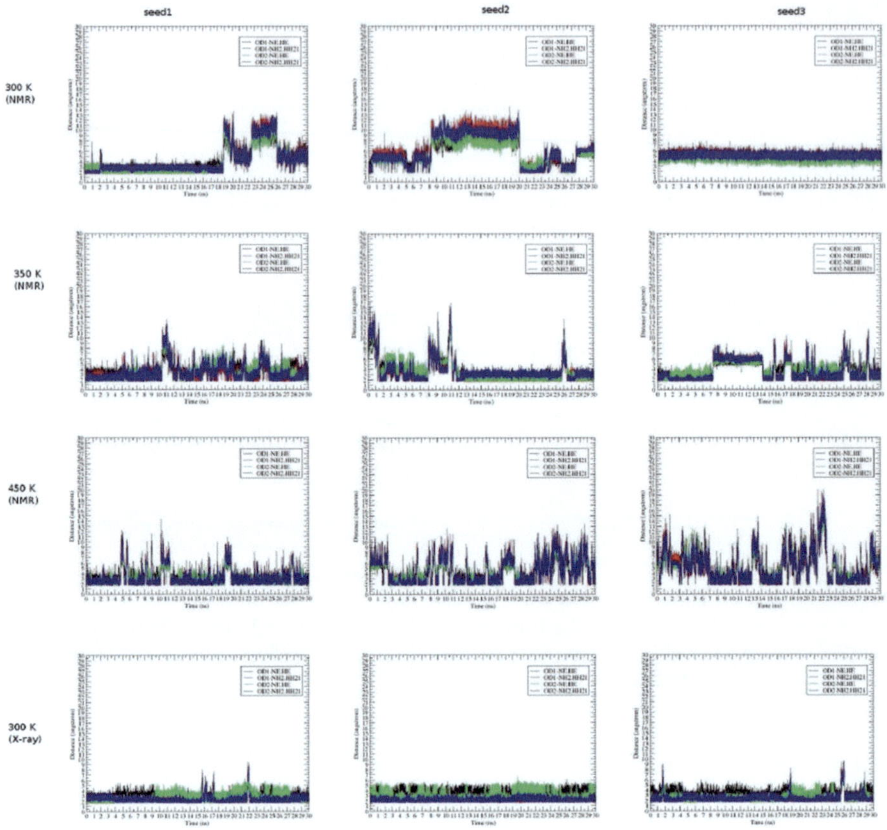

Fig. 8.4 The occupying of HB ASP177-ARG163 during the 30 ns of 300 K (NMR), 350 K (NMR), 450 K (NMR), and 300 K (X-ray) for seed1∼seed3 *[seed1 to seed3: from left to right; 300 K (NMR), 350 K (NMR), 450 K (NMR), and 300 K (X-ray): from up to down]*

of the role of SBs that plays in the structural stability of RaPrP. The survey shows to us that rabbits were not challenged directly in vivo with other known prion strains and the saPMCA result did not pass the test of the known BSE strain of cattle. Thus, we might still look rabbits as a prion resistant species (and the concept of species resistance to prion disease should be re-evaluated using new powerful tools because it is not reasonable to attribute species-specific prion disease resistance just based purely on the absence of natural cases and incomplete in vivo challenges [294]). MD results indicate that the three α-helices of the wild-type are stable under the neutral pH environment, and the three α-helices of the mutants (I214V and S173N) are unfolded into rich β-sheet structures under the same pH environment. In addition, we found an interesting result that the SBs such as ASP201–ARG155, ASP177–ARG163, HIS186-ARG155 contribute greatly to the structural stability of RaPrP (Fig. 8.5).

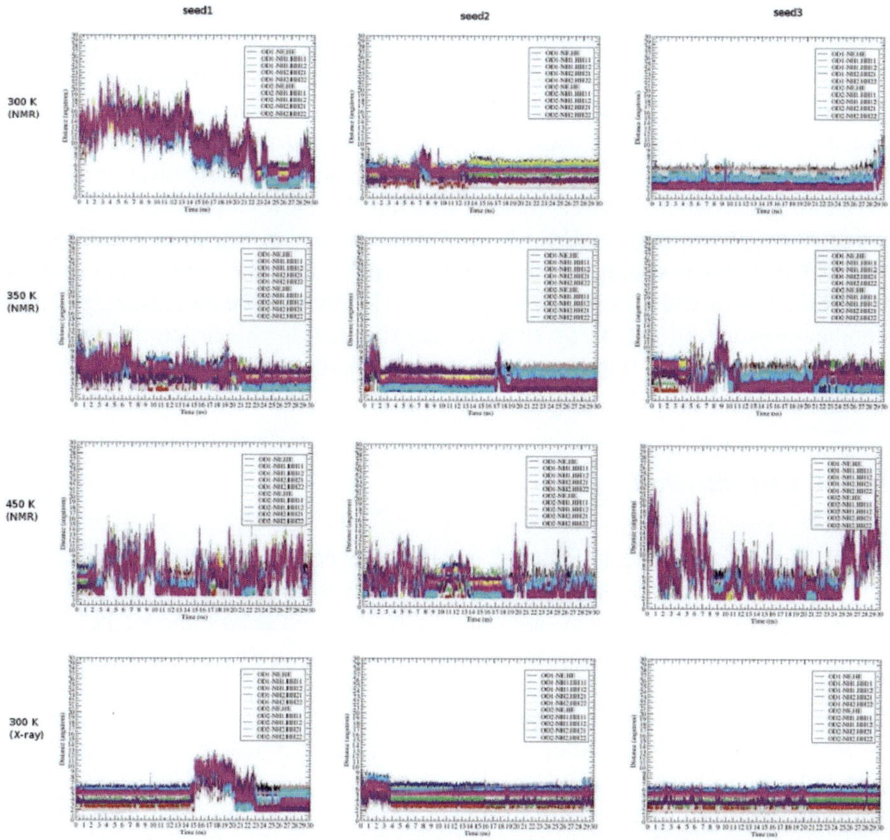

Fig. 8.5 The occupying of HB ASP201-ARG155 during the 30 ns of 300 K (NMR), 350 K (NMR), 450 K (NMR), and 300 K (X-ray) for seed1∼seed3 *[seed1 to seed3: from left to right; 300 K (NMR), 350 K (NMR), 450 K (NMR), and 300 K (X-ray): from up to down]*

8.6 Review on the Salt Bridge Asp177-Arg163 (O-N) of RaPrP

Prion diseases are invariably fatal and highly infectious neurodegenerative diseases that affect a wide variety of mammalian species such as sheep and goats, cattle, deer, elks, humans and mice etc., but rabbits have a low susceptibility to be infected by prion diseases with respect to other species. The stability of rabbit prion protein is due to its highly ordered β2-α2 loop (PLoS One 5(10) e13273 (2010); Journal of Biological Chemistry 285(41) 31682–31693 (2010)) and a hydrophobic staple helix-capping motif (PNAS 107(46) 19808–19813 (2010); PLoS One 8 (5) e63047 (2013)). The β2-α2 loop and the tail of Helix 3 it interacts with have been a focus in prion protein structure studies. For this loop we found a SB linkage ASP177-ARG163 (O–N) (Journal of Theoretical Biology 342 (7 February 2014) 70–82 (2014)). Some scientists said *on the 2FJ3.pdb NMR file of the rabbit prion protein, the distance of ASP177-ARG163 (O–N) gives the SB of about 10 Å which is nearly null in terms of energy and such a SB is not observed in their work*. But, from the 3O79.pdb X-ray file of the rabbit prion protein, we can clearly observe this SB. This article analyses the NMR and X-ray structures and gives an answer to the above question: *the SB presents at pH 6.5 in the X-ray structure is simply gone at pH 4.5 in the NMR structure* is simply due to the different pH values that impact electrostatics at the SB and hence also impact the structures. Moreover, some molecular dynamics simulation results of the X-ray structure are reported in this article to reveal the secrets of the structural stability of rabbit prion protein.

8.6.1 Introduction

It has been a challenge to rational whether the contagious TSE is caused by prions [203, 485, 486, 540, 541]. As a misfolded protein, prion is neither a virus, nor a bacterium, and nor any microorganism. Prion diseases cannot be caused by the vigilance of the organism immune system but it can freely spread from one species to another species. Humans TSEs (for example, CJD and vCJD) can happen randomly through a number of processes, such as infections of transplanted tissue, blood transfusions and/or consumption of infected beef products, etc. Many mammals such as cat, mink, deer, elk, moose, sheep, goat, nyala, oryx, greater kudu and ostrich etc. are also susceptible to TSEs. However, a small group of other animals such as rabbits, horses, dogs and buffaloes seem to be little affected by prions [336, 474, 591, 680, 697]. As a result, it is important to understand and to identify the specific causes why these animals are unlikely to be affected by prions, as it will provide insight to prion diseases and help to resolve the prion diseases issue.

The role of PrP^{Sc} infection in animals such as rabbit has been subject to a heated debating. A number of previous studies showed that a few animals such as rabbits exhibit low susceptibility to be infected by the PrP^{Sc} [29, 160, 204, 336, 350, 407, 441, 587, 591, 625, 626, 717].

Now experimental structural data for rabbit PrP^C ($RaPrP^C$) is available from the Protein Data Bank (PDB bank: http://www.rcsb.org/). For example, the structures of $RaPrP^C$ obtained from NMR [381] (with a PBD entry of 2FJ3) and X-ray [336] (with a PBD entry of 3O79) measurements. As a result, it is desirable to reveal the properties and specific mechanisms of the $RaPrP^C$ and the conversion process of $PrP^C \rightarrow PrP^{Sc}$ of rabbits from limited experimental results. Here PrP^C is a soluble normal cellular prion protein and PrP^{Sc} is insoluble abnormally folded infectious and diseased prions. The present study will base on the X-ray and NMR structure of $RaPrP^C$ using MD simulation techniques. The information from the present MD studies is able to provide valuable insight for the $PrP^C \rightarrow PrP^{Sc}$ conversion.

The information will provide useful rational in the design of novel therapeutic approaches and drugs that stop the conversion and disease propagation.

This section can capture the α-helices→ β-sheets conversion of PrP^C →PrP^{Sc} under pH environments from neutral to low. The removing of SBs (SBs) under low pH environment can lead to this conversion. The 2FJ3.pdb of the $RaPrP^C$ has the distance of ASP177-ARG163 (O–N) about 10 Å which is nearly null in terms of energy. Thus, this SB does not exist and will not at all contribute to the PrP^C →PrP^{Sc} conversion. This SB just links the $\beta 2$-$\alpha 2$ loop, which has been a focus in PrP studies [50, 144–147, 156, 166, 251, 252, 338, 402, 421, 463, 503, 504, 531, 533, 534, 544, 555, 556, 561, 625, 626]. Hence, our SB has caused much more debate, and it is very necessary to us to specially organize a section to address this problem.

The rest of this section is organized as follows. The MD simulation materials and MD reproducible methods for the X-ray and NMR structures of $RaPrP^C$ wild-type are provided in next subsection, followed by the analysis and discussion focusing on the MD trajectory results of the SB between ASP177 and ARG163 linking the $\beta 2$-$\alpha 2$ loop and their discussions. Numerous new MD results of X-ray $RaPrP^C$ structure will be reported in the Results and Discussion subsection. Finally, a concluding remark on this SB, the $\beta 2$-$\alpha 2$ loop and the interaction of this loop with the tail of H3 of PrP^C is summarized.

8.6.2 Materials and Methods

The materials, e.g., data used in the present study are based on the laboratory NMR and X-ray PDB files of 2FJ3.pdb, 3O79.pdb respectively. For the NMR structure $RaPrP^C$(124–228), the experimental temperature is 298 K, pH value is 4.5, and pressure is 1 ATM; for the X-ray structure $RaPrP^C$(126–230), the resolution is 1.60 Å, R-value is 0.161, R-free is 0.218, temperature is 100 K, pH value is 6.5 (where, possibly owing to the different pH conditions, in X-ray structure 3O79.pdb we can observe the SB Asp177-Arg163 (O–N) but in NMR structure 2FJ3.pdb we cannot).

The MD methods employed are the same as the previous studies [679, 682, 703, 704]. Briefly, all simulations used the ff03 force field of the AMBER 11 package [91]. The systems were surrounded with a 12 Å layer of TIP3PBOX water molecules and neutralized by sodium ions using the XLEaP module of AMBER 9. To remove the unwanted bad contacts, the systems of the solvated proteins with their counter ions had been minimized mainly by the steepest descent method and followed by a small number of conjugate gradient steps on the data, until without any amino acid clash checked by the Swiss-Pdb Viewer 4.1.0 (http://spdbv.vital-it.ch/). Next, the solvated proteins were heated from 100 to 300 K in 1 ns duration. Three sets of initial velocities denoted as seed1, seed2 and seed3 are performed in parallel for stability (this will make each set of MD starting from different MD initial velocity, implemented in Amber package we choose three different odd-real-number values for "ig") – but for the NMR structure and the X-ray structure of

RaPrP^C, each set of the three has the same "ig" value in order to be able to make comparisons. The thermostat algorithm used is the Langevin thermostat algorithm in constant NVT ensembles. The SHAKE algorithm (only on bonds involving hydrogen) and PMEMD (Particle Mesh Ewald Molecular Dynamics) algorithm with non-bonded cutoff of 12 Å were used during heating. Equilibrations were reached in constant NPT ensembles under Langevin thermostat for 5 ns. After equilibrations, production MD phase was carried out at 300 K for 30 ns using constant pressure and temperature ensemble and the PMEMD algorithm with the same non-bonded cutoff of 12 Å during simulations. The step size for equilibration was 1 and 2 fs in the MD production runs. The structures were saved to file every 1000 steps. During the constant NVT and then NPT ensembles of PMEMD, periodic boundary conditions have been applied.

In order to obtain the low pH (acidic) environment, the residues HIS, ASP, GLU were changed into their zwitterion forms of HIP, ASH, GLH, respectively, and Cl- ions were added by the XLEaP module of the AMBER package. Thus, the SBs of the system (residues HIS, ASP, GLU) under the neutral pH environment were broken in the low pH environment (zwitterion forms of HIP, ASH, GLH).

8.6.3 Results and Discussion

8.6.3.1 Confirmation from the X-Ray Structure 3O79.pdb File

The 3O79.pdb file of RaPrP^C X-ray structure consists of A and B structures. We can confirm as follows there is the SB ASP178-ARG164 (O–N) in both the A and B molecule structures. In A structure, we find the SBs illuminated in the 1st graph of Fig. 8.6 and listed in Table 8.2, where the cut-off distance for a SB is 4 Å [353] according to the definition of a SB.

In B structure, we find the SBs illuminated in the 2nd graph of Fig. 8.6 and listed in Table 8.3. Our MD experiences showed that the underlined (in real and dash lines) SBs are two basic and fundamental SBs of RaPrP^C.

8.6.3.2 Confirmation from the MD of X-Ray 3O79.pdb Structure at 300 K Room Temperature

We denote the three a-helices and the two β-strands of PrP^C as $\beta 1$, $\alpha 1$, $\beta 2$, $\alpha 2$ and $\alpha 3$. The MD results from the simulations are summarized in Figs. 8.7∼8.12, under the parallel conditions of seed1, seed2 and seed3 in the duration of 30 ns.

From Figs. 8.7 and 8.8, we cannot see gigantic changes of secondary structures under pH environments from neutral to low; but we found the following changes: (i) the 'tails' (i.e. C-terminus) of $\alpha 2$ and $\alpha 3$ have α-helix unfolded into HB turns, bends and 3_{10}-helices, for seed2 a little part of $\alpha 2$ unfolded into β-bridge, (ii) some parts of the yellow coloured 3_{10}-helices between $\alpha 1$ and $\beta 2$ become into purple HB

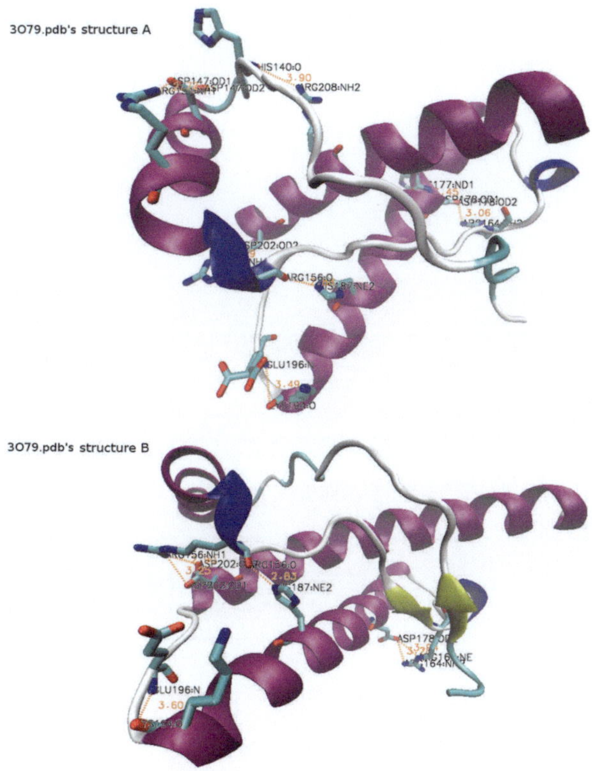

Fig. 8.6 All the SBs in the A structure and B structure of the X-ray PDB file of RaPrP^C (3O79.pdb). The *orange dashed lines* denote the SBs and the *yellow data* denote the distances of the SBs

Table 8.2 All the SBs in the A structure of the X-ray PDB file of RaPrP^C (3O79.pdb)

O atoms	N atoms	Distance (Å)
ASP178.OD2	ARG164.NH2	3.06
ASP202.OD2	ARG156.NH1	2.89
ARG156.O	HIS187.NE2	2.88
LYS194.O	GLU196.N	3.49
ASP178.OD1	HIS177.ND1	3.45
HIS140.O	ARG208.NH2	3.90
ASP147.OD1	ARG151.NH1	2.77
ASP147.OD2	ARG151.NH1	3.71

turns, (iii) for seed1 and seed3 some parts of the purple coloured HB turns between $\beta 2$ and $\alpha 2$ become into yellow coloured 3_{10}-hleices, (iv) for seed1 the head of $\alpha 2$ becomes into bends, (v) for some snapshots their two β-strands $\beta 1$ and $\beta 2$ (i.e. the β-sheet) become longer. Figure 8.8 shows the snapshots of Fig. 8.7 at 5, 10, 15, 20, 25 and 30 ns. In Fig. 8.8, we can see there is a short 3_{10}-helix in the $\beta 2$-$\alpha 2$ loop under both neutral and low environments, but for seed1 the S174N mutation (4HMM.pdb)

Table 8.3 All the SBs in the B structure of the X-ray PDB file of RaPrP^C (3O79.pdb)

O atoms	N atoms	Distance (Å)
ASP178.OD2	ARG164.NE	3.94
ASP178.OD2	ARG164.NH2	3.28
ASP202.OD1	ARG156.NH1	3.25
ASP202.OD2	ARG156.NH1	2.90
ARG156.O	HIS187.NE2	2.83
LYS194.O	GLU196.N	3.60

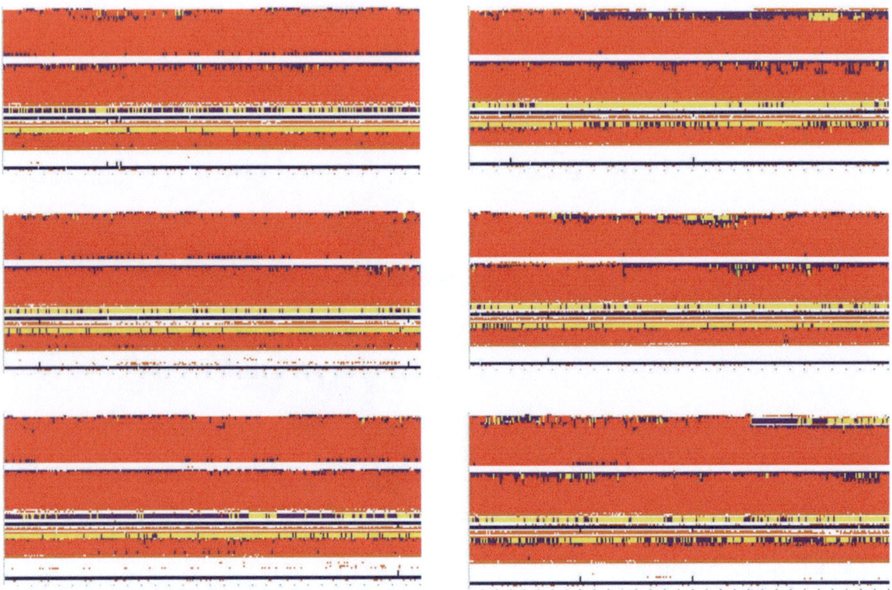

Fig. 8.7 Secondary Structure graphs for X-ray RaPrP^C wild-type at 300 K (x-axis: time (0–30 ns), y-axis: residue number (126–230); *left column*: neutral pH, *right column*: low pH; up to down: seed1–seed3). H is the α-helix, I is the π-helix, G is the 3-helix or 3_{10} helix, B is the residue in isolated β-bridge, E is the extended strand (participates in β-ladder), T is the HBed turn, and S is the bend

made this 3_{10}-helix disappear at 30 ns; during the long 30 ns of MD simulations in neutral pH environment, for both the wild-type and the S174N mutant, we found in the β2-α2 loop there are 3 HYDs/cores Pro165-Val166-Phe175-Val176 with occupied rate greater than 95 %, and 2 HYDs Pro165-Met129, Val166-Ala225 with high occupied rate. Being different from wild-type, the S174N mutation produced two new HYDs Val176-Val180, Val176-Ile215 with 100 % occupied rate during the whole 30 ns of seed1∼seed3 MD simulations. In the β2-α2 loop, for the wild-type at 300 K under neutral pH environment, there always exist 3 HBs Asp178-Arg164 (see Table 8.4 for the occupied rates), Cys179-Thr183 (with occupied rates 82.83 %, 72.79 %, 84.29 % for seed1∼seed3 respectively), Asp178-Tyr169 (with occupied rates 27.44 %, 33.93 %, 5.09 % for seed1∼seed3 respectively); there also exist some

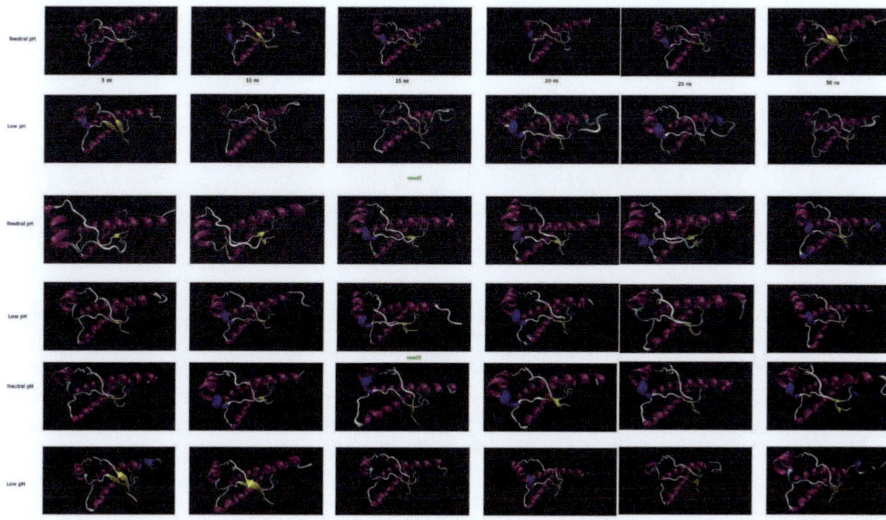

Fig. 8.8 Snapshots at 5, 10, 15, 20, 25 and 30 ns (from left to right in turns) for MD of X-ray structure of RaPrP^C wild-type at 300 K. The 1st and 2nd rows are for seed1, the 3rd and 4th rows are for seed2, and the 5th and 6th rows are for seed3. The 1st, 3rd and 5th rows are for neutral pH environment, and the 2nd, 4th, and 6th rows are for low pH environment

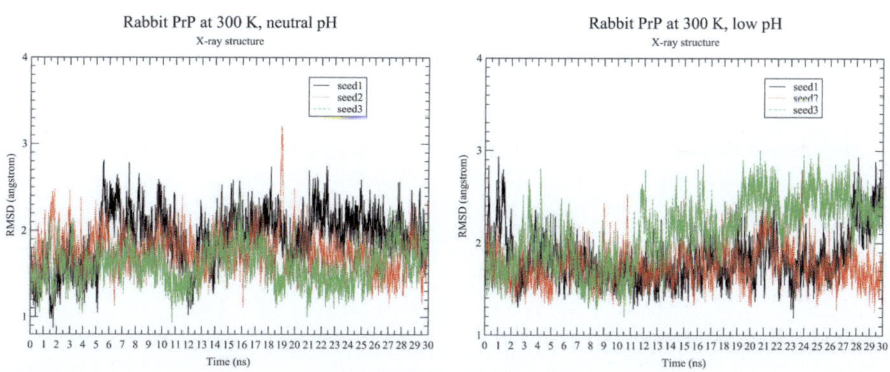

Fig. 8.9 RMSD graphs for X-ray RaPrP^C wild-type at 300 K (x-axis: time (0–30 ns), y-axis: RMSD value (angstrom); *left column*: neutral pH, *right column*: low pH; seed1: *black real line*, seed2: *red dot line*, seed3: *green dashed line*)

special HBs Gln172-Gln219 (with occupied rates 8.07 %, 7.39 % for seed1, seed3 respectively), Val166-Ser170 (with occupied rate 7.01 % for seed3), Tyr169-Tyr218 (with occupied rate 5.51 % for seed3), Tyr218-Ser170 (with occupied rate 48.13 % for seed3), Ser170-Asn171 (with occupied rate 16.76 % for seed2), and Asn171-Ser174 (with occupied rate 23.92 % for seed2). We found the S174N mutation made the loss of the strong HB Cys179-Thr183. Other HBs such as Asp202-Tyr149, Asp202-Arg156, Asp202-Tyr157, and Lys204-Glu146 are keeping H3 and

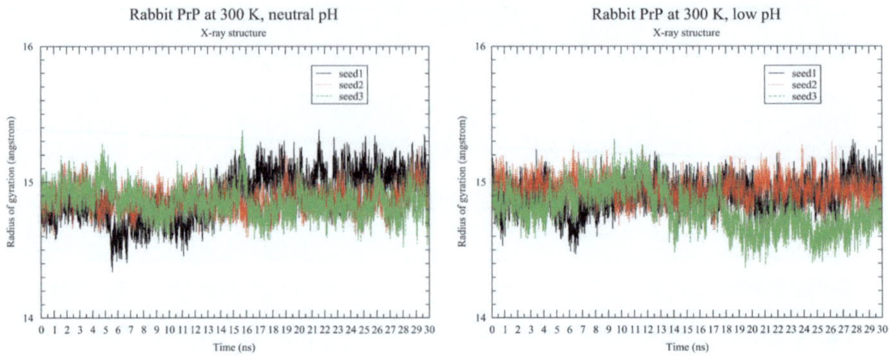

Fig. 8.10 Radius of Gyration graphs for X-ray RaPrP^C wild-type at 300 K (x-axis: time (0–30 ns), y-axis: Radius of gyration value (angstrom); *left column*: neutral pH, *right column*: low pH; seed1: *black real line*, seed2: *red dot line*, seed3: *green dashed line*)

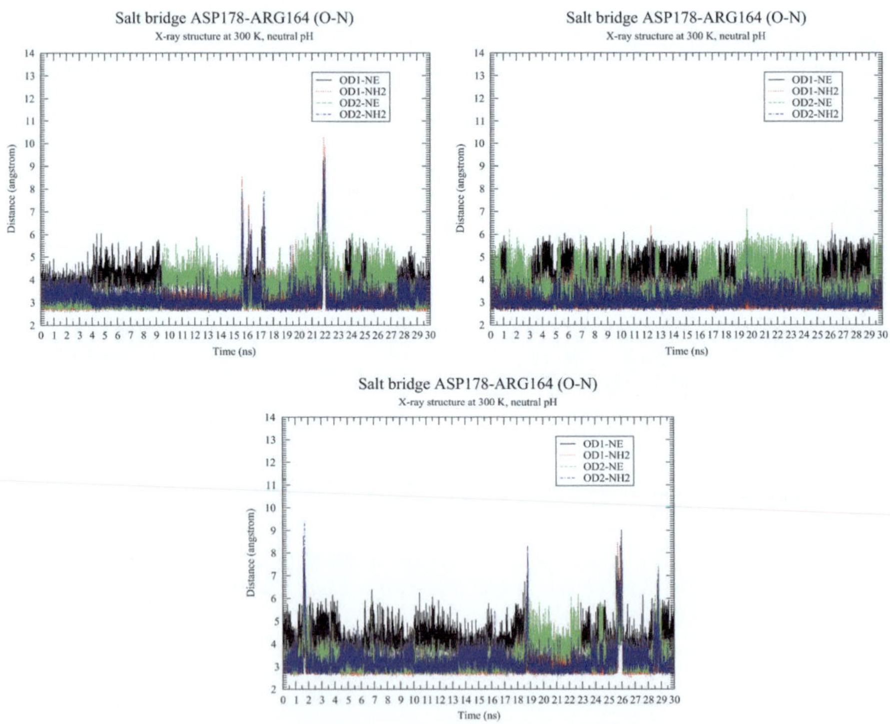

Fig. 8.11 The occupying of SB ASP178-ARG164 (O–N) during the 30 ns of MD simulations of the wild-type X-ray structure RaPrP at 300 K under neutral pH environment for seed1–seed3 [seed1–seed3: from left to right in turns]

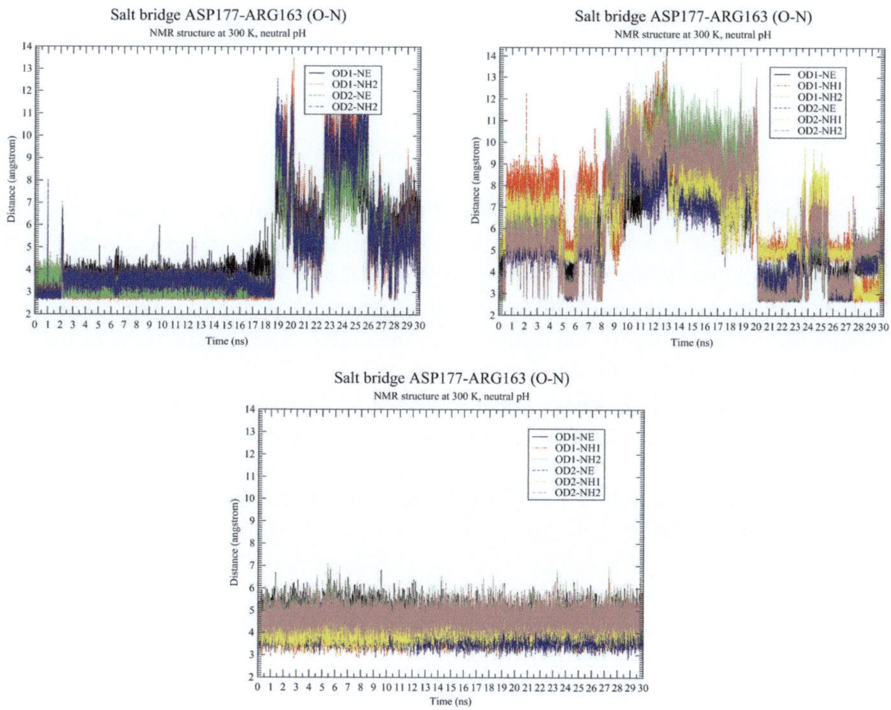

Fig. 8.12 The occupying of SB ASP177-ARG163 (O–N) during the 30 ns of MD simulations of the wild-type NMR structure RaPrP at 300 K under neutral pH environment for seed1–seed3 [seed1–seed3: from left to right in turns]

Table 8.4 The occupied rates for HB Asp178-Arg164 of the wild-type (X-ray structure) at 300 K, neutral pH value during the 30 ns of MD simulations

HB	Seed1 (%)	Seed2 (%)	Seed3 (%)
ASP178@OD1-ARG164@NH2.HH21	51.01	52.47	59.76
ASP178@OD2-ARG164@NH2.HH21	44.65	53.73	38.06
ASP178@OD1-ARG164@NE.HE	30.57	13.92	10.00
ASP178@OD2-ARG164@NE.HE	31.73	14.91	45.49

H1 linked; at the same time, H1 linking with H2 and the H2-H3 loop are kept by two strong SBs Arg156-His187 and Arg156-Glu196. For SBs, we found the S174N mutation has not greatly changed SBs of the wild-type; the occupied rates of SB Asp178-Arg164 for the wild-type are 70.59 %, 26.27 %, 61.89 % for seed1∼seed3 respectively.

The above analysis means that for the wild-type X-ray RaPrP structures we can capture the α-helices→ β-sheets conversion of PrP^C →PrP^{Sc} under pH environments from neutral to low, because of the removing of SB networks of neutral pH environment. The SB ASP178-ARG164 (O–N) (Fig. 8.12) is one of these important SBs and seeing Fig. 8.12 we know it always exists during 30 ns. For seed1∼seed3, SBs ASP178.OD1-ARG164.NE and ASP178.OD2-ARG164.NE are always existing during the whole 30 ns, SBs ASP178.OD1-ARG164.NH2 and ASP178.OD2-ARG164.NH2 are always strongly existing during the whole 30 ns, but SBs ASP178.OD1-ARG164.NH1 and ASP178.OD2-ARG164.NH1 are not existing at all during the whole 30 ns.

If we set the distance cut-off 3.0 Å and the angle cut-off 20° for the HBs as in the VMD package (www.ks.uiuc.edu/Research/vmd/), seeing the last row of Fig. 4 in [704], we know that the HB between ASP178 and ARG164 is almost 100 % there under neutral pH environment, and it becomes very much weaker under low pH environment because of the removing of the electrostatic interactions of the SB between ASP178 and ARG164.

Regarding the RMSD (Fig. 8.9) and Radius Of Gyration (Fig. 8.10) of the MD simulations of X-ray structure, we can see from Figs. 8.9∼8.10 that the 30 ns simulations is short but good enough for a small protein like the prion protein. We also can see from Figs. 8.9∼8.10 that the three seeds are valid and there are not great differences among them.

8.6.3.3 Confirmation from the MD of the NMR 2FJ3.pdb Structure at 300 K Room Temperature

The secondary structure changes for the NMR wild-type RaPrP at room temperature 300 K can be seen in Fig. 2 of [704]. As we can be seen from Fig. 2 of [704], the left panels are dominated by red colour, indicating that under the neutral pH conditions, the three a-helices ($\alpha 1$, $\alpha 2$ and $\alpha 3$) of the wild-type prion RaPrP^C, remain dominant the α-helices without significant changes during the period of 30 ns, regardless the seed conditions. For example, the α-helices ($\alpha 1$, $\alpha 2$ and $\alpha 3$ in red colour) of the top-left panel does not experience any apparent colour changes, indicating that under the neutral pH condition, the α-helices of the wild-type prion resist structural changes. This is particular the case in $\alpha 3$ and $\alpha 2$, although small noticeable changes in $\alpha 1$ has been observed. However, on the right hand side of the figure, under acidic condition (i.e. low pH environment) the seed1–seed3 cases of the wild-type prion protein show changes indicated by their changes of colour codes. In low pH environment, the wild-type prion protein turns into a colourful panel but (i) large presentation of HB bends for $\alpha 3$ in seed1–seed3 and $\alpha 2$ in seed2, (ii) 3^{10}-helix for $\alpha 2$ in seed3 and $\alpha 1$ in seed1–seed3, and (iii) $\beta 1$ and $\beta 2$ become longer under acidic condition for seed2 and seed3, for seed1 the $\beta 2$ becomes into β-bridge structures. Therefore, under acidic condition (low pH environment), the SB network of the wild-type (RaPrP^C) is broken thus leads to the unfolding of the stable α-helical structures of RaPrP^C. Hence, it suggests that the structural distributions of the wild-type (RaPrP^C) protein depends on the pH. Generally, we could see the clear secondary structure changes from the unfolding of the α-helical structures conversion into β-sheets from Fig. 2 of [704]. If we set the HB distance cut-off be 3.0 Å, seeing the first row of Fig. 4 of [704], we know that the HB between ASP177 and ARG163 does not exist for seed3, exists occasionally for seed2, and will not exist after 19 ns for seed1. However, the following Fig. 8.11 can still confirms there is a SB ASP177-ARG163 (O–N).

By Fig. 8.11, (i) for seed1 (Fig. 8.11 graph 1), the SBs D177.OD1-R163.NE and D177.OD1-R163.NH2 exist until 18.742 ns, the SB D177.OD2-R163.NE exists until 18.724 ns and the SB D177.OD2-R163.NH2 exists until 18.744 ns; (ii) for seed2 (Fig. 8.11 graph 2), generally speaking, except for during 9∼20 ns the SBs of D177-R163 (O–N) exist from 0 ns to 30 ns, SB D177.OD1-R163.NE exists during 0∼8.03 ns and 20.108∼30 ns (where the gap might be due to the SBs D177-HIS176 and D166-R163 drawing D177 and R163 away respectively), SB D177.OD1-R163.NH1 exists during 7.678∼8.032 ns and 27.572∼30 ns and occasionally during 23.38∼23.88 ns, SB D177.OD1-R163.NH2 exists during 0∼8.052 ns and 20.11∼27.628 ns, SB D177.OD2-R163.NE exists during 0∼8.168 ns and 20.108∼29.986 ns, SB D177.OD2-R163.NH1 exists during 7.672∼7.908 ns and 27.572∼29.992 ns, and SB D177.OD2-R163.NH2 exists during 0∼8.054 ns and 20.11∼27.69 ns; and (iii) for seed3 (Fig. 8.11 graph 3), weak SB D177.OD1-R163.NE always exists during 0∼30 ns, very strong SB D177.OD1-R163.NH1 always exists during 0∼30 ns, strong SB D177.OD1-R163.NH2 always exists during 0∼30 ns, very strong SB D177.OD2-R163.NE always exists during 0∼30 ns, very strong SB D177.OD2-R163.NH1 always exists during 0∼30 ns, and strong SB D177.OD2-R163.NH2 always exists during 0∼30 ns.

From the NMR structure 2FJ3.pdb file, RaPrP^C has no SB D177-R163 (O–N) because of the following large distances (10.2 Å) of D177-R163 (O–N) (Table 8.5): 10.73 Å for D177.OD1-R163.NE, 12.13 Å for D177.OD1-R163.NH1, 10.63 Å for D177.OD1-R163.NH2, 8.70 Å for D177.OD2-R163.NE, 10.26 Å for D177.OD2-R163.NH1, and 8.85 Å for D177.OD2-R163.NH2. For the 2FJ3.pdb, we found other SBs of RaPrP^C (Table 8.6): D201.OD1-R155.NH1 with distance 2.87 Å, D201.OD1-R155.NH2 with distance 2.55 Å, D146.OD2-K139.O with distance 2.82 Å, D143.O-E145.N with distance 3.76 Å, E151.OE2-R155.NH1 with distance 3.67 Å, E151.OE2-R155.NE with distance 3.25 Å, E195.OE2-K193.NZ with distance 2.93 Å, R155.O-K186.O with distance 3.06 Å, R155.O-K186.ND1 with distance 2.79 Å, E210.OE1-K176.ND1 with distance 3.97 Å, and E210.OE2-K176.ND1 with distance 2.60 Å.

Table 8.5 The O–N distances of ASP177-ARG163 of the NMR PDB file of RaPrP^C (2FJ3.pdb)

O atoms	N atoms	Distance (Å)
ASP177.OD1	ARG163.NE	10.73
ASP177.OD1	ARG163.NH1	12.13
ASP177.OD1	ARG163.NH2	10.63
ASP177.OD2	ARG163.NE	8.70
ASP177.OD2	ARG163.NH1	10.26
ASP177.OD2	ARG163.NH2	8.85

Table 8.6 All the SBs in the NMR structure of the PDB file of $RaPrP^C$ (2FJ3.pdb)

O atoms	N/O atoms	Distance (Å)
ASP201.OD1	ARG155.NH1	2.87
ASP201.OD1	ARG155.NH2	2.55
ASP146.OD2	LYS139.O	2.82
ASP143.O	GLU145.N	3.76
GLU151.OE2	ARG155.NH1	3.67
GLU151.OE2	ARG155.NE	3.25
GLU195.OE2	LYS193.NZ	2.93
ARG155.O	LYS186.O	3.06
ARG155.O	LYS186.ND1	2.79
GLU210.OE1	LYS176.ND1	3.97
GLU210.OE2	LYS176.ND1	2.60

8.6.3.4 Confirmation from the MD of the NMR 2FJ3.pdb Structure at 350 and 450 K

Seeing Tab. 1 of [704], for the SB ASP177-ARG163 (O–N), we know its occupied rates for seed1–seed3 are 47.80 %, 40.38 %, 21.92 % at 450 K, and 19.54 %, 6.09 %, 38.69 % at 350 K. We observed from the second and third rows of Fig. 4 of [704] that ASP177.OD1/2-ARG163.NE/NH2 should exist at 350 K and 450 K under neutral pH environment for $RaPrP^C$. In conclusion, by Tab. 1 and Fig. 4 of [704] the SB ASP177-ARG163 (O–N) exists at 350 K & 450 K for $RaPrP^C$.

In Fig. 3 of [704],we can see that percentages of β-sheet are clearly increasing under low pH environment at 350 K for seed1 and seed3 (for seed1 the loop between $\alpha2$ and $\alpha3$ becomes into β-ladders and for seed3 the $\beta2$-$\alpha2$ loop becomes into β-ladders and β-bridges), and the three a-helices are unfolded into a colourful panel of almost all of the seven colours but large presentation of yellow 3_{10}-helices, purple HB turns and orange bends for seed1–seed3. In Fig. 1 of [704], under low pH environment for seed2 the two β-strands of β-sheet clearly become longer, for seed1–seed3 some parts of the three a-helices become into β-sheet structures, and the three a-helices are unfolded into a colourful panel of almost all of the seven colours but large presentation of orange bends for seed1–seed3. This reason to cause the above α-helices$\rightarrow$ β-sheets conversion is due to the removing of SBs (by changing residues HIS, ASP, GLU into their zwitterion forms of HIP, ASH, GLH in order to obtain the low pH (acidic) environment). SB ASP177-ARG163 (O–N) is one of the SBs.

8.6.3.5 Confirmation from the MD of the Homology Structure of $RaPrP^C$ at 500 K

For $RaPrP^C$, its X-ray structure (3O79.pdb) was released into PDB bank on 2010-11-24 and its NMR structure (2FJ3.pdb) was released into PDB bank on 2006-12-31. Earlier in 2004, CSIRO scientists built a homology model for $RaPrP^C$ (denoted as 6EPA.pdb). The homology structure is for RaPrP^C(120–229) made by mutations using the NMR structure of human PrP^C(125–228) (1QLX.pdb) [700]. We did MD for the homology structure at 500 K and found that SB ASP177-ARG163 (O–N) is conserved through a large part of the 30 ns' simulations [700].

Interestingly, the SB ASP177-ARG163 (O–N) does not affect the structures of human and mouse prion proteins very much [680]. Kuwata et al. (2007) presented an anti-prion drug GN8 fixing the distance between N159 and E196 being 1.54 Å [358]. Thus, we might propose to fix the distance between ASP177 and ARG163 in a SB distance to design an anti-prion drug (this proposal might be wrong but this SB at $\beta 2$-$\alpha 2$ loop plays a very important role for the stability of RaPrP^C).

8.6.4 Conclusion

This chapter mainly discusses the presence/influence of a SB in RaPrP in the context of a possible role in PrP toxicity. We compare the X-ray crystal structure and NMR structure. Of note, both structures were obtained under different pH conditions, which affect electrostatics and structure. It has to be considered that presence of a SB depends on the environmental condition. Hence, the topic itself is interesting us and some furthermore study directions (i)–(iii) are pointed to us. (i). Given the dependence of SBs on the environment, the statement made on some scientists claiming a low energy contribution for a 10 Å SB at pH 4.5 thereby challenging us results might be wrong is not valid if justified by a structure obtained at different pH environment – MD studies of many and dynamic structures will be a way to confirm some observations. (ii). Because NMR and X-ray structures were got under different environments, in order to make comparisons we put them into the same environment to make MD calculations, this might be scientific in theory but not completely agreeing with the X-ray environment – we will furthermore do MD studies along the way of the X-ray environment. To get a clue on the protonation states at the different pH values the electrostatics should be calculated for the starting structure, as geometry may result in rather strong pKa deviations of individual groups. (iii). For the X-ray structure and NMR structure, different starting geometries sampled or different starting velocities should be furthermore studied for the MD theory.

There was a big lot of controversy over "prion" theory and recently on saPMCA produced rabbit prions. We have found one focus of prion protein structures is at the $\beta 2$-$\alpha 2$ loop and its interacted C-terminal of H3 [54–56, 83, 145, 146, 166, 252, 301, 349, 355, 356, 371, 463, 466, 556, 625, 626, 672, 699]. This article found there is a SB ASP177-ARG163 (O–N) in RaPrP^C, which just keeps this loop being linked. The function of this SB linkage should be furthermore confirmed in experimental laboratories.

Chapter 9
Surface Electrostatic Charge Distributions

9.1 Wild-Type RaPrP Compared with its Mutants, Humans, Mice, Horses and Dogs

The S2-H2 loop was reported to play an important role to stabilize the structural stability of rabbit and horse prion proteins [56, 166, 206, 329, 335, 336, 390, 463, 557, 558, 625, 626], and recently Wen et al. (2010) reported that rabbit prion protein has a unique distribution of large-continuous-positive-charge-surface electrostatic potential [390, 625, 626], which is with the highly ordered and well NMR-signal-recognized S2-H2 loop contributing to the structural stability of rabbit prion protein. The section will study (by MD techniques) the dynamics of the surface electrostatic potential distributions of the 3D NMR structure of rabbit prion protein.

To describe the potential energy associated with a charge distribution the concept of the electrostatic potential is introduced. The electrostatic potential at a given position is defined as the potential energy of a test particle divided by the charge of this object. When the charge distribution is over a particular area then the distribution is called as surface charge distribution. Electrostatic potential surfaces are valuable in computer-aided drug design because they help in *optimization* of electrostatic interactions between the protein and the ligand. These surfaces can be used to compare different inhibitors with substrates or transition states of the reaction.

First, we show the surface electrostatic charge distributions of 3D NMR structures of rabbit, horse, dog, human, mouse, rabbit I214V mutant and rabbit S173N mutant prion proteins (Fig. 9.1). We may see in Fig. 9.1 that wild-type rabbits have a larger land of positive charge distributions than horses, dogs, humans and mice. Fig. 9.1 just illuminates to us the surface charge distribution of one snapshot of rabbit prion protein. In this section we will study the dynamics of the surface charge distributions during long MD simulations of 30 ns to confirm the above observation.

The MD snapshots of RaPrP^C(124–228) (PDB entry 2FJ3) at 350 K and 450 K for both *seed1* and *seed2* used in this section were generated in Chaps. 3, 4 and 5. The surface charge distributions of RaPrP^C of the snapshots at 5, 10, 15, 20, 25, 30

J. Zhang, *Molecular Structures and Structural Dynamics of Prion Proteins and Prions*, Focus on Structural Biology 9, DOI 10.1007/978-94-017-7318-8_9

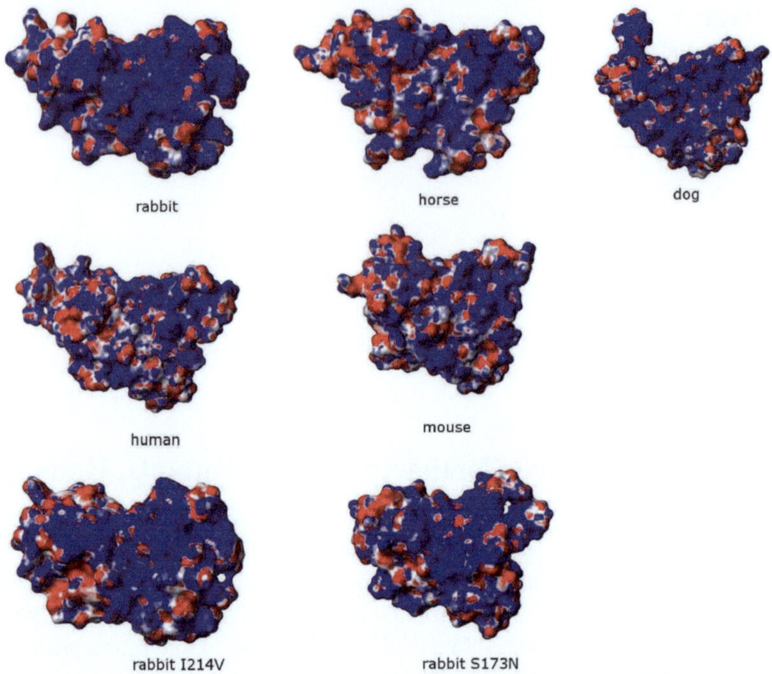

Fig. 9.1 Distributions of the surface electrostatic potential of rabbit, horse, dog, human, mouse, rabbit I214V mutant, and rabbit S173N mutant prion protein NMR structures (PDB IDs 2FJ3, 2KU4, 1XYK, 1QLX, 1AG2, 2JOM, 2JOH). *Blue* is for positive charge whereas *red* is for negative charge. This figure and all other figures in the below were generated using MolMol (ftp://ftp.mol.biol.ethz.ch/software/MOLMOL/win/, http://www.csb.yale.edu/userguides/graphics/molmol/tutorial/example4.html)

of 350 and 450 K for different initial velocities *seed1* and *seed2*, in neutral and low pH environments will be illuminated (Figs. 9.2 and 9.3).

We may see in Figs. 9.2∼9.3 that the continuation of blue or red colors for low pH value is less for neutral pH value; this might imply to us RaPrP has less structural stability in low pH environment. Similarly, we may list the snapshots of surface charge distributions of HoPrP^C(119–231) (PDB entry 2KU4), DoPrP^C(121–231) (PDB entry 1XYK), HuPrP^C(125–228) (PDB entry 1QLX) and MoPrP^C(124–226) (PDB entry 1AG2) during the long MD simulations of 30 ns. However, we cannot find their clear differences from RaPrP^C(124–228). In the below, we only analyze the surface charge distributions of special residues ARG163 (Fig. 9.4) and ASP177 (Fig. 9.5) and of the S2-H2 loop RaPrP(163–171) (Fig. 9.6).

By observations from Fig. 9.6 we cannot see much differences among these snapshots at the 30th nanosecond. Figures 9.4 and 9.5 illuminate to us that the NMR snapshot has large land of uniform distributions around the residues R163 and D177 respectively, but we cannot find the real difference among all the snapshots at the 30th nanosecond.

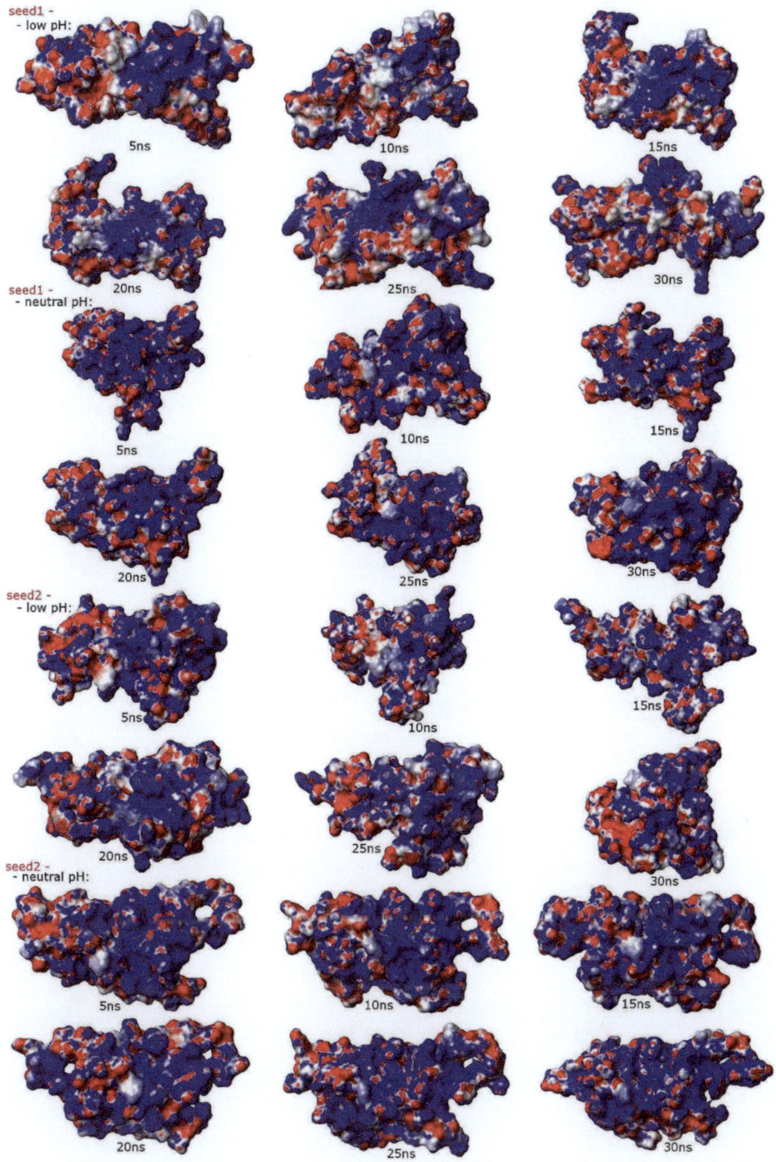

Fig. 9.2 Distributions of the surface electrostatic potential of *rabbit* prion protein structure at *350 K*

The surface of NMR RaPrP^C(124–228) molecule has a large land of *continuous* positive charge distribution (Fig. 9.1a). This agrees with the observation of [390, 625, 626]. To check whether this continuous positive charge distribution is still reserved during the long MD simulations of 30 ns under many different conditions,

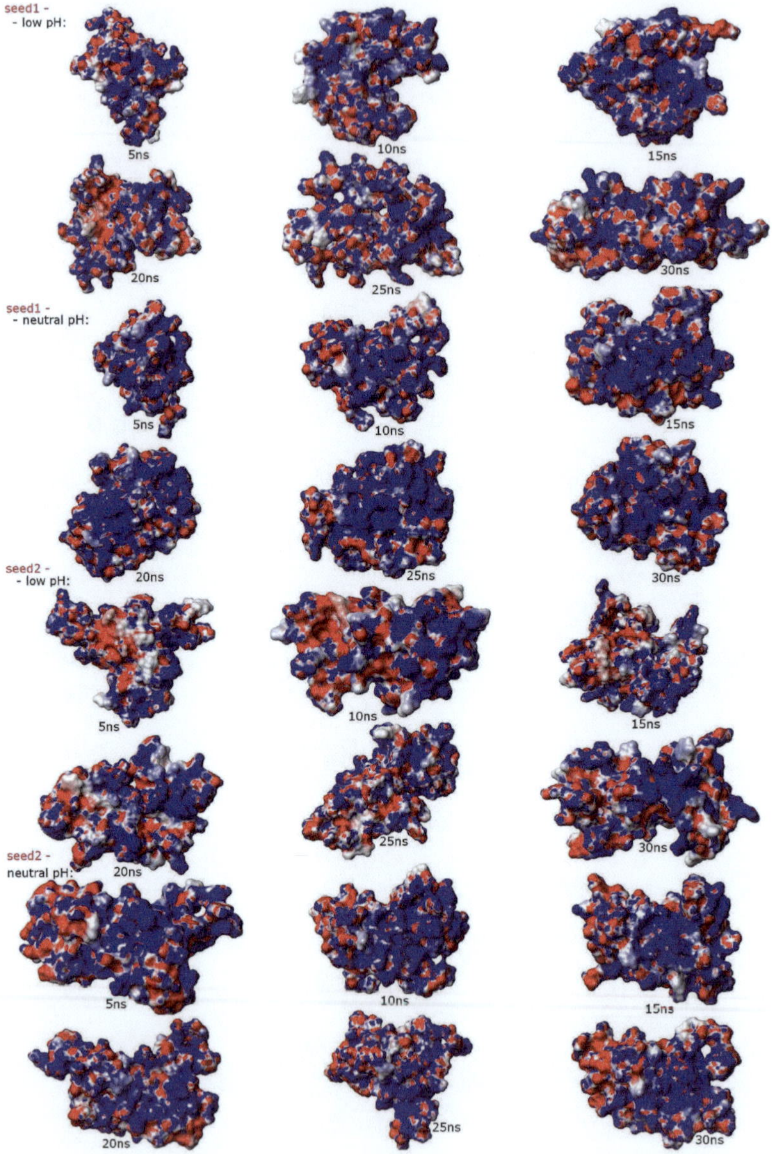

Fig. 9.3 Distributions of the surface electrostatic potential of *rabbit* prion protein structure at *450 K*

this section presents numerous snapshots of graphs for NMR RaPrP(124–228). These graphs should have a value acted as a rich database to study the 'unique distribution of surface electrostatic potential' of NMR RaPrP^{C}(124–228) molecule. Generally, we find that, in neutral pH environment, rabbit prion protein has a *large*

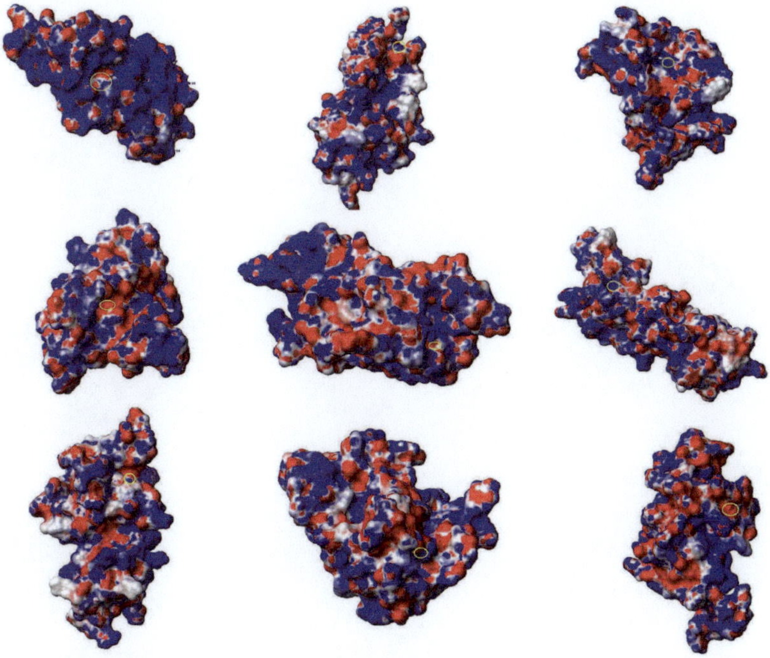

Fig. 9.4 Distributions of the surface charges around residue R163 (labeled into the yellow circle) of RaPrP^C(124–228) at snapshots of 2JF3.pdb, the 30th nanosecond of low pH – seed 1 – 350 K, low pH – seed 2 – 350 K, neutral pH – seed 1 – 350 K, neutral pH – seed 2 – 350 K, low pH – seed 1 – 450 K, low pH – seed 2 – 450 K, neutral pH – seed 1 – 450 K, neutral pH – seed 2 – 450 K

land of positive charge distribution; this agrees with the finding of Wen et al. [390, 625, 626]. In low pH environment, the *large continuous* land disappears and the structural stability of rabbit prion protein is broken (formulated α-helices into β-sheets). In conclusion, the surface electrostatic potential charge distribution, more or less, contributes to the structural stability of rabbit prion protein and contributes to the study of the species barrier of rabbit prion protein to prion diseases.

9.2 From the MD Results of NMR RaPrP and Mutants

Prion is a misfolded protein found in mammals that causes infectious diseases of the nervous system in humans and animals. Prion diseases are invariably fatal and highly infectious neurodegenerative diseases that affect a wide variety of mammalian species such as sheep and goats, cattle, deer, elk and humans etc. Recent studies have shown that rabbits have a low susceptibility to be infected by prion diseases with respect to other animals including humans. The present study employs MD means to unravel the mechanism of rabbit prion proteins (RaPrP^C)

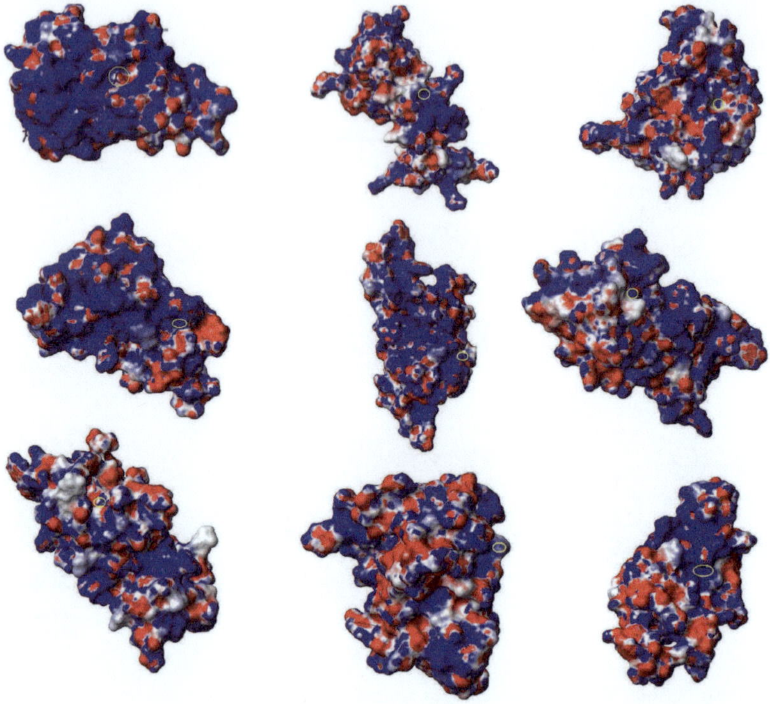

Fig. 9.5 Distributions of the surface charges around residue D177 (labeled into the yellow circle) of $RaPrP^C$(124–228) at snapshots of 2JF3.pdb, the 30th nanosecond of low pH – seed 1 – 350 K, low pH – seed 2 – 350 K, neutral pH – seed 1 – 350 K, neutral pH – seed 2 – 350 K, low pH – seed 1 – 450 K, low pH – seed 2 – 450 K, neutral pH – seed 1 – 450 K, neutral pH – seed 2 – 450 K

based on the recently available rabbit NMR structures (of the wild-type and its two mutants of two surface residues). The electrostatic charge distributions on the protein surface are the focus when analysing the MD trajectories. It is found that we can conclude that surface electrostatic charge distributions indeed contribute to the structural stability of wild-type $RaPrP^C$; this may be useful for the medicinal treatment of prion diseases.

9.2.1 Introduction

It has been a challenge to rational whether the contagious TSE is caused by prions [203, 485, 486, 540, 541]. As a misfolded protein, prion is neither a virus, nor a bacterium, and nor any microorganism. Prion diseases cannot be caused by the vigilance of the organism immune system but it can freely spread from one species to another species. Humans TSEs (for example, CJD and vCJD) can happen randomly through a number of processes, such as infections of transplanted

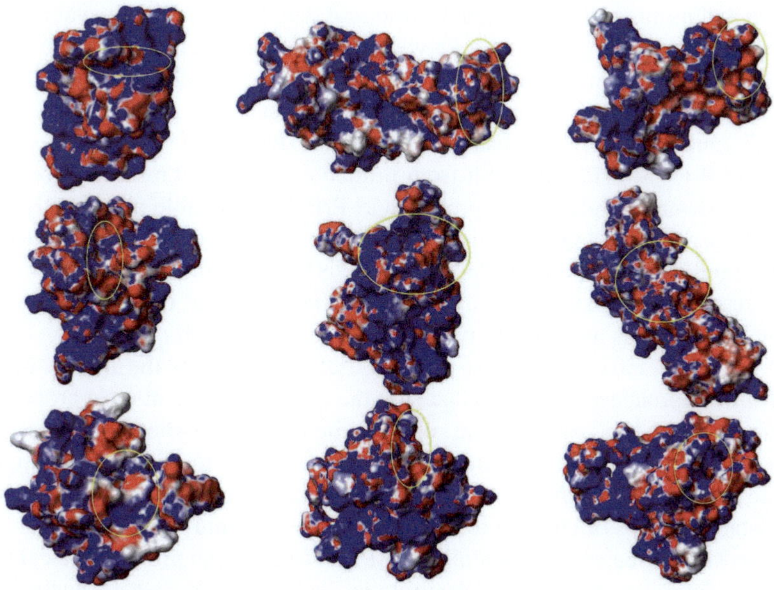

Fig. 9.6 Distributions of the surface charges around the S2-H2 loop RaPrP(163–171) (labeled into the *yellow circle*) at snapshots of 2JF3.pdb, the 30th nanosecond of low pH – seed 1 – 350 K, low pH – seed 2 – 350 K, neutral pH – seed 1 – 350 K, neutral pH – seed 2 – 350 K, low pH – seed 1 – 450 K, low pH – seed 2 – 450 K, neutral pH – seed 1 – 450 K, neutral pH – seed 2 – 450 K

tissue, blood transfusions and/or consumption of infected beef products, etc. Many mammals such as cat, mink, deer, elk, moose, sheep, goat, nyala, oryx, greater kudu and ostrich etc. are also susceptible to TSEs. However, a small group of other animals such as rabbits, horses, dogs and buffaloes seem to be little affected by prions [336, 474, 591, 681, 697, 708]. As a result, it is important to understand and to identify the specific causes why these animals are unlikely to be affected by prions, as it will provide insight to prion diseases and help to resolve the prion diseases issue.

The role of PrP^{Sc} infection in animals such as rabbit has been subject to a heated debating. Although a number of previous studies showed that a few animals such as rabbits exhibit low susceptibility to be infected by the PrP^{Sc} [29, 160, 204, 336, 350, 407, 441, 587, 591, 625, 626, 717]. However, other recent studies do not agree. Rather, recent studies suggested that (i) the rabbit prion may be produced through serial automated Protein Misfolding Cyclic Amplification (saPMCA) in vitro [106, 203], and (ii) the rabbit prion is infectious and transmissible [106]. In their most recently studies, Zhang et al. (2014) and Vidal et al. (2013) suggested that rabbits were not challenged directly in vivo by other known prion strains [583, 704]. In addition, the saPMCA results did not pass the test of the known BSE strain of cattle. As a result, rabbits can be considered as a prion resistant species.

A limited number of experimental structural data for RaPrP^C are available from the Protein Data Bank (PBD: http://www.rcsb.org/). For example, the structures of RaPrP^C obtained from NMR [384] (with a PBD entry of 2FJ3) and X-ray [336] (with a PBD entry of 3O79) measurements. As a result, it is desirable to reveal the properties and specific mechanisms of the rabbit PrP^C (RaPrP^C) and the conversion process of PrP^C →PrP^{Sc} of rabbit from limited experimental results. Here PrP^C is a soluble normal cellular prion protein and PrP^{Sc} is insoluble abnormally folded infectious and diseased prions. The present study will base on the NMR structure of RaPrP^C (and its I214V and S173N mutants: NMR structure PDB entries 2JOM, 2JOH, respectively) using MD simulation techniques. The information from the present MD studies is able to provide valuable insight for the PrP^C →PrP^{Sc} conversion. The information will provide useful rational in the design of novel therapeutic approaches and drugs that stop the conversion and disease propagation.

This section can capture the clear α-helices→ β-sheets conversion of PrP^C →PrP^{Sc} under pH environments from neutral to low. SB changes of charged residues under these two environments can lead to this conversion. Thus, electrostatic charges of charged residues are focused to study in this section.

In this study, electrostatic potential (ESP) surfaces of the proteins are calculated in order to determine the shape of the proteins, which helps to reveal how the ligands (drugs) dock into the proteins, the lock and key model in computer-aided drug design [689]. That is, the ESP surfaces are used as indicators for different inhibitors with substrates or transition states of the reaction. These bioinformatics results may be useful to understand the mechanisms and to suggest helpful information in the medicinal treatment of prion diseases. As a result, the analysis of surface electrostatic charge distributions of MD trajectories will be presented. The rest of this section is organized as follows. The MD simulation materials and methods for NMR structures of RaPrP^C wild-type and mutants are provided in next section, followed by the analysis and discussion focusing on the MD trajectory results of the surface electrostatic charge distributions and their discussions. Finally, concluding remarks on RaPrP are summarized.

9.2.2 Materials and Methods

The materials, e.g., data used in the present study are based on the NMR PDB files of 2FJ3.pdb, 2JOM.pdb, 2JOH.pdb. The MD methods employed are the same as the previous studies [679, 704]. Briefly, all simulations used the ff03 force field of the AMBER 11 package [91]. The systems were surrounded with a 12 Å layer of TIP3PBOX water molecules and neutralized by sodium ions using the XLEaP module of AMBER 9. Then 14 Cl-, 14 Cl-, 13 Cl- and 5909, 4185 and 4729 water molecules were added in the rabbit prion wild-type, S173N mutant and I214V mutant, respectively for pH less than 7. To remove the unwanted HB contacts, the systems of the solvated proteins with their counter ions were minimized mainly by the steepest descent method and followed by a small number of conjugate gradient steps on the data. Next, the solvated proteins were heated from 100 K to 450 K step by step in a 3 ns duration (prions are not completely removed even at a high temperature of 600 °C [493]. For prions many experimental works and MD works have done at 500 K [523]; simulation results on protein structures and their dynamics of this paper showed that the force field parameters are suitable to allow simulations at 450 K). Three sets of initial velocities denoted as seed1, seed2 and seed3 are performed in parallel for stability. The thermostat algorithm used is the Langevin thermostat algorithm in constant NVT ensembles. The SHAKE algorithm (only on bonds involving hydrogen) and PMEMD (Particle Mesh Ewald MD) algorithm with non-bonded cutoff of 12 Å were used during heating. Equilibrations were reached in constant NPT ensembles under Langevin thermostat for 5 ns. After equilibrations, production MD phase was carried out at 450 K for 30 ns using constant pressure and temperature ensemble and the PMEMD algorithm with the same non-bonded cutoff of 12 Å during simulations. The step size for equilibration was 0.5 and 1 fs in the MD production runs. The structures were saved to file every 1000 steps. The MD simulations are repeated under neutral pH environment.

In order to obtain the low pH (acidic) environment, the residues HIS, ASP, GLU were changed into their zwitterion forms of HIP, ASH, GLH, respectively, and Cl- ions were added by the XLEaP module of the AMBER package. Thus, the SBs of the system (residues HIS, ASP, GLU) under the neutral pH environment were broken in the low pH environment (zwitterion forms of HIP, ASH, GLH). At acidic pH, our MD simulations can also capture the misfolding of prion proteins from RaPrP^C α-helices to RaPrP^{Sc} β-sheets [101].

This section is focusing to study the surface electrostatic charge distributions [702]. Here, in the below, we will specially discuss some excellent dielectric constant [96] and dielectric constant [264] for the calculations of electrostatic potential surface charges.

9.2.2.1 Discussion on the MD Methods

The pmemd.MPI is faster and scales better in parallel than sander.MPI, but pmemd is less accurate than sander. The main reason for choosing PMEMD [513] in our case is that (i) the solvated systems are large, (ii) the MD simulation time are long, and (iii) the parallel supercomputer resources of VLSCI (http://www.vlsci.org.au).

The use of cutoffs in GB (Generalized Born) simulations as implemented in PMEMD does not conserve energy, and their use involves an approximation with an unknown effect on accuracy. For this reason, we chose not to implement vdW and electrostatic cutoffs in the GPU version of this code. Cutoffs for the nonbonded interactions are implemented for explicit solvent simulations with periodic boundary conditions using the PME method, as described in [253]. However, cutoffs in calculating the effective Born radii are supported.

We explain a bit more about PME and the explicit water with counter ions distribution as described in [597]. In the periodic boundary method the particles being simulated are enclosed in a box which is then replicated in all three dimensions to give a periodic array. During the simulation only one of the particles is represented, but the effects are reproduced over all the image particles with each particle not only interacting with the other particles but also with their images in neighbouring boxes. Particles that leave one side of the box re-enter from the opposite side as their image. In this way the total number of particles in the central box remains constant. The periodic boundary method would appear to be very computationally intensive. However, by employing the PME method it is possible to obtain the infinite electrostatics in a way that scales as NlnN time. This involves dividing the calculation up between a real space component and a reciprocal space component. The use of PME means we are calculating all of the 'infinite' electrostatics and so are not actually truncating the electrostatics. The vdW interactions are still needed though which means we cannot make the cut-off too small less than 8 Å for periodic boundary PME calculations. Prior to MD, we use optimization (first Holding the solute fixed to just minimize the positions of the water and ions and then minimizing the entire system) to relax the system to remove bad vdW and electrostatic interactions and the gaps between the solvent and solute and solvent and box edges, etc. During the equilibration period, we first run short MD at constant volume and then use periodic boundaries to keep the pressure constant and so allow the volume of the box to change. This will allow the water and ions to equilibrate around the solute and come to an equilibrium density. The visual description of colours for charges should be averaged from residue charge.

We have chosen 450 K, which is high temperature to SB. This means that the Cp is low enough to keep the initial structure (ΔG<0) at 450 K [96]; but it is not simulated using the Gibbs-Helmholtz equation [96] by all versions of the Amber package (http://ambermd.org/).

We use FirstGlance in Jmol (http://bioinformatics.org/firstglance/fgij/) to detect all the charges of RaPrP^C wild-type, I214V mutant and S173N mutant: 10+ (7 ARG, 3 LYS) (3HIS) and 13- (5 ASP, 7 GLU, 1 C-terminus), in defining the following SBs that we have calculated: ASP146-ARG147, GLU210-ARG207, GLU206-LYS203, GLU206-ARG207, ASP146-HIS139, GLU151-ARG150, GLU151-ARG147, GLU151-ARG155, ASP177-ARG163, GLU145-ARG135, ASP143-HIS139, GLU145-HIS139, GLU195-ARG155, ASP146-ARG150, ASP177-HIS176, ASP143-ARG147, GLU195-LYS193, HIS186-ARG155, HIS186-LYS184, ASP166-ARG163, GLU210-HIS176, HIS139-ARG135, GLU199-LYS193, GLU206-HIS176 [704].

This section is focusing to study the ESP surface charge distributions. Here in the below we will specially discuss some excellent dielectrique constant [96] and dielectric constant [264] for the calculations of ESP surface charges.

9.2.2.2 Discussion on Dielectrique Constant and Free Energy

The dielectric response for a particular region of the protein is obtained at the microscopic level by employing a TI/MD (Thermodynamic Integration/MD) method; MD simulations and "TI" are used to determine how the protein responds to charge insertion at a particular site at the atomistic level. In this paper we presented the external polarisation in terms of charges (positives, neutral, negatives). This paper uses Maestro 9.7 2014-1 (Academic use only) free package (http://www.schrodinger.com/) to draw the Poisson-Boltzmann ESP surface charges of the external polarisation: we choose indi = 1.0 as the solute dielectric constant, exdi=80.0 as the solvent dielectric constant, and 12 Å as the solvent radius, 450 K as the Temperature, and EPS mapped on molecular surface is chosen. We know that the delta ΔCp => ΔGu=f(T) is a good method to show how a molecular structure is more stable with a SB (Fig. S4 of [96]; using the delta method on the 2FJ3.pdb file of the rabbit PrP, the distance of ASP177-ARG163 (O-N) gives a SB of 10 Å which is nearly null in terms of energy. But this is just the result of one snapshot under special conditions; during the whole MD simulations at temperatures such as 300 K, 350 K, 400 K, 450 K, and 500 K the distance of ASP177-ARG163 (O-N) is at the SB distance during most of the 30 ns' MD simulation time.

9.2.2.3 About Dielectric Constant

In the method using a mesoscopique dielectric constant a nice figure (Fig. 4B of [264]) shows the change of e which gives what is under the coloured surface. The transparent surfaces shows contours of equal dielectric as determined from heterogeneous mesoscopique dielectric theory [264]. SBs into the core of the protein will show a better energy for stabilization (weak e, therefore better ΔGu) than all SBs rather on the external protein. All the charges we are showing as red and blue are not just on the periphery. SBs are calculated by oppositely-charged atoms that are within 6.5 Å and are not directly hydrogen-bonded. It should be the average charge calculated per residue or the specific atom charge of the residue. The donor residues involved are ASP, GLU, HID and the acceptor residues involved are LYS, ARG, HID, and the real computed distance is within 6.5 Å in Amber package. On prion proteins, Guest et al. (2011) declared "SBs known to be absent in disease causing human mutants of the prion protein were found to be among the strongest present in the protein, so that the human mutants were electrostatically the least stable of those proteins studied. Conversely, the prion protein with the most stable SBs belonged to a species known to be resistant to prion disease (frog)" [263, 264].

9.2.3 Results and Discussion

The MD results from the simulations are summarized in Figs. 9.7, 9.8 and 9.9, respectively, under the parallel conditions of seed1, seed2 and seed3 in the duration of 30 ns. Each figure, say, Fig. 9.7, it composes of six panels – three on the left for the system under the neutral pH condition and three on the right for the systems under the acidic (low pH) solutions. The variations of the secondary molecular structures

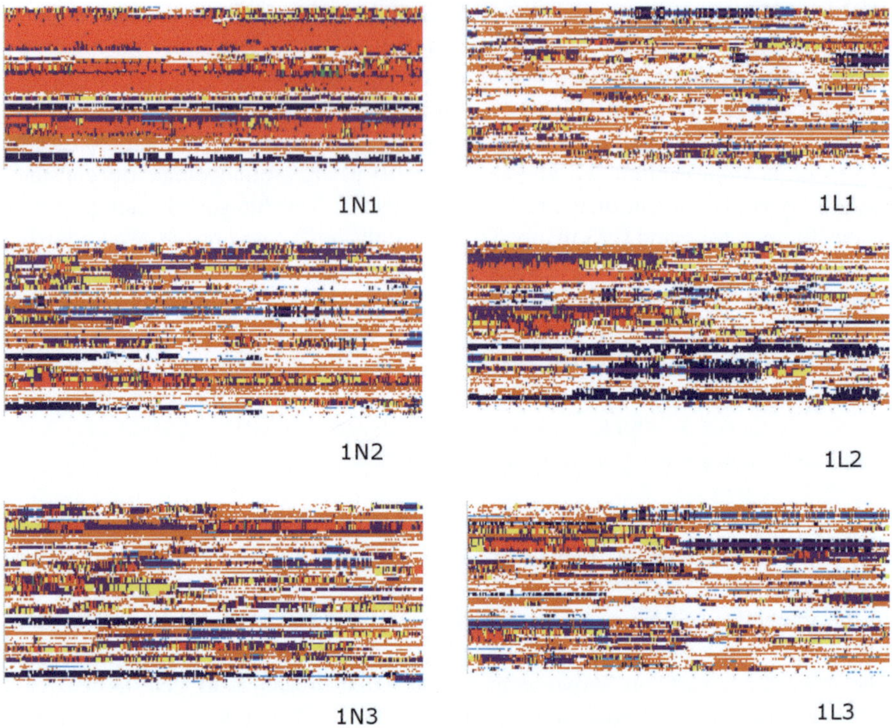

Fig. 9.7 Seed1. Secondary Structure graphs for RaPrP^C wild-type, I214V mutant and S173N mutant (from up to down) at 450 K (x-axis: time (0-30 ns), y-axis: residue number (124-228); left column: neutral pH, right column: low pH). H is the α-helix, I is the π-helix, G is the 3-helix or 3/10 helix, B is the residue in isolated β-bridge, E is the extended strand (participates in β-ladder), T is the HBed turn, and S is the bend

of the wild-type (RaPrP^C) and its I214V and S173N mutants are given on the top row, middle row and bottom row, respectively. As a result, the six panels in Fig. 9.7 can be called as 1N1, 1N2, 1N3, 1L1, 1L2 and 1L3, for the results under seed1 conditions, in neutral pH environment for RaPrP^C, I214V mutant and S173 mutant for 1N1, 1N2 and 1N3; in low pH environment (acidic) for RaPrP^C, I214V mutant and S173 mutant for 1L1, 1L2 and 1L3. For example, 1L2 is the results for seed1 conditions in low pH solution for I214V mutant. Within each panel, say, 1N1 (the top-left corner panel), three subpanels named H1, H2 and H3 from bottom to top represents the α-helix 1, α-helix 2 and α-helix 3 of the prion protein, respectively.

Under the convention, coloured codes represent various helices. The information in the figures was produced using the DSSP program [326]. For example, red, green, yellow, blue, dark blue, purple and orange represent for H, I, G, B, E, T and S, accordingly. The coloured codes represent various helices. For example, letter H refers to the α-helix, I refers to the π-helix, G refers to the 3-helix or 3/10 helix, B refers to the residue in isolated β-bridge, E refers the extended strand (participates

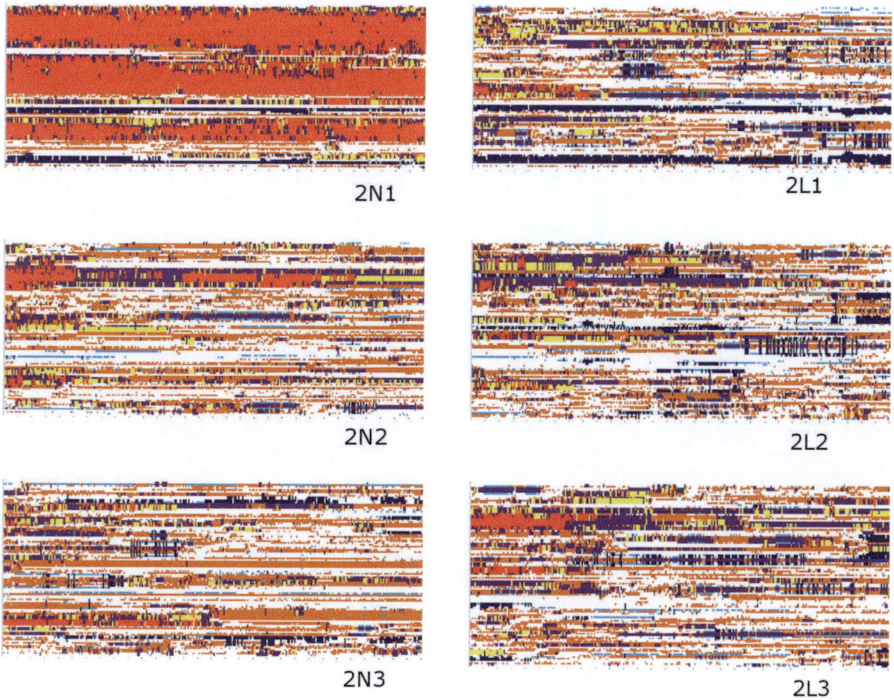

Fig. 9.8 Seed2. Secondary Structure graphs for RaPrP^C wild-type, I214V mutant and S173N mutant (from up to down) at 450 K (x-axis: time (0-30 ns), y-axis: residue number (124-228); left column: neutral pH, right column: low pH). H is the α-helix, I is the π-helix, G is the 3-helix or 3/10 helix, B is the residue in isolated β-bridge, E is the extended strand (participates in β-ladder), T is the HBed turn, and S is the bend

in β-ladder), T refers to the HBed turn, and S refers to the bend. The colour coded letters are given in the figures.

As can be seen from Figs. 9.7, 9.8 and 9.9, the top-left corner panels, i.e, 1N1, 2N1 and 3N1 are dominated by red colour, indicating that under the neutral pH conditions, the three α-helices (H1, H2 and H3) of the wild-type prion RaPrP^C, remain dominant the α-helices without significant changes during the period of 30 ns, regardless the seed conditions. For example, the α-helices (H1, H2 and H3 in red colour) of 1N1 in Fig. 9.7 (top-left panel) does not experience any apparent colour changes, indicating that under the neutral pH condition, the α-helices of the wild-type prion resist structural changes. This is particular the case in H3 and H2, although small noticeable changes in H1 has been observed. However, the I214V (second rows) and S173N (third rows) mutants of the wild-type prion protein show significant changes indicated by their changes of colour codes. This is also the cases under acidic condition (i.e., low pH environment) on the right hand side of the figures, which shows colourful panels – indication very little similarities can be seen in the corresponding bands. As Figs. 9.7, 9.8 and 9.9 are under parallel conditions

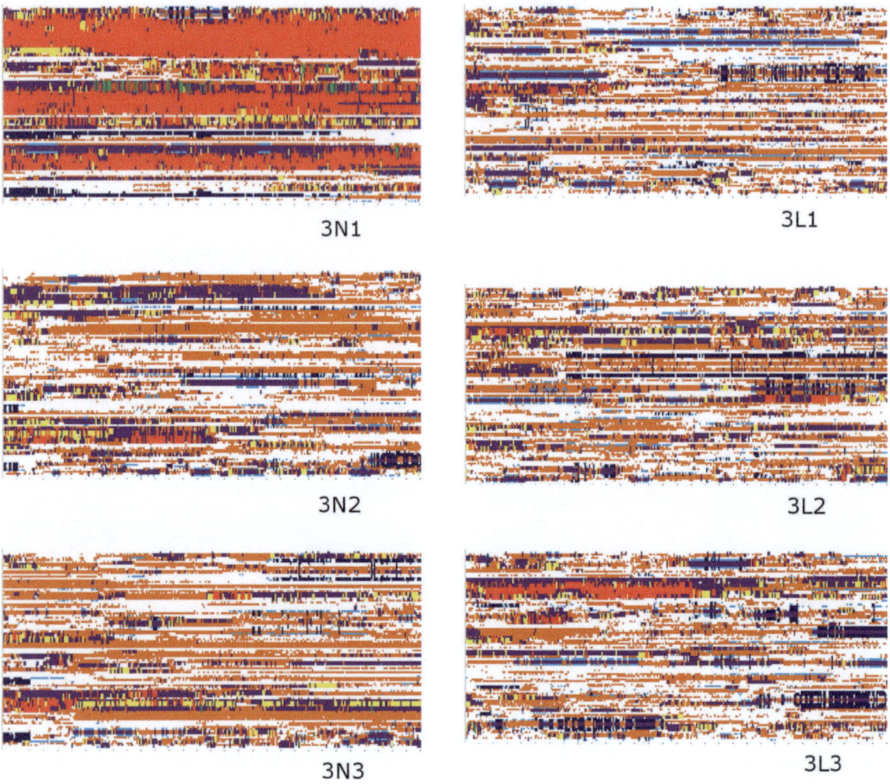

Fig. 9.9 Seed3. Secondary Structure graphs for RaPrP^C wild-type, I214V mutant and S173N mutant (from up to down) at 450 K (x-axis: time (0-30 ns), y-axis: residue number (124-228); left column: neutral pH, right column: low pH). H is the α-helix, I is the π-helix, G is the 3-helix or 3/10 helix, B is the residue in isolated β-bridge, E is the extended strand (participates in β-ladder), T is the HBed turn, and S is the bend

with similar changes patterns, the following discussion will focus on Fig. 9.7 under seed1 conditions.

In Fig. 9.7, the 1N1 panel is unique – it is dominant by red, indicating little mutational changes, whereas all other panels in this figure results in colourful changes. It reveals that the three α-helices of the wild-type (RaPrP^C) prion protein are unfolded and remain in the same structures under neutral pH conditions. However, in low pH environment, this wild-type prion protein (1L1) turns into a colourful panel of almost all of the seven colours but large presentation of orange (HBed bend). Therefore, under acidic condition (low pH environment), the SB network of the wild-type (RaPrP^C) is broken thus leads to the unfolding of the stable α-helical structures of RaPrP^C [704]. Hence, it suggests that the structural distributions of the wild-type (RaPrP^C) protein depends on the pH. On the other hand, the I214V and S173N mutants under either neutral or low pH environment,

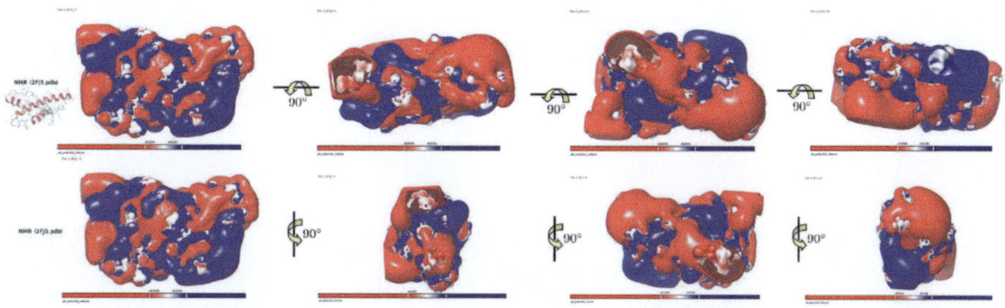

Fig. 9.10 Surface electrostatic charge distributions for the NMR structure of $RaPrP^C$ wild-type, where blue is for positive charge whereas red is for negative charge. The pb_potential_volume is ±43.8195

such as 1N2/1L2 and 1N3/1L3, are unfolded into rich β-sheet structures. Generally, we could see the clear secondary structure changes from (α-, π-, 3_{10}-) helix structures to β-sheet (ladder and bridge) structures from Figs. 9.7∼9.9.

Residue at 214 within the C terminus could potentially influence the conversion of $RaPrP^C$ to $RaPrP^{Sc}$ [591]. For residue at 173 in the $\beta 2$-$\alpha 2$ loop, Wen et al. (2010) reported a large-continuous-positive-charge-surface ESP, which is unique to the $RaPrP^C$ prion protein [626]. In the studies, the solution structures (backbones) of the $RaPrP^C$ protein and its S173N variant are determined for their structured C-terminal domains. It suggested that the highly ordered $\beta 2$-$\alpha 2$ loop has been well recognized by the NMR-signals contribute to the structural stability of $RaPrP^C$ [625, 626] at one snapshot of 300 K and 1 ATM condition. In Fig. 9.10, the surface EPS charge distributions are calculated based on the NMR structure of the $RaPrP^C$ wild type.

Figures 9.11∼9.12 show the surface ESP charge distributions of the wild-type $RaPrP^C$ protein at a temperature of 450 K for the neutral pH environment obtained in the present study. Figure 9.11 gives the snapshots at 5 ns, 10 ns, 15 ns, 20 ns, 25 ns, 30 ns, for the conditions of seed1 (first row), seed2 (second row) and seed3 (3rd row). In these figures, the charges coloured by red and blue represent negative and positive respectively. As shown in Fig. 9.11, it is discovered that a large region of positive charge (blue) distribution has still been reserved during the long MD simulation of 30 nanoseconds. However, the continuous of the land has not been always reserved yet (where continuous means connect to each other, not discreted), as indicated by Fig. 9.12 with rotational views.

The I214V mutation does not change the ESP due to the non-charged feature of the two amino acids at site 214 [625]. Figure 9.13 shows the surface ESP charge distributions of the mutation of the wild-type $RaPrP^C$ protein at a temperature of 450 K in the neutral pH environment obtained in the present study. However, it reveals that the structures of I214V mutant do not unfold from the α-helices (refer to Figs. 9.7∼9.9) in the 30 ns duration of the MD simulation. This suggests that the ESP may be not the reason leading to the structure unfolding of the I214V mutant of the $RaPrP^C$ protein.

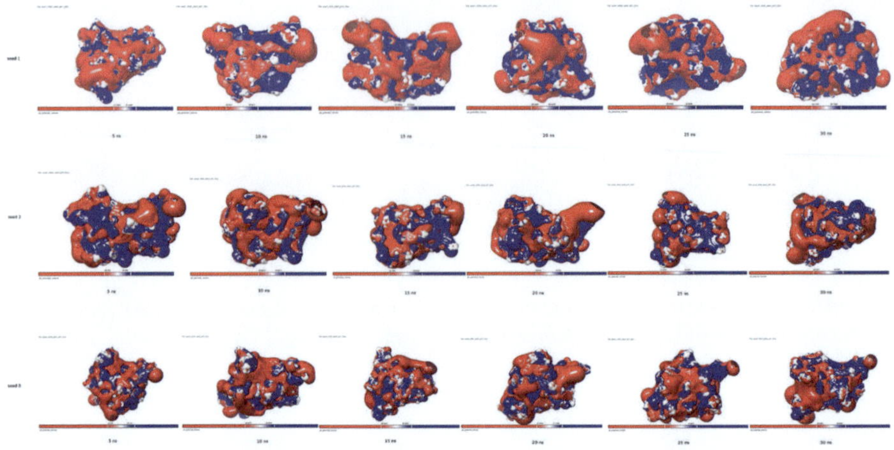

Fig. 9.11 Surface electrostatic charge distributions for RaPrP^C wild-type at 450 K in neutral pH environment at 5 ns, 10 ns, 15 ns, 20 ns, 25 ns, 30 ns (left to right), where blue is for positive charge whereas red is for negative charge. Up to down: seed1∼seed3. The pb_potential_volumes for seed1 are ±21.3561 (5 ns), ±27.637 (10 ns), ±23.9485 (15 ns), ±28.2425 (20 ns), ±32.4692 (25 ns), ±25.1583 (30 ns), for seed2 are ±26.452 (5 ns), ±19.6673 (10 ns), ±28.1642 (15 ns), ±25.912 (20 ns), ±31.4185 (25 ns), ±29.0557 (30 ns), and for seed3 are ±24.138 (5 ns), ±28.8575 (10 ns), ±26.5422 (15 ns), ±21.7444 (20 ns), ±31.0833 (25 ns), ±28.0295 (30 ns)

The S173 mutation, however, just lies in the $\beta 2$-$\alpha 2$ loop. Thus this mutation S173N leads to the changes of the network of the ESP in this loop, which is indicated in Fig. 9.14 under seed1 and seed2 conditions, which is very different from seed3 conditions in the same figure. It indicates that the ESP may lead to the structure unfolding of the S173N mutant of the RaPrP^C protein in seed1 and seed2, whereas the ESP may be not the reason for the structure unfolding of the S173N mutant of the RaPrP^C protein. From Figs. 9.12∼9.14, we may see that the pb_potential_volume for wild-type is less than for mutants.

9.2.4 Conclusions

There are still a number of challenges remain in "prion" theory. The present study using the MD simulation strategies has found that surface electrostatic charge distributions play an important role in the structural stability of RaPrP^C but are clearly not the only reason contributing to the structural stability the RaPrP^C protein. This finding only partially agrees to the conclusion that "large-continuous-positive-charge-surface ESP contributes to the structural stability of RaPrP^C" (especially in the $\beta 2$-$\alpha 2$ loop region) [625, 626]. Further studies to understand the specific mechanism of RaPrP^C are still needed.

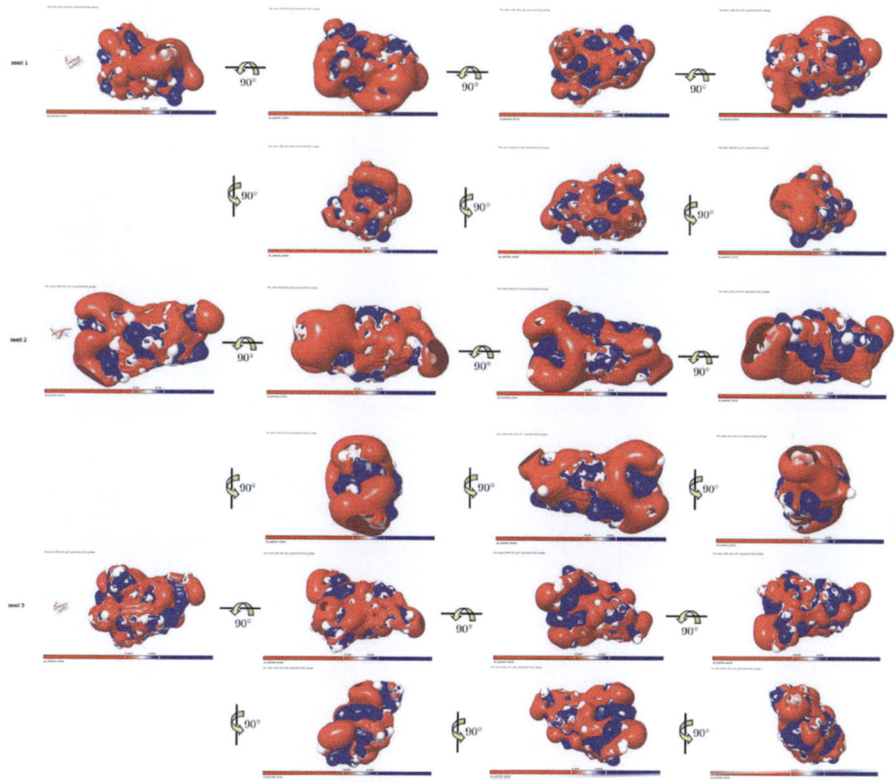

Fig. 9.12 Surface electrostatic charge distributions for the average structure of 30 ns' MD of RaPrP^C wide-type at 450 K in neutral pH environment, where blue is for positive charge whereas red is for negative charge. Up to down: seed1∼seed3. The pb_potential_volumes for seed1, seed2, seed3 are ±25.8555, ±25.328, ±33.5938 respectively

9.3 Other Clues

There are many clues to reveal prion diseases; for example, copper bindings, π-interactions, poplar contacts, etc.

9.3.1 Weak and Strong Copper Bindings

The Cu^{2+}, Cu^{+} bindings should also be a clue of prion diseases. Here we just illuminate two pictures of the weak-binding HGGGW octarepeat region PrP(57–89) and the strong-binding GGGTH region PrP(92–96) Fig. 9.15 [161]; the author has done QM/MM studies on all prion copper binding sites whether in structured region or unstructured region of prion proteins.

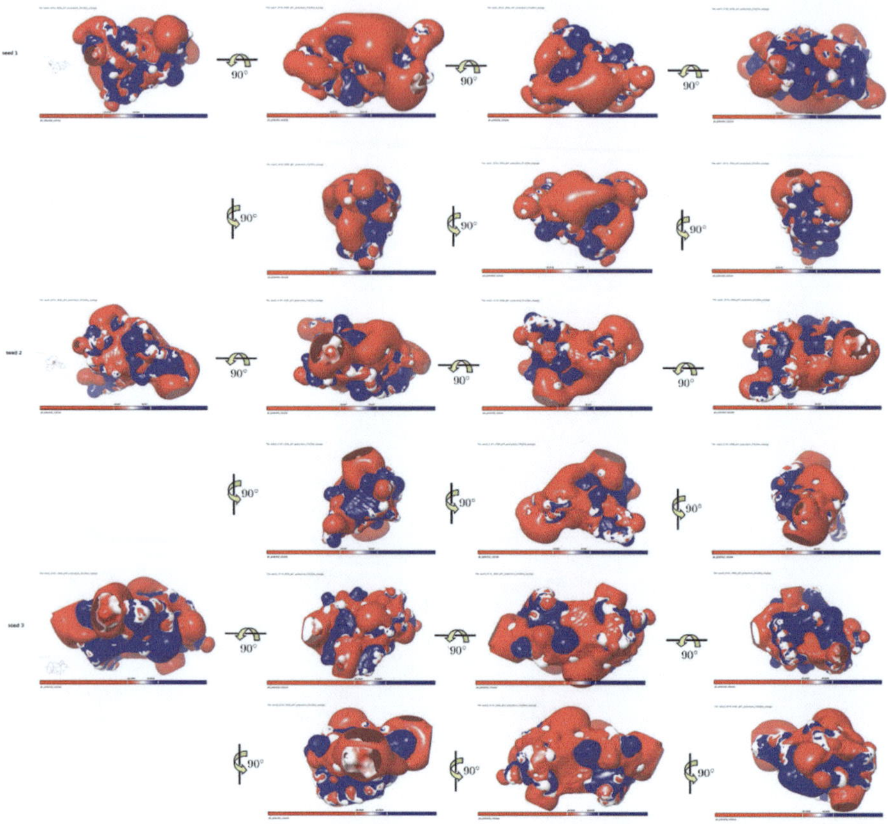

Fig. 9.13 Surface electrostatic charge distributions for the average structure of 30 ns' MD of RaPrP^C I214V mutant at 450 K in neutral pH environment, where blue is for positive charge whereas red is for negative charge. Up to down: seed1∼seed3. The pb_potential_volumes for seed1, seed2, seed3 are ±34.9142, ±38.007, ±40.6668 respectively

These two copper binding sites both are in the unstructured region, in the next chapter we will briefly study the molecular modeling on the AGAAAAGA of this region without structures.

9.3.2 *π-π Stackings and π-Cations in All PrP^C Structures*

π-π stackings and π-cations clearly do some contributions to maintain the structural stability of a normal cellular prion protein (PrP). This short article is to do a survey on the π-π stackings and π-cations in all the PrP structures listed in the PDB (www.rcsb.org) Bank. We find the following important π-π stackings: Y218–F175–Y169 (around the β2-α2 loop), Y162–Y128 (linking the two β-strands), F141–Y150–Y157 (in α-helix 1), H187–F198 (linking

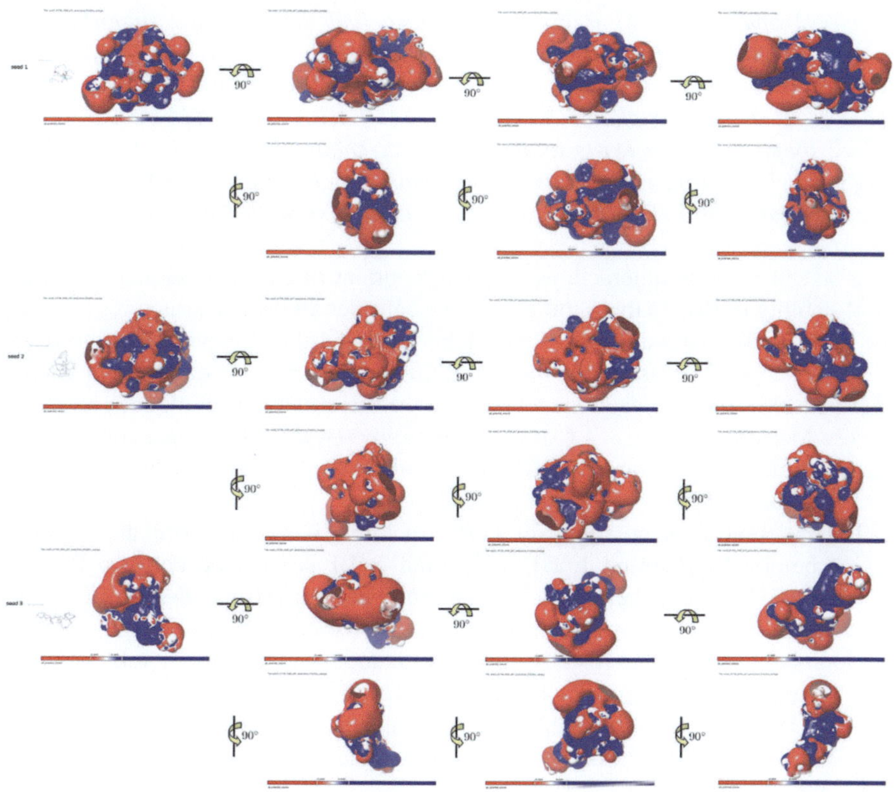

Fig. 9.14 Surface electrostatic charge distributions for the average structure of 30 ns' MD of RaPrP^C S173N mutant at 450 K in neutral pH environment, where blue is for positive charge whereas red is for negative charge. Up to down: seed1$\sim$seed3. The pb_potential_volumes for seed1, seed2, seed3 are $\pm$42.8397, $\pm$39.629, $\pm$31.4955 respectively

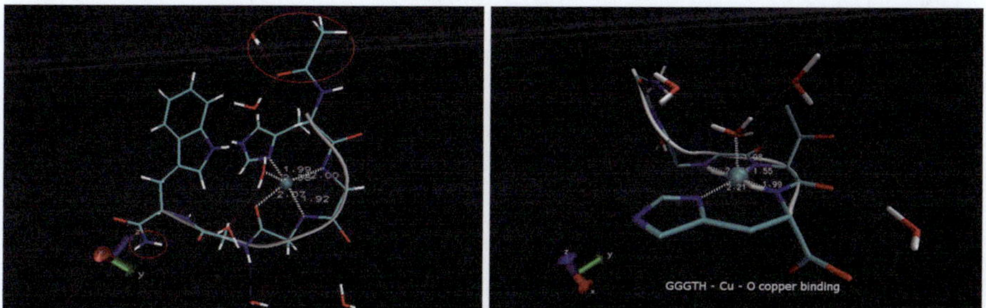

Fig. 9.15 Two copper binding examples for prion proteins

α-helix 2 and the α2-α3 loop); and we also find the following important π-cations: F141–R208.(N)NH2 (linking the β1-α1 loop and α-helix 3), Y128–R164.(N)NH2–Y169 (linking β-strand 1 and the β2-α2 loop). Thus, for PrPs, there exists a long "π-chain" Y218–F175–Y169–R164–Y128–Y162 covering the β2-α2 loop, and there exists another long "π-chain" R208–Y141–Y150–Y157–F198–H187 covering the α-helix 1. This short article can be acted as a "quick reference card" for PrP protein structure π-interaction studies in laboratories or in theories.

π-π and π-cation interactions play an important role in maintaining the structural stability of PrP. In this short article, we consider all the PrP structures listed in the PDB Bank: first we use the Swiss-PdbViewer 4.1.0 (spdbv.vital-it.ch) to relax (i.e. do Energy Minimisation in the use of Steepest Descent – Conjugate Gradients -Steepest Descent optimization methods) all the PrP structures (Table 9.1), and then we use Maestro 10.1 2015–1 (Academic use only) (www.schrodinger.com) to find the π interactions (see Table 9.2, where the code in the ( ) bracket is the PBD ID for each PrP species).

Let us denote some notations for a PrP structure: L0 is the N-terminal structured region before the β-strand 1, B1 is β-strand 1, L1 is the loop between B1 and α-helix 1, H1 is the α-helix 1, L2 is the loop between H1 and β-strand 2, B2 is β-strand 2, L3 is the loop between B2 and α-helix 2, H2 is the α-helix 2, L4 is the loop between H2 and α-helix 3, and H3 is the α-helix 3. From Table 9.2, we may see that there are π-π stackings (Fig. 9.16): Y218–F175–Y169 (linking H3–H2–L3), H187–F198 (linking H2–L4), F141–Y150–Y157 (in H1), Y225–Y226 (in H3), Y162–Y128 (linking B2–B1), etc. From Table 9.2, we also see the following

Table 9.1 Energy variations during the energy minimizations (number of iterations are in the brackets) for each PrP

Species	SD	CG	SD
mousePrP (1AG2)	−7234.507 (391)	−7243.597 (13)	−7275.160 (45)
mousePrP (1XYX)	−7812.781 (498)	−7892.577 (123)	−8002.255 (218)
humanPrP (1QLX)	−7460.885 (296)	−7610.384 (195)	−7640.157 (54)
humanPrP (1QLZ)	−7458.125 (492)	−7464.300 (7)	−7494.678 (45)
humanPrP (1QM0)	−7403.133 (474)	−7475.931 (118)	−7501.340 (49)
humanPrP (1QM1)	−7249.798 (350)	−7345.074 (114)	−7410.470 (99)
humanPrP (1QM2)	−7451.833 (410)	−7590.378 (284)	−7631.895 (106)
humanPrP (1QM3)	−7271.444 (326)	−7474.208 (309)	−7591.563 (285)
humanPrP (2LSB)	−8274.376 (741)	−8336.196 (121)	−8423.878 (202)
humanPrP(118–224) (4N9O)	−7496.459 (333)	−7634.974 (308)	−7693.850 (165)
humanPrP-pH7 (1HJM)	−7517.054 (745)	−7534.802 (45)	−7581.247 (130)
humanPrP-pH7 (1HJN)	−7160.900 (263)	−7370.024 (242)	−7474.183 (190)
humanPrP-M129	−7249.798 (350)	−7345.074 (114)	−7410.470 (99)

(continued)

Table 9.1 (continued)

Species	SD	CG	SD
bovinePrP (1DWY)	−7698.001 (544)	−7809.687 (255)	−7819.315 (26)
bovinePrP (1DWZ)	−7335.884 (463)	−7518.091 (372)	−7586.896 (173)
bovinePrP (1DX0)	−7468.698 (555)	−7593.610 (298)	−7673.004 (240)
bovinePrP (1DX1)	−7592.867 (904)	−7623.991 (116)	−7635.851 (47)
SyrianHamsterPrP (1B10)	−7418.702 (225)	−7653.251 (258)	−7688.044 (50)
SyrianHamsterPrP (2LH8)	−7524.862 (540)	−7555.382 (68)	−7568.822 (31)
dogPrP	−7148.517 (549)	−7225.084 (151)	−7251.191 (60)
catPrP	−6935.915 (192)	−7186.961 (186)	−7382.361 (229)
sheepPrP	−8066.183 (291)	−8066.300 (22)	−8204.010 (179)
mousePrP[N174T]	−7418.443 (211)	−7657.886 (206)	−7923.308 (419)
humanPrP[Q212P]-M129	−7662.032 (464)	−7688.836 (49)	−7846.123 (424)
rabbitPrP[S173N]-NMR	−7173.492 (271)	−7666.456 (492)	−7684.357 (36)
rabbitPrP[I214V]-NMR	−7785.640 (774)	−7802.710 (39)	−7835.354 (88)
rabbitPrP[S170N]-X-ray	−8682.104 (414)	−8753.827 (178)	−8827.220 (254)
rabbitPrP[S174N]-X-ray	−8551.921 (286)	−8659.515 (160)	−8739.475 (178)
rabbitPrP[S170N,S174N]-X-ray	−8864.615 (363)	−8899.511 (65)	−8945.262 (106)
mousePrP - at 37 °C	−7679.395 (212)	−8031.103 (397)	−8134.173 (213)
mousePrP[V166A]	−8040.172 (436)	−8153.529 (174)	−8164.996 (22)
mousePrP[D167S] - at 20 °C	−7938.545 (594)	−7987.788 (83)	−8051.072 (121)
mousePrP[D167S,N173K]	−7615.915 (546)	−7751.804 (205)	−7906.153 (290)
mousePrP[Y169G]	−7713.983 (249)	−7885.343 (147)	−7913.277 (27)
mousePrP[Y169A]	−7948.267 (507)	−7949.964 (1)	−7955.516 (3)
	−7972.101 (17)	−8064.175 (142)	−8143.807 (179)
mousePrP[S170N]	−7341.668 (74)	−7790.935 (262)	−7947.893 (224)
mousePrP[S170N,N174T]	−7988.545 (468)	−8126.978 (241)	−8149.948 (55)
mousePrP[F175A]	−7660.438 (680)	−7859.632 (381)	−7873.355 (34)
mousePrP[Y225A,Y226A]	−7457.356 (244)	−7588.338 (109)	−7757.797 (213)
mousePrP[Y169A,Y225A,Y226A] - at 20 °C	−7609.773 (173)	−7690.231 (45)	−7700.162 (4)
elkPrP	−7894.305 (875)	−7959.371 (160)	−7978.686 (53)
pigPrP	−6354.886 (141)	−6735.155 (321)	−6813.048 (109)
bankVolePrP	−7727.260 (478)	−7799.991 (118)	−7951.502 (376)
tammarWallabyPrP	−8028.082 (393)	−8195.817 (248)	−8238.248 (69)
rabbitPrP-NMR	−7712.972 (814)	−7730.035 (36)	−7790.620 (154)
rabbitPrP-X-ray	−8643.309 (602)	−8658.327 (40)	−8717.344 (187)
horsePrP	−7335.273 (226)	−7614.526 (433)	−7633.569 (45)
humanPrP-V129 (3HAK)	−7928.211 (287)	−7969.004 (56)	−8013.225 (78)
humanPrP-M166V (1E1G)	−6977.549 (69)	−7612.355 (568)	−7851.550 (568)
humanPrP-M166V (1E1J)	−7332.531 (320)	−7499.970 (270)	−7540.679 (92)
humanPrP-M166C/E221C (1H0L)	−7104.881 (374)	−7346.396 (366)	−7363.282 (36)
humanPrP-D178N (2K1D)	−6891.865 (369)	−7227.109 (630)	−7275.960 (130)
humanPrP-S170N (1E1P)	−7865.426 (875)	−7964.758 (242)	−7985.477 (59)
humanPrP-S170N (1E1S)	−7173.158 (165)	−7608.837 (354)	−7655.861 (69)

(continued)

Table 9.1 (continued)

Species	SD	CG	SD
humanPrP-D178N-M129 (3HEQ)	−7798.315 (291)	−7952.650 (246)	−8075.597 (398)
humanPrP-D178N-V129 (3HJX)	−7277.354 (230)	−7431.112 (165)	−7483.991 (111)
humanPrP-F198S-M129 (3HES)	−7345.292 (606)	−7368.778 (87)	−7371.617 (9)
humanPrP-F198S-V129 (3HER)	−7356.892 (572)	−7409.278 (158)	−7418.585 (33)
humanPrP-R200K (1FKC)	−7751.933 (536)	−7834.431 (156)	−7850.802 (36)
humanPrP-R200K (1FO7)	−7565.328 (298)	−7651.469 (89)	−7704.544 (71)
humanPrP-V209M-M129 (2M8T)	−7508.455 (425)	−7555.117 (117)	−7638.879 (251)
humanPrP-V210I-M129 (2LEJ)	−7629.392 (210)	−7782.742 (176)	−7832.093 (83)
humanPrP-V210I-M129 (2LV1)	−7874.923 (424)	−7915.883 (78)	−8041.248 (304)
humanPrP-E219K-M129 (2LFT)	−7958.208 (397)	−8063.400 (193)	−8156.185 (245)
humanPrP-R220K (1E1U)	−7096.946 (506)	−7139.047 (60)	−7210.754 (136)
humanPrP-R220K (1E1W)	−7056.962 (420)	−7173.744 (190)	−7240.287 (137)
chickenPrP (1U3M)	−6873.529 (769)	−6912.738 (115)	−6998.128 (316)
turtlePrP (1U5L)	−8269.539 (917)	−8273.342 (11)	−8317.274 (145)
xenopusLeavisPrP (1XU0)	−7516.396 (626)	−7553.854 (65)	−7595.688 (84)
sheepPrP-H168 (1XYU)	−7627.764 (465)	−7726.488 (147)	−7770.972 (80)
ovinePrP-R168 (1Y2S)	−7456.458 (236)	−7668.949 (204)	−7682.468 (18)

Table 9.2 π-π-stackings and π-cations for each (optimized) PrP

Species	**π-π-stacking**	**π-cation**
mousePrP (1AG2)	F175–Y218,Y162–Y128,H187–F198	F141–R208.(N)NH2
(1XYX)	Y169–Y218	
humanPrP (1QLX)		Y169–R164.(N)NH2
(1QLZ)		
(1QM0/1/2/3)		
(2LSB)	F175–Y218	Y128–R164.(N)NH2,F198–R156.(N)NH2,
		H155–R136.(N)NH2
bovinePrP (1DWY)		Y128–R164.(N)NH2
(1DWZ)		
(1DX0/1)		
SyrianHamsterPrP (1B10)	Y169–F175–Y218	
(2LH8)	Y169–F175–Y218	
caninePrP (1XYK)		
catPrP (1XYJ)		Y150–R156.(N)NH2
sheepPrP (1UW3)	F141–Y150,Y169–F175–Y218	
mousePrP[N174T] (1Y15)	F141–Y150,Y169–F175–Y218	Y128–R164.(N)NH2
humanPrP[Q212P]-M129 (2KUN)		H237–R228.(N)NH2
rabbitPrP[S173N]-NMR (2JOH)		Y127–L124.(N)N,Y144–R147.(N)NH2
rabbitPrP[I214V]-NMR (2JOM)	H139–Y149	Y148–R155.(N)NH2
rabbitPrP[S170N]-X-ray (4HLS)	Y169–F175	F141–R208.(N)NH2
rabbitPrP[S174N]-X-ray (4HMM)	Y169–F175	F141–R208.(N)NH2
rabbitPrP[S170N,S174N] (4HMR)	Y169–F175	F141–R208.(N)NH2

(continued)

Table 9.2 (continued)

Species	π-π-stacking	π-cation
mousePrP - at 37°C (2L39)		Y169-R164.(N)NH2
mousePrP[V166A] (2KFO)	Y169–F175	
mousePrP[D167S] (2KU5)	F175–Y218	
mousePrP[D167S,N173K] (2KU6)	F175–Y218,H187–F198	
mousePrP[Y169G] (2L1D)	F141–Y150,F175–Y218,Y225–Y226	Y128–R164.(N)NH2
mousePrP[Y169A] (2L40)	W145–Y149,H187–F198	
mousePrP[S170N] (2K1O)	Y225–Y226	
mousePrP[S170N,N174T] (1Y16)		Y169–R164.(N)NH2
mousePrP[F175A] (2L1E)	Y163–Y218	F141–R208.(N)NH2
mousePrP[Y225A,Y226A] (2KFM)	Y169–F175–Y218	
mousePrP[Y169A,Y225A,Y226A] (2L1K)		
elkPrP (1XYW)	Y169–F175–Y218	
pigPrP (1XYQ)		
bankVolePrP (2K56)	Y169–F175–Y218	Y169–R164.(N)NH2
tammarWallabyPrP (2KFL)		F198–R156.(N)NH2
rabbitPrP-NMR (2FJ3)	F140–Y149	Y127–L124.(N)N
rabbitPrP-X-ray (3O79)	Y169–F175	
horsePrP (2KU4)		F198–R156.(N)NH2
humanPrP-pH7 (1HJM) (1HJN)	H187–F198,Y218–F175–Y169	F141–R208.(N)NH2
humanPrP(118–224) (4N9O)	F175–Y218	Y128–R164.(N)NH2
humanPrP-V129 (3HAK)	Y225–Y226,F175–Y218	F141–R208.(N)NH2
humanPrP-M166C/E221C (1H0L)	F175–Y218–Y169	
humanPrP-M166V (1E1G)		Y225–R228.(N)NH2,Y128–L125.(N)N
(1E1J)		Y128–L125.(N)N
humanPrP-S170N (1E1P)	H187–F198	
(1E1S)	F141–Y150	T128–R164.(N)NH2
humanPrP-D178N (2K1D)	Y128–Y162,Y150–F141–	F141–R208.(N)NH2
humanPrP-D178N-M129 (3HEQ)	Y225–Y226,F175–Y218	
humanPrP-D178N-V129 (3HJX)		
humanPrP-F198S-M129 (3HES)	F175–Y218	F141–R208.(N)NH2
humanPrP-F198S-V129 (3HER)	F175–Y218,Y225–Y226	
humanPrP-R200K (1FKC)	F175–Y218,Y150–Y157	Y162–L125.(N)N
(1FO7)	F175–Y218,Y150–Y157	Y162–L125.(N)N
humanPrP-V209M-M129 (2M8T)		
humanPrP-V210I-M129 (2LEJ)	Y150–Y157	H187–R136.(N)NH2
(2LV1)	Y225–Y226	F141–R208.(N)NH2,W99=K101.(N)NZ
humanPrP-E219K-M129 (2LFT)	Y157–Y198,F175–Y218	
humanPrP-R220K (1E1U)	Y225–Y226	H155–R136.(N)NH2
(1E1W)	Y225–Y226,Y218–F175–Y169	H155–R136.(N)NH2
chickenPrP (1U3M)	Y235–Y238,F148–W156	
turtlePrP (1U5L)	F141–W150,Y162–Y188,Y166–F176	
xenopusLeavisPrP (1XU0)	Y149–Y153	
sheepPrP-H168 (1XYU)	F175–Y218,F141–Y150,H187–F198	Y128–R164.(N)NH2–Y169
ovinePrP-R168 (1Y2S)	Y150–Y157,H187–F198	

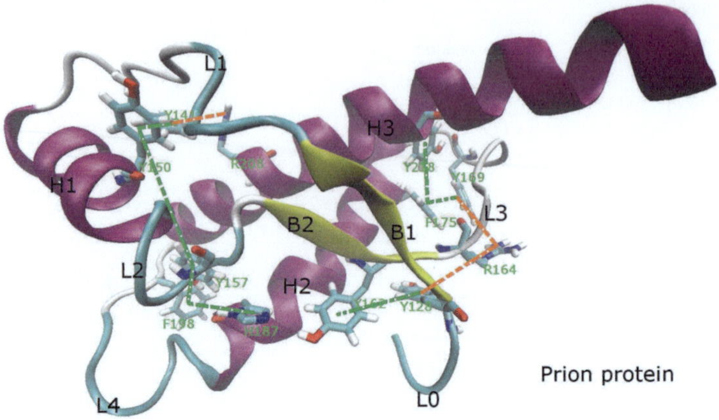

Fig. 9.16 Some selected π-interactions of PrPs, where the green dashed-lines denote the π-π stackings and the orange colored dashed-lines denote the π-cations

π-cations (Fig. 9.16): F141–R208.(N)NH2 (linking L1–H3), Y162–L125.(N)N–Y128–R164.(N)NH2–Y169 (linking B2–L0–B1–L3), F198–R156.(N)NH2 (linking L4–L2), H155–R136.(N)NH2 (linking L2–L1), Y127–L125.(N)N (in L0), etc. The above bioinformatics might be acted as a "quick reference card" for PrP protein structure π-interaction studies [701].

Chapter 10
The Hydrophobic Region PrP(109–136)

10.1 Introduction

Prion diseases such as CJD in humans and BSE or 'mad-cow' disease in cattle are invariably fatal neurodegenerative diseases. Prions differ from conventional infectious agents in being highly resistant to treatments that destroy the nucleic acids found in bacteria and viruses. The infectious prion is thought to be an abnormally folded isoform (PrP^{Sc}) of a host protein known as the prion protein (PrP^{C}). The conversion of PrP^{C} to PrP^{Sc} occurs post-translationally and involves conformational change from a predominantly α-helical protein to one rich in β-sheet amyloid fibrils. Much remains to be understood about how the normal cellular isoform of the prion protein PrP^{C} undergoes structural changes to become the disease associated amyloid fibril form PrP^{Sc}. The hydrophobic domain of PrP^{C}(109–136) is highly conserved, containing a palindrome and the repeats of the GxxxG protein-protein interaction motif (two glycines separated by any three residues; please note that the minimum number of residues to form fibrils should be 5 [67]). It is reported that the palindrome AGAAAAGA is an inhibitor/blocker of prion diseases [67, 288] and the Glycine-xxx-Glycine motif GAVVGGLGGYMLG is also an inhibitor of prion diseases [102, 274, 373]. The alterations of residues in AGAAAAGA and GAVVGGLGGYMLG will drastically affect the ability of cells and lead to the amyloid fibril formations (e.g. A117V will cause the GSS prion disease, and the numerous mutants in [274]). Our computational results also confirm the amyloid fibril formation ability of the PrP(109–136) hydrophobic region (Fig. 10.1). As shown in Fig. 10.1, if energy is less than the threshold energy −26 kcal/mol then amyloid fibril is formed in the corresponding region of residues, thus, the palindrome segment PrP(113–120) and the GLGGY segment PrP(124–128) can be firmly confirmed having a strong amyloid fibril formation property. This chapter will give detailed reviews on the PrP(109–136) region from the 3D molecular structure (Fig. 10.2) point of view and presents the studies of its molecular structural dynamics. The rest of this chapter is arranged as follows. In Sect. 10.2, the

J. Zhang, *Molecular Structures and Structural Dynamics of Prion Proteins and Prions*, Focus on Structural Biology 9, DOI 10.1007/978-94-017-7318-8_10

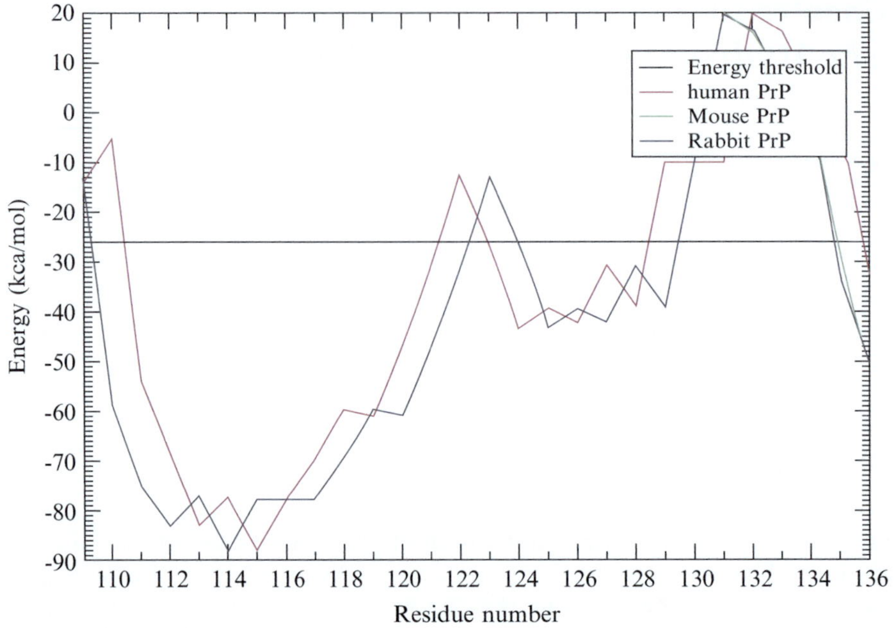

Fig. 10.1 Identifying the amyloid fibril formation property in PrP(109–136) region by the fibril prediction program of [695]

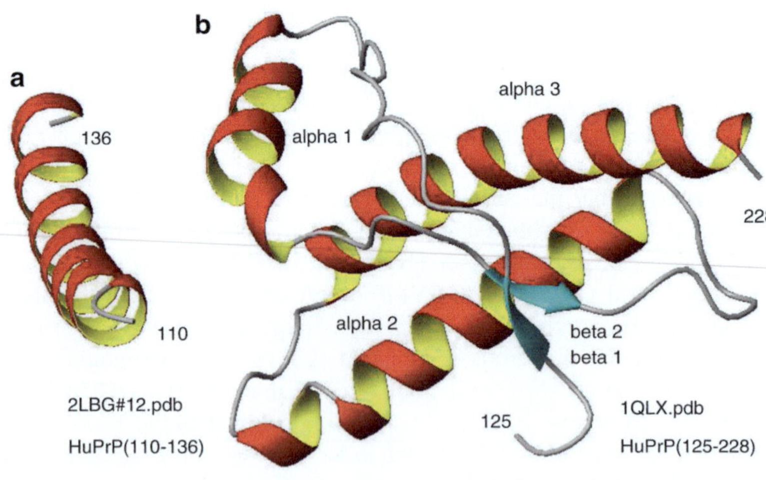

Fig. 10.2 The 3D molecular structure of human PrP. (**a**) The 3D molecular NMR structure of human PrP(110–136) in dodecylphosphocholine micelles. (**b**) The 3D molecular NMR structure of human PrP(125–228)

analysis of the Glycine-xxx-Glycine motif GAVVGGLGGYMLG inhibiting prion diseases will be done from the molecular structural point of view. Section 10.3 will make some concluding remarks on the hydrophobic PrP(109–136) region.

10.2 Structural Studies on the PrP(119–131) GAVVGGLGGYMLG

Some 30 segments from the Alzheimer's amyloid-β (Aβ) and tau proteins, the PrP prion protein, insulin, etc form amyloid-like fibrils, microcrystals that reveal steric zipper structures [517]. Harrison et al (2006) reported there are similarities between Aβ and PrP in the segment of the three GxxxG repeats (where both Aβ and PrP have the crucial residue Methionine located in the middle (GxMxG) of the last repeat) [30, 273] that controls prion formation [274]. Harrison et al (2010) used cell biological approaches of investigating numerous mutants in this region to reveal the mechanism of prion inhibition, and mutagenesis studies demonstrate that minor alterations to this highly conserved region of PrP^C drastically affect the ability of cells to uptake and replicate prion infection in both cell and animal bioassay [274]. This section presents some explanations for the biological experimental performance of [274] from the molecular structural point of view.

First, we do the alignments of the structured region of mouse, human, dog, rabbit, horse and elk PrP^C (with PDB IDs 1AG2, 1QLX, 1XYK, 2FJ3, 2KU4, 1XYW respectively) (Fig. 10.3). The PrP(125–136) 3D structures of human, dog, rabbit, horse and elk superposed onto mouse are with backbone-atom-RMSD values

```
CLUSTAL 2.1 multiple sequence alignment

2FJ3_rabbit.pdb     ------LGGYMLGSAMSRPLIHFGNDYEDRYYRENMYRYPNQVYYRPVDQYSNQNSFVHD 54
1XYW_elk.pdb        --VVGGLGGYMLGSAMSRPLIHFGNDYEDRYYRENMYRYPNQVYYRPVDQYNNQNTFVHD 58
1XYK_dog.pdb        --VVGGLGGYMLGSAMSRPLIHFGNDYEDRYYRENMYRYPDQVYYRPVDQYSNQNNFVRD 58
2KU4_horse.pdb      GSVVGGLGGYMLGSAMSRPLIHFGNDYEDRYYRENMYRYPNQVYYRPVSEYSNQKNFVHD 60
1AG2_mouse.pdb      -----GLGGYMLGSAMSRPMIHFGNDWEDRYYRENMYRYPNQVYYRPVDQYSNQNNFVHD 55
1QLX_human.pdb      ------LGGYMLGSAMSRPIIHFGSDYEDRYYRENMHRYPNQVYYRPMDEYSNQNNFVHD 54
                          *************:****.*:*********:***:******:.:*.**:.**:*
                    901234567890123456789012345678901234567890123456789012345678
                   119        130       140       150       160       170
                               β1                 α1            β2

2FJ3_rabbit.pdb     CVNITVKQHTVTTTTKGENFTETDIKIMERVVEQMCITQYQQESQAAYQRA-- 105  (-1)
1XYW_elk.pdb        CVNITVKQHTVTTTTKGENFTETDIKMMERVVEQMCITQYQRESEAYYQRGAS 111
1XYK_dog.pdb        CVNITVKQHTVTTTTKGENFTETDMKIMERVVEQMCVTQYQKESEAYYQRGAS 111
2KU4_horse.pdb      CVNITVKQHTVTTTTKGENFTETDVKIMERVVEQMCITQYQKEYEAFQQRGAS 113
1AG2_mouse.pdb      CVNITIKQHTVTTTTKGENFTETDVKMMERVVEQMCVTQYQKESQAYY----- 103
1QLX_human.pdb      CVNITIKQHTVTTTTKGENFTETDVKMMERVVEQMCITQYERESQAYYQR--- 104
                    *****:******************:*:*********:***::* :*
                    90123456789012345678901234567890123456789012345678901
                   180       190       200       210       220       230
                         α2                          α3
```

Fig. 10.3 The alignments of the structured region of mouse, human, dog, rabbit, horse and elk proteins with PDB IDs 1AG2, 1QLX, 1XYK, 2FJ3, 2KU4, 1XYW respectively. The residues of PrP^C(125–136) were underlined

2.579427, 2.228940, 2.745877, 2.532690, 2.877734 Å, respectively. In Fig. 10.3, the alignment of sequences was generated by the online CLUSTAL 2.1 program at *www.ebi.ac.uk/Tools/msa/clustalw2/*, where "*" means that the residues or nucleotides in that column are identical in all sequences in the alignment, ":" means that conserved substitutions have been observed, "." means that semi-conserved substitutions are observed, the red colored residues are Small (small+ hydrophobic (incl.aromatic -Y)), the blue colored residues are Acidic, the green colored residues are Hydroxyl + sulfhydryl + amine + G, and the gray colored residues are Unusual amino/imino acids etc; the residue numbers are from 119 to 231 (for rabbit the numbers are 1 less than others); the last column of numbers denotes the number of residues accounted from 1; the PrP 125–136 LGGYMLGSAMSR, β-strand 1 ($\beta\mathbf{1}$), α-helix 1 ($\alpha\mathbf{1}$), β-strand 2 ($\beta\mathbf{2}$), α-helix 2 ($\alpha\mathbf{2}$) and α-helix 3 ($\alpha\mathbf{3}$) of a PrP structure (Fig. 10.2b) were underline denoted.

In Fig. 10.3, we can see the PrP(125–136) LGGYMLGSAMSR is highly conserved among all species, strongly suggesting it has functional and evolutionary significance. To understand the functions of proteins at a molecular level, in protein structures, the non-covalent interactions such as HBing, ionic interactions (SB), vdW forces, and HYDs are driving the proteins to be able to perform their biological functions (*en.wikipedia.org/wiki/Protein_structure*). Thus, in the below, we will investigate the HBs, SBs, vdWs, HYDs in the structures and their structural dynamics of LGGYMLGSAMSR of human and rabbit, GLGGYMLGSAMSR of mouse, VVGGLGGYMLGSAMSR of elk and dog, GSVVGGLGGYMLGSAMSR of horse (where S120 is special for horse, instead of A120) and structural connections with other residues/loops/sheets/helices in the C-terminal. Mutations will destroy these non-covalent interactions that well maintain the structure so the function of the prion protein. This will give clear explanations to the mutants in the Glycine-rich region of [108, 274], which affect the uptake of prion infectivity very much.

In 2010, horses were reported to be resistant to prion diseases [336]. First, we analyze the role of GSVVGGLGGYMLGSAMSR (PrP(119–136)) in horse PrP^C(119–231) (PDB entry 2KU4). In our study, the MD simulation conditions are at 350 K in explicit solvent under neutral and low pH environments, heatings are using the Langevin thermostat algorithm in constant NVT ensembles and the equilibrations and productions are using Langevin thermostat algorithm in constant NPT ensembles. We know the following HBs of GSVVGGLGGYMLGSAMSR (PrP(119–136)):

ARG136-TYR157 (linking β1-to-α1 loop with α1-to-β2 loop, 68.27 %, 71.04 %),
MET134-ASN159 (linking β1-to-α1 loop with α1-to-β2 loop, 29.83 %, 21.47 %),
GLY131-GLN160 (linking β1 with β2, 26.67 %, 37.03 %),
SER132-GLN217 (linking β1 with α3, 29.70 %, 12.90 %),
ARG136-PRO158 (linking β1-to-α1 loop with α1-to-β2 loop, 9.53 %, 6.01 %),
ARG136-TYR157 (linking β1-to-α1 loop with α1-to-β2 loop, 68.27 %, 71.04 %),
GLY126-ARG164 (linking bend before β1 with β2-to-α2 loop, 33.31 %[1]),
TYR128-ASP178 (linking coil before β1 with α2, 11.81 %[1]),
SER120-LEU125 (in the peptide, 9.72 %[1]),

LEU125-10TYE128 (in the peptide, 5.58 $\%^2$),
GLY119-4VAL122 (in the peptide, 6.51 $\%^2$),
GLY127-ARG164 (link bend before $\beta1$ with $\beta2$-to-$\alpha2$ loop, 5.87 $\%^2$),

where the first percentage is for seed1 ($\%^1$) and the second percentage is for seed2 ($\%^2$) and the two seeds mean two different initial velocities of 30 ns MD, and the following HYDs of GSVVGGLGGYMLGSAMSR (PrP(119–136)):

In PrP(119–136) of 2LBG.pdb, between the two adjacent residues there are always occupied by HYDs with rate of 100 %, except for between GLY123 and GLY124, and between GLY126 and GLY127,
Among the residues in PrP(119–136), there are HYDs in

TYR128–LEU130 (where LEU130 is a residue in $\beta1$),
VAL121–GLY119, GLY123, TYR128, MET129, LEU130 (where MET129 and LEU130 are in $\beta1$),
VAL122–SER120, GLY124, LEU125, TYR128,
GLY123–SER120, LEU125,
TYR128–GLY126,
MET129–GLY131 (where both MET129 and GLY131 are in $\beta1$),
MET134–SER132, ARG136 (where SER132 is in $\beta1$),

The HYDs between a residue in PrP(119–136) and a residue in PrP(137–231):

SER135–PRO137 (where SER135, PRO137 is in $\beta1$-to-$\alpha1$ loop),
ARG136–MET154, TYR157 (where ARG136 is in $\beta1$-to-$\alpha1$ loop, MET154 and TYR157 are in $\alpha1$-to-$\beta2$ loop),
ASN159–SER135, ALA133, ARG136, MET134 (where ASN159 is in $\alpha1$-to-$\beta2$ loop),
TYR162–TYR128, MET129, LEU130 (where MET129, LEU130 are in $\beta1$, and TYR162 is in $\beta2$, there is no HB between TYR162 and LEU130),
GLN217–GLY131 (where is in $\beta1$, GLN217 is in $\alpha3$),
ARG164–GLY127, MET129, TYR128 (where MET129 is in $\beta1$, ARG164 is in $\beta2$-to-$\alpha2$ loop),
TYR163–TYR128, LEU130, MET129 (where MET129, LEU130 are in $\beta1$, TYR163 is in $\beta2$),
GLN162–ALA133, LEU130, MET134, GLY131, SER132 (where ALA133, LEU130, GLY131, SER132 are in $\beta1$, GLN162 is in $\beta2$),
VAL161–GLY131, LEU130 (where GLY131, LEU130 are in $\beta1$, VAL161 is in $\beta2$),
MET213–MET134 (MET213 is in $\alpha3$, MET134 is in the $\beta1$-to-$\alpha1$ loop),

with more than 50 % occupancy rate over the long MD trajectory of 30 ns. The mutations listed in [274] will break these HBs and HYDs to lose the inhibition to prion diseases of horses.

Rabbits are also resistant to infection from prion diseases from some species [381, 390, 591, 625, 626] and the outbreak of "mad rabbit disease" is unlikely [106, 203]. Nisbet et al (2010) reported there is 87 % sequence homology between mouse PrP and rabbit PrP, approximately 9 of the 33 (i.e. 1/3) of the difference in the region DGRRSSSTV of mouse and QRAAGVL of rabbit, and residues surrounding the glycosylphosphatidylinositol anchor attachment site of PrP modulate prion infection [441]. Under the MD simulation conditions of 450 K in explicit solvent, neutral and low pH environments, with heatings using the Langevin thermostat algorithm in constant NVT ensembles and the equilibrations and productions using Langevin thermostat algorithm in constant NPT ensembles, we have found rabbit PrP has the C-terminal residue R227 forming a HB/SB network with inner residues and the beginning residues of rabbit homology structure PrP(120–229) (6EPA.pdb) and NMR structure PrP(124–228) (2FJ3.pdb) but mouse PrP has no such an Arginine residue at the end of C-terminal owning this property.

Dogs were reported in 2008 to be resistant to infection from prion diseases from other species [474]. In our recent work, MD studies were done under 450 K in explicit solvent, neutral and low pH environments, with heatings using the Langevin thermostat algorithm in constant NVT ensembles and the equilibrations and productions using Langevin thermostat algorithm in constant NPT ensembles. In the hydrophobic region 121–136 of dog PrP, MD studies find that there are strong HBs S132–Y163–Y128 linking the two antiparallel β-strands, and strong HYDs M134–A133, M129–L130, V121–V122 residing in the core of the hydrophobic region 121–136. We should note that residue 129 for dog PrP is L129 (www.uniprot.org/uniprot/O46501) but for many others is M129.

For the hydrophobic region PrP(109–136), we should also notice the following points:

(1) the segment GYMLGS or GYVLGS of HuPrP(127–132) can form Class 8 antiparallel amyloid fibrils (3NHC.pdb and 3NHD.pdb) [17, 517],
(2) the mutants A117V and M129V cause GSS [100, 165, 418] and CJD [342, 593] respectively,
(3) the β-sheet core of PrP^{Sc} consists of three layers of β-strands E1(116–119), E2(129–132) and E3(160–164) [175], where E1 and E2 are in the hydrophobic region (109–136),
(4) H111 is a residue of copper binding sites [276, 321, 343, 495],
(5) Y128 is in the center of HB-and-SB-network of HB(Y128–D178), SB(D178–R164), HB(Y128–R164), HB(Y128–H177), HB(H177-N154) [13, 27, 28, 246],
(6) A133 and S132 have HBs with R220 and a water binding site with G131 [180, 345], in PrP(113–132) the hydrophobic cluster with vdWs rendering of atoms in residues 113–127 interacts with the first β-strand (see Fig. 10.2b of [313]), M129 makes interactions with the side chain of V122 and pulls the N-terminus into the β-sheet [13] and M129 is very close to Y163 [30],
(7) conservation of the Gly-rich region PrP(119–131) is required for uptake of prion infectivity [274, 465, 669] is shown by the physical or chemical properties of numerous mutations in the PrP(119–131) Gly-rich region [274],
(8) one O-linked sugar at Ser135 can affect the coil-to-β structural transition of the prion peptide, but at Ser132 the effect is opposite [98], etc., and
(9) the NMR structure of HuPrP(110–136) in dodecylphosphocholine micelles was known (2LBG.pdb) [516] and we do MD simulations for it as follows.

We used the ff03 force field of the AMBER 11 package [91], in a neutral pH environment. The systems were surrounded with a 12 Å layer of TIP3PBOX water molecules and with 2 Cl- ions added using the XLEaP module of AMBER 11. The templates used are the 2LBG.pdb from the Protein Data Bank and its 17 mutants at G114V, A117V, G119A, G119L, G119P, A120P, G123A, G123P, G124A, L125A, G126A, G127A, G127L, M129V, G131A, G131L, G131P, which are got by the mutate module of the free package Swiss-PdbViewer Version 4.1.0 (spdbv.vital-it.ch). These 18 models were firstly optimized by SD method and then CG method for 2 stages. Minimization Stage 1 is holding the solute fixed with a force constant of 500.0 kcal mol^{-1} A^{-2} for 500 steps of SD minimization followed by 500 steps of CG minimization. Minimization Stage 2 is minimizing the entire system for 1000 steps

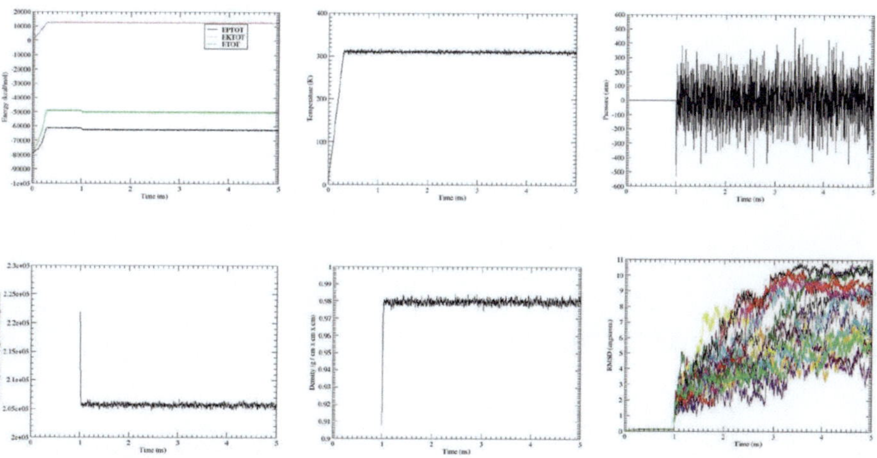

Fig. 10.4 Variations of the potential energy (EPTOT), kinetic energy (EKTOT), total energy (ETOT), temperature, pressure, volume, density, and RMSDs during the 5 ns of MD simulations for all the models. In the RMSD graph, different colors stand for HuPrP(110–136) and its 17 mutants

of SD minimization followed by 1500 steps of CG minimization. The minimized models were checked by Swiss-PdbViewer and found there is not any bad contact that makes amino acids clashed. Then, the solvated proteins were quickly heated from 0 K to 310 K linearly for 300 ps and then systems were kept at 310 K for 700 ps. The systems were in constant NVT ensembles using Langevin thermostat algorithm with weak restraints (a force constant of 10.0 kcal mol^{-1} A^{-2}) on the solvated proteins. The SHAKE and PMEMD algorithms with nonbonded cutoffs of 12 Å were used. Next, the systems were done MD simulations in constant NPT ensembles (with constant pressure 1 atm and constant temperature 310 K) under the Langevin thermostat for 4 ns and the PRESS, VOLUME (DENSITY) and RMSDs were sufficiently stable for each of the 18 models (Fig. 10.4). A step size of 2 fs was used for all the MD simulations, the structures were saved to file every 1000 steps and the Metropolis criterion was used. These MD simulation conditions are completely consistent with the experimental work of NMR structure of HuPrP(110–136) (2LBG.pdb).

We picked out the snapshots at 0, 1, 2, 3, 4 and 5 ns for each model (Fig. 10.5) and found the α-helix structure of each model has been unfolding and longer MD simulations might make the α-helix structure unfolded completely (Figs. 10.5∼10.6). This shows that the structure in PrP(110–136) region is structurally unstable and might be critical to the conversion from the predominantly α-helical PrP^{C} into the rich β-sheet PrP^{Sc}.

The preliminary findings to explain the above performance during the 5 ns of MD simulations may be described as follows. At the beginning of the simulations, the 3D structures of all the 13 models are α-helices (Figs. 10.5∼10.6). with rich HBs. In an ideal α-helix, there are 3.6 residues per complete rotating so a rotation of 100

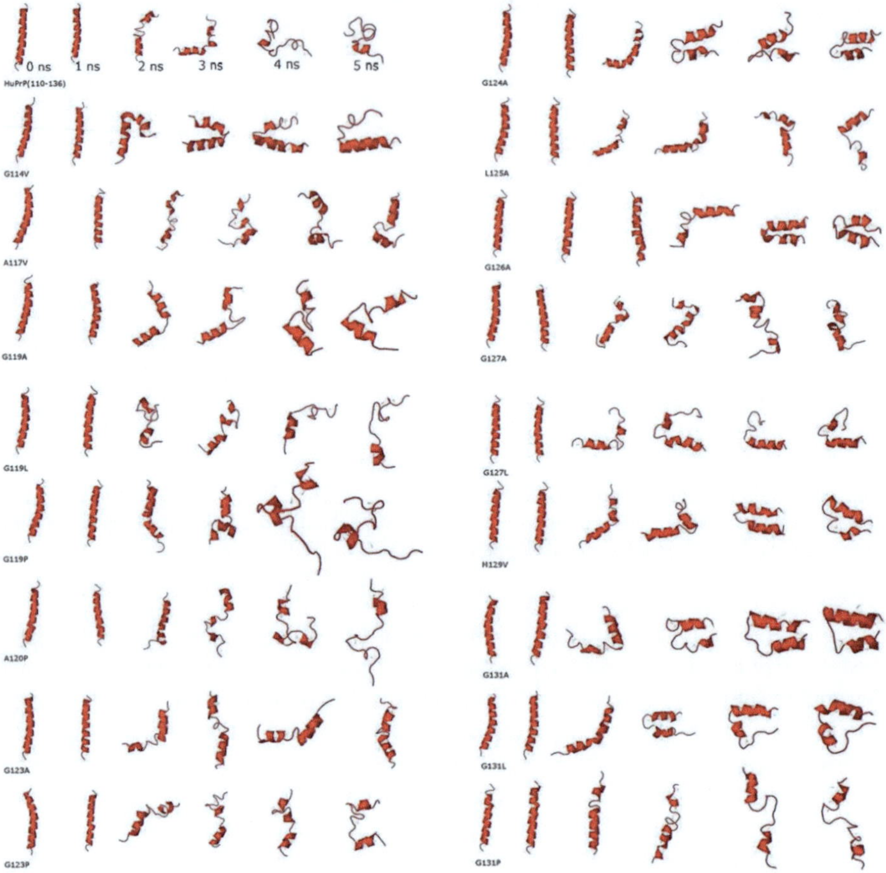

Fig. 10.5 The respective snapshots at 0, 1, 2, 3, 4 and 5 ns of the MD simulations for the 18 models. The *dashed lines* denote HBs

degrees per residue. Thus, in all the 13 models, there are about 7 turns (Fig. 10.5) in each α-helix. A α-helix is maintained by HBs. Clearly, the disappearance of HBs is a reason for the unfolding of all the 13 models. As the progress of MD simulations (Fig. 10.6), most of them have disappeared, except for the three main HBs GLY/ALA/LEU/PRO131–SER135–ARG136, TYR128–SER132, which have different occupied rate for each model (Table 10.1) and are just in the N-terminal of the structural region of HuPrP(125–228) (1QLX.pdb). This might show that PrP(110–124) region of PrP(110–136) (2LBG.pdb) is very unstable. We also found that there is a SB between HIS111–LYS110 with the occupied rate 100 % for the 13 models. Specially for the mutants G127L, M129V, G131A, and G131L, there is another SB between HIS111–136ARG linking the head and tail of HuPrP(110–136). Seeing the snapshots of 3, 4 and 5 ns of mutant G127L in Fig. 10.5, 4 and 5 ns of mutant M129V in Fig. 10.5, 3, 4 and 5 ns of mutant G131A

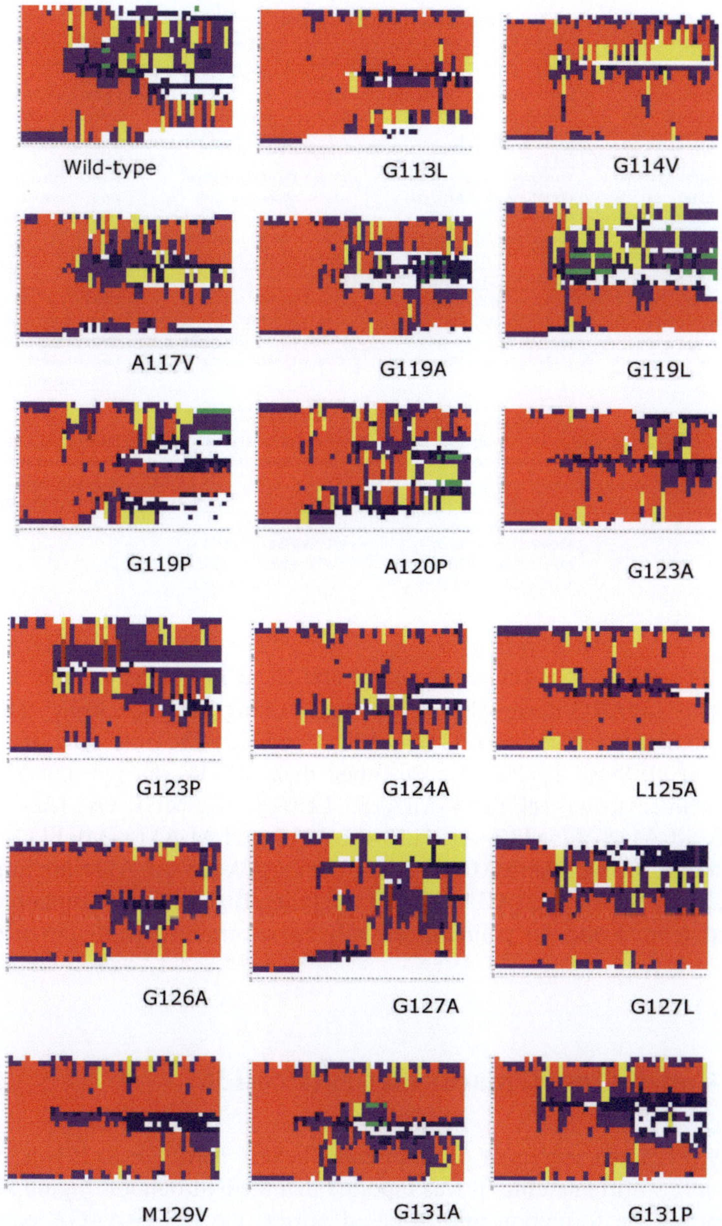

Fig. 10.6 Variations of the secondary structures of the 18 models. H is the α-helix, B is the residue in isolated β-bridge, E is the extended strand, G is the 3-helix or 3/10 helix, I is the π-helix, T is the HBed turn and S is the bend as described in the DSSP program [326]

Table 10.1 Occupied rate (%) of the HBs for each model

HuPrP(110–136)	G114V	A117V	G119A
131@O–135@OG.HG (25.48)	131@O–135@OG.HG (26.28)	131@O–135@OG.HG (40.52)	131@O–135@OG.HG (49.24)
125@O–135@OG.HG (17.68)	135@O–136@NH1.HH11 (21.72)	128@O–132@OG.HG (8.88)	135@O–136@NH1.HH11 (14.6)
117@O–136@NH1.HH11 (7.32)	129@O–132@OG.HG (7.76)	135@O–136@NH1.HH11 (8.12)	128@O–132@OG.HG (5.68)
116@O–136@NH1.HH12 (5.76)	132@O–135@OG.HG (7.72)	132@O–135@OG.HG (7.36)	
G119L	G119P	A120P	G123A
135@O–136@NH1.HH11 (23.96)	131@O–135@OG.HG (31.24)	135@O–136@OG.HG (28.08)	131@O–135@OG.HG (28.08)
131@O–135@OG.HG (22.8)	135@O–136@NH1.HH11 (24.4)	132@O–135@OG.HG (8.6)	135@O–136@NH1.HH11 (17.76)
132@O–135@OG.HG (15.08)	128@O–132@OG.HG (10.08)	135@O–136@NH1.HH11 (7.72)	128@O–132@OG.HG (6.16)
	130@O–135@OG.HG (7.2)	112@O–111@ND1.HD1 (5.6)	
		128@O–132@OG.HG (5.04)	
G123P	G124A	L125A	G126A
131@O–135@OG.HG (25.52)	131@O–135@OG.HG (41.4)	131@O–135@OG.HG (36.24)	131@O–135@OG.HG (46.36)
135@O–136@NH1.HH11 (21.92)	135@O–136@NH1.HH11 (21.4)	135@O–136@NH1.HH11 (17.92)	135@O–136@NH1.HH11 (14.12)
	128@O–132@OG.HG (5.56)	132@O–135@OG.HG (8.48)	128@O–132@OG.HG (7.04)
	134@O–136@NH1.HH11 (5.44)	128@O–132@OG.HG (5.6)	133@O–136@NE.HE (5.6)
			132@O–135@OG.HG (5.08)
G127A	G127L	M129V	G131A
131@O–135@OG.HG (26.2)	135@O–136@NH1.HH11 (23.6)	131@O–135@OG.HG (34.4)	131@O–135@OG.HG (47.76)
135@O–136@NH1.HH11 (14.88)	131@O–135@OG.HG (21.2)	135@O–136@NH1.HH11 (13.2)	128@O–132@OG.HG (17.08)
112@O–111@ND1.HD1 (13.08)	133@O–135@OG.HG (8.04)	128@O–132@OG.HG (8.92)	135@O–136@NH1.HH11 (13.48)
			132@O–136@NH1.HH11 (7.72)
			134@O–111HIS@ND1.HD1 (6.2)
	G131L	G131P	
	131@O–135@OG.HG (70.4)	131@O–135@OG.HG (32.48)	
	135@O–136@NH1.HH11 (18.4)	135@O–136@NH1.HH11 (16.72)	
	128@O–132@OG.HG (7.88)	128@O–132@OG.HG (15.2)	

in Fig. 10.5, and 3, 4 and 5 ns of mutant G131L in Fig. 10.5, we may know the SB HIS111–136ARG makes these snapshots looking like a "hairpin". HIS111 is a very important residue in PrP(110–136). Along with the unfolding of α-helical structure of all these 13 models, we found many HYDs disappeared except for some fundamental ones MET134–ALA133, LEU130–129MET, VAL122–VAL121-ALA120, ALA118–ALA117–ALA116–ALA115, and ALA113–MET112 with the occupied rate 100 %, where ALA118–ALA117–ALA116–ALA115 are in the core of the palindrome AGAAAAGA and this might imply to us the hydrophobic core is very hard to break and this palindrome really has enormous potential to be amyloid fibrils.

10.3 Concluding Remarks on PrP(109–136)

To really reveal the secrets of prion diseases is very hard. For us it is a long shot but certainly worth pursuing. It was reported that the hydrophobic region PrP(109–136) controls the formation into diseased prions: the AGAAAAGA palindrome and Glycine-xxx-Glycine repeats (both being the inhibitor of prion diseases) are just in this region. This chapter gives some investigations and explanations on the PrP(109–136) region in view of its 3D structures and molecular dynamics studies. The structural bioinformatics presented in this chapter can be acted as a reference in 3D images for laboratory experimental works. This presents some clue or hints for the author to study prion proteins and prions.

10.4 Prion AGAAAAGA Amyloid Fibril Structures

Protein misfolding into structures such as amyloids is a key feature of many disorders of humans. Neurodegenerative diseases, including more than 20 kinds of Parkinson's disease, Alzheimer's disease (Aβ), Huntington's disease, Parkinson's disease, Prions' diseases, β-Secretase, etc., are protein misfolding diseases. Neurodegenerative diseases are also called conformational diseases in that a functional native or mutant protein undergoing a conformational transition will lead to protein aggregation or formation of amyloid fibrils. This conformational transition may be attempted to study by MD techniques. We also use QM/MM approaches (a small active site / substrate region is treated with QM with the rest of the system treated with classical MM) to predict the structure and relative stability of mutant proteins associated with neurodegenerative diseases. Neurodegeneratively diseased amyloid fibrils are unstable, noncrystalline and insoluble so that their structures are very difficult to be determined in laboratory by experimental X-ray crystallography, NMR spectroscopy, dual polarization interferometry. MM (Molecular Modeling) skills such as mathematical optimization can allow us to construct many amyloid fibril models for neurodegenerative diseases and unstructured regions of proteins. This is to say MD, QM/MM, MM computing can allow us to study protein misfolding neurodegenerative diseases with the amyloid fibril structures. Particularly, this section will study the PrP(113–120) region (AGAAAAGA) of prion diseases to get its MD and MM computational molecular structures. Because there is little X-ray or NMR structural data available to date on AGAAAAGA, the studies of this chapter may be useful for the goals of medicinal chemistry.

Firstly, in computational theory, by the fibril prediction program [695] we can identify that prion AGAAAAGA (PrP(113–120)) owns the amyloid fibril formation property (see Fig. 10.7).

Secondly, the prion AGAAAAGA region has also been identified to own the amyloid fibril formation property in the following laboratory works (in each year):

- Abskharon RN, Giachin G, Wohlkonig A, Soror SH, Pardon E, Legname G, Steyaert J (2014) Probing the N-terminal β-sheet conversion in the crystal structure of the human prion protein bound to a nanobody. J Am Chem Soc 136(3):937–944.
- Cheng HM, Tsai TWT, Huang WY, Lee HK, Lian HY, Chou FC, Mou Y et al (2011) Steric zipper formed by hydrophobic peptide fragment of Syrian hamster prion protein. Biochem 50:6815–6823.
- Jones EM, Wu B, Surewicz K, Nadaud PS, Helmus JJ, Chen S, Jaroniec CP et al (2011) Structural polymorphism in amyloids: new insights from studies with Y145Stop prion protein fibrils. J Biol Chem 286:42777–42784.
- Wagoner VA, Cheon M, Chang I, Hall CK (2011) Computer simulation study of amyloid fibril formation by palindromic sequences in prion peptides. Proteins 79:2132–2145.
- Lee SW, Mou Y, Lin SY, Chou FC, Tseng WH, Chen CH, Lu CY et al (2008) Steric zipper of the amyloid fibrils formed by residues 109–122 of the Syrian hamster prion protein. J Mol Biol 378:1142–1154.
- Sasaki K, Gaikwad J, Hashiguchi S, Kubota T, Sugimura K, Kremer W, Kalbitzer HR et al (2008) Reversible monomer-oligomer transition in human prion protein. Prion 2:118–122.

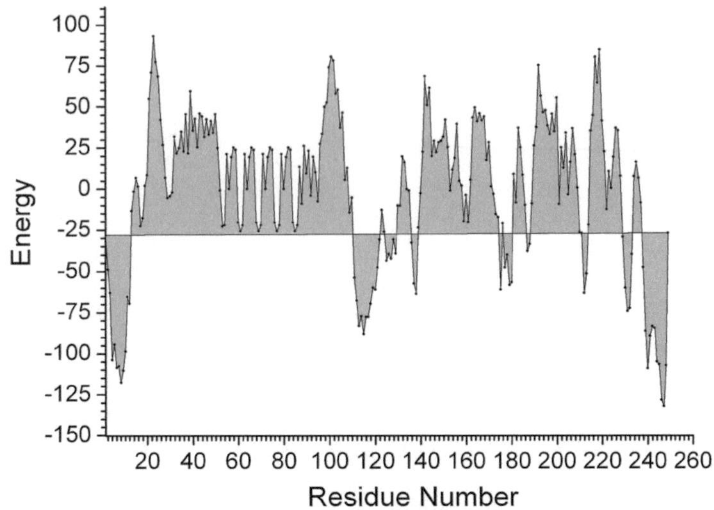

Fig. 10.7 Identifying the amyloid fibril formation property PrP (113–120) AGAAAAGA region is identified to own the amyloid fibril formation property

- Sawaya MR, Sambashivan S, Nelson R, Ivanova MI, Sievers SA, Apostol MI, Thompson MJ et al (2007) Atomic structures of amyloid cross-beta spines reveal varied steric zippers. Nature 447:453–457.
- Kajava A, Squire JM, Parry DAD (2006) Fibrous Proteins: Amyloids, Prions and Beta Proteins, Advances in Protein Chemistry, vol 73. Elsevier.
- Haigh CL, Edwards K, Brown DR (2005) Copper binding is the governing determinant of prion protein turnover. Mol Cell Neurosci 30:186–196.
- Norstrom EM, Mastrianni JA (2005) The AGAAAAGA palindrome in PrP is required to generate a productive PrP^{Sc}-PrP^{C} complex that leads to prion propagation. J Biol Chem 280:27236–27243.
- Govaerts C, Wille H, Prusiner SB, Cohen EE (2004) Evidence for assembly of prions with left-handed beta-helices into trimers. Proc Natl Acad Sci USA 101:8342–8347.
- Kourie JI, Kenna BL, Tew D, Jobling MF, Curtain CC, Masters CL, Barnham KJ et al (2003) Copper modulation of ion channels of PrP[106–126] mutant prion peptide fragments. J Membr Biol 193:35–45.
- Zanuy D, Ma B, Nussinov R (2003) Short peptide amyloid organization: stabilities and conformations of the islet amyloid peptide NFGAIL. Biophys J 84:1884–1894.
- Ma BY, Nussinov R (2002) Molecular dynamics simulations of alanine rich β-sheet oligomers: insight into amyloid formation. Protein Sci 11:2335–2350.
- Wegner C, Romer A, Schmalzbauer R, Lorenz H, Windl O, Kretzschmar HA (2002) Mutant prion protein acquires resistance to protease in mouse neuroblastoma cells. J Gen Virol 83:1237–1245.
- Kourie JI (2001) Mechanisms of prion-induced modifications in membrane transport properties: implications for signal transduction and neurotoxicity. Chem Biol Interact 138:1–26.
- Brown DR (2000) Prion protein peptides: optimal toxicity and peptide blockade of toxicity. Mol Cell Neurosci 15:66–78.
- Holscher C, Delius H, Burkle A (1998) Overexpression of nonconvertible PrPC delta114–121 in scrapie-infected mouse neuroblastoma cells leads to trans-dominant inhibition of wild-type PrP^{Sc} accumulation. J Virol 1998, 72:1153–1159.

- Jobling MF, Stewart LR, White AR, McLean C, Friedhuber A, Maher F, Beyreuther K et al (1999) The hydrophobic core sequence modulates the neurotoxic and secondary structure properties of the prion peptide 106–126. J Neurochem 73:1557–1565.
- Chabry J, Caughey B, Chesebro B (1998) Specific inhibition of in vitro formation of protease-resistant prion protein by synthetic peptides. J Biol Chem 273:13203–13207.
- Gasset M, Baldwin MA, Lloyd DH, Gabriel JM, Holtzman DM, Cohen F, Fletterick R et al (1992) Predicted alpha-helical regions of the prion protein when synthesized as peptides form amyloid. Proc Natl Acad Sci USA 89:10940–10944.

Thirdly, we answer the question "What does the prion amyloid fibril structure in AGAAAAGA look like?" Gasset et al. (2004) [254] reported there are similarities between the PrP sequence AGAAAAGA and that of silkworm fibroin, and the homology between PrP sequence AGAAAAGAVVGGLGG and that of spider fibroin. Thus, we may say that the hydrophobic region of PrP(109–136) should be a region to form β-sheets and amyloid polymers, instead of α-helices of the Garnier-Robson analysis [234]. Our MD and MM (where the β-sheet structure is maintained by vdWs, HYDs and HBs in 3D protein structures, which can be represented as mathematical optimization problems of LJ potential energy and HB potential energy) computations show that prion AGAAAAGA amyloid fibrils should have the Class 7 (β-strand antiparallel, face=back, up–up) or Class 1 (β-strand parallel, face-to-face, up–up) structure of [102, 408, 517].

In conclusion, this chapter solves a problem on the amyloid fibril structures of prion AGAAAAGA region; it has an amyloid fibril β-sheet structure Class 1 or 7. This shows that MD, QM/MM, MM computations play a very important role to reveal the protein misfolding conformational diseases.

10.5 Recent Research Advances in the Mammalian PrP Glycine-xxx-Glycine Motif

Prion diseases such as CJD in humans and bovine spongiform encephalopathy ('mad-cow' disease) in cattle are invariably fatal neurodegenerative diseases. Prions differ from conventional infectious agents in being highly resistant to treatments that destroy the nucleic acids found in bacteria and viruses. The infectious prion is thought to be an abnormally folded isoform (PrP^{Sc}) of a host protein known as the prion protein (PrP^C). The highly conserved glycine-zipper region PrP^C(119–131) of mammalian prion proteins, consisting of 3 repeats GAVVGGLGGYMLG of the GxxxG protein-protein interaction motif (two glycines separated by any three residues), plays a crucial role in the formation or conversion of PrP^{Sc} from PrP^C. This section will briefly review recent research advances in this motif.

This motif contains part of the β-strand 1 (i.e. $\beta 1$: amino acids 128–131) and a PrPctm putative transmembrane domain (amino acids 112–135) [30, 108]. A portion of the PrP^C(118–135) exhibits a nonfibrillar property and in vivo cytotoxicity, and PrP amyloid formation may be associated with self-assembly by the GxxxG motif of prion protein [108].

- In 2006, Barnham et al. (2006) [30] reported both Aβ and PrP have three GxxxG repeats, the crucial residue methionine (Aβ35, PrP129) located in the middle (GxMxG) of the last repeat, and disruption of GxxxG motifs will alter properties of Aβ and PrP [30]. In 2006, Choi et al. (2006) [108] generated a monoclonal antibody 1C5 (IgG1) recognized by the GxxxG motif of PrPC.
- In 2007, Harrison et al. (2007) [273] summarized the GxxxG PrPctm motif: "these motifs are commonly found in TM α-helices where they act to allow close packing and binding between helices. The GxxxG motif of PrP is highly conserved and antibodies raised against this region detect PrP from many mammalian species. Both PrP and β-amyloid peptide, causative agent of Alzheimer's disease, contain this motif within regions that are capable of crossing the membrane, thus presenting a potential common mode of action for these two proteins which are both involved in neurodegenerative diseases" and pointed out "G131V disrupts one of the GxxxG motifs found within the hydrophobic region, leading to the possibility that this motif may be relevant to the actions of TM-PrP".
- In 2010, Harrison et al. (2010) [274] reported that "mutagenesis studies demonstrate that minor alterations to this highly conserved region of PrPC(119–131) drastically affect the ability of cells to uptake and replicate prion infection in both cell and animal bioassay" and concluded that "these residues provide conformational flexibility".
- In 2014, Coleman et al. (2014) [153] studied mutations G114V and A117V lied next to the glycine rich region of PrPC(119–131) that can abrogate prion infection and concluded that "small, protease sensitive prion species have a significant role in the progression of prion disease and that the hydrophobic domain is an important determinant of PrP conversion". Seeing Fig. 10.6, we know that, because of the mutations, the structure will unfolded / misfolded ("broken") at amino acids 114 and 117 respectively after 1 ns (where the PRESS, VOLUME (DENSITY) and RMSD in the MD were sufficiently stable to reach the equilibrations in the NPT systems). The A117V mutation breaks the strong hydrophobic core linkage ALA118–ALA117–ALA116–ALA115.

In conclusion, we might be able to say that the PrP glycine rich region GAVVGGLGGYMLG can abrogate prion infection and it should be an important determinant of PrP conversion.

10.6 A Survey on the (Central) Hydrophobic Region of PrP

10.6.1 Introduction

The "mad cow" disease (i.e. Bovine Spongiform Encephalopathy (BSE)) is a contagious type of Transmissible Spongiform Encephalopathies (TSEs). It can be easily transmitted to human beings by consumption of contaminated food from animal products. Research indicates that the BSE can be caused by prions (the misfolding prion proteins), although a lot work needs to be done before a resolution for the "mad cow disease". A prion is not a virus, not a bacteria, nor any microorganism. As a result, prion diseases cannot be caused by the vigilance of the organism immune system as it can freely spread from one species to another species, such as from animals to humans. For example, the human version of "mad cow" diseases are named CJD and vCJDs. Unfortunately, such diseases in human beings could happen randomly through infections of transplanted tissue or blood transfusions or consumption of infected animal products. A number of animals such as cats, minks, deer, elks, moose, sheep, goats, nyalas, oryxes, greater kudus, ostriches and many other species are also susceptible to TSEs. The origin of such TSEs remains unknown so far.

However, it has been reported that the hydrophobic region 109–136 of prion proteins (PrP) controls the formation of affected prions. The hydrophobic region of the PrP(109–136) is found to be the control the formation of the diseased prions (PrP^C). Furthermore, the normal PrP(113–120) AGAAAAGA palindrome is likely an inhibitor/blocker of prion diseases [67, 288]. In addition, the highly conserved Glycine-xxx-Glycine motif PrP(119–131) is able to inhibit the formation of infectious prion proteins in cells [102, 274, 373]. In this section, we first provide a detailed survey focusing on experimental evidences and findings in laboratories (not on the computers) of the hydrophobic region of PrP (listed in the PubMed of NCBI http://www.ncbi.nlm.nih.gov/pubmed).

10.6.2 A Detailed Review on PrP Central Hydrophobic Region

The highly and evolutionarily conserved PrP hydrophobic region has been considered essential to PrP conformational conversion. This section does a survey of the research work on *prion hydrophobic region* listed in the PubMed database mainly from the molecular structure point of view.

- 2015:
  - Huang & Caflisch (2015) reported that the evolutionary/strictly conserved Tyr169 stabilizes the $\beta 2$–$\alpha 2$ loop of the prion protein through MD simulations of the wild-type MoPrP and its Y169G mutant [293].
  - Ning et al. (2015) did REMD of the monomer, 2- and 4-peptide systems of PrP(113–120) and its A117V mutant and provided structural insights into the impacts of the A117V mutation on the folding and assembly of the palindrome [440].
- 2014:
  - Chu et al. (2014) equipped Δ105–125, ΔCR_PrP (i.e. the PrP missing central hydrophobic region, a variant that is known to be highly neurotoxic in transgenic mice) with a C-terminal membrane anchor via a semisynthesis

strategy and found the importance of the central hydrophobic domain in the membrane anchor in PrP-lipid interactions [148].
- Daskalov et al. (2014) studied the HET-s prion (which owns a β-solenoid with a triangular hydrophobic core) and identified a region that modulates prion formation [168, 515]. Solid-state NMR data showed the hydrophobic core of HET-s(218–289) is rigid [364].
- Xu et al. (2014) presented the structures of prion-like MAVS and found there are "electrostatic interactions between neighboring strands and hydrophobic interactions within each strand" [651].
- Baral et al. (2014) did promazine binding to mouse PrP and found the "binding induces structural rearrangement of the unstructured region proximal to β1, through the formation of a 'hydrophobic anchor'" and promazine "stabilizes the misfolding initiator-motifs such as the C terminus of α2, the α2-α3 loop, as well as the polymorphic β2-α2 loop" [26].
- Coleman et al. (2014) studied the murine homologues (G113V and A116V, which lie in the hydrophobic domain of PrP) and concluded that "the hydrophobic domain is an important determinant of PrP conversion" [153].
- Mays et al. (2014) studied the C1 cleavage that occurs amino-terminal of PrP^C's hydrophobic domain [419].

- 2013:
  - Wang et al. (2013) used [Au(bpy)Cl_2]PF_6 and [Au(dien)Cl]Cl_2 (where bpy is 2,2'-bipyridine and dien is diethylenetriamine), six prion peptides with either a His111-mutated or a Met109/112-mutated residue to investigate interaction and peptide aggregation [603].
  - Pimenta et al. (2013) described the NMR structure of ovine prion-like Doppel peptide (1–30) (PDB ID: 2M1J) and the interaction of this peptide with the conserved SRP54M hydrophobic groove [471].
  - The central domain (CD, 95–133) of PrP^C comprises the charge cluster (CC, 95–110) and a hydrophobic region (HR, 112–133); Vilches et al. (2013) reported that among CC, HR and CD only the CD peptide is neurotoxic and this peptide is able to activate caspase-3 and disrupt the cell membrane, leading to cell death [585].
  - "Among mutants spanning the region 95–135, a construct lacking solely residues 105–125 ($\triangle$105–125) had distinct properties when compared with the full-length prion protein 23–231 or other deletions" [461].
  - Zweckstetter (2013) reported that "prion stop mutants that accumulate in amyloidogenic plaque-forming aggregates fold into a β-helix" and residue 129 is located in the hydrophobic core of the β-helix, the trimer interface of a trimeric left-handed β-helix model is formed by residues L125, Y128 and L130 [723].

- 2012:
  - Lau et al. (2012) demonstrated that Shadoo (Sho) binds DNA and RNA in vitro via the arginine-rich (or alanine-rich) hydrophobic region of five tandem

A/LAAG amino residue repeats R1-R5 (including tandem positively charged "RGG boxes"), where is the most homologous region of Sho and PrP (but Sho sequences showed variability in the number of alanine residues) [170, 366, 547].

- Zhao et al. (2012) reported that "two fixed missense mutations (102Ser→Gly and 119Thr→Ala), and three missense mutations (92Pro > Thr/Met, 122Thr > Ile and 139Arg > Trp) in the coding region presenting different ($P < 0.05$) genotypic and allelic frequency distributions between cattle and buffalo" [708].
- The 14–3–3 beta protein (highly abundant in brain, a biomarker for sCJD) interacts with the central hydrophobic amino acids 106–126 of prion protein [314].
- The α-cleavage within the central hydrophobic domain not only disrupts a region critical for both PrP toxicity and PrP^C to PrP^{Sc} conversion but also produces the N1 fragment that is neuroprotective and the C1 fragment that enhances the proapoptotic effect of staurosporine in one report and inhibits prion in another [385].
- Zhang et al. (2012) found a new haplotype in a Sunite sheep and the sheep of Inner Mongolia in China have several haplotypes with the similar results of Stewart and Daude's: SPRN (shadowof prion protein homology) contained an alanine-rich sequence, which is homologous to a hydrophobic core with amyloidgenic characteristics in PrP [705].
- Sauve et al. (2012) published the NMR structure of HuPrP(110–136) in dodecylphosphocholine (DPC) micelles (PDB ID: 2LBG) [516].

• 2011:

- Biljan et al. (2011) published the NMR structure of HuPrP-M129 mutant V210I (85–231) (PDB ID: 2LEJ) including the unstructured N-terminal part (residues 90–124) [56].

  Julien et al. (2011) reported the relative and regional stabilities of the hamster, mouse, rabbit, and bovine PrPs towards urea unfolding, and also investigated the effect of the S174N mutation in rabbit PrP^C [324].
- Shi et al. (2011) reported that point-mutations within the hydrophobic transmembrane region increase the amount of ^{Ctm}PrP (a kind of PrP^{Sc}) in cells, such as human homologue A117V which is associated with GSS and G114V associated with gCJD, while the mutations outsides transmembrane region do not [526].

• 2010:

- Wang et al. (2010) reported that "the hydrophobic domain deletion mutant significantly weakened the hydrophobic rPrP-lipid interaction and abolished the lipid-induced C-terminal PK-resistance", "both disease-associated P105L and P102L mutations, localized between lysine residues in the positively charged region, significantly affected lipid-induced rPrP conversion" and "the hydrophobic domain localized 129 polymorphism altered the strength of hydrophobic rPrP(recombinant mouse PrP)-lipid interaction" [617].

- Ile et al. (2010) published the NMR structure of HuPrP-M129 mutant Q212P (PrP(90–231)) (PDB ID: 2KUN) and concluded that the Q212P mutation caused GSS syndrome might be due to the disruptions of the hydrophobic core consisting of $\beta 2$–$\alpha 2$ loop and $\alpha 3$ helix [301].
- Biasini et al. (2010) revealed through the deleting residues 114–121 (△114–121) that the hydrophobic core region governs mutant prion protein aggregation and intracellular retention [52].
- Harrison et al. (2010) reported that there are similarities between Aβ (Alzheimer's amyloid-β) and PrP in the segment of the three GxxxG repeats (where both Aβ and PrP have the crucial residue Methionine located in the middle (GxMxG) of the last repeat) that controls prion formation, and found that minor alterations to this highly conserved region of PrP^C drastically affect the ability of cells to uptake and replicate prion infection in both cell and animal bioassay [274].
- Oliveira-Martins et al. (2010) reported that "PrP^C undergoes extensive proteolysis at the α site (109K↓H110)" and "C1 prevalence was unaffected by variations in charge (PrP(90–110)) and hydrophobicity (PrP(110–130)) of the region neighbouring the α-cleavage site, and by substitutions of the residues in the palindrome that flanks this site, instead, α-cleavage was size-dependently impaired by deletions within the domain PrP(106–119)" [449].
- The hydrophobic core (HC), a transmembrane domain, harbors the most highly conserved regions of PrP^C. A mutation in HC is associated with prion disease resulting in an enhanced formation of a transmembrane form (^{Ctm}PrP). Lutz et al. (2010) created a set of mutants carrying microdeletions of 2-8 amino acids within HC between position 114 and 121 (△114–121), and showed that these mutations display reduced propensity for transmembrane topology and HC might function as recognition site for the protease(s) responsible for PrP^C α-cleavage [404]. They also found that the mutant G113V displayed increased ^{Ctm}PrP topology and decreased α-cleavage in their in vitro assay and concluded that HC represents an essential determinant for transmembrane PrP topology, whereas the high evolutionary conservation of this region is rather based upon preservation of PrP^C α-cleavage, thus highlighting the biological importance of this cleavage.
- Valensin et al. (2010) reported that HuPrP possesses two copper binding sites localized at His-96 and His-111 in the so called "amylodogenic" or neurotoxic region (residues 91–126) and chicken PrP possesses a similar region (PrP(105–140)) containing two His (His-110 and His-124) and an identical hydrophobic tail of 15 amino acids rich in Ala and Gly [579].
- Turi et al. (2010) studied HuPrP(91–115), HuPrP(76–114) – H85A and HuPrP(84–114) – H96A and found that His96 predominates almost completely for Ni^{2+} ions, while His85 and His111 predominate Cu^{2+} ions [162, 323, 575].

- 2009:
  - Julien et al. (2009) reported that "strong hydrophobic interactions between helices $\alpha 1$ and $\alpha 3$, and between $\alpha 2$ and $\alpha 3$, stabilize these regions even at very high concentrations of urea" but this result is for PrP structural region 121–230 of bovine [325].
  - Fei et al. (2009) reported Preceding with a hydrophobic residue cysteine, instead of a charged residue, "caused the rate of assembly into fibrils to increase greatly for both peptides and full-length Ure2p", and concluded that "disulfide bond formation significantly accelerates the assembly of Ure2p fibrils because of the proximity of a potential amyloid stretch" [202].
  - Tseng et al. (2009) reported that hydrophobic does not play a major role in the unfolding of α-helix 1 of PrP^C [572] but In the mouse PrP (PDB ID: 1AG2) there are nine hydrophobic residues and five non-polar Glycines in $\beta 1$-strand: GLGGYMLGSAMSRPMIHFGN(PrP(124–142)), five hydrophobic residues in $\beta 2$-strand: N153MYRYPNQV YYRPVD167 (PrP(153–167)).
- 2008:
  - Margittai et al.(2008) found that, to maximize the hydrophobic contact surface, most fibrils (that contain an extensive core region of about 20 amino acids or more) share a common parallel in-register arrangement of β-strands [415].
  - Ciccotosto et al. (2008) reported "the major region of neurotoxicity has been localized to the hydrophobic domain located in the middle of the PrP sequence" [150].
- 2007:
  - Owen et al. (2007) reported "the protease thermolysin cleaves at the hydrophobic residues Leu, Ile, Phe, Val, Ala, and Met, residues that are absent from the protease accessible aminoterminal region of PrP^{Sc}" [454].
  - Ott et al. (2007) reported "the presence of small hydrophobic amino acids such as Val and Ile was insufficient to promote re-orientation. Only Met and Leu, large hydrophobic amino acids, could promote transmembrane domain inversion" [452].
  - Sakudo et al. (2007) reported that neurotoxic signals of aged PrP(106–126) are mediated by N-terminal half of the hydrophobic region (HR) but not the octapeptide repeat (OR) of PrP [511].
  - Berti et al. (2007) reported the copper(II) binding result of the 91–120 region of HuPrP: "room-temperature NMR spectroscopy data were consistent with the binding site centered on His111 being approximately fourfold stronger than that centered on His96, low-temperature EPR spectroscopy results yielded evidence for the opposite trend", and the MD results showed that "Met112 approaching the metal at room temperature, a process that is expected to stabilize the His111-centered binding site through hydrophobic shielding of the metal coordination sphere" [49].

  - Harrison et al. (2007) summarized the GxxxG PrPctm motif GAVVGGLGGYMLG and made a mutation G131V leading to the possibility that this motif may be relevant to the actions of TM-PrP [273].

- 2006:

  - Lee et al. (2006) found that "the OR (octapeptide repeat region, PrP(51–90)) and N-terminal half of HR (hydrophobic region, PrP(112–145)) of PrP retains anti-apoptotic activity similar to full-length PrP" [374].

- 2005:

  - Sakudo et al. (2005a) reported that "OR and N-terminal half of the HR were required for the inhibitory effect of PrP(113–132) but not STI1 (stress-inducible protein 1) pep.1" and "mediate the action of STI1 upon cell survival and upregulation of SOD (superoxide dismutase) activity" [509]. In 2005, Sakudo et al. (2005b) also reported that "Removal of the OR (PrP(53–94)) enhances apoptosis and decreases SOD activity. Deletion of the N-terminal half of HR (PrP(95–132)) abolishes its ability to activate SOD and to prevent apoptosis, whereas that of the C-terminal half of HR (PrP(124–146)) has little if any effect on the anti-apoptotic activity and SOD activation" [510].
  - Gaggelli et al. (2005) did NMR and EPR studies on the interaction of the HuPrP(106–126) (KTNMKHMAGAAAAGAVVGGLG) with copper(II), manganese(II) and zinc(II) at pH 5.7 and concluded that "the hydrophobic C-terminal region was not affecting the copper-binding (of His111) properties of the peptide and this hydrophobic tail is left free to interact with other target molecules" [218].

- 2004:

  - Kuznetsov et al. (2004) reported that "PrP(114–125) and the C terminus of helix B may be considered as primary candidates for sites involved in conformational transition from PrP^C to PrP^{Sc}" and "most PrP mutations associated with neurodegenerative disorders increase local hydrophobicity" [359].
  - Ott et al. (2004) reported that "the N terminal and hydrophobic regions of the signal sequence affect (membrane) integration (of priton protein) most significantly. Mutations in either region result in a considerable increase in the number of chains that integrate" [453].
  - Haire et al. (2004) reported that the strand of residues 129–131 in sheep PrP crystal structure is involved in lattice contacts about a crystal dyad to generate a four-stranded intermolecular β-sheet between neighbouring molecules [268].

- 2003:

  - Susceptibility to scrapie is primarily controlled by polymorphisms in the ovine PrP gene (PRNP) and Seabury et al. (2003) reported a polymorphism of ovine

PrP at Proline 116, flanking the cleavage site between Lys112 and His113, and A117V (causing Gerstmann-Strausler-Scheinker syndrome) [521].

- Saez-Cirion et al. (2003) reported that the hydrophobic internal region 130GAVVGGLGGYMLGSAMSR147 of bovine PrP shares structural and functional properties with HIV type 1 fusion peptide [508]. The 121KHVAGAAAAGAVVGGLGGYMLGSAMSR147 transmembrane region (BovinePrP(tm)) has mainly a helical structure but also containing some random coil (upon addition of calcium, the random coils disappear while the helical conformation remains) [508]. Amyloid-type fibers PrP^{Sc} will be formed, if in the absence of membrane prion sequence [508].
- Tcherkasskaya et al. (2003) reported that hydrophobic interactions between side chains of the peptide variants PrP(104–122) (containing a polar head KTNMKN followed by a long hydrophobic tail MAGAAAAAGAVV) and PrP(109–122) prevent the formation of the rigid β-sheet structures [567].
- Premzl et al. (2003) reported that the alignment of the hydrophobic segment shows strong conservation across all PrPs and Shadoos; 12 of the 20 residues of the hydrophobic region are identical or almost identical, and another 6 are conserved hydrophobic [480].
- Cui et al. (2003) investigated the deletions of the hydrophobic domain (PrPΔ(112–119), PrP(112–136), PrP(135–150)) and found that the conserved hydrophobic core region is a critical domain for the activity of PrP [162].
- Kourie et al. (2003) reported that copper modulates the ion channels of PrP[106–126] mutant prion peptide fragments, the hydrophobic core AGAAAAGA is not a ligand Cu(2+) site but the Cu(2+)-binding site is located at M(109) and H(111) of prion fragment PrP[106–126] [352].
- In mouse PrP, the selected rabbit-specific mutations Asn99Gly, Leu108Met, Asn173Ser, and Val214Ile significantly interfere with the conversion of PrP^C to PrP^{Sc} [591].

- 2002:
  - Suzuki et al. (2002) reported that Fugu PrP-like has the PrP-conserved hydrophobic region of the Xenopus PrP (but doppel gene lacks PrP conserved hydrophobic region) [554].
  - The amino acid sequence AGAAAAGA, comprising residues 112–119 of the murine PrP^C, has been shown to be amyloidogenic and evolutionarily conserved. To assess the effect of mutations at and around this hydrophobic sequence on protease resistance, Wegner et al. (2002) replaced the sequence either by alanines or by glycines and, in a third mutant, a large part surrounding this region was removed. At last Wegner et al. (2002) concluded that "mutations in the central hydrophobic region lead to immediate alterations in PrP structure and processing" [619].
  - Chen et al. (2002) reported "one O-linked sugar at Ser135 can affect the coil-to-beta structural transition of the prion peptide" but at Ser132 the effect is opposite [98], where PrP^C(132–140) domain is closely related to the prion

propagation and 132–140 portion of PrP^C is a logical target for development of anti-prion drugs [464].

- 2001:
  - Dragani et al. (2001) showed that, in the hydrophobic core of all GST (glutathione S-transferase) and related protein (including the yeast prion protein Ure2), at the beginning of α6-helix of GST domain II, a N-capping box (S/TxxD) and an hydrophobic staple motif are strictly conserved in all GSTs and GST-related proteins [192, 336].
  - Holscher et al. (2001) reported that "deletion of each of the two hydrophobic regions in PrP (i.e. ΔPrP(112–121), ΔPrP(231–254)) revealed that the C-terminally located hydrophobic region (transmembrane2, PrP(213–254)) can function as second signal sequence in PrP" [287].
  - A direct correlation between the PrP amino acid sequence and TSE incubation time was demonstrated in transgenic mice expressing mouse PrP with amino acid substitutions Val111Met, Leu108Met, Pro101Leu [33, 34, 553].
  - Knaus et al. (2001) observed the N-terminal residues PrP(119–124) in the electron density of the crystal structure of HuPrP dimer (but the PrP(119–124) region is disordered in the NMR structure of PrP monomers) [344].
  - Laws et al. (2001) reported the solid-state NMR structure of MoPrP(89–143, P101L) [368].
- 1999:
  - Ragg et al. (1999) obtained the two-dimensional ^{1}H NMR structure of PrP(106–126) under the following solvent conditions: deionized water/2,2,2-trifluoroethanol 50:50 (v/v) and dimethylsulfoxide; and the data were analyzed by restrained MD calculations [489]. "In deionized water at pH 3.5, the peptide adopted a helical conformation in the hydrophobic region spanning residues Met112–Leu125, with the most populated helical region corresponding to the Ala115–Ala119 segment (about 10 %). In trifluoroethanol/H_2O, the α-helix increased in population especially in the Gly119–Val122 tract (about 25 %). The conformation of this region was found to be remarkably sensitive to pH, as the Ala120–Gly124 tract shifted to an extended conformation at pH 7. In dimethylsulfoxide, the hydrophobic cluster adopted a prevalently extended conformation. For all tested solvents the region spanning residues Asn108–Met112 was present in a 'turn-like' conformation and included His111, situated just before the starting point of the α-helix. Rather than by conformational changes, the effect of His111 is exerted by changes in its hydrophobicity, triggering aggregation. The amphiphilic properties and the pH-dependent ionizable side-chain of His111 may thus be important for the modulation of the conformational mobility and heterogeneity of PrP(106–126)" [489].
  - Kanyo et al. (1999) determined the X-ray crystallographic atomic-resolution structures of Syrian hamster PrP(104–113) binding Fab 3F4 [330]. "The

peptide binds in a U-shaped groove on the Fab surface, with the two specificity determinants, Met109 and Met112, penetrating deeply into separate hydrophobic cavities formed by the heavy and light chain complementarity-determining regions, and two intrapeptide HBs are observed" [330].
- Liu et al. (1999) reported some NMR information for the Syrian hamster PrP(90–123): "the N-terminus (residues 90–119) is largely unstructured despite some sparse and weak medium-range NOEs implying the existence of bends or turns. The transition region between the core domain and flexible N-terminus, i.e., residues 113–128, consists of hydrophobic residues or glycines and does not adopt any regular secondary structure in aqueous solution. There are about 30 medium- and long-range NOEs within this hydrophobic cluster, so it clearly manifests structure" in 1B10.pdb [391].

- 1997:
  - Smith et al. (1997) compared CD (circular dichroism) data from two peptides corresponding to the hydrophobic region PrP(106–136) which contained either methionine or valine at position 129 and concluded that "there was no detectable difference between the CD spectra of these peptides irrespective of the solvent conditions we used" [537].
- 1996:
  - Inouye et al. (1996) revealed the hydrophobic core in prion by X-ray diffraction as the beta-silk structure [307].
- 1995:
  - Priola et al. (1995) found that "a single hamster PrP amino acid Met138 blocks conversion to protease-resistant PrP in scrapie-infected mouse neuroblastoma cells" (while a mouse specific isoleucine Ile138 facilitated the conversion of mouse PrP^{C} into PrP^{Sc}) [483]. They also found homology at PrP(112–138) was required for mouse $PrP^{C} \rightarrow PrP^{Sc}$ conversion [432, 483].
- 1994:
  - De Fea et al. (1994) reported that "a charged, extracytoplasmic region, termed the Stop Transfer Effector (STE) sequence, has been shown to direct then ascent translocating chain to stop at the adjoining hydrophobic domain to generate the first membrane-spanning hydrophobic region PrP(112–143)" [173].
- 1993:
  - Harris et al. (1993) proposed that the cleavage is lying within a region of 24 amino acids that is identical in chicken PrP and mammalian PrP and is representing a major processing event that may have physiological as well as pathological significance [272].

- 1992:
  - Muramoto et al. (1992) found that ΔPrP(95–107), ΔPrP(108–121), and ΔPrP(122–140) inhibit the $PrP^C \rightarrow PrP^{Sc}$ conversion [437]. Kuwata et al. (2003) furthermore reduced the region of residues and found that mouse PrP(106–126) can form amyloid-like fibrils ("the fibrils contain $\approx 50\,\%$ β-sheet structure, and strong amide exchange protection is limited to the central portion of the peptide spanning the palindromic sequence VAGAAAAGAV") [357]. PrP(90–140) might be the region of interactions between PrP^C and PrP^{Sc} [586].
- 1987:
  - Liao et al. (1987) cloned and sequenced rat prion-related protein (PrP) cDNA, and showed a 23 amino acids hydrophilic region that extends to amino acid position 122, and the numerous G-G-G-X repeats, where X is a hydrophobic amino acid, may function in β-pleated sheet amyloid formation [387].
  - Bazan et al. (1987) analyzed the primary PrP sequence hydrophobicities to detect potential amphipathic regions and several hydrophobic segments, a proline- and glycine-rich repeat region and putative glycosylation sites are incorporated into a PrP model [41].

Throughout the above review on the recent research advances of PrP hydrophobic region, we noticed that hydrophobic region PrP(109–136) controls the formation of diseased prions PrP^{Sc} conformationally changed from PrP^C.

Part II
Molecular Modeling of Prions

Introduction

The non-covalent interactions of HB, SB, vdW, and HYD (Fig. 1) are driving the proteins to be able to perform their biological functions. In Part I, the author has found some SBs play a key role in prion diseases revealed from RaPrP. Neurodegenerative diseases are amyloid fibril diseases. Part II will study the HB, vdW, HYD interactions for MM of amyloid fibrils in view of GO computational algorithms. In the following chapters, the author will focus on the studies on MM structures of AGAAAAGA - the unstructured 113–120 region of prions, and vdWs, HBs will be used to construct the MM structures of AGAAAAGA.

"Neurodegeneration is the umbrella term for the progressive loss of structure or function of neurons, including death of neurons. Many neurodegenerative diseases including Parkinson's, Alzheimer's, and Huntington's occur as a result of neurodegenerative processes. As research progresses, many similarities appear which relate these diseases to one another on a sub-cellular level. Discovering these similarities offers hope for therapeutic advances that could ameliorate many diseases simultaneously" (en.wikipedia.org/wiki/Neurodegeneration). One similarity is they all featured amyloids, which "are insoluble fibrous protein aggregates sharing specific structural traits" (en.wikipedia.org/wiki/Amyloid). "Amyloid is characterized by a cross-β sheet quaternary structure" and "recent x-ray diffraction studies of microcrystals revealed atomistic details of core region of amyloid" (en.wikipedia.org/wiki/Amyloid and references [238, 245, 434, 439, 462, 517, 552, 636] therein). All the quaternary structures of amyloid cross-β spines can be reduced to the one of 8 classes of steric zippers of [517], with strong vdW interactions between β-sheets and HBs to maintain the β-strands.

A prion is a misshapen protein that acts like an infectious agent (hence the name, which comes from the words protein and infection). Prions cause a number of fatal diseases such as mad cow disease in cattle, scrapie in sheep and kuru and CJDs in humans. Prion diseases (being rich in β-sheets (about 43 % β-sheet) [85, 169, 257, 447, 457, 491]) belong to neurodegenerative diseases. Many

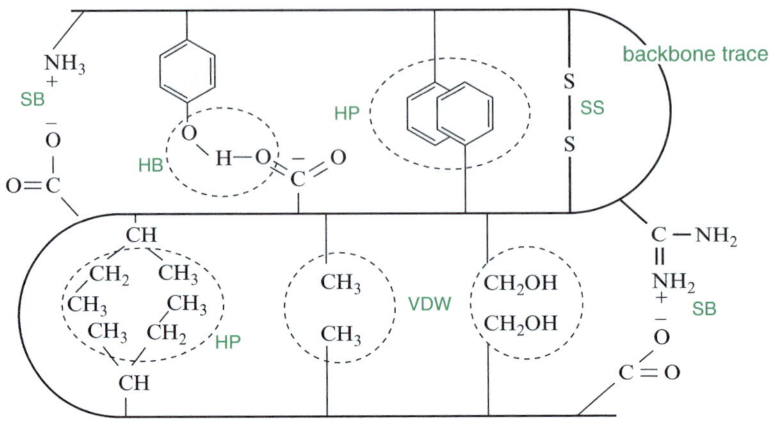

Fig. 1 The non-covalent HB, SB, vdW and HYD interactions to stabilize 3D structure of a protein

experimental studies such as [67, 68, 70, 85, 274, 288, 315, 316, 357, 445, 619] have shown that the normal hydrophobic region (113–120) AGAAAAGA of prion proteins is an inhibitor/blocker of prion diseases. PrP lacking this palindrome could not convert to prion diseases. However, to the best of the author's knowledge, there is little X-ray or NMR structural data available to date on AGAAAAGA, which falls just within the N-terminal unstructured region (1–123) of prion proteins, due to its unstable, noncrystalline and insoluble nature. This book studies the MM structures of this region of prions. Chapters 11, 12, 13, 14, 15, 16, and 17 will show the role of vdWs/HYDs and HBs in the molecular modeling of prion AGAAAAGA amyloid fibrils.

A new era in the structural analysis of amyloids started from the 'steric zipper'-β-sheets [439]. As the two sheets zip up, HYDs (& vdWs) have been formed. The extension of the 'steric zipper' above and below (i.e. the β-strands) is maintained by HBs (but there is no HB between the two β-sheets). This is the common structure associated with some 20 neurodegenerative amyloid diseases, ranging from Alzheimer's and type 2 diabetes to prion diseases. For prion AGAAAAGA amyloid fibril structure [327], basing on the common property of HYDs, vdWs, and HBs, we presents a series of computational GO algorithms in Chaps. 11, 12, 13, 14, 15, 16, and 17.

X-ray crystallography finds the X-ray final structure of a protein, which usually need refinements using a simulated annealing protocol in order to produce a better structure. thus, it is very amenable to use simulated annealing to format the models constructed. Chapter 11 presents a hybrid method of global search simulated annealing with local steepest descent, conjugate gradient search. Chapter 12 presents a hybrid method of simulated annealing with local discrete gradient search. Because the canonical duality theory in optimization can get global optimal solution(s) with zero duality gap, and the interactions of vdWs and HBs reach the optimal at the bottom of potential well, Chap. 13 presents a novel canonical dual global optimization search computational approach for solving molecular distance geometry problem (MDGP). Inspired by natural evolution such as inheritance, mutation, selection and crossover, genetic algorithm mimics the process of natural evolution. Chapter 12 also presents a hybrid method of genetic algorithm with local optimization searches; furthermore, Chap. 14 presents a hybrid method of evolutionary computations with simulated annealing, and the hybrid methods are applied to the computational potential energy minimization studies on the prion AGAAAAGA amyloid fibril molecular structures. At the end of Chap. 14, a new thinking about the construction of 3D-structure of a protein is given. Replica exchange (or so-called parallel tempering) can be used as a super simulated annealing; Chap. 15 presents a simulated annealing refined replica exchange global search algorithm. Chapter 16 presents a LBFGS quasi-Newtonian method for molecular modeling prion AGAAAAGA amyloid fibrils. The author found a package CALYPSO which has a successful promise in crystal structure determination/identification or in designing the multi-functional materials. This package was designed based on the particle swarm global search optimization strategy. Chapter 17 briefly presents the particle swarm global optimization search algorithm. All these global optimization computing methods presented in Chaps. 11, 12, 13, 14, 15, 16, and 17 are very practical and useful for the computational studies of the molecular structures of prion amyloid fibrils. Chapter 18 summarized three models for prion amyloid fibrils at AGAAAAGA region. Each chapter of Part II can also be read independently.

Chapter 11
The Hybrid Method of Steepest Descent: Conjugate Gradient with Simulated Annealing

11.1 Steepest Descent Local Search Method

"Steepest descent (SD) is based on the observation that if the real-valued function $f(x)$ is defined and differentiable in a neighborhood of a point x_0 then $f(x)$ decreases fastest if one goes from x_0 in the direction of the negative gradient of $f(x)$ at x_0" and SD local search method (en.wikipedia.org/wiki/Gradient_descent, [20, 435, 481]) converges fast. But this method is usually very difficult to reach real global minimums – at this moment, we may use the efficient conjugate gradient (CG) described as follows. The author also found a SD iteration method using a fixed iteration stepsize mathematical formula like the one described in the next section.

11.2 Conjugate Gradient Local Search Method

In this section, the author introduces his CG method designed with his collaborator Sun J. (National University of Singapore) [551]. Global convergence results are derived for well-known conjugate gradient methods in which the line search step is replaced by a step whose length is determined by a formula. The results include the following cases: (1) The Fletcher-Reeves method, the Hestenes-Stiefel method, and the Dai-Yuan method applied to a strongly convex LC^1 objective function; (2) The Polak-Ribiere method and the Conjugate Descent method applied to a general, not necessarily convex, LC^1 objective function.

J. Zhang, *Molecular Structures and Structural Dynamics of Prion Proteins and Prions*, Focus on Structural Biology 9, DOI 10.1007/978-94-017-7318-8_11

11.2.1 Overview

The conjugate gradient method is quite useful in finding an unconstrained minimum of a high-dimensional function $f(x)$ because it does not require the storage and computation of any matrices. In general, the method has the following form:

$$x_{k+1} = x_k + \alpha_k d_k \tag{11.1}$$

$$d_k = -g_k \quad for \quad k = 1, \tag{11.2}$$

$$-g_k + \beta_k d_{k-1} \quad for \quad k > 1, \tag{11.3}$$

where g_k denotes the gradient $\nabla f(x_k)$, α_k is a stepsize obtained by a line search or by a given formula, d_k is the search direction, and β_k is chosen so that d_k becomes the kth conjugate direction when the function is quadratic and the line search is exact. Varieties of this method differ in the way of selecting β_k. Some well-known formulae for β_k are given by

$$\beta_k^{FR} = \frac{||g_k||^2}{||g_{k-1}||^2}, \quad \text{(Fletcher-Reeves [209])} \tag{11.4}$$

$$\beta_k^{PR} = \frac{g_k^T(g_k - g_{k-1})}{||g_{k-1}||^2}, \quad \text{(Polak-Ribière [473])} \tag{11.5}$$

$$\beta_k^{HS} = \frac{g_k^T(g_k - g_{k-1})}{d_{k-1}^T(g_k - g_{k-1})}, \quad \text{(Hestenes-Stiefel [283])} \tag{11.6}$$

and

$$\beta_k^{CD} = \frac{||g_k||^2}{-d_{k-1}^T g_{k-1}}, \quad \text{(The Conjugate Descent Method [209])} \tag{11.7}$$

where $|| \cdot ||$ is the Euclidean norm and "T" stands for the transpose. Recently Dai and Yuan [164] also introduced a formula for β_k:

$$\beta_k^{DY} = \frac{||g_k||^2}{d_{k-1}^T(g_k - g_{k-1})}. \tag{11.8}$$

For ease of presentation we call the methods corresponding to (11.4), (11.5), (11.6), (11.7), and (11.8) the FR method, the PR method, the HS method, the CD method, and the DY method, respectively.

The global convergence of methods (11.4), (11.5), (11.6), (11.7), and (11.8) has been studied by many authors, including Al-Baali [11], Gilbert and Nocedal [243], Hestenes and Stiefel [283], Hu and Storey [292], Liu, Han and Yin [392], Powell [477], Touati-Ahmed and Storey [570], and Zoutendijk [722], etc. A common feature of the studies is to determine the value of α_k by a line search process. For examples, in [164] Dai and Yuan showed that both the CD method and the FR method are globally convergent if the following line search conditions for α_k are satisfied:

$$\begin{cases} f(x_k) - f(x_k - \alpha_k d_k) \geq -\delta \alpha_k g_k^T d_k, \\ \sigma_1 g_k^T d_k \leq g(x_k + \alpha_k d_k)^T d_k \leq -\sigma_2 g_k^T d_k, \end{cases} \tag{11.9}$$

where $0 < \delta < \sigma_1 < 1$ and $0 < \sigma_2 < 1$. In fact the above conditions represent a kind of generalized Wolfe line search [633]. Gilbert and Nocedal [243] conducted an elegant analysis on conjugate gradient methods and showed that by suitably selecting β_k the methods are globally convergent if α_k is determined by a line search step satisfying a Wolfe-like condition. Their work was inspired by a deep study of Powell [478] on the conjugate gradient methods. With the resolution of global convergence under line searches it is then interesting to ask the following question: What if the stepsize α_k is calculated by a given formula, as is done in the quadratic case, without using a line search procedure? In this section we use a steplength formula given by Wang [607] to substitute for the line search step and show that after such a revision the well-known conjugate gradient methods are still globally convergent.

11.2.2 The Proof for Convergence

We adopt the following assumption on function f which are commonly used in the literature.

Assumption 1. The function f is LC^1-smooth in a neighborhood N of the level set $L := \{x \in R^n | f(x) \leq f(x_1)\}$ and L is bounded. Here, by LC^1-smooth we mean that the gradient $\nabla f(x)$ is Lipschitz continuous with modulus μ, i.e., there exist $\mu > 0$ such that $||\nabla f(x_{k+1}) - \nabla f(x_k)|| \leq \mu ||x_{k+1} - x_k||$ for any $x_{k+1}, x_k \in N$.

Assumption 1 is sufficient for the global convergence of the PR method and the CD method, but seems to be not enough for the convergence of the FR method, the HS method, and the DY method without line search. Thus, for these three methods we impose the following stronger assumption.

Assumption 2. The function f is LC^1 and is strongly convex on N. In other words, there exists $\lambda > 0$ such that $(\nabla f(x_{k+1}) - \nabla f(x_k))^T (x_{k+1} - x_k) \geq \lambda ||x_{k+1} - x_k||^2$ for any $x_{k+1}, x_k \in N$.

Note that Assumption 2 implies Assumption 1 since a strongly convex function has bounded level sets.

We analyze conjugate gradient methods using the following steplength formula:

$$\alpha_k = -\delta g_k^T d_k / ||d_k||^2, \tag{11.10}$$

where $\delta \in (0, \kappa/\mu)$ and $\kappa \in (0, 1)$.

Lemma 11.2.1. *Suppose that x_k is given by (11.1), (11.3), and (11.10). Then*

$$g_{k+1}^T d_k = \rho_k g_k^T d_k \tag{11.11}$$

holds for all k, where

$$\rho_k = 1 \quad \textit{if} \quad \alpha_k = 0, 1 - \delta\phi_k \quad \textit{if} \quad \alpha_k \neq 0 \tag{11.12}$$

and

$$\phi_k = \frac{(g_{k+1} - g_k)^T (x_{k+1} - x_k)}{||x_{k+1} - x_k||^2}. \tag{11.13}$$

Proof. The case of $\alpha_k = 0$ is trivial. We now prove for the case of $\alpha_k \neq 0$. From (11.1) and (11.10) we have

$$\begin{aligned}
g_{k+1}^T d_k &= g_k^T d_k + (g_{k+1} - g_k)^T d_k \\
&= g_k^T d_k + \alpha_k^{-1} (g_{k+1} - g_k)^T (x_{k+1} - x_k) \\
&= g_k^T d_k + \alpha_k^{-1} \frac{(g_{k+1} - g_k)^T (x_{k+1} - x_k)}{||x_{k+1} - x_k||^2} ||x_{k+1} - x_k||^2 \\
&= g_k^T d_k + \alpha_k \phi_k ||d_k||^2 \\
&= g_k^T d_k - \left(\delta g_k^T d_k / ||d_k||^2\right) \phi_k ||d_k||^2 \\
&= (1 - \delta\phi_k) g_k^T d_k \\
&= \rho_k g_k^T d_k.
\end{aligned}$$

The proof is complete. □

Corollary 11.2.2. *In formulae (11.7) and (11.8) we have $\beta_k^{DY} = \beta_k^{CD} / (1 - \rho_{k-1})$. Under Assumption 1 we have $1 - \delta\mu \leq \rho_k \leq 1 + \delta\mu$ for all k, and under Assumption 2 we have $0 < \rho_k \leq 1 - \delta\lambda$ for all k with $\alpha_k \neq 0$.*

Proof. The first part is trivial. By Assumption 1 or 2 we have

$$||g_{k+1} - g_k|| \leq \mu ||x_{k+1} - x_k||$$

or

$$(g_k - g_{k-1})^T (x_k - x_{k-1}) \geq \lambda ||x_k - x_{k-1}||^2.$$

then from (11.1) and (11.13) we have

$$\rho_k = 1 - \delta\phi_k = 1 - \frac{\delta(g_k - g_{k-1})^T(x_k - x_{k-1})}{||x_k - x_{k-1}||^2},$$

which leads to the corresponding bounds for ρ_k. □

Lemma 11.2.3. *Suppose that Assumptions 1 (or Assumption 2) holds and that x_k is given by (11.1), (11.3), and (11.10). Then*

$$\sum_{d_k \neq 0} \frac{(g_k^T d_k)^2}{||d_k||^2} < \infty.$$

Proof. By the mean-value theorem we have

$$f(x_{k+1}) - f(x_k) = \bar{g}^T(x_{k+1} - x_k)$$

where $\bar{g} = g(\bar{x})$ for some $\bar{x} \in [x_k, x_{k+1}]$. Now by the Cauchy-Schwartz inequality, (11.1), (11.10), and Assumption 1 we obtain

$$\begin{aligned} \bar{g}^T(x_{k+1} - x_k) &= g_k^T(x_{k+1} - x_k) + (\bar{g} - g_k)^T(x_{k+1} - x_k) \\ &\leq g_k^T(x_{k+1} - x_k) + ||\bar{g} - g_k||\,||x_{k+1} - x_k|| \\ &\leq g_k^T(x_{k+1} - x_k) + \mu||x_{k+1} - x_k||^2 \\ &= \alpha_k g_k^T d_k + \mu\alpha_k^2||d_k||^2 \\ &= \alpha_k g_k^T d_k + \mu\alpha_k\delta g_k^T d_k/||d_k||^2 \\ &= \alpha_k g_k^T d_k + \mu\delta\alpha_k g_k^T d_k \\ &= -\delta(1 - \mu\delta)(g_k^T d_k)^2/||d_k||^2; \end{aligned}$$

i.e.,

$$f(x_{k+1}) - f(x_k) \leq -\delta(1 - \mu\delta)(g_k^T d_k)^2/||d_k||^2, \tag{11.14}$$

which implies $f(x_{k+1}) \leq f(x_k)$. It follows by Assumption 1 that $\lim_{k\to\infty} f(x_k)$ exists. Thus, from (11.14) we obtain

$$\frac{(g_k^T d_k)^2}{||d_k||^2} \leq \frac{1}{\delta(1 - \mu\delta)}\left[f(x_k) - f(x_{k+1})\right].$$

This finishes our proof. □

Lemma 11.2.4. *Suppose that Assumption 1 (or Assumption 2) holds and that x_k is given by (11.1), (11.3), and (11.10). Then*

$$\liminf_{k\to\infty} ||g_k|| \neq 0 \quad \textit{implies} \quad \sum_{d_k \neq 0} \frac{||g_k||^4}{||d_k||^2} < \infty.$$

Proof. If $\liminf_{k\to\infty} ||g_k|| \neq 0$, then there exists $\gamma > 0$ such that $||g_k|| \geq \gamma$ for all k. Let

$$\lambda_k = |g_k^T d_k|/||d_k||.$$

The by Lemma 11.2.3 there holds

$$\lambda_k \leq \frac{\gamma}{2(1+\kappa)}$$

for all large k. By Lemma 11.2.1 and Corollary 11.2.2 we have

$$|g_{k+1}^T d_k| = |\rho_k g_k^T d_k| \leq (1+\delta\mu)|g_k^T d_k| \leq (1+\kappa)|g_k^T d_k|; \tag{11.15}$$

Considering (11.3), we have

$$g_k = \beta_k d_{k-1} - d_k. \tag{11.16}$$

By multiplying g_k on both sides of (11.16), we obtain

$$||g_k||^2 = \beta_k g_k^T d_{k-1} - g_k^T d_k. \tag{11.17}$$

From (11.17) and (11.15) it follows that

$$\begin{aligned}
\frac{||g_k||^2}{||d_k||} &= \frac{\beta_k g_k^T d_{k-1} - g_k^T d_k}{||d_k||} \\
&\leq \frac{|\beta_k|\ |g_k^T d_{k-1}| + |g_k^T d_k|}{||d_k||} \\
&= (1+\kappa)\lambda_{k-1}\frac{||\beta_k d_{k-1}||}{||d_k||} + \lambda_k \\
&= \lambda_k + (1+\kappa)\lambda_{k-1}\frac{||d_k + g_k||}{||d_k||} \\
&\leq \lambda_k + (1+\kappa)\lambda_{k-1} + (1+\kappa)\lambda_{k-1}\frac{||g_k||^2}{||d_k||\ ||g_k||} \\
&= \lambda_k + (1+\kappa)\lambda_{k-1} + (1+\kappa)\lambda_{k-1}\frac{||g_k||^2}{\gamma||d_k||} \\
&\leq \lambda_k + (1+\kappa)\lambda_{k-1} + (1+\kappa)\frac{\gamma}{2(1+\kappa)}\frac{||g_k||^2}{\gamma||d_k||} \\
&= \lambda_k + (1+\kappa)\lambda_{k-1} + \frac{||g_k||^2}{2||d_k||};
\end{aligned}$$

i.e.

$$\frac{||g_k||^2}{||d_k||} \leq \lambda_k + (1+\kappa)\lambda_{k-1} + \frac{||g_k||^2}{2||d_k||}.$$

The above relation can be re-written as

$$\frac{||g_k||^2}{||d_k||} \leq 2\lambda_k + 2(1+\kappa)\lambda_{k-1} \leq 2(1+\kappa)(\lambda_k + \lambda_{k-1}).$$

Hence we have

$$||g_k||^4/||d_k||^2 \leq 4(1+\kappa)^2(\lambda_k + \lambda_{k-1})^2 \leq 8(1+\kappa)^2(\lambda_k^2 + \lambda_{k-1}^2). \tag{11.18}$$

By Lemma 11.2.3 we can see that $\sum \lambda_k^2 < \infty$. Hence we finally obtain from (11.18)

$$\sum_{d_k \neq 0} \frac{||g_k||^4}{||d_k||^2} < \infty. \square$$

Lemma 11.2.5. *Suppose Assumption 2 holds. Then the FR method satisfies*

$$\frac{||d_k||^2}{||g_k||^4} \leq \frac{||d_{k-1}||^2}{||g_{k-1}||^4} + \frac{\Omega}{||g_k||^2}, \tag{11.19}$$

where Ω is a constant.

Proof.

$$\begin{aligned}
-g_k^T d_{k-1} &= -\rho_{k-1} g_{k-1}^T d_{k-1} \\
&= -\rho_{k-1} g_{k-1}^T(-g_{k-1} + \beta_{k-1}^{FR} d_{k-2}) \\
&= -\rho_{k-1} g_{k-1}^T \left(-g_{k-1} + \frac{||g_{k-1}||^2}{||g_{k-2}||^2} d_{k-2}\right) \\
&= \rho_{k-1}||g_{k-1}||^2 - \frac{\rho_{k-1}||g_{k-1}||^2}{||g_{k-3}||^2} g_{k-1}^T d_{k-2}.
\end{aligned}$$

Thus, we have a recursive equation which leads to

$$\begin{aligned}
-g_k^T d_{k-1} &= \rho_{k-1}||g_{k-1}||^2 - \frac{\rho_{k-1}||g_{k-1}||^2}{||g_{k-3}||^2} g_{k-1}^T d_{k-2} \\
&= \rho_{k-1}||g_{k-1}||^2 + \rho_{k-1}\rho_{k-2}||g_{k-1}||^2 - \rho_{k-1}\rho_{k-2}\frac{||g_{k-1}||^2}{||g_{k-3}||^2} g_{k-2}^T d_{k-3} \\
&= \ldots = ||g_{k-1}||^2(\rho_{k-1} + \rho_{k-1}\rho_{k-2} + \cdots + \rho_{k-1}\rho_{k-2}\cdots\rho_2),
\end{aligned}$$

which implies that

$$\begin{aligned}
-g_k^T d_{k-1} &= ||g_{k-1}||^2(\rho_{k-1} + \rho_{k-1}\rho_{k-2} + \cdots + \rho_{k-1}\rho_{k-2}\cdots\rho_2) \\
&\le ||g_{k-1}||^2 \sum_{k=1}^{\infty}(1-\delta\lambda)^k =: \nu||g_{k-1}||^2.
\end{aligned}$$

Hence we finally obtain

$$\begin{aligned}
\frac{||d_k||^2}{||g_k||^4} &= \left(||g_k||^2 - 2\beta_k^{FR} g_k^T d_{k-1} + (\beta_k^{FR})^2||d_{k-1}||^2\right)/||g_k||^4 \\
&\le \left(||g_k||^2 + 2\nu\beta_k^{FR}||g_{k-1}||^2 + (\beta_k^{FR})^2||d_{k-1}||^2\right)/||g_k||^4 \\
&= (1+2\nu)/||g_k||^2 + ||d_{k-1}||^2/||g_{k-1}||^4.
\end{aligned}$$

By defining $\Omega := 1 + 2\nu$, the proof is complete. □

Lemma 11.2.6. *The CD method satisfies (11.19) under Assumption 1 while the DY method does the same under Assumption 2.*

Proof. For the CD method we have

$$\begin{aligned}
g_k^T d_k &= g_k^T\left(-g_k + \frac{||g_k||^2}{-d_{k-1}^T g_{k-1}} d_{k-1}\right) \\
&= -||g_k||^2 - ||g_k||^2 \frac{g_{k-1}^T d_{k-1}}{d_{k-1}^T g_{k-1}} \\
&= -||g_k||^2 - ||g_k||^2 \frac{\rho_{k-1} g_{k-1}^T d_{k-1}}{d_{k-1}^T g_{k-1}} \\
&= -(1+\rho_{k-1})||g_k||^2 < 0.
\end{aligned}$$

Note that from the inequality above we have

$$(g_k^T d_k)^2 > ||g_k||^4. \tag{11.20}$$

Applying (11.3), (11.7), Lemma 11.2.1, and (11.20), we obtain

$$\begin{aligned}
||d_k||^2 &= ||-g_k + \beta_k^{CD} d_{k-1}||^2 \\
&= \left\| -g_k + \frac{||g_k||^2}{-g_{k-1}^T d_{k-1}} d_{k-1} \right\|^2 \\
&= \|g_k\|^2 + \frac{||g_k||^4}{(g_{k-1}^T d_{k-1})^2}||d_{k-1}||^2 + \frac{2||g_k||^2}{g_{k-1}^T d_{k-1}} g_k^T d_{k-1}
\end{aligned}$$

$$= \|g_k\|^2 + \frac{||g_k||^4}{(g_{k-1}^T d_{k-1})^2}||d_{k-1}||^2 + \frac{2\rho_{k-1} g_{k-1}^T d_{k-1}}{g_{k-1}^T d_{k-1}}||g_k||^2$$

$$= (1 + 2\rho_{k-1})||g_k||^2 + \frac{||g_k||^4}{(g_{k-1}^T d_{k-1})^2}||d_{k-1}||^2$$

$$\leq [1 + 2(1 + \delta\mu)]||g_k||^2 + \frac{||g_k||^4}{||g_{k-1}||^4}||d_{k-1}||^2$$

Dividing both sides by $||g_k||^4$ we get

$$\frac{||d_k||^2}{||g_k||^4} \leq \frac{||d_{k-1}||^2}{||g_{k-1}||^4} + \frac{[1 + 2(1 + \delta\mu)]}{||g_k||^2}. \tag{11.21}$$

By defining $\Omega := 1 + 2(1 + \delta\mu)$, we complete the proof for the CD method. Changing β_k^{FR} to β_k^{DY} in the above proof and noting Corollary 11.2.2, the proof for the DY method can be obtained similarly. □

Theorem 11.2.7. *Suppose that Assumption 1 holds. Then the CD method will generate a sequence* $\{x_k\}$ *such that* $\liminf_{k\to\infty} ||g_k|| = 0$. *If Assumption 2 holds, then the same conclusion holds for the FR and DY methods as well.*

Proof. If $\liminf_{k\to\infty} ||g_k|| \neq 0$, then there exists $\gamma > 0$ such that $||g_k|| \geq \gamma$ for all k, and by Lemma 11.2.3 we have

$$\sum_{d_k \neq 0} \frac{||g_k||^4}{||d_k||^2} < \infty. \tag{11.22}$$

In order to deduce a contradiction we obtain from Lemmae 11.2.5 and 11.2.6

$$||d_k||^2/||g_k||^4 \leq ||d_{k-1}||^2/||g_{k-1}||^4 + \Omega/||g_k||^2. \tag{11.23}$$

Hence we get

$$\begin{aligned} ||d_k||^2/||g_k||^4 &\leq ||d_{k-1}||^2/||g_{k-1}||^4 + \Omega/\gamma^2 \\ &\leq ||d_{k-2}||^2/||g_{k-2}||^4 + 2\Omega/\gamma^2 \leq \cdots \\ &\leq ||d_1||^2/||g_1||^4 + (k-1)\Omega/\gamma^2 \end{aligned} \tag{11.24}$$

Thus, (11.24) means that

$$||g_k||^4/||d_k||^2 \geq 1/(ka - a + b), \tag{11.25}$$

where $a = \Omega/\gamma^2, b = ||d_1||^2/||g_1||^4 = 1/||g_1||^2$ are constants. From (11.25) we have

$$\sum_{d_k \neq 0} ||g_k||^4/||d_k||^2 = +\infty, \tag{11.26}$$

which is contradictory to (11.22). Hence the theorem is valid. □

Now we turn to the PR and the HS methods.

Theorem 11.2.8. *Suppose that Assumption 1 holds. Then the PR method will generate a sequence* $\{x_k\}$ *such that* $\liminf_{k\to\infty} ||g_k|| = 0$. *If Assumption 2 holds, then the same conclusion holds for the HS method as well.*

Proof. We consider the PR method first. Assume in contrary that $||g_k|| \geq \gamma$ for all k. From Lemma 11.2.3 we have

$$\sum_{d_k \neq 0} \frac{(g_k^T d_k)^2}{||d_k||^2} = \sum ||\alpha_k d_k||^2 = \sum ||x_{k+1} - x_k||^2 < \infty.$$

Hence

$$|\beta_k^{PR}| = \left|g_{k+1}^T (g_{k+1} - g_k)/||g_k||^2\right| \to 0$$

as $k \to \infty$. Since L is bounded, both $\{x_k\}$ and $\{g_k\}$ are bounded. By using

$$||d_k|| \leq ||g_k|| + |\beta_k| \, ||d_{k-1}|| \tag{11.27}$$

one can show that $||d_k||$ is uniformly bounded. Thus we have

$$\begin{aligned} |g_k^T d_k| &= |g_k^T(-g_k + \beta_k^{PR} d_{k-1})| \\ &\geq ||g_k||^2 - |\beta_k| \, ||g_k|| \, ||d_{k-1}|| \\ &\geq ||g_k||^2/2 \ \text{(since } ||\beta_k d_{k-1}|| \leq ||g_k||/2) \end{aligned}$$

for large k. Thus we have

$$\frac{(g_k^T d_k)^2}{(||d_k||^2 ||g_k||^2)} \geq \frac{1}{2} \frac{||g_k||^2}{||d_k||^2}. \tag{11.28}$$

Since $||g_k|| \geq \gamma$ and $||d_k||$ is bounded above, we conclude that there is $\epsilon > 0$ such that

$$\frac{(g_k^T d_k)^2}{(||d_k||^2 ||g_k||^2)} \geq \epsilon,$$

which implies

$$\sum_{d_k \neq 0} ||g_k||^2 \frac{(g_k^T d_k)^2}{(||d_k||^2 ||g_k||^2)} = \infty.$$

This is a contradiction to (11.22). The proof for the PR method is complete. We now consider the HS method. Again, we assume $||g_k|| \geq \gamma$ for all k and note that $g_k - g_{k-1} \to 0$. Since

$$
\begin{aligned}
g_k^T d_k &= g_k^T(-g_k + \beta_k^{HS} d_{k-1}) \\
&= g_k^T\left(-g_k - \frac{g_k^T(g_k - g_{k-1})}{(1-\rho_{k-1})d_{k-1}^T g_{k-1}} d_{k-1}\right) \\
&= -||g_k||^2 - \frac{\rho_{k-1}}{1-\rho_{k-1}} g_k^T(g_k - g_{k-1}),
\end{aligned}
$$

we have

$$
\begin{aligned}
\left(\frac{g_k^T d_k}{||g_k||^2}\right)^2 &= \left(1 + \frac{\rho_{k-1}}{1-\rho_{k-1}} \frac{g_k^T(g_k - g_{k-1})}{||g_k||^2}\right)^2 \\
&\geq \left(1 - \frac{1-\delta\lambda}{\delta\lambda} \frac{|g_k^T(g_k - g_{k-1})|}{||g_k||^2}\right)^2 \geq \frac{1}{2}
\end{aligned}
$$

for large k. This is the same as (11.28). Moreover, from

$$
|\beta_k^{HS}| = \left|\frac{g_k^T(g_k - g_{k-1})}{(1-\rho_{k-1})d_{k-1}^T g_{k-1}}\right| \leq \frac{\sqrt{2}}{\delta\lambda} \left|\frac{g_k^T(g_k - g_{k-1})}{||g_{k-1}||^2}\right| \to 0
$$

we conclude that $||d_k||$ is bounded through (11.27). The rest of the proof is the same as the proof for the PR method. □

11.2.3 Concluding Remarks

We show that by taking a "fixed" steplength α_k computed by formula (11.10), the conjugate gradient method is globally convergent for several popular choices of β_k. The result discloses an interesting property of the conjugate gradient method – its global convergence can be guaranteed by taking a pre-determined steplength rather than following a set of line search rules. This steplength might be practical in cases that the line search is expensive or hard. Our proofs require that the function is at least LC^1 (sometimes strongly convex in addition) and the level set L is bounded. We point out that the latter condition can be relaxed to that the function is lower bounded on L for the CD method because in the corresponding proof an upper bound of $||g_k||$ is not assumed. Computational results showed that the CG methods without line search have similar numerical behavior to these CG methods with line search [377].

11.3 Simulated Annealing Global Optimization Search Algorithm

SA, "the name and inspiration come from annealing in metallurgy, a technique involving heating and controlled cooling of a material to increase the size of its crystals and reduce their defects. The heat causes the atoms to become unstuck from their initial positions (a local minimum of the internal energy) and wander randomly through states of higher energy; the slow cooling gives them more chances of finding configurations with lower internal energy than the initial one" ([311, 410, 574, 580], en.wikipedia.org/wiki/Simulated_annealing and references [94, 167, 181, 255, 339, 424, 482, 622] therein).

11.4 The Application of Hybrid SDCG-SA-SDCG Method

To date the hydrophobic region AGAAAAGA palindrome of prion proteins (113–120) has little existing experimental structural data available; but many experimental works showed this region has amyloid fibril forming property. This region belongs to the N-terminal unstructured region (1–123) of prions that is not easy to determine the structure using NMR or X-ray crystallography. This chapter successfully constructs three amyloid fibril models for this region. The models are formatted by standard SA and refined by the traditional SD-CG optimization methods. Because the NMR or X-ray structure of prion AGAAAAGA peptide has not been determined yet, these models can be used as a reference for the experimental studies on the hydrophobic region AGAAAAGA palindrome of prion proteins (113–120). Standard SA is effective in molecular modeling as is also shown by this chapter. All in all, this chapter describes construction of the experimentally unknown 3D structures of the hydrophobic region AGAAAAGA of prion proteins by using suitable templates from Protein Data Bank [47] and refinement of the model with several optimization techniques within the computational chemistry package AMBER [89]. Three constructed models for amyloid fibrils may be useful for the goals of medicinal chemistry.

11.4.1 Overview

Prion diseases are invariably fatal and highly infectious neurodegenerative diseases that affect humans and animals. These neurodegenerative diseases are caused by the conversion from a soluble into an insoluble fibrillar form of the prion protein (PrP) [484, 486]. The infectious prion (PrP^{Sc}) is an abnormally folded form of the normal cellular prion (PrP^{C}) [257, 484, 486]. The conversion of PrP^{C} to PrP^{Sc} is believed to involve conformational change from a predominantly α-helical protein (42 % α-helix, 3 % β-sheet) to a protein rich in β-sheets (30 % α-helix, 43 % β-sheet) [85, 169, 257, 447, 457, 486, 491]. Prion amyloid fibrils are believed rich in β-sheet structure and contain a cross-β core. Many experimental works (e.g. [67, 68, 288, 315, 316, 357, 445, 619]) show that the hydrophobic region AGAAAAGA of prion proteins (113–120) plays an important role in the conversion of PrP^{Sc} to PrP^{C}. PrP lacking / deleting the palindrome (PrP 113–120) neither converted to PrP^{Sc} nor generated proteinase K-resistant PrP [68, 288, 445, 619]. Brown et al. pointed out that the AGAAAAGA peptide was found to be necessary (though not sufficient) for blocking the toxicity and amyloidogenicity of PrP 106–126 [67, 70]. The peptide AGAA did not form fibrils but the peptide AGAAAAGA formed fibrils in both water and PBS [67]. Thus, the minimum sequence necessary for fibril formation should be AGAAA, AGAAAA, AGAAAAG, GAAAAGA or AGAAAAGA (in this chapter AGAAAA, AGAAAAG and GAAAAGA will be used). According to Brown [67], AGAAAAGA is important for fibril formation and is an inhibitor of PrP^{Sc} neurotoxicity. Studies the atomic-resolution structures of the AGAAAAGA peptide will be useful for the goals of medicinal chemistry for controlling prion diseases. The motivation of this chapter is to study the atomic-resolution structures of the AGAAAAGA peptide of prions.

Many other experimental works (for examples, see [315, 357]) also showed that the hydrophobic region AGAAAAGA of prion proteins (113–120) has an amyloid fibril forming property. However, to date the hydrophobic region AGAAAAGA of prion proteins (113–120) has little existing experimental structural data available. It is well known that the N-terminal residues (1–123) of prions are unstructured (the AGAAAAGA (113–120) region just belongs to this region); NMR and X-ray crystallography cannot easily get the molecular structure of the N-terminal region. Due to the noncrystalline and insoluble nature of the amyloid fibril, it is difficult to obtain atomic-resolution structures with traditional experimental methods.

Computational approaches, however, allow us to obtain a description of the peptide at a microscopic level. This chapter computationally constructs some amyloid fibril models for the AGAAAAGA (113–120) region, which are significantly useful for the experimental study of the hydrophobic region AGAAAAGA palindrome of prion proteins (113–120). The studies on the structure or the dynamics of the AGAAAAGA region should play a role in the aggregation process and may be useful for the goals of medicinal chemistry for controlling prion diseases. Some studies on prion have done for PrP 106–126 region [159, 281, 315, 316, 357, 448, 584, 711]. After the breakthrough work of Sawaya et al. [517] of 2007, many computational studies have been done on prion or yeast prion amyloid fibrils [97, 373, 535, 589, 657, 694]; this chapter computationally studies PrP(113–120) region of PrP(106–126).

11.4.2 Materials and Methods

Recently common structural features shared by all amyloid fibrils were revealed by [517]; atomic structures of all the microcrystals revealed are steric zippers, with strong vdW interactions between β-sheets and HBs to maintain the β-strands. Basing on these steric zippers, three prion AGAAAAGA palindrome amyloid fibril models will be constructed. Model-chapt11-1 belongs to Class 7 of [517], Model-chapt11-2 and Model-chapt11-3 belong to Class 1 of [517] (numerical experience of the author shows that, for prion AGAAAAGA palindrome, the β-sheet structure of other Classes of [517] cannot be preserved after optimizations or simulated annealing).

The models were built on the LYQLEN peptide derived from human insulin residues 13–18 (PDB entry 2OMP [47, 517]) and the GNNQQNY peptide from the yeast prion protein Sup35 (PDB entry 1YJP [47, 517]). A 12 chains AGAAAA model (Model-chapt11-1), a 10 chains AGAAAAG model (Model-chapt11-2) and a 10 chains GAAAAGA model (Model-chapt11-3) were successfully constructed. Models were separately constructed by the unmerge, mutate, and merge modules of Insight II of Accelrys Inc. Model-chapt11-1 (with EFABIJ LKDCHG chains) was derived from 2OMP.pdb, its AB chains were simply gotten from the AB chains of 2OMP.pdb by the unmerge, mutate and merge modules of Insight II and then its other chains can be gotten by mathematical formulas C = A + (-1.885 0 17.243), D = B + (−1.885 0 17.243), I = A + (9.666 0 0), J = B + (9.666 0 0), E = A + (−9.666 0 0), F = B + (−9.666 0 0), H = D + (9.666 0 0), G = C + (9.666 0 0), L = D + (−9.666 0 0), K = C + (−9.666 0 0). The B chain of 1YJP.pdb can be gotten by a mathematical formula

$$B = \begin{pmatrix} -1 & 0 & 0 \\ 0 & 1 & 0 \\ 0 & 0 & -1 \end{pmatrix} A + \begin{pmatrix} 0 \\ 2.433 \\ 0 \end{pmatrix}$$

from its A chain. AB chains of Model-chapt11-2 and Model-chapt11-3 were constructed separately from AB chains of 1YJP.pdb by the unmerge, mutate, and merge modules of Insight II; other chains of Model-chapt11-2 and Model-chapt11-3 were gotten from AB chains by the parallelization of AB chains. The initial structures of Model-chapt11-1 to Model-chapt11-3 are shown in Fig. 11.1.

The models were firstly optimized by steepest descent (SD) method and conjugate gradient (CG) method [89]. These two methods are traditional optimization methods. The former has nice convergence but is slow when close to minimums. The latter is efficient but its gradient RMS and GMAX gradient [89] do not have a good convergence. We use the SD method and then the CG method to optimize our models. When the models cannot be optimized further, we employ standard simulated annealing (SA) [89, 691]. In this chapter numerical potential energy results show that SA is very effective for further optimizations and the Model-chapt11-1 to Model-chapt11-3 constructed in this chapter should be very useful to study the hydrophobic region AGAAAAGA palindrome of prion proteins (113–120). These models were optimized by SD and CG methods, formatted by SA and

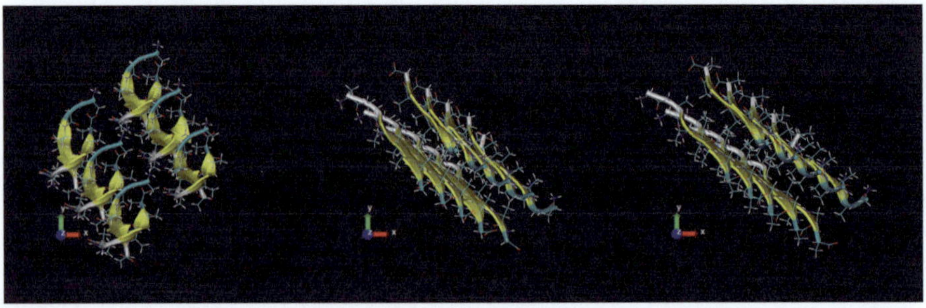

Fig. 11.1 Initial structures of prion AGAAAAGA amyloid fibril Model-chapt11-1 to Model-chapt11-3

finally refined by SD and CG methods. The work was performed with the AMBER 10 package [89] and the graphs were drawn by XMGRACE of Grace 5.1.21 and VMD 1.8.7beta5 [297].

We use the ff03 force field of the AMBER 10 package, in a neutral pH environment. The systems were surrounded with a 12 Å layer of TIP3PBOX water molecules using XLEaP module of AMBER 10.

The solvated proteins were minimized by the SD method and then CG were performed. Model-chapt11-1 were optimized by 57,350 steps of SD and 26,784 steps of CG; Model-chapt11-2 by 57,950 steps of SD and 23,943 steps of CG; Model-chapt11-3 by 59,500 steps of SD and 23,109 steps of CG.

Then the solvated proteins were quickly heated from 0 to 300 K linearly during 20 ps. The systems were kept at 300 K for 80 ps. The systems then were slowly cooled from 300 to 100 K linearly for 400 ps. At 100 K, the systems were kept for 100 ps. All the systems were in constant NVT ensembles using Langevin thermostat algorithm with weak restraints (a force constant of 10.0 KCal mol^{-1} Å^{-2} was used) on the solvated proteins. The SHAKE algorithm and SANDER (Simulated Annealing with NMR-Derived Energy Restraints) algorithm with nonbonded cutoffs of 9 Å were used during the heating, cooling and the 100 ps at 100 K. The equilibration was done in constant NPT ensembles under a Langevin thermostat for 4400 ps and the RMSD, PRESS, and VOLUME (DENSITY) were sufficiently stable [89] for each of the models (Fig. 11.2), where the jump in RMSD around 0.2 Å correlates with removing restraints for the change from NVT to NPT, but it did not change the structures at 100 K. The equilibration was under constant pressure 1 atm and constant temperature 100 K in neutral pH environment (Why was the equilibration performed at low temperature 100 K? It is in order to be consistent with the experimental works of Sawaya et al. [517]). Step size is 2 fs for the whole SA runs. The structures were saved to file every 100 steps. During the SA, the Metropolis criterion was used. The changes of temperatures along with time are shown in Fig. 11.3.

After the SA, Model-chapt11-1 was refined by 10,000 steps of SD and 2347 steps of CG; 10,000 steps of SD and 1774 steps of CG for Model-chapt11-2; 10,000 steps of SD and 619 steps of CG for Model-chapt11-3. All the above works were performed on the Tango facilities of the Victorian Partnership for Advanced Computing (www.vpac.org) of Australia. Longer simulations for the equilibration can slightly further refine the RMSD, PRESS, and VOLUME (DENSITY) values and longer optimizations will also improve the values of ENERGY, RMS and GMAX very slightly.

11.4.3 Results and Discussion

For the sake of clarity, we use a program in Zhang et al. [695] to theoretically confirm that prion AGAAAAGA (113–120) segment has an amyloid fibril forming property. The theoretical computation results are shown in Fig. 11.4, from which we can see that prion AGAAAAGA (113–120) region is clearly identified as the amyloid fibril formation region because the energy is less than the amyloid fibril

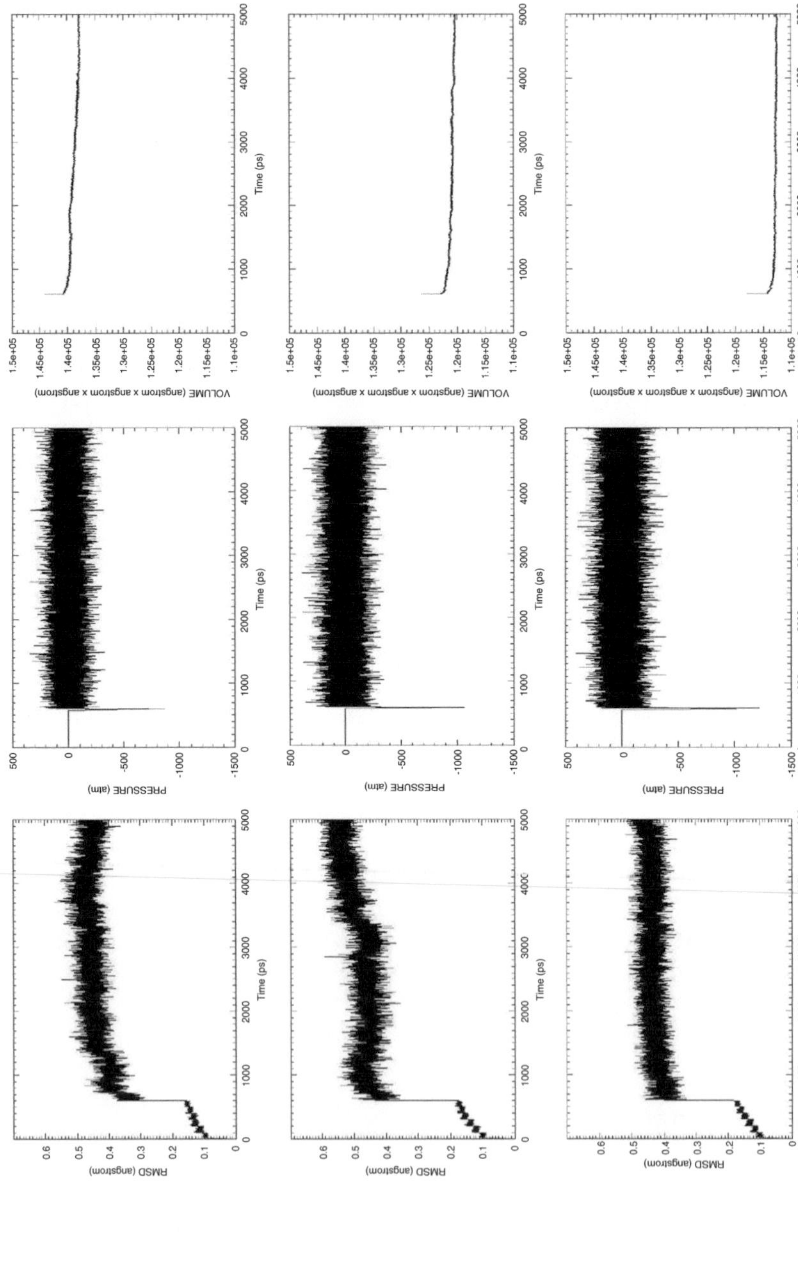

Fig. 11.2 RMSD, PRESSURE, and VOLUME values for Model-chapt11-1 to Model-chapt11-3 in 5ns (1st row: Model-chapt11-1, 2nd row: Model-chapt11-2, 3rd row: Model-chapt11-3)

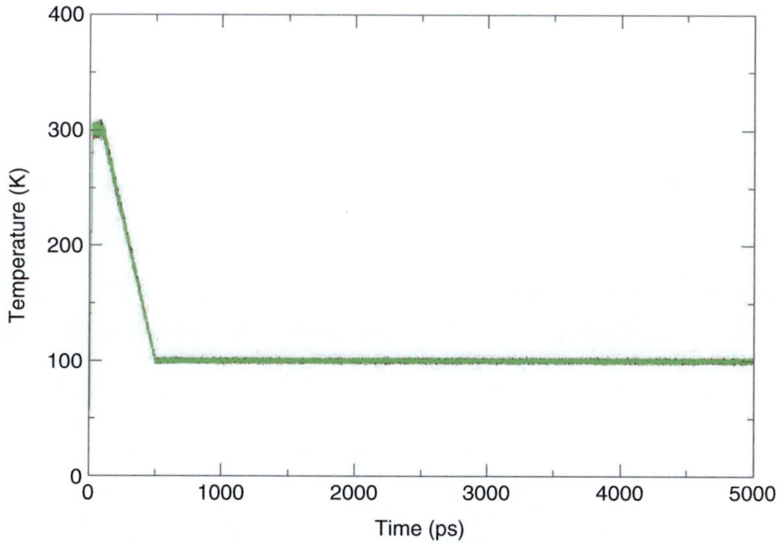

Fig. 11.3 Temperature-time graph of the SA

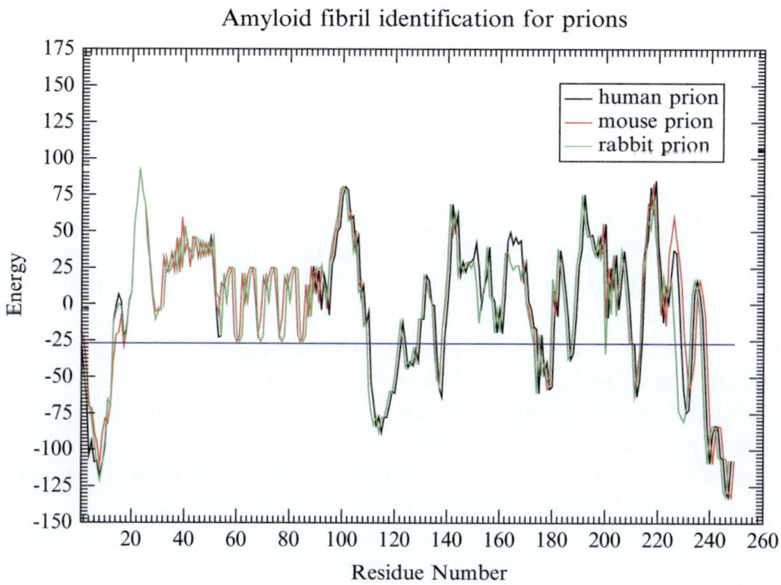

Fig. 11.4 Prion AGAAAAGA(113–120) is clearly identified

formation threshold energy -26 [695]. In this section we will illuminate the optimal molecular structures of Model-chapt11-1 to Model-chapt11-3. Firstly how to get the optimal models is described as follows.

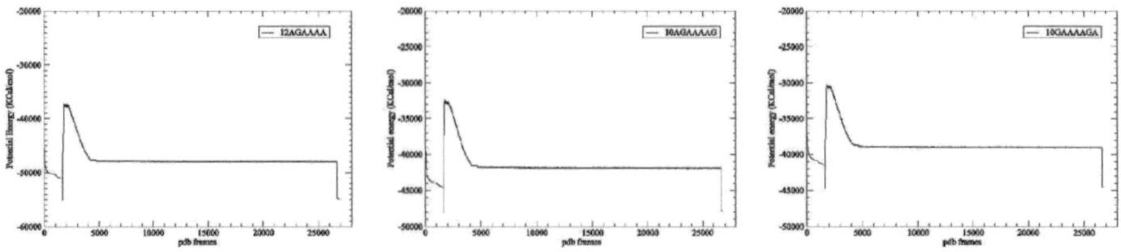

Fig. 11.5 Potential energy graph

Table 11.1 Potential energy values

Model	OPT1 1st step	OPT1 last step	SA 1st step	SA last 1000 steps (average value)	OPT2 1st step	OPT2 last step
1	−2233.5	−5111.9	−55157.8	−48016.6	−48027	−54845
2	−2725.0	−4458.5	−48056.1	−41894.3	−41896	−47827
3	−2603.8	−4150.6	−44747.5	−39022.4	−39049	−44533

Figure 11.5 shows the potential energy development for the three Models. We can see that the potential energy goes down during the SD and CG optimization phase (OPT1), suddenly drops down and quickly goes up and then slowly goes down and levels off during the SA phase (SA), and at last quickly goes down and then levels off during a short phase (OPT2). At the beginning of SA, the energy quickly drops off is due to the temperatures of the systems being suddenly changed from 100 K [517] to 0 K. This is a case of so called "quenching". Some energy values are listed in Table 11.1.

In Table 11.1, the first column of energies (OPT1 First Step) are the ones that derive from the LYQLEN and GNNQQNY templates by mutations. The two β-sheets are very close with many bad vdW contacts and the distance between β-strands is too short for the vdW contacts and HB contacts. OPT1 removes these bad contacts and makes the structures of the models become much more stable with lower potential energies. OPT1 makes Model-chapt11-1 to Model-chapt11-3 separately lose 2878.4, 1733.5, 1546.8 KCal/mol of potential energy. However, OPT1 is a local search optimization method which cannot thoroughly optimize the models into their most stable structures.

In Fig. 11.5 we see that models are trapped into their local optimal structures. SA is a global search optimization method and it can make OPT1 jump out of the local trap, even accepting very bad cases with a probability according to the Metropolis criterion. Thus, in Table 11.1 we see that SA is quickly quenching the molecular structures to escape from the local traps and at last SA makes the three systems separately lose 7141.2, 6161.8, 5725.1 KCal/mol.

After SA, OPT2 can safely bring the molecular structures of the models to the most stable states. OPT2 makes the molecules lose 6818, 5931, 5484 KCal/mol of potential energies separately for Model-chapt11-1 to Model-chapt11-3. OPT2 can lose the energies of Model-chapt11-1 to Model-chapt11-3 nearly as the same

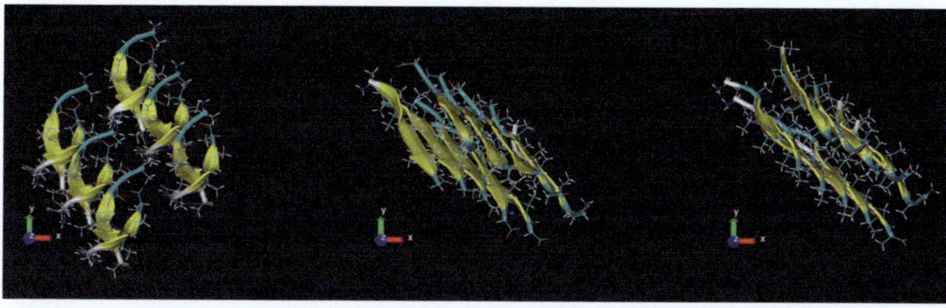

Fig. 11.6 Optimized prion AGAAAAGA amyloid fibril Model-chapt11-1 to Model-chapt11-3

numbers as that of SA (i.e. the decrease in energy in OPT2 can be significant compared to the decrease in energy in SA between its 1st step and the average of the last 1000 steps). Compared with OPT1, OPT2 made greater optimization; this demonstrates the effectiveness of SA. For *each* of the models, the above analyses were done; however, these three models are not comparable because each of the models is independent of the others.

The final optimal molecular structures of Model-chapt11-1 to Model-chapt11-3 after OPT2 are illuminated in Fig. 11.6 (the RMSDs from the initial structures illuminated in Fig. 11.1 are 0.84, 1.25, 0.98 Å separately for Model-chapt11-1 to Model-chapt11-3). The HBs between the two closet adjacent β-strands and the vdw contacts between the two inner closest adjacent alanines can be seen from Fig. 11.6.

For Model-chapt11-1, there are several HBs in each of the antiparallel β-strands. For Model-chapt11-2 and Model-chapt11-3, in each β-sheet the β-strands are parallel to each other. It is very interesting to see in Model-chapt11-2 two chains in one sheet are tightly bonded by four HBs, but in another sheet three chains are bonded to one by one as Model-chapt11-3.

Model-chapt11-1 is β-strand antiparallel, face=back, up-up [517], Model-chapt11-2 and Model-chapt11-3 are β-strand parallel, face-to-face, up-up [517] and for all the models about 5 Å between the two closet adjacent β-sheets are maintained by HYDs, about 4.5 Å between the two closet adjacent β-strands which are linked by HBs. Model-chapt11-1 and Model-chapt11-2 & Model-chapt11-3 belong to Class 7 and 1 of Sawaya et al. [517] separately. There is a hydrophobic core in each of the models. These amyloid fibrils are rich in β-sheet structure and contain a cross-β core form of PrP^{Sc}, which causes prion diseases.

At last, the numerical computational experience of the author is introduced as follows. In Sawaya et al. [517], there are 8 Classes of the steric zipper structures for peptide segments of fibril-forming proteins. For each Class, this chapter constructed the molecular structures for the hydrophobic region AGAAAAGA palindrome of prion proteins (113–120). Computer numerical experience shows that:

(1) For Class 1, basing on the NNQQNY peptide from yeast prion protein Sup35 (PDB entry 1YJO) a hexamer model with 6 AAAAGA chains can be constructed by OPT1 but cannot pass SA and OPT2, however, basing on the

GNNQQNY peptide from the yeast prion protein Sup35 (PDB entry 1YJP) Model-chapt11-2 with 10 AGAAAAG chains and Model-chapt11-3 with 10 GAAAAGA chains were successfully constructed by OPT1-SA-OPT2;

(2) For Class 2 (i.e. β-strand parallel, face-to-back, up-up), a tetramer model with 4 AGAAAA chains can be constructed basing on the SNQNNF peptide of human prion 170–175 (PDB entry 2OL9) by OPT1 but cannot pass SA and OPT2;

(3) For Class 7, basing on the LYQLEN peptide derived from human insulin residues 13–18 (PDB entry 2OMP), Model-chapt11-1 with 12 AGAAAA chains was successfully constructed by OPT1-SA-OPT2;

(4) For other cases of Sawaya et al. [517] the β-sheet structure of prion AGAAAAGA palindrome cannot be preserved after OPT1;

(5) Basing on the NNQNTF peptide of elk prion 173–178 (PDB entry 3FVA) [251, 252, 631] which belongs to Class 1 of Sawaya et al. [517], new models for prion AGAAAAGA palindrome might be able to be constructed because 6 NNQNTF chains of elkPrP (173–178) are still preserved after OPT1. This will be further studied by the author.

11.4.4 Concluding Remarks

To date the hydrophobic region AGAAAAGA palindrome PrP(113–120) of the unstructured N-terminal region (1–123) of prions has little existing experimental structural data available. This chapter successfully constructs three molecular structure models for AGAAAAGA palindrome (113–120). These models should be very helpful for the experimental studies of the hydrophobic region AGAAAAGA palindrome of prion proteins (113–120) when the NMR or X-ray molecular structure of prion AGAAAAGA peptide has not been easily determined yet. This chapter also shows that the standard simulated annealing is very effective in molecular modeling. In conclusion, this chapter describes construction of the experimentally unknown 3D structures of the hydrophobic region AGAAAAGA of prion proteins by using suitable templates from Protein Data Bank [47] and refinement of the model with several optimization techniques within AMBER. Three constructed models for amyloid fibrils may be useful for the goals of medicinal chemistry.

11.5 Implementing Simulated Annealing in Global Optimization Computations

SA was inspired from annealing in metallurgy, a technique involving heating and controlled cooling of a material to increase the size of its crystals and reduce their defects, both are attributes of the material that depend on its thermodynamic free energy. In this section, firstly we will study SA in details on its initial feasible solution choosing, initial temperature selecting, neighbourhood solution searching, efficient way of calculating for the difference of objective function values of two neighbourhood solutions, acceptance function (Metropolis function), temperature cooling, and the criteria of inner and outer loops' stopping, etc.

SA, as well as Tabu Search (TS) and Genetic Algorithm (GA), is one of the successful heuristic computational methods. It simulates the annealing process with Monte Carlo (MC) property. The works of Metropolis, Kirkpatrick, Johnson, Aarts, et al. are well-known. In discrete optimization, simulated annealing method has found a lot of applications. The book for example [582] is good collections of its applications to discrete optimization problems. For continuous optimization problems, there are a lot of references on it. However, we still rarely see a very successful simulated annealing method for large scale continuous optimization problems in very high dimensions, especially in the constrained case.

The SA method appeared as early as in 1953 [424] as a MC method and was firstly investigated and used in 1983 by Kirkpatrick et al. [339]. SA is a stochastic method. It differs from the traditional descent methods (see, for example, [490] and references therein) in that a local search algorithm for a neighborhood solution search, whether it randomly descents or steeply descents, allows downhill moves only, while in an attempt to escape local optima SA algorithm allows occasional uphill moves as well. SA techniques are based upon the physical analogy of cooling crystal structures (including the case of so called 'quenching') that spontaneously arrives at a stable configuration, characterized by-globally or locally-minimal potential energy. Starting with an initial solution x, and an initial "temperature" T_0, which is a parameter, we obtain a neighboring solution x' and compare its cost with that of x. If the cost of x' is smaller than that of x, i.e. $f(x') < f(x)$, we accept the new solution x'. The same thing would happen if we are applying the local search method by random descent method [490]. On the other hand, if $f(x')$ is greater than $f(x)$, (in which case local search algorithms (see, for example, [490]) will not accept x'), the SA algorithm may accept x', but with a probability $e^{\Delta_{x'x}/T_0}$ where $\Delta_{x'x}$ is the difference in the costs of x' and x, i.e., $\Delta_{x'x} = f(x') - f(x)$. This process is carried out for a certain number of times, which we call iterations, for each temperature. Then we reduce the temperature according to a particular schedule, and repeat. The convergence of SA algorithms are studied, for example, in [1, 396].

An essential element of the SA algorithm is the probability $e^{-\Delta_{x'x}/T}$ of an uphill move of size $\Delta_{x'x}$ being accepted when the current temperature is T. This is dependent on both $\Delta_{x'x}$ and T. For a fixed temperature T, smaller uphill moves $\Delta_{x'x}$ have a higher probability of being accepted. On the other hand, for a particular uphill move $\Delta_{x'x}$, a higher temperature will result in a larger probability for that uphill move being accepted. As stated in [241], "at a high temperature any uphill move might be indiscriminately accepted with large probability so that the objective function and the tumbles around the space are not very important; as T is lowered the objective function becomes more and more significant; until as T goes to zero the search becomes trapped in the lowest minima that it has reached."

The SA algorithm for solving a practical problem is typically implemented in two nested loops: the outer loop and the inner loop. The outer loop controls temperatures, while the inner loop iterates a fixed number of times for the given temperature. The inner loop is for the problem specific decisions. The decisions of the outer loop involve the setting of initial temperature (T_0), the cooling schedule, the temperature length which is the number of outer loop iterations performed at each temperature, and the stopping condition of the outer loop. The inner loop of SA typically considers the following aspects: feasible solution space, initial feasible solution, neighborhood move, objective function values (and efficient calculation of their difference), and the decision which decides whether the move is found acceptable or probably acceptable according the so-called Metropolis criterion.

In this section, we will study SA in details on its initial feasible solution choosing, initial temperature selecting, neighborhood solution searching, efficient way of calculating for the difference of objective function values of two neighborhood solutions, acceptance function (Metropolis function), temperature cooling, and the criteria of inner and outer loops' stopping, etc.

11.5.1 Overview

In this subsection we consider problem

$$\text{Minimize}\, f(x) \text{ subject to } x \in X,$$

where X is a subset of R^n is a compact set and f is continuous, being solved by SA algorithm.

The word *renew* denotes the counts of the solution being accepted in the inner loop. The pseudo-code (referred, for example, to [413]) of the SA algorithm is listed as follows:

Algorithm 11.1. *Simulated annealing algorithm*:

Initialization:

Define the objective function f and its feasible solution space.
Call **Initial feasible solution producing procedure** to produce an initial feasible solution x.
Call **Procedure of selecting initial temperature** to produce the initial temperature T_0.
Calculate the size of neighborhood N_size.
Calculate $f(x)$, and set $x\_best = x$ and $f\_best = f(x)$.
Set best_count = frozen_count = 0, and value of δ.

Cooling (outer loop procedure):

Repeat (outer loop)
  Call **Inner loop procedure**.
  Call **cooling schedule** $T = \alpha(T)$ to decrease to a new temperature.
  If best_count > 0 **then** set frozen_count = 0
  If renew/iteration count < 1/N_size **then**
    set frozen_count = frozen_count + 1
Until outer loop stopping criterion is met

Inner loop Procedure:
Set iteration_count = 0.
Repeat (inner loop)

Call **Neighborhood solution search procedure** to generate a feasible neighborhood solution x'.
Calculate $f(x')$.
Call **Efficient procedure** of calculating the cost difference $\triangle_{x'x} = f(x') - f(x)$.
If $\triangle_{x'x} < -\delta$ **then**
Set $x = x'$, renew = renew + 1.
Set $f(x) = f(x')$.
If $f(x) < f$_best **then**
x_best = x
f_best = $f(x)$
best_count = best_count+1
Record results on "Best So Far"
Endif
else
If random[0, 1] < $\exp(-\triangle_{x'x}/T)$ **then**
$x = x'$
$f(x) = f(x')$
renew = renew + 1
Endif
Set iteration_count = iteration_count + 1.
Endif

Until stopping criteria of inner loop is met

In implementing the SA algorithm described above, *initial feasible solution producing procedure, the procedure of selecting initial temperature, neighborhood solution search procedure, efficient way to calculate the difference of objective function values of two neighborhood solutions, acceptance function (here it is the Metropolis function), cooling scheduling of the temperature, and stopping criterions of inner and outer loops* are its important components. Different definitions of those are discussed in the literature on SA methods, which will be discussed in the following subsections.

SA algorithm corresponds to a Markov chain. For each temperature T fixed, if the variation of Markov chain arrives at a stable state and then T goes down, we call the SA algorithm *homogeneous SA algorithm*; if not, it is a *non-homogeneous SA algorithm*.

11.5.2 Initial Feasible Solution Producing Procedure

For a convex function the initial feasible solution can be chosen anywhere, from which the global minimum is reached by moving towards to the lower values, in the feasible region; however, for a non-convex function, it depends on the initial solution very much either to find a local minimum of it or to find its global minimum [545]. For real projects, usually there are many requirements, i.e. constraints, for reaching its aims.

A good feasible initial solution producing procedure is clearly needed. Numerical experiments show that, without the sensitive procedure of choosing the initial simplex (see § 3.1.1 of [279]), the Simplex Simulated Annealing (SSA) method of paper [279] is very difficult to make it work. However, for many problems solved by SA method, there, often, is a simple way of producing initial solution: randomly taking a feasible solution from the feasible region as the initial solution. Constraint programming is a new high-level language paradigm for satisfaction and optimization problems. To produce a feasible solution by constraint programming strategy as the initial solution is also a very popular way; for example, see [422]. Using a local/global search optimization method to quickly get a solution as the initial solution of SA is also a good way.

11.5.3 Initial Temperature Selecting Procedure

Numerical experiments tell us that proper initial temperature T_0 can make the SA method quickly get the optimal value of the objective function. If at initial temperature we accept almost all the solution (i.e. acceptance rate $\chi_0 \approx 1$), then in theory, by Metropolis criterion $\exp(\Delta_{x'x}/T_0 \approx 1)$, where T_0 should be "sufficiently" large. Johnson, Kirkpatrick, Aarts et al. [2, 318, 319, 339, 360] presented several initial temperature selecting procedures. The idea of Kirkpatrick is: to choose a large T_0, give χ_0 in advance (for example χ_0=0.8), generate many solutions, if the acceptance rate χ is less than χ_0 then increase T_0, repeat until $\chi > \chi_0$ to get a T_0. Johnson's formula is

$$T_0 = \frac{\Delta^- f^+}{\ln(\chi_0^{-1})}$$

where $\Delta^- f^+$ is the average increase of objective function values of randomly generated solutions. The one of [2] is frequently used; for example, in [279, 413]. Aarts' formula is

$$T_0 = \frac{\Delta^- f^+}{\ln(\frac{m_2}{m_2\chi - m_1(1-\chi)})}$$

where m_1 is the number of solutions making the objective function value decrease, m_2 is the number of solutions making the objective function value increase, and χ, for example, may be set as χ_0. However, those procedures are not definitely working well for all problems. Fixed temperature schedule is studied and applied in [152, 174, 208]. In homogeneous SA, T_0 chosen should be properly large enough to sufficiently accept all the candidate feasible solutions possibly produced. We also often use the following ways:

(1) Uniformly sample some solutions, calculate their objective function values, and take the variance of those objective function values as T_0.
(2) Randomly generate some solutions, determine $|\triangle_{\max}|$ which is the maximal difference of each pair of solutions, and calculate $T_0 = -\triangle_{\max}/lnp_r$, where p_r is the initial acceptance probability and in theory it should be close to 1. For non-homogeneous SA, in theory we have formulas [598] for calculating T_0.

11.5.4 Neighborhood Solution Searching Procedure

This is one key element in implementing SA. For discrete optimization problem, for instance, in the book [582], there are several successful neighborhood solution searching procedures. All those procedures should be at least based on two basic ideas: (a) neighbor means "nearby", (b) SA method is a stochastic method so that the neighborhood solution should be randomly taken. We may take those ideas for developing neighborhood solution searching procedure for continuous optimization problems.

First we review some neighborhood solution searching procedures of continuous optimization problems. Miki et al. (1999) presented a formula $x_i^{k+1} = x_i^k + r * m$, where r is a random number with uniform distribution in [−1,1], m is the neighborhood range which makes the rate between accepted and rejected moves approximate 0.5 [425]. In [62], first generate a random direction vector $\theta_k \in R^n$, with $||\theta_k|| = 1$, then find a fixed step size $\triangle r$, thus get a neighborhood solution x_{k+1} of $x_k : x_{k+1} = x_k + \triangle r\theta_k$. The choice of $\triangle r$ is thoroughly discussed in [66, 599]. In [599], the direction vector θ_k is defined in a new way. It is suggested to take into account the point $x_h, h < k$, generated by the algorithm and different from x_k, if $f(x_h) < f(x_k)$ then $\theta_k = x_h - x_k$, otherwise $\theta_k = x_k - x_h$. Contrary to the results in [62, 66], Corana et al. (1987), Siarry et al. (1997) and Vanderbilt et al. (1984) search through the space of feasible region in an anisotropic way [158, 578]. In [158], at each iteration k a single variable of x_k is modified, and iterations are subdivided into cycles of n iterations during which each variable is modified; i.e. $x_{k+1} = x_k + rv_{i+1}e_{i+1}$, where r is a uniform random number in [-1,1], $i \in \{0, \ldots, n-1\}$ is such that $k + 1 = h * n + i$ for some nonnegative integer h, and v_{i+1} (that is anisotropic) is the maximum allowed step along the direction e_{i+1} of the $(i+1)$-st axis. Instead of varying a single variable in x_k in each iteration, Siarry et al. (1997) varied p variables [529]. Another concept to simulated annealing method is *adaptive* (see [247, 303–306, 320]). This means SA method should possess the ability of adapting itself to the problem it solves, the objective function f and the temperature, etc. whether globally or locally. The code of Ingber's ASA (Adaptive Simulated Annealing) algorithm [303–306] can be retrieved from the web site www.ingber.com, and many techniques such as "fast annealing", "re-annealing", "quenching", "multistart strategy", and "priori information" are used.

Romeijn et al. (1994, 1999) proposed a two-phase generator: *"first generating a random direction θ_k, with $||\theta_k|| = 1$, and then generating a random point λ_k in the set $\Lambda_k = \Lambda_k(\theta_k) = \{\lambda : x_k + \lambda\theta_k \in X\}$, thus $x_{k+1} = x_k + \lambda_k\theta_k$"*, and if x_{k+1} is not in X or if there is a jamming problem, i.e. Λ_k is very small, then use the *'reflections'* technique [498, 499]. Employing computer science theory is also useful for the neighborhood solution searching procedure; for example Bilbro et al. (1991) gave a tree annealing approach [53]: *divide the feasible regions in the form of a tree, and x_{k+1} is sampled from a distribution which selects a random path from the root of the tree to a leaf of the tree in a way that the sub-regions with a high percentage of accepted points are favored.* Employing a local search method into simulated annealing method is also very popular. Desai et al. (1996) proposed a technique [179]: *randomly perturb the current solution x_k to get a new point $\bar{x}_{k+1}$, and start a local search from $\bar{x}_{k+1}$ to get a new local minimum x_{k+1}*, which attempts to combine the robustness of annealing in rugged terrain with the efficiency of local optimization methods in simple search spaces. The parallel version of [179] may be seen in [654]. Lucidi et al. (1989) presented a random tunneling technique by means of acceptance-rejection sampling [400]. Over the unit hypercube feasible region, Fox (1995) gave a special neighborhood solution searching procedure [212]. First the objective function f is evaluated at the points of a net. Then, the unit hypercube is subdivided into many boxes of a set C, which are with different widths along different axes. Over C, generate a probability mass function p by intersecting the net with each box to generate many points and find the minimum value of those points. Then sample a box B from C according to the probability mass function p, sample a uniform point y from B and apply some local search steps starting from y. Repeat the sampling of B and y a finite number of times, and get the set $F(x_k, k)$, which is a finite set of candidate neighborhood points of x_k at iteration k. And then the acceptance probabilities are applied to define the distribution of the next iteration $L \geq k$, and according to the acceptance probabilities the algorithm randomly selects a point in $F(x_k, k)$ and moves to it. For getting neighborhood solution, there is an idea: simultaneously perturbing all of the variables of the problem in a "proper" random fashion. In [279], instead of the reflect-expand-contract-shrink Nelder-Mead method (seen, for example, in [280]), the reflects-shrink Simplex Direct Search (SDS) method is given. The SDS method uses various numbers of reflections in a flexible manner and follows a shrinking if after reflecting all the n worst vertices of the initial simplex better movement is still failed to be gotten.

In Fast Simulated Annealing method of [560], Cauchy distribution is used to produce new solution. Greenes et al. (1986) used the probability of fitness function, which is based on objective function, to produce new solution [256].

From the ideas of all those reviewed above, we present two versions of the neighborhood solution search procedure for SA algorithm. In the SDS algorithm, for the objective function value of each vertex we add a random fluctuation: $f(x_i) + k_B T \log(z)$, where k_B is the Boltzmann constant in appropriate units and z is a random number of the interval (0,1). We might carry out a multiple shrinking, in which the highest vertex is simultaneously moved along all the coordinates towards the lowest-energy vertex.

Noticing that the neighborhood solution search for simulated annealing method should be at least based on ideas (a) and (b), we may simply give a neighborhood solution search procedure for simulated annealing algorithm: *Uniformly randomly keeping $n-1$ elements of x, and making the left one element of x uniformly randomly take a value such that the new solution x' is still feasible. This gives x'*. When the feasible region of the optimization problem is the unit simplex S, the neighborhood solution search procedure should be modified: *Uniformly randomly keeping $n-2$ elements of x, and making one element from the two elements left to x uniformly randomly take a value from [0,1] such that the value of the sum of the $n-1$ elements is not greater than 1. Another left element of x' is given the value 1-sum. This gives x'*.

Another version of neighborhood solution procedure may be found in the following SA pseudo-code:
New version of the simulated annealing algorithm.

```
X_best = x & f_best = f; q := q0
DO j := 1 to J
   T := T0(j)
   Repeat
      DO k := 1 to L
         Repeat
            Randomly generate the search direction d ∈ (−1, 1)
            Let x'_i = x_i + qd and x'_l = x_l when l ≠ i
         Until x' is feasible
         Calculate Δx'x
         IF (Δx'x < 0) or (exp(−Δ_{x'x}/T) > random[0, 1]) THEN
            Accept x'
            IF f < f_best THEN x_best = x & f_best = f
         Calculate next annealing temperature T
         Adjust the step length q := g(Acc) ∗ q (where g() is a function given)
      END DO
   Until outer loop stopping criterion is satisfied
   IF f_best < f THEN x = x_best & f = f_best
END DO
```

where $g()$ is an adjustment function, for example $g(x) = (x-0.5)^3+1$. As a whole, the new solution generating procedure composes two parts: the way to generate candidate solution, and how to generate the probability distribution of the candidate solution. Hence, we may replace our uniform distribution by Cauchy distribution, Gauss normal distribution, or their combined distribution, and get some new results for comparisons.

11.5.5 Efficient Calculation of Cost Difference

Since a very large number of iterations are performed, it is essential to calculate efficiently the cost differences between a solution and its neighborhood solution. Take a simple instance, if $f(x) = Ax + b$, it is clearly $\triangle_{x'x} = A(x' - x)$ is much efficient than $\triangle_{x'x} = (Ax' + b) - (Ax + b)$, especially when the computational effort is very much for the computer.

Note: There is a subtle difference in the meaning between *objective function* and *fitness function*. *Objective function* measures the variable's performance in the search region; whereas *fitness function* provides a measure of variable's relative performance by transforming the objective function f into $F(x) = g(f(x))$. For example, proportional fitness assignment ($F(x_i) = f(x_i)/\sum f(x_i)$), linear transformation ($F(x) = af(x) + b$) are some simple transformations.

11.5.6 Acceptance Function

The (Markov chain state) acceptance function, generally, is given in a probability form that should meet the following criteria:

(1) At each fixed temperature, the acceptance function should maintain the average percentage of accepted moves at about 1/5 of the total number of moves, which cannot make the objective function value decrease;
(2) With the decrease of temperatures, the probability of accepting an increasing move decreases; and
(3) When temperature becomes zero, only the solutions that make the objective function value decrease can be accepted.

The usual acceptance function for the SA algorithm is the so-called Metropolis function

$$A(x, x', T) = \min\{1, exp\{-\triangle_{x'x}/T\}\}.$$

Note here we let the acceptance function depend on the difference of the function values of x and x' instead of depending directly on x and x'. More generally, we may rewrite the above formula as follows:

$$A(x, x', T) = \min\{1, exp\{-\triangle_{x'x}/\gamma(T)\}\},$$

if $\gamma : (0, +\infty) \rightarrow (0, +\infty)$ is a strictly increasing function under some balance conditions [520]. Barker's function

$$A(x, x', T) = 1/(1 + exp\{-\triangle_{x'x}/T\})$$

is another popular acceptance function. Johnson et al. (1987) suggested

$$A(x, x', T) = \min\{1, 1 - \triangle_{x'x}/T\}.$$

and made the speed of SA algorithm increase by 30 % [317]. Sechen (1988) used TS to reduce the time wasted on calculating $\exp\{-\Delta_{x'x}/T\}$ [522].

11.5.7 Cooling Scheduling of Temperature

During the SA iterations, the temperature sequence $\{T_k\}$ is being produced. If $\lim_{k\to+\infty} T_k = 0$, we say that $\{T_k\}$ is a cooling schedule. In this subsection, we review some successful cooling schedules.

Aarts and Laarhoven (1985) presented a cooling scheme [2]

$$T_{k+1} = T_k/(1 + T_k \log(1+\epsilon)/(3\sigma_k)),$$

where σ_k is the standard deviation of the observed value of the cost function, and ϵ is 0.1 in [152].

$$T_{k+1} = T_k exp(-T_k(f_{T_k} - f_{T_{k-1}})/\sigma^2 T_k)$$

is another cooling schedule [5]. Reeves (1995) described a cooling schedule of Lundy and Mees [490], where the temperature is reduced according to

$$T_{k+1} = T_k/(1 + \beta T_k),$$

or equivalently,

$$T_k = T_0/(1 + k\beta T_0),$$

where β is suitably small, and only one iteration is performed in each inner loop. For the convergence of non-homogeneous SA method, in 1984 Geman et al. (1984) gave the Boltzmann annealing or called classical SA [239], in which the temperature is calculated by

$$T_k = T_0/\ln(k+c), k = 1, \ldots, \infty,$$

where $c = 1$. A little modification of c is used in [545]:

$$T_k = T_0/\ln(k+c), k = 0, \ldots, \infty$$

with $c = e = 2.7183$. For the above two formulas c should not be less than 1. In 1987, Szu et al. (1987) proposed the Fast Annealing method [560]. The cooling schedule of this method is with a faster decrease:

$$T_k = T_0/(k+1), k = 1, \ldots, \infty$$

that decreases sharper than $T_k = T_0/\ln(k+c)$. However, we should match the rate of temperature decrease with the neighborhood solution generating procedure. Nahar et al. (1987) divided $[0, T_0]$ into K intervals and find $T_k, k = 1, \ldots, K$ [360]. The Very Fast Simulated Re-annealing method [303] was presented in 1989 by Ingber. Its cooling schedule is

$$T_k = T_0 exp(-ck^{1/n}), k = 1, \ldots, \infty$$

where c is a scale factor. Ingber (1989) also used a slower schedule [303] of $T_k = T_0/(k+1)$, which is

$$T_k = T_0/(k+1)^{1/n}, k = 1, \ldots, \infty.$$

Although many cooling schedules are mentioned above, the geometric cooling scheme [339] proposed by Kirkpatrick et al. (1983)

$$T_{k+1} = \alpha T_k, k = 0, \ldots, \infty,$$

where $\alpha \in (0, 1)$ is a constant, is still a widely used and a popular SA cooling schedule (refer to [2, 279, 318, 319, 360, 490]) because it compromises the quality and CPU time of optimization. Kirkpatrick et al. (1983) take $\alpha = 0.95$; and Johnson et al. take $\alpha \in [0.5, 0.99]$. Our numerical experiments also shows that (0.8, 0.99) is a good interval chosen for α. Given T_0, T_f and the number of outer loop iterations, the webpage www.btluke.com/simanf1.html lists many kind of cooling schedules.

11.5.8 Stopping Criterion of Inner Loops

The number of iterations in each inner loop is also called the temperature length. In many forms of simulated annealing method, a fixed number of iterations are performed for each temperature. Usually the fixed number is detected by a long sequence of iterations in which no new solutions have been accepted. This fixed number depends on the size of the neighborhood Nsize, which is defined to be the total number of possible distinct neighborhood solutions that can be obtained, and its mathematical form is $Nfactor * Nsize$, where $Nfactor$ is some multiplying factor, for example $Nfactor = 10$. In our pseudo-code of the simulated annealing method, we also introduce a symbol renew which records the number of times the solutions are accepted at a temperature. We may also terminate the inner loop if this number has exceeded $Cut*Nfactor*Nsize$ where Cut is another multiplying factor. Section 4.2.3 of [545] describes this fixed number in view of stochastic process terminologies. In homogeneous SA, from the view of objective function values two stopping criteria may be presented:

(1) Checking stability of the expectation value of objective function values; and
(2) The change of objective function values is lower than some tolerance in a certain amount of iterations.

11.5.9 Stopping Criterion of Outer Loops

The choice of *final temperature Tf* determines a stopping criterion of outer loops. At the end of each inner loop, if the best solution obtained in that inner loop has not been changed and at the same time we are not having many changes in the current solution, we reduce the temperature and start another inner loop. If a solution $x*$ has been consecutively "frozen" at many current temperatures, then we stop and say $x*$ is the best solution found by the simulated annealing method. Usually we set a proper small temperature as Tf as the stopping condition of outer loops. Our numerical experiments show that we may get the small temperature T when "Floating point exception (core dumped)" is reminded by computer. In our pseudo-code, we also count the number *best_count* of times at which the best feasible solution is not replaced again, and calculate the proportion of solutions accepted, by *frozen_count*. We may halt the algorithm when *frozen_count* reaches a predetermined number. As a whole, based on $\lim_{k\to\infty} T_k = 0$, we find a "frozen" temperature for the stopping criterion of outer loops. Nahar et al. (1987) used the number of temperatures, in other words, the number of Markov chains or iterations, as stopping criterion of outer loops [360]. Notice here the lengths of Markov chains, L_k, may be upper bounded by a constant $\bar{L}$. Johnson et al. (1987) used the acceptance rate to terminate the outer loops [317]: current acceptance rate $\chi_k > \chi_f$ given, where χ_f is the final acceptance rate.

From the point of view of the objective function values, we may also give the terminating criterion for the outer loops. If $|\Delta_{x'x}| \leq \epsilon$ or $f(x) - f_{best} \leq \epsilon$, $|\frac{f_{best} - f_{opt}}{f_{opt}}| \leq \epsilon$ (where $|f_{best} - f_{opt}| < \epsilon$ if $f_{opt} = 0$), where ϵ is a sufficiently small positive number and f_{opt} is the optimal value known, we stop the method. This simply means when the objective function values cannot be improved we may stop the algorithm.

In another form, we use the information not only on the temperature but also on the objective function values; then we can also give a stopping criterion for the simulated annealing method. Suppose P_F is a proper number given, if $A(x, x', T) = P_F$, we stop the simulated annealing method. If in many successive Markov chains the solution has not changed, we can stop the method.

Setting an upper limit of executing time is also a way to stop the algorithm. The user may terminate the method manually according to a user-defined aim.

Chapter 12
Hybrid Method of Discrete Gradient with Simulated Annealing or Genetic Algorithm

12.1 Overview

Prion diseases include CJD, vCJDs, GSS syndrome, FFI, Kuru in humans, scrapie in sheep, BSE (or 'mad-cow' disease) and CWD in cattle. These diseases are invariably fatal and highly infectious neurodegenerative diseases affecting humans and animals. However, for threating all these diseases, there is no effective therapeutic approaches or medications [8, 486, 624]. Prion diseases are caused by conversion of prion proteins from a soluble into an insoluble fibrillar form [484, 486]. The normal cellular prion protein (PrP^C) is rich in α-helices but the infectious prions (PrP^{Sc}) are rich in β-sheets [85, 169, 257, 447, 457, 486, 491]. For PrP^C, the N-terminal residues (1–123) are unstructured, but the C-terminal residues (124–231) are well structured with three α-helices, two short anti-parallel β-strands, and a buried disulfide bond between the 2nd and 3rd α-helix (residues number 179 and number 214). The prion protein molecular structure and dynamics can be seen from [679, 681, 697] etc. For PrP^{Sc}, the rich β-sheets formulate into prion amyloid fibrils.

The hydrophobic region (113–120) AGAAAAGA palindrome of prion proteins falls just within the N-terminal unstructured region (1–123) of prion proteins which is hard to determine its molecular structure using NMR spectroscopy or X-ray crystallography [494]. However, many experimental studies such as [67, 68, 70, 288, 315, 316, 357, 445, 619] have shown that: (1) the hydrophobic region (113–120) AGAAAAGA of prion proteins plays an important role in the conversion of PrP^C to the abnormally folded form PrP^{Sc}; and (2) AGAAAAGA is important for amyloid fibril formation and is an inhibitor of prion diseases. Zhang (2009) also confirmed through computer molecular dynamics simulations that the stability of prion proteins might be attributable mainly to the N-terminal unstructured region (1–123) [678]. Due to the noncrystalline and insoluble nature of the amyloid fibril, it is very difficult to obtain atomic-resolution structures of AGAAAAGA using traditional experimental methods [571, 710]. For the sake of clarity on computers again, we use the program used by [695] to theoretically

J. Zhang, *Molecular Structures and Structural Dynamics of Prion Proteins and Prions*, Focus on Structural Biology 9, DOI 10.1007/978-94-017-7318-8_12

confirm that prion (113–120) AGAAAAGA segment has an amyloid fibril forming property. The theoretical computation results are shown in Fig. 11.4 of Chap. 11, from which we can see that the prion AGAAAAGA region (113–120) is clearly identified as the amyloid fibril formation region because the energy is less than the amyloid fibril formation threshold energy of −26 KCal/mol [695]. Thus, we got confidence in constructing the atomic-resolution molecular structures of prion (113–120) AGAAAAGA amyloid fibrils by computer computational approaches or introducing novel mathematical formulations and physical concepts.

Many studies have indicated that computational approaches or introducing novel mathematical formulations and physical concepts into molecular biology, such as Mahalanobis distance [115, 141], pseudo amino acid composition [116, 125], graphic rules [15, 113, 114, 124, 715], complexity measure factor [645, 646], homology modeling [117], cellular automaton [641, 643, 644, 648], molecular docking [139], grey theory [642], geometric moments [647], low-frequency (or Terahertz frequency) phonons [110, 111, 129], solitary wave [142, 536], and surface diffusion-controlled reaction [143], can significantly stimulate the development of biological and medical science. Various computer computational approaches were used to address the problems related to "amyloid fibril" [86, 118, 120, 130, 609, 620]. Here, we would like to use the hybrid local and GO search methods to investigate the optimal atomic-resolution amyloid fibril models in hopes that the findings thus obtained may be of use for controlling prion diseases. Using the traditional local search steepest descent and conjugate gradient methods hybridized with the standard global search simulated annealing method, Chap. 11 successfully constructed three optimal atomic-resolution structures of prion AGAAAAGA amyloid fibrils. These structures were constructed based on the breakthrough work of [517]. In Chap. 11, the author pointed out that basing on the NNQNTF peptide of elk prion 173–178 (PDB entry 3FVA that was released into Protein Data Bank (www.rcsb.org) on 30-JUN-2009, deposition date 15-JAN-2009) we might also be able to construct amyloid fibril models for prion AGAAAAGA palindrome; this chapter is doing this homology model construction work. The homology models were built using an improved hybrid SA [339] Discrete Gradient (DG [21–23]) method. Then the models were optimized/solved using the traditional SD and CG local search methods of [89]; the former has nice convergence but is slow when close to minimums and the latter is efficient but its gradient RMS and GMAX gradient do not have a good convergence [89]. We used the SD method followed by the CG method to optimize our models. When the models could not be optimized further, we employed the standard global search SA method of [89] to escape from the stationary point calculated by the local search SD & CG methods. Through the further refinement of SD and CG local search methods, at last two optimal models were successfully got. Numerical results in this chapter show that the hybridization of local and global search optimization methods is very effective. X-ray crystallography finds the X-ray final structure of a protein, which usually need refinements using a SA protocol in order to produce a better structure; this chapter also correctly illustrates the SA protocol of crystallography.

X-ray crystallography is a powerful tool to determine the protein 3D structure. However, it is time-consuming and expensive, and not all proteins can be successfully crystallized, particularly for membrane proteins. Although NMR spectroscopy is indeed a very powerful tool in determining the 3D structures of membrane proteins [81, 82, 455, 467–470, 519, 606], it is also time-consuming and costly. To the best of the author's knowledge, there is little structural data available on the AGAAAAGA palindrome in the hydrophobic region (113–120) of prion proteins, although many experimental studies have shown that this region has amyloid fibril forming properties. In view of this, the present study was devoted to address this problem from computational approaches such as global energy optimization [131, 137], simulated annealing [112, 126], and structural bioinformatics [120]. Thus, the optimization computational approaches (such as SADG, GADG, SDCG etc) presented in this chapter are very necessary and important to study amyloid fibrils, nanotubes, etc. The prion AGAAAAGA optimal atomic-resolution structures of this chapter might have a value for finding treatments for prion diseases.

12.2 The Optimization Model Building

Recently the protein fibril structure of NNQNTF (173–178) segment from elk prion protein was released [631]. Its PDB entry ID is 3FVA in the Protein Data Bank. This fibril has six chains, belonging to Class 1 of [517]. The atomic structure is a steric zipper, with vdW interactions between β-sheets and HBs to maintain the β-strands.

Basing on this steric zipper, two prion AGAAAAGA palindrome amyloid fibril models – a six chains AAAAGA model (Model-chapt12-1) and a six chains GAAAAG model (Model-chapt12-2) – will be successfully constructed. The minimum sequence necessary for fibril formation should be AGAAA, AAAGA, AGAAAA, GAAAAG, AAAAGA, AGAAAAG, GAAAAGA or AGAAAAGA (also for Chap. 11), which are important for fibril formation and are an inhibitor of PrP^{Sc} neurotoxicity [67]. Because the peptide NNQNTF has six residues and the six chains AGAAAA model could not successfully pass SA, AAAAGA and GAAAAG were picked out of the eight possible sequences. The D chain (i.e. β-sheet 2) of 3FVA.pdb can be obtained from A Chain (i.e. β-sheet 1) using the mathematical formula

$$D = \begin{pmatrix} -1 & 0 & 0 \\ 0 & 1 & 0 \\ 0 & 0 & -1 \end{pmatrix} A + \begin{pmatrix} -14.31482 \\ 2.42 \\ -21.03096 \end{pmatrix}. \tag{12.1}$$

AD chains of Model-chapt12-1 and Model-chapt12-2 (Figs. 12.1∼12.2) were respectively got from AD chains of 3FVA.pdb using the mutate module of the free package Swiss-PdbViewer (SPDBV Version 4.01) (spdbv.vital-it.ch), but the vdW contacts are too far, very bad at this moment (Figs. 12.1∼12.2). To get good vdW interactions will be an optimization problem described as follows.

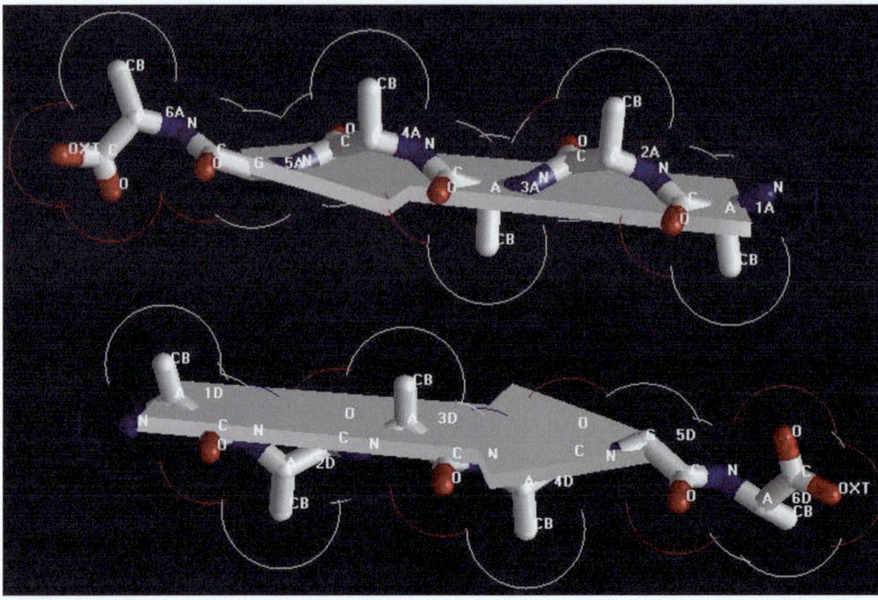

Fig. 12.1 Model-chapt12-1 – bad vdW contacts of AD chains of the AAAAGA model

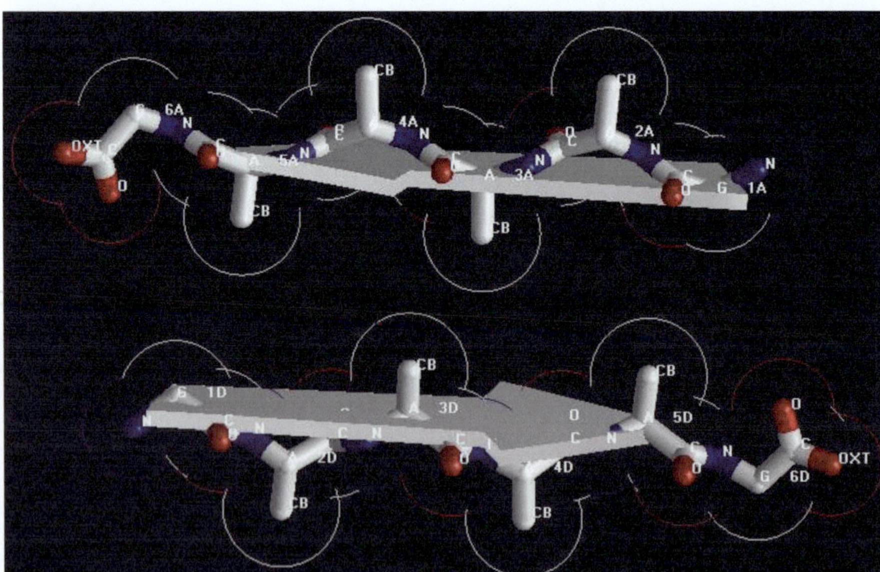

Fig. 12.2 Model-chapt12-2 – bad vdW contacts of AD chains of the GAAAAG model

Neutral atoms are subject to two distinct forces in the limit of large distance and short distance: a dispersion force (i.e. attractive vdW force) at long ranges, and a repulsion force, the result of overlapping electron orbitals. The LJ potential represents this behavior (en.wikipedia.org/wiki/Lennard-Jones_potential, or [397] and references therein). The LJ potential is of the form

$$V(r) = 4\varepsilon\left[\left(\frac{\sigma}{r}\right)^{12} - \left(\frac{\sigma}{r}\right)^{6}\right], \tag{12.2}$$

where ε is the depth of the potential well and σ is the atom diameter; these parameters can be fitted to reproduce experimental data or deduced from results of accurate quantum chemistry calculations. The $(\frac{\sigma}{r})^{12}$ term describes repulsion and the $(\frac{\sigma}{r})^{6}$ term describes attraction. If we introduce the coordinates of the atoms whose number is denoted by N and let $\varepsilon = \sigma = 1$ be the reduced units, the form (12.2) becomes

$$f(x) = 4\sum_{i=1}^{N}\sum_{j=1,j<i}^{N}\left(\frac{1}{\tau_{ij}^{6}} - \frac{1}{\tau_{ij}^{3}}\right), \tag{12.3}$$

where $\tau_{ij} = (x_{3i-2} - x_{3j-2})^2 + (x_{3i-1} - x_{3j-1})^2 + (x_{3i} - x_{3j})^2$, $(x_{3i-2}, x_{3i-1}, x_{3i})$ is the coordinates of atom i, $N \geq 2$. The minimization of LJ potential $f(x)$ on R^n (where $n = 3N$) is an optimization problem:

$$\min_{s.t. x\in R^{3N}} f(x). \tag{12.4}$$

For solving the optimization problem (12.4), many studies, for examples [154, 189, 295, 296, 370, 460, 500, 634, 635, 652–656], have been done. A brief review on some results and computational algorithms before the year 2003 for solving (12.4) can be referred to [676], for the year 2004 can be referred to [638, 639]; the website www-wales.ch.cam.ac.uk/index.html should not be ignored. Basing on these studies, we present the following two successful algorithms.

Algorithm 12.1. Hybrid SADG method

Initialization:

Define the objective function f *and its feasible solution space.*
Call the *initial feasible solution generating procedure* to get x.
Call *initial temperature selecting procedure* to get T.
Inialization of f: $f = f(x)$.
Initialize the neigbourhood feasible solution: $x\_neighbour = 0$.
Initialize x_best: $x\_best = x$.
Initialize f_best: $f\_best = f$.

do {

DG local search part:

f_best_local = **local_search**(x_best, x_new_gotten);
$x = x\_new\_gotten$;

SA global search part:

do {
  do {
    $x\_neighbour = randomly\_perturb(x)$ [24];
    $f\_neighbour = f(x\_neighbour)$;
    Calculate the difference $\Delta = f\_neighbour - f$;
    If $(\Delta \leq 0)$ **or** (random[0,1] $<$ exp($-\Delta/T$))
      $x = x\_neighbour \quad f = f\_neighbour$;
    If $(f \leq f\_best)$ $\quad x\_best = x \quad f\_best = f$;
  } while (equilibrium has not been reached);
  Temperature annealing
} while (SA stop criterion has not been met);

} while ($f\_best - f\_best\_local \leq -0.001$);

The local search DG method is efficient and effective but it is also always trapped in a local solution. SA is a global search method but sometimes just gets a global solution with low probability according to the Metropolis criterion. SA can make DG jump out of the local trap and then DG can make SA reach an optimal solution with the 100 % full probability. The limit of Algorithm 12.1 might be the large numbers of iterations in SA when searching the global solution with low probability.

Algorithm 12.2. Hybrid GADG (Genetic Algorithm [211] DG) method

Step 0. Set the seeding of the initial parental population. We set the cluster of 98 atoms as the base of the seed because the 98 atoms cluster has a tetrahedral structure.
Step 1. Apply the discrete gradient method on all individuals of the initial parental population to relax them to their nearest local minimal energy positions.
Step 2. Call the mating procedure of [171] to get the center of the mass of the parental population. Then set a random number for this mating procedure to mate more parents with each other. Thus the offspring population is produced.
Step 3. Select from the parental population and offspring population to get the best combination of a new population. Take the new population as parental population.
Step 4. Run the Newton method (where the Hessian matrix is calculated explicitly) [178] to relax all the individuals of parental population to local minimal energy positions.
Step 5. Run discrete gradient method to refine the local minimum positions. The refined local minimal energy positions are set as the parental population.
Step 6. Call the twinning mutations of [632]. Then set a random number as the mutation rate for these mutation schemes to make mutations to the parental population. Then offspring population is produced.
Step 7. Make a best combination of the parental population and the offspring population to get a new population. Take the new population as parental population.
Step 8. Run the explicit Newton method to relax all the individuals of parental population to local minimal energy positions.
Step 9. Repeat Step 5.
Step 10. Repeat Step 7.
Step 11. If the algorithm reaches its convergence, then terminates, otherwise, goto Step 2.

In Algorithm 12.2 the DG method and the explicit Newton method are local search optimization methods, and genetic algorithm is a global search method that brings the local solution to jump out of the trap. The shortcomings of the Algorithm 12.2 lie in its initial solution specially chosen; whereas Algorithm 12.1 can start from any initial solution. The computational load of Algorithm 12.2 is very heavy, compared with that of Algorithm 12.1.

SA and GA methods both are stochastic heuristic global search methods. But there are three basic differences between SA and GA: (1) SA simulates the annealing process of crystal materials with Monte Carlo property, GA simulates the process of natural competitive selection, crossover, and mutation of species; (2) GA is a parallel computing (population) algorithm and SA algorithm is a sequential computing (individual) algorithm; (3) The choice of initial solution is different. In Step 0 of Algorithm 12.2, we may choose clusters of 55, 75, 76, 85, 97, 98, or 147 atoms as the base of seed, because of their well-known structures of the best known solutions. For example, the 98 atoms cluster has a tetrahedral structure; the 55 atoms cluster has an icosahedral structure; the cluster of 75 or 85 atoms has an octahedral structure; and the cluster of 76 atoms has a decahedral structure. For 48 atoms, we use the structures of 13, 23, 19, 10 atoms. In brief, all those 38 atoms face-centred-cubic (fcc) truncated octahedron, 75–77 and 102–104 atoms decahedron, and 98 atoms tetrahedron structures are based on polytetrahedral packing, which is structurally more similar to the icosahedral packing of most Lennard-Jones clusters. Because of different background of SA and GA, the acceptance probability of solution is different very much. The computation load of GADG is clearly large than that of SADG.

In Algorithm 12.1 the DG method [22] is a derivative-free local search method for nonsmooth optimization with the continuous approximations to the Clarke subdifferential [151] and the Demyanov-Rubinov quasidifferential [176], and the SA algorithm is using the neighborhood search procedure of [24]. The convergence of the proposed hybrid method directly follows from the convergence of SA and DG methods. The hybrid method starts from an initial point, first executes DG method to find local minimum, then carries on SA method in order to escape from this local minimum and to find a new starting point for the DG method. Then we again apply the DG method starting from the last point and so on until the sequence of the optimal objective function values gotten is convergent. Similarly for Algorithm 12.2, the DG and Newton methods are local search optimization methods, GA is global search optimization method, and the convergence of the hybrid GADG method simply follows from that of these three optimization methods. Algorithm 12.2 can successfully reproduce all the best L-J potential energy values known (physchem.ox.ac.uk/~doye/jon/structures/LJ.html) nearly up to 310 atoms and fortunately some more precise energy values and better solution structures are got (Table 12.1 and Figs. 12.3∼12.4), but it is not easy to be applied to our model construction work of this chapter.

In implementing Algorithm 12.1, we use T = 0.9*T as the temperature annealing schedule and the initial temperature is taken large enough according to the rule in [339], where we take $T_0 = 300\,\text{K}$. We restrict the number of iterations for the outer procedure by 100 and number of iterations for the inner procedure by 1,000. The DG method part is terminated when the distance between the approximation to the subdifferential and origin is less than a given tolerance $\epsilon > 0$ ($\epsilon = 10^{-4}$). The proposed hybrid method fails to solve (12.4) when number of atoms $N \geq 20$

Table 12.1 Our precise LJ potential energy best values

Number of atoms	Our precise best value	Best value known*
39	−180.033185202447	−180.033185140508
40	−185.249838614238	−185.249838598471
42	−196.277533506901	−196.277533404920
48	−232.199531999140	−232.199529316227
55	−279.248470461822	−279.248470308143
75	−397.492330708363	−397.492330681104
76	−402.894865906469	−402.894865881675
97	−536.681382651509	−536.681382483245

*http://doye.chem.ox.ac.uk/jon/structures/LJ.html

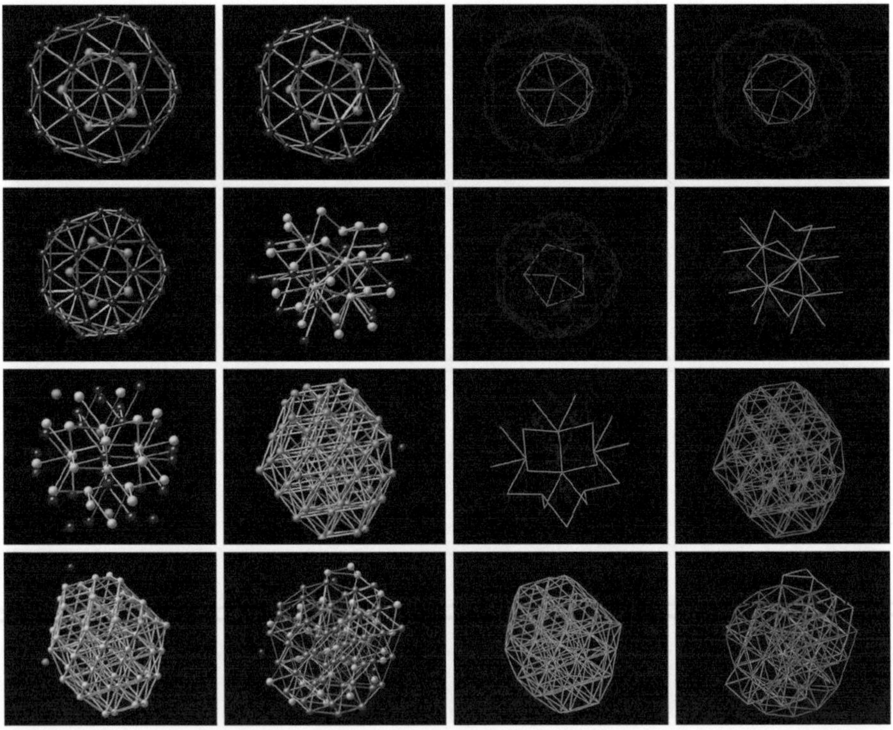

Fig. 12.3 N = 39, 40, 42, 48, 55, 75, 76, 97 (got by Algorithm 12.2)

(however, the AD Chains of Model-chapt12-1 and Model-chapt12-2 have 60/58 atoms). In order to reduce the number of local minima we suggest to approximate the function (where this idea was contributed by Birgov A.M. of Ballarat University and implemented by the author)

$$\varphi(\tau) = 4\left(\frac{1}{\tau^6} - \frac{1}{\tau^3}\right) \tag{12.5}$$

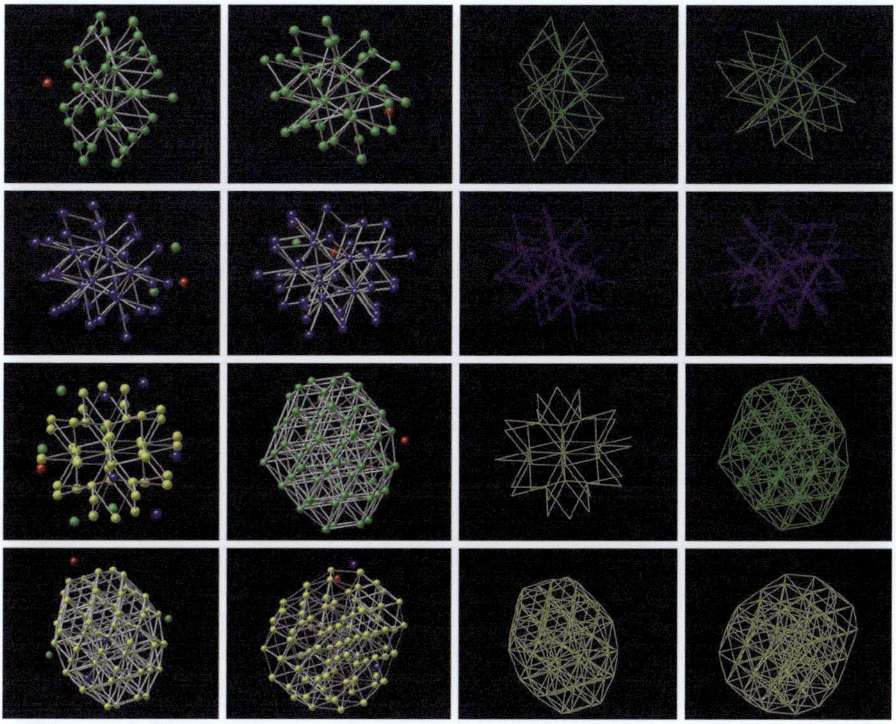

Fig. 12.4 N = 39, 40, 42, 48, 55, 75, 76, 97 (best structures known)

(which is neither convex nor strictly the difference of two convex functions) by the following function [677]:

$$g(\tau) = \max(g_1(\tau), \min(g_2(\tau), g_3(\tau))) \tag{12.6}$$

where $g_1(\tau)$ is the piecewise linear approximation of the function $\varphi(\tau)$ in segment $(0, r_0]$, $g_2(\tau)$ is the piecewise linear approximation of this function over segment $[r_0, r_1]$, and finally $g_3(\tau)$ is the piecewise linear approximation over $[r_1, b]$ and b is an enough large number. Here

$$r_0 = \sqrt[3]{2}, \quad r_1 = 1/\sqrt[3]{2/7}. \tag{12.7}$$

Such an approximation of the function $\varphi(\tau)$ allows us to remove many local minima of the LJ potential function and to get a good approximation to the global minimum of the objective function f in problem (12.4) (a note: recently GO algorithm has been designed at Ballarat University for piecewise linear functions based on the new codifferential theory). In numerical experiments we take $b = 16$ and divide the segment $[0.001, r_0]$ into 100 segments, the segment $[r_0, r_1]$ into 100 segments and the $[r_1, 16]$ into 50 segments which allows one to get good approximations for the function $\varphi(\tau)$. The replacement of the function $\varphi(\tau)$ by the function $g(\tau)$ makes the objective function nonsmooth. On the other side such a replacement significantly reduce the number of local minima. Since the discrete gradient method is a method of nonsmooth optimization the proposed hybrid method can be applied for solving this transformed problem. When solving the L-J problem (12.4), first we use the DG method with build-up technique to relax to an initial solution. Then we apply the hybrid method, with the above approximation for the objective function, to get another initial solution. Starting from this initial solution we again apply the derivative-free DG method and at last get the global solution. Results of numerical experiments (Table 12.2) show that our techniques can effectively solve the L-J problem (12.4) when number of atoms is not greater than 310. For Model-chapt12-1,

Table 12.2 Our numerical results for the LJ Potential Problem

Number of atoms	Best value obtained	Best value known*
19	−72.659782	−72.659782
20	−77.177043	−77.177043
21	−81.684571	−81.684571
22	−86.573675	−86.809782
23	−92.844461	−92.844472
24	−97.348815	−97.348815
25	−102.372663	−102.372663
27	−112.825517	−112.873584
30	−128.096960	−128.286571
34	−150.044528	−150.044528
44	−207.631655	−207.688728
49	−239.091863	−239.091864
56	−283.324945	−283.643105
65	−334.014007	−334.971532
67	−347.053308	−347.252007
84	−452.267210	−452.6573
93	−510.653123	−510.8779
148	−881.072948	−881.072971
170	−1,024.791771	−1,024.791797
172	−1,039.154878	−1,039.154907
268	−1,706.182547	−1,706.182605
288	−1,850.010789	−1,850.010842
293	−1,888.427022	−1,888.427400
298	−1,927.638727	−1,927.638785
300	−1,942.106181	−1,942.106775
301	−1,949.340973	−1,949.341015
304	−1,971.044089	−1,971.044144
308	−1,999.983235	−1,999.983300

*http://doye.chem.ox.ac.uk/jon/structures/LJ.html

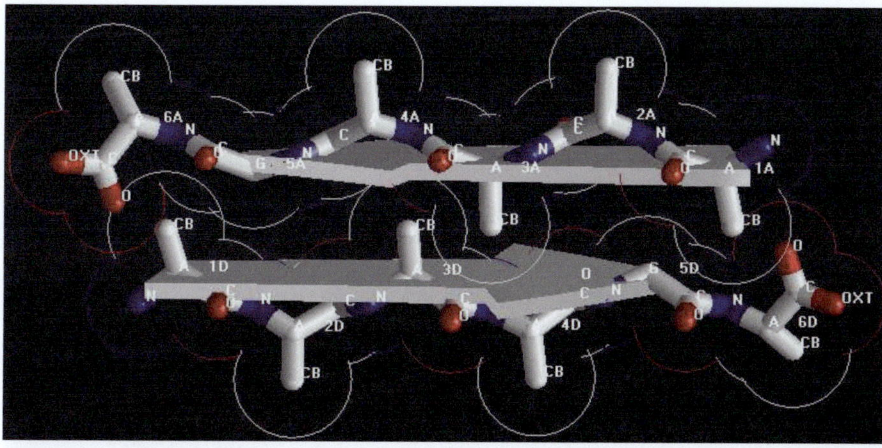

Fig. 12.5 Model-chapt12-1 – good vdW interactions of AD chains of the AAAAGA model

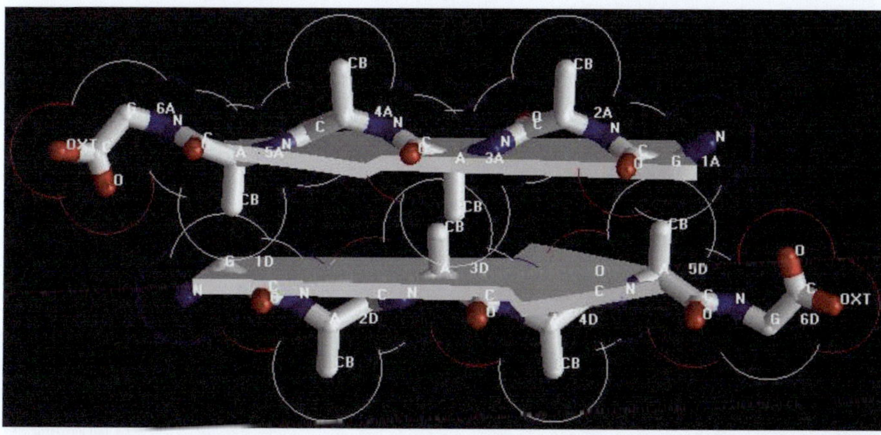

Fig. 12.6 Model-chapt12-2 – good vdW interactions of AD chains of the GAAAAG model

seeing Fig. 12.1 we may know that vdW interactions such as between 1D.CB–6A.O, 3D.CB–3A.CB, 6D.O–1A.CB, etc. should be maintained. Solving the optimization problem (12.4) can get A Chain and D Chain, where D Chain should have good vdW interactions with A Chain. Similarly for Model-chapt12-2, vdW interactions should be maintained between 3D.CB-3A.CB, etc. (Fig. 12.2). AD Chains in all have 60/58 atoms. Thus, we may use the above improved hybrid SADG algorithm to easily get the optimal coordinates of AD Chains of Model-chapt12-1 and Model-chapt12-2, where D Chain has good vdW interactions with A Chain now (Figs. 12.5∼12.6). Other chains (i.e. β-strands) of Model-chapt12-1 and Model-chapt12-2 are got from AD Chains by the parallelization of AD Chains. The initial structures of Model-chapt12-1 and Model-chapt12-2 are shown in Figs. 12.7∼12.8.

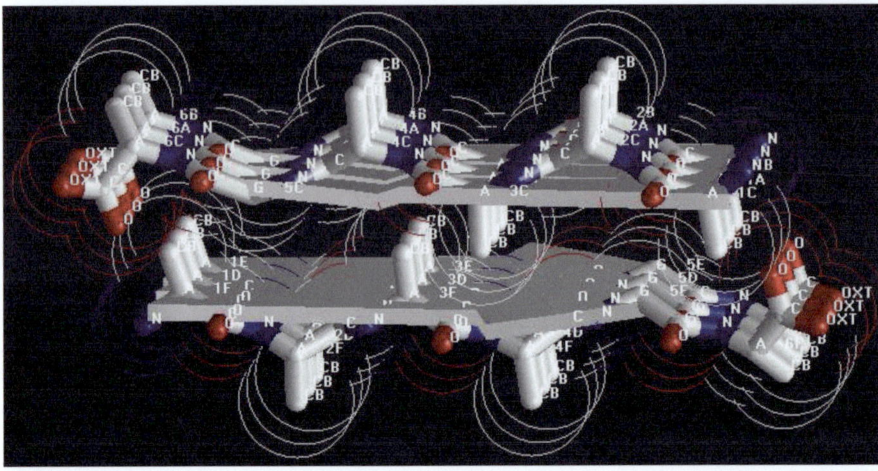

Fig. 12.7 Model-chapt12-1 – initial structure of prion AAAAGA amyloid fibril

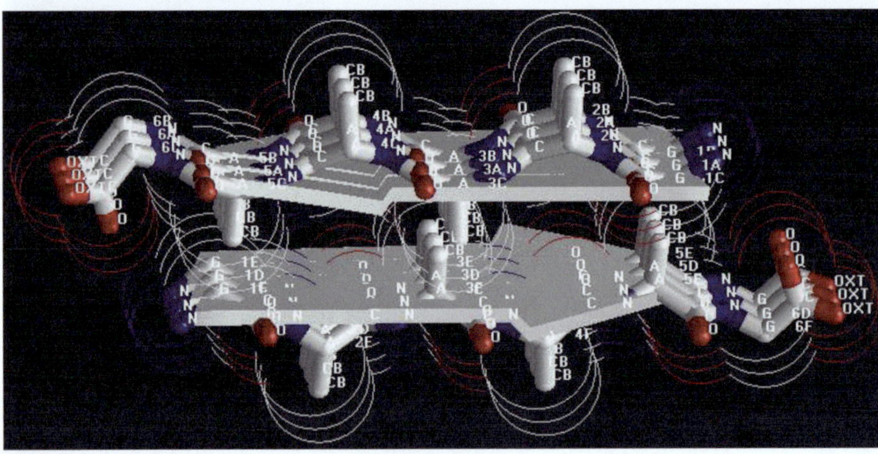

Fig. 12.8 Model-chapt12-2 – initial structure of prion GAAAAG amyloid fibril

12.3 Model Solving/Optimization

The LJ potential (12.2) energy of atoms' vdW interactions is just a part of the total potential energy of a protein [89, 397]:

$$E_{\text{total}} = \sum\nolimits_{\text{bonds}} K_r (r - r_{eq})^2 + \sum\nolimits_{\text{angles}} K_\theta (\theta - \theta_{eq})^2 + \sum\nolimits_{\text{dihedrals}} \frac{V_n}{2}[1 + \cos(n\phi - \gamma)]$$
$$+ \sum\nolimits_{i<j}^{\text{vdW}} \left[\frac{A_{ij}}{R_{ij}^{12}} - \frac{B_{ij}}{R_{ij}^{6}} \right] + \sum\nolimits_{i<j}^{\text{electrostatic}} \left[\frac{q_i q_j}{\epsilon R_{ij}} \right] + \sum\nolimits_{\text{H-bonds}} \left[\frac{C_{ij}}{R_{ij}^{12}} - \frac{D_{ij}}{R_{ij}^{10}} \right]. \quad (12.8)$$

The initial structures of Model-chapt12-1 and Model-chapt12-2 illuminated in Figs. 12.7∼12.8 are not the optimal structures with the lowest total potential energies. The initial structures also have no hydrogen atoms (so no hydrogen bonds existed) and water molecules added. For each Chain, the C-terminal and N-terminal atoms also have problems. Clearly there are a lot of close/bad contacts between β-strand atoms as illuminated in Figs. 12.7∼12.8. Thus, we still use the hybrid techniques of SD, CG, SA optimization methods within AMBER [89] as Chap. 11 to optimize the above Model-chapt12-1 and Model-chapt12-2 in order to get the most stable structures. Each of the most stable structures will have its lowest total potential energy, i.e.

$$\min E_{\text{total}}. \tag{12.9}$$

We used the ff03 force field of AMBER 10, in a neutral pH environment. The amyloid fibrils were surrounded with a 8 Å layer of TIP3PBOX water molecules using the XLEaP module of AMBER 10. 1,360, 1,372 waters and 180, 168 hydrogen atoms were added separately for Model-chapt12-1 and Model-chapt12-2 by the XLEaP module. The solvated amyloid fibrils were minimized by the SD method and then the CG method were performed (OPT1). Model-chapt12-1 were optimized by 95,016 steps of SD and 27,751 steps of CG; Model-chapt12-2 by 95,016 steps of SD and 24,418 steps of CG. Then the solvated amyloid fibrils were quickly heated from 0 to 300 K linearly during 20 ps. The systems were kept at 300 K for 80 ps. The systems then were slowly cooled from 300 to 100 K linearly for 400 ps. At 100 K, the systems were kept for 100 ps, and then for 4,400 ps until the systems reach sufficient equilibration at 100 K (where the RMSD, PRESS, and VOLUME (DENSITY) were sufficiently stable though their variations are very large). The SANDER (Simulated Annealing with NMR-Derived Energy Restraints) algorithm with nonbonded cutoffs of 12 Å were used during the heating, cooling and the 100 ps at 100 K. Step size is 2 fs for the whole SA runs. During the SA, the Metropolis criterion was used. After the SA, the models were refined by SD and CG methods again (OPT2), Model-chapt12-1 was refined by 20,000 steps of SD and 597 steps of CG, and 20,000 steps of SD and 1,921 steps of CG for Model-chapt12-2. All the above works were performed on the Tango facilities of the Victorian Partnership for Advanced Computing (www.vpac.org) of Australia.

12.4 Results and Discussion

Figure 12.9 shows the potential energy development for the two Models (where the OPT1-SA-OPT2 of AMBER 10 were used to generate the potential energy and the figures were drawn with XMGRACE of Grace 5.1.21). We can see that the potential energy goes down during the SD and CG optimization phase OPT1, suddenly drops down and quickly goes up and then slowly goes down and levels off during the SA phase, and at last quickly goes down and then levels off during a short

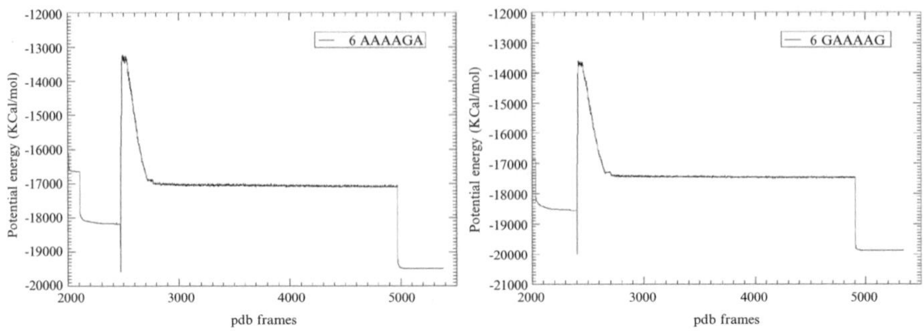

Fig. 12.9 Potential energy of Model-chapt12-1 (left) and Model-chapt12-2 (right)

Table 12.3 Potential energy values (1st = 1st step, last = last step, Last 10,000 = Last 10,000 steps)

Model	OPT1 1st	OPT1 50th	OPT1 last	SA 1st	SA Last 10,000 (average value)	OPT2 1st	OPT2 last
1	5.0294E+12	117,540	−18,179	−19,583.8427	−17,074.54472	−17,053	−19,482
vdW	******	36,804.4917					
2	1.8427E+16	419,080	−18,564	−19,992.2174	−17,461.46262	−17,474	−19,887
vdW	******	232,373.1312					

phase of OPT2. At the beginning of SA, the energy quickly drops off is due to the temperatures of the systems being suddenly changed from 100 K [631] to 0 K. This is a case of so called "quenching". Some energy values are listed in Table 12.3. In Table 12.3, the first column of energies (OPT1 1st-step) are the ones of the initial structures of Model-chapt12-1 and Model-chapt12-2. The distance between β-strands is too short for the vdW contacts so that Amber 10 cannot show the large LJ potential values (in Table 12.3 Column 1). This also implies the initial structures (Figs. 12.7∼12.8) are far from their optimal structures. OPT1 removes these bad vdW and hydrogen bond contacts and makes the structures become much better with lower potential energies. However, OPT1 is a local search optimization method which cannot thoroughly optimize the models into their most stable structures.

In Fig. 12.9 we see that models are trapped into their local optimal structures. SA is a global search optimization method that can make OPT1 jump out of the local trap, even accepting very bad cases with low probability according to the Metropolis criterion. Thus, in Table 12.3 we see that SA rapidly quenches the molecular structures, allowing escape from the local traps; SA finally results in the loss of 2,509.29798, 2,530.75478 KCal/mol for the two systems, respectively.

After SA, OPT2 can safely bring the molecular structures of the models to the most stable states. OPT2 makes the molecules in Model-chapt12-1 and Model-chapt12-2 lose 2,429, 2,413 KCal/mol of potential energy, respectively. OPT2 results in a loss of energy from Model-chapt12-1 and Model-chapt12-2 of nearly the same magnitude as that of SA (i.e. the decrease in energy in OPT2 is significant compared to the change of energy in SA between the 1st step and the average of the last 10,000 steps). OPT1 could not make further optimization, but OPT2 could make further optimization after SA; this demonstrates the effectiveness of SA (shown in Table 12.3 by comparing the values of OPT2 last-step with OPT1 last-step).

The final optimal molecular structures of Model-chapt12-1 and Model-chapt12-2 after OPT2 are shown in Figs. 12.10∼12.11, where the snapshots after OPT2 were drawn by the free package Molecular Visualisation & Modeling (MVM) (zmm-soft.com). The RMSDs (root mean square deviations) from the initial structures shown in Figs. 12.7∼12.8 (where the initial structures were drawn by MVM and were generated by the improved SADG Algorithm 12.1) are 2.71, 2.95 Å, respectively, for Model-chapt12-1 and Model-chapt12-2. The hydrogen bonds between the two closet adjacent β-strands and the vdW contacts between the two inner closest adjacent Alanines can be seen in Figs. 12.10∼12.11. In both models, there is about 5 Å between the two closet adjacent β-sheets, maintained by hydrophobic bonds, and about 4.5Å between the two closet adjacent β-strands, which are linked by hydrogen bonds such as Ala2.H-Ala7.O, Ala6.H-Gly11.O for Model-chapt12-1 and Ala2.O-Ala8.H, Ala21.H-Ala32.O for Model-chapt12-2. There is a hydrophobic core in each of the models. These amyloid fibrils are rich in β-sheet structure and contain a cross-β core form of infectious prions, which causes prion diseases.

Numerical results of this chapter showed that a six chains AGAAAA model could not successfully pass SA. However, two prion AGAAAAGA palindrome amyloid fibril models – a six chains AAAAGA model (Model-chapt12-1) and a six chains GAAAAG model (Model-chapt12-2) – were successfully passing OPT1-SA-OPT2 and got at last.

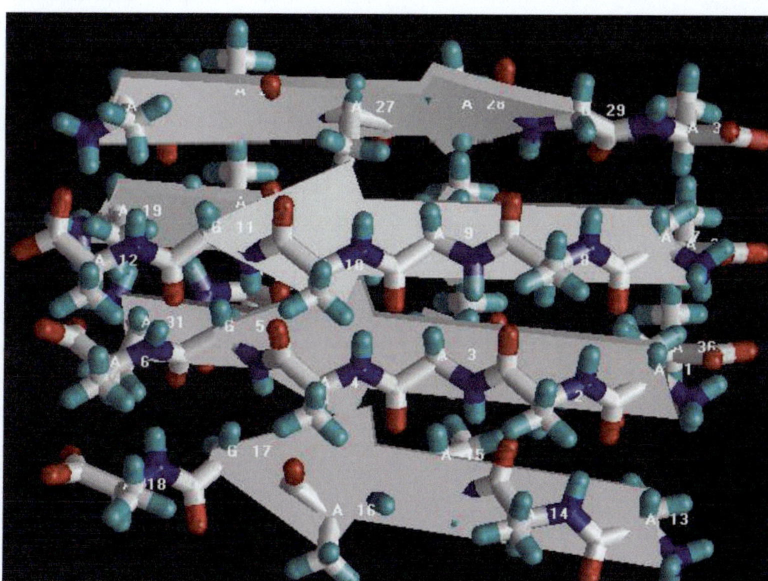

Fig. 12.10 Model-chapt12-1 – optimal structure of prion AAAAGA amyloid fibril

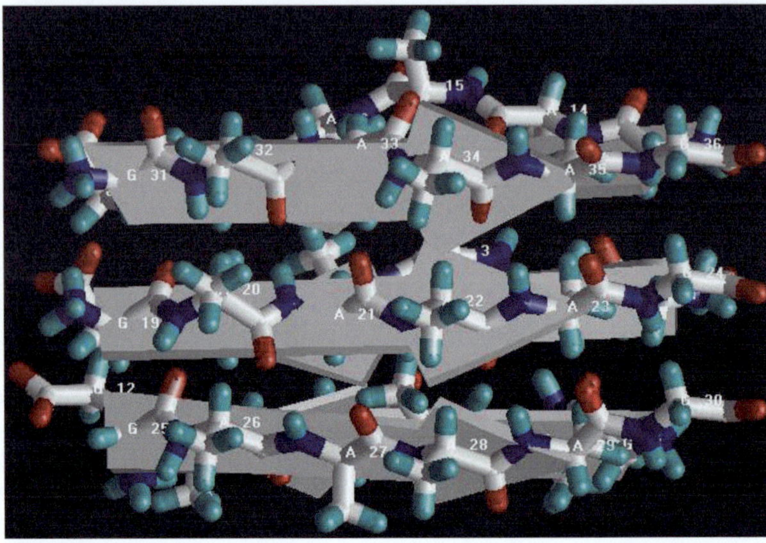

Fig. 12.11 Model-chapt12-2 – optimal structure of prion GAAAAG amyloid fibril

12.5 Concluding Remarks

In recent years large-scale GO problems have drawn considerable attention. These problems have many applications, in particular in biochemistry and data mining. Numerical methods for GO are often very time consuming and could not be applied for high-dimensional non-convex and/or non-smooth optimization problems. The study of new algorithms which allow one to solve large-scale GO problem is very important. One technique is to use hybrid of global and local/global search algorithms. This chapter presents two hybrid methods for solving the large-scale L-J potential GO problem. The methods do not guarantee the calculation of a global solution; however results of numerical experiments show that they, as a rule, calculate a solution which is global one or close to it. The improved hybrid SADG method can be successfully applied to the construction work of optimal atomic-resolution structures of prion AGAAAAGA amyloid fibrils. As the three models constructed for amyloid fibrils in Chap. 11, the two amyloid fibril models gained in this chapter may be useful in furthering the goals of medicinal chemistry.

Chapter 13
A Novel Canonical Dual Global Optimization Computational Approach

13.1 Overview

According to a recent comprehensive review [125], to develop a useful model for biological systems, the following things were usually needed to consider: (i) the material of benchmark used to develop and test the model, (ii) the formulation of modeling method, (iii) operating procedures during the modeling process, (iv) properly perform the cross-validation tests to objectively evaluate the anticipated accuracy of the model, and (v) web-server establishment. Below, let us elaborate some of these procedures. In this chapter, the material used to develop the model is 3NHC.pdb and its 3D-crystal structures; the modeling method is the Mathematical Optimization methods of the canonical dual theory (CDT) [199, 219–230, 232, 233] (*procedure 1*) and of the Amber 11 package's SD method [91] and CG method [91] (*procedure 2*); and the test to the accuracy of the model is performed by the RMSD value of last snapshots between *procedures 1*$\sim$*2*.

Various computational molecular dynamics approaches have been used to study PrP (106–126) [357, 595] but, to the best of our knowledge, to predict molecular structures of prion AGAAAAGA amyloid fibrils the computational approaches are few (Chaps. 11 and 12). Chapter 11 successfully constructed three AGAAAAGA amyloid fibril models by the standard simulated annealing (SA) method and several traditional optimization methods within AMBER 10 package. In Chap. 12, the hybrid simulated annealing discrete gradient (SADG) method was successfully used for modeling two AGAAAAGA amyloid fibril models (instead of the Insight II (accelrys.com) package used in Chap. 11), and then the models were refined/optimized by the SDCG methods, SA method and SDCG methods again. In this chapter, all the optimization approaches of Chaps. 11 and 12 will be replaced by the optimization theory of CDT. Numerical computational results show that the optimization approaches of CDT have a very perfect performance. It is even no need to do furthermore SDCG refinements by the AMBER package. We could not do comparisons (for example, the angstrom values between adjacent β-sheets and

J. Zhang, *Molecular Structures and Structural Dynamics of Prion Proteins and Prions*, Focus on Structural Biology 9, DOI 10.1007/978-94-017-7318-8_13

β-strands) for the models of Chaps. 11 and 12 and of this chapter, because these models have different number of chains and different structural Classes listed in [357].

As we all know, the disease prions PrP^{Sc} are rich in β-sheets amyloid fibrils (about 43 % β-sheet) [257]. There are some classical works on the β-sheets and β-barrels [127, 128, 133, 135]. X-ray crystallography and NMR spectroscopy are two powerful tools to determine the protein 3D structure. However, not all proteins can be successfully crystallized, particularly for membrane proteins. Although NMR is indeed a very powerful tool in determining the 3D structures of membrane proteins (see, e.g., [82, 455, 468, 470, 519, 606] and a recent review [469]), it is also time-consuming and costly. To acquire the structural information in a timely manner, one has to resort to various structural bioinformatics tools (see, e.g., [117, 119, 121, 123] and a comprehensive review [117]). Particularly, computational approaches allow us to obtain a description of the protein 3D structure at a submicroscopic level. Under many circumstances, due to the unstable, noncrystalline and insoluble nature of the β-sheets amyloid fibrils, it is very difficult to use traditional X-ray and NMR experimental methods to obtain atomic-resolution structures of β-sheets amyloid fibrils [571, 710]. Although X-ray and NMR techniques cannot determined the 3D structures of some proteins and their binding interactions with ligands in a timely manner that are important for drug design and basic research, many structural bioinformatics tools can play a complementary role in this regard as demonstrated by a series papers published recently (see, e.g. [80, 118, 122, 139, 193, 194, 250, 386, 601, 615, 621]). This chapter, in some sense, presents a structural bioinformatics tool in view of the CDT-based mathematical optimization theory.

The accuracy of the models presented in this chapter is tested by the RMSD value. The last snapshot of *procedure 2* will be superposed onto the last snapshot of *procedure 1*, and the RMSD value is zero after the alignment by VMD 1.8.7beta5 [297]. This implies to us that the CDT strategy can accurately built the prion AGAAAAGA amyloid fibril models. To test the accuracy of their model, some examination validation methods are always used. In developing a prediction model or algorithm, the following three cross-validation methods are often used for examining its effectiveness in practical application: independent dataset test, subsampling (5-fold or 10-fold cross-validation) test, and jackknife test [141]. However, as demonstrated by Equations 28–32 of [125], among the three cross-validation methods, the jackknife test is deemed the least arbitrary that can always yield a unique result for a given benchmark dataset, and hence has been increasingly used and widely recognized by investigators to examine the accuracy of various models and predictors (see, e.g. [99, 140, 185, 278, 328, 389, 429, 675, 714]; all these papers reflect the current trend of increasingly and widely using the jackknife test to examine varieties of models or predictors).

There is another criteria to evaluate the models. To avoid homology bias and remove the redundant sequences from the benchmark dataset, a cutoff threshold of 25 % was recommended [125, 140] to exclude those proteins from the benchmark datasets that have equal to or greater than 25 % sequence identity to any other. However, in this study we did not use such a stringent criterion because the currently available data do not allow us to do so. Otherwise, the numbers of proteins left would be too few to have statistical significance. As it is well known, the more stringent of a benchmark dataset in excluding homologous sequences, the more reliable the results derived based on their model. According to a recent comprehensive review paper [125], an ideal cutoff threshold should be set at 25 %, meaning that none of protein sequences included in the benchmark dataset has greater than 25 % pairwise sequence identity to any other.

The last procedure to develop a useful model for biological systems is a web-server establishment. Since user-friendly and publicly accessible web-servers represent the future direction for developing practically more useful models, simulated methods, or predictors [138], we shall make efforts in our future work to provide a web-server for the method presented in this chapter.

This chapter addresses an important problem on neurodegenerative amyloid fibril or plaque diseases. The rest of this chapter is arranged as follows. In the next section, i.e. Sect. 13.2, the CDT will be briefly introduced and its effectiveness will be illuminated by applying the CDT-based optimization approach to a well-known system of minimizing the Double Well Potential function and Rosenbrock's function. In Sect. 13.3, the molecular modeling works of prion AGAAAAGA amyloid fibrils will be done. Section 13.3 also successfully gains the optimal prion AGAAAAGA amyloid fibril models by the applications of the CDT-based optimization theory. Furthermore refinement/optimization to these models by the SDCG methods of the package Amber 11 will also be done in Sect. 13.3. The zero RMSD value implies to us that the CDT optimization strategy can accurately obtain the prion AGAAAAGA amyloid fibril models. Thus, when using the time-consuming and costly X-ray crystallography or NMR spectroscopy we still cannot determine the protein 3D structure, we may introduce computational approaches or novel mathematical formulations and physical concepts into molecular biology to study molecular structures. This concluding remark will be made in the last section, i.e. Sect. 13.4. Other models will be studied in Sect. 13.5 in the use of CDT approach.

13.2 The Canonical Dual Global Optimization Approach

We briefly introduce the CDT of [221, 232, 233] specially for solving the following minimization problem of the sum of fourth-order polynomials:

$$(P) : \min_x \left\{ P(x) = \sum_{i=1}^{m} W_i(x) + \frac{1}{2}x^T Qx - x^T f : x \in R^n \right\}, \tag{13.1}$$

$$\textit{where} \quad W_i(x) = \frac{1}{2}\alpha_i \left(\frac{1}{2}x^T A_i x + b_i^T x + c_i \right)^2, A_i = A_i^T, Q = Q^T \in R^{n \times n},$$

$$b_i, f \in R^n, c_i, \alpha_i \in R^1, i = 1, 2, \ldots, m, x \in \mathcal{X} \subset R^n.$$

The dual problem of (P) is

$$(P^d): \max_{\varsigma} \left\{ P^d(\varsigma) = \sum_{i=1}^{m} \left(c_i \varsigma_i - \frac{1}{2} \alpha_i^{-1} \varsigma^2 \right) - \frac{1}{2} F^T(\varsigma) G^+(\varsigma) F(\varsigma) : \varsigma \in S_a \right\}, \quad (13.2)$$

$$\text{where} \quad F(\varsigma) = f - \sum_{i=1}^{m} \varsigma_i b_i, G(\varsigma) = Q + \sum_{i=1}^{m} \varsigma_i A_i, S_a = \{\varsigma \in R^m | F(\varsigma) \in Col(G(\varsigma))\},$$

G^+ denotes the Moore-Penrose generalized inverse of G, and $Col(G(\varsigma))$ is the column space of $G(\varsigma)$. The prime-dual Gao-Strang complementary function of CDT [221, 232, 233] is

$$\Xi(x, \varsigma) = \sum_{i=1}^{m} \left[\left(\frac{1}{2} x^T A_i x + b_i^T x + c_i \right) \varsigma_i - \frac{1}{2} \alpha_i^{-1} \varsigma_i^2 \right] + \frac{1}{2} x^T Q x - x^T f. \quad (13.3)$$

For (P) and (P^d) we have the following CDT:

Theorem 13.2.1 ([221, 232, 233]). *The problem (P^d) is canonically dual to (P) in the sense that if $\bar{\varsigma}$ is a critical point of $P^d(\varsigma)$, then $\bar{x} = G^+(\bar{\varsigma})F(\bar{\varsigma})$ is a critical point of $P(x)$ on R^n, and $P(\bar{x}) = P^d(\bar{\varsigma})$. Moreover, if $\bar{\varsigma} \in S_a^+ = \{\varsigma \in S_a | G(\varsigma) \succ 0\}$, then $\bar{\varsigma}$ is a global maximizer of $P^d(\varsigma)$ over S_a^+, $\bar{x}$ is a global minimizer of $P(x)$ on R^n, and*

$$P(\bar{x}) = \min_{x \in R^n} P(x) = \Xi(\bar{x}, \bar{\varsigma}) = \max_{\varsigma \in S_a^+} P^d(\varsigma) = P^d(\bar{\varsigma}). \quad (13.4)$$

It is easy to prove that the canonical dual function $P^d(\varsigma)$ is concave on the convex dual feasible space S_a^+. Therefore, Theorem 13.2.1 shows that the nonconvex primal problem (P) is equivalent to a concave maximization problem (P^d) over a convex space S_a^+, which can be solved easily by well-developed methods. Over $S_a^- = \{\varsigma \in S_a | G(\varsigma) \prec 0\}$ we have the following theorem:

Theorem 13.2.2 ([233]). *Suppose that $\bar{\varsigma}$ is a critical point of (P^d) and the vector $\bar{x}$ is defined by $\bar{x} = G^+(\bar{\varsigma})F(\bar{\varsigma})$. If $\bar{\varsigma} \in S_a^-$, then on a neighborhood $\mathcal{X}_o \times \mathcal{S}_o \subset \mathcal{X} \times S_a^-$ of $(\bar{x}, \bar{\varsigma})$, we have either*

$$P(\bar{x}) = \min_{x \in \mathcal{X}_o} P(x) = \Xi(\bar{x}, \bar{\varsigma}) = \min_{\varsigma \in \mathcal{S}_o} P^d(\varsigma) = P^d(\bar{\varsigma}), \quad (13.5)$$

or

$$P(\bar{x}) = \max_{x \in \mathcal{X}_o} P(x) = \Xi(\bar{x}, \bar{\varsigma}) = \max_{\varsigma \in \mathcal{S}_o} P^d(\varsigma) = P^d(\bar{\varsigma}). \quad (13.6)$$

By the fact that the canonical dual function is a d.c. function (difference of convex functions) on S_a^-, the double-min duality (13.5) can be used for finding the biggest

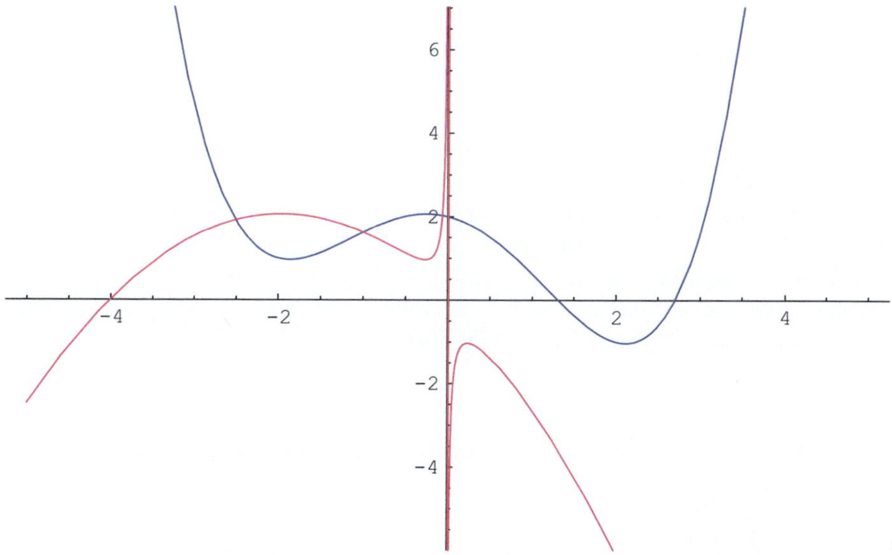

Fig. 13.1 The prime and dual double-well potential functions (Prime: *blue*, Dual: *red*)

local minimizer of (P) and (P^d), while the double-max duality (13.6) can be used for finding the biggest local maximizer of (P) and (P^d). In physics and material sciences, this pair of biggest local extremal points play important roles in phase transitions.

To illuminate that the CDT works, we minimize the well-known Double Well Potential (DWP) function [221] (blue colored in Fig. 13.1):

$$P(x) = \frac{1}{2}(\frac{1}{2}x^2 - 2)^2 - \frac{1}{2}x. \tag{13.7}$$

We can easily get $\Xi(x, \varsigma) = (\frac{1}{2}x^2 - 2)\varsigma - \frac{1}{2}\varsigma^2 - \frac{1}{2}x$,

$$P^d(\varsigma) = -\frac{1}{8\varsigma} - \frac{1}{2}\varsigma^2 - 2\varsigma \tag{13.8}$$

(red colored in Fig. 13.1) and $S_a^+ = \{\varsigma \in R^1 | \varsigma > 0\}$. Let $\Xi(x, \varsigma)' = 0$, we get three critical points of $\Xi(x, \varsigma)$:

$$(\bar{x}^1, \bar{\varsigma}^1) = (2.11491, 0.236417),$$
$$(\bar{x}^2, \bar{\varsigma}^2) = (-1.86081, -0.268701),$$
$$(\bar{x}^3, \bar{\varsigma}^3) = (-0.254102, -1.96772).$$

By Theorem 13.2, we know $\bar{x}^1 = 2.11491$ is the global minimizer of (13.7), $\bar{\varsigma}^1 = 0.236417$ is the global maximizer of (13.8) over S_a^+, and $P(\bar{\varsigma}^1) = \Xi(\bar{x}^1, \bar{\varsigma}^1) = P^d(\bar{\varsigma}^1) = -1.02951$. By Theorem 5.2.2, we know that the local minimizers: $\bar{x}^2 = -1.86081, \bar{\varsigma}^2 = -0.268701$ (over S_a^-), $P(\bar{\varsigma}^2) = \Xi(\bar{x}^2, \bar{\varsigma}^2) = P^d(\bar{\varsigma}^2) = 0.9665031$ and the local maximizers: $\bar{x}^3 = -0.254102, \bar{\varsigma}^3 = -1.96772$ (over S_a^-), $P(\bar{\varsigma}^3) = \Xi(\bar{x}^3, \bar{\varsigma}^3) = P^d(\bar{\varsigma}^3) = 2.063$.

Thus, by Fig. 13.1 illuminating the application of CDT to the DWP problem, we may see that the canonical dual approach works. The powerful of CDT is shown in Tables 1–3 of arXiv 1105.2270v3 [501]. In the next section, we will apply this successful canonical dual approach to the molecular model building and solving problem of prion AGAAAAGA amyloid fibrils.

13.3 Prion AGAAAAGA Amyloid Fibril Model Building & Solving

Many experimental studies such as [67, 68, 70, 288, 315, 316, 357, 445, 619] have shown two points: (1) the hydrophobic region (113–120) AGAAAAGA of prion proteins is critical in the conversion from a soluble PrP^C into an insoluble PrP^{Sc} fibrillar form; and (2) normal AGAAAAGA is an inhibitor of prion diseases. Various computational approaches were used to address the problems related to "amyloid fibril" [86, 117, 119, 130, 609, 620, 678]. By introducing novel mathematical canonical dual formulations and computational approaches, in this chapter we may construct atomic-resolution molecular structures for prion (113–120) AGAAAAGA amyloid fibrils.

The atomic structures of all amyloid fibrils revealed steric zippers, with strong vdW interactions between β-sheets and HBs to maintain the β-strands [517]. About β-sheets and β-barrels, there are various interactions and motions, such as the interactions between β-strands [133, 134, 136, 137], interaction between two β-sheets [132], as well as the low-frequency accordion-like motion in a β-sheet and breathing motion in a β-barrel [109] and their biological functions [110]. The "amyloid fibril" problem can be looked as a molecular distance geometry problem (MDGP) [262], which arises in the interpretation of NMR data and in the determination of protein structure [as an example to understand MDGP, the problem of locating sensors in telecommunication networks is a DGP. In such a case, the positions of some sensors are known (which are called anchors) and some of the distances between sensors (which can be anchors or not) are known: the DGP is to locate the positions of all the sensors. Here we look sensors as atoms and their telecommunication network as a molecule]. The three dimensional structure of a molecule with n atoms can be described by specifying the 3-Dimensional coordinate positions $x_1, x_2, \ldots, x_n \in R^3$ of all its atoms. Given bond lengths d_{ij} between a subset S of the atom pairs, the determination of the molecular structure is

$$(P_0) to \quad find \quad x_1, x_2, \ldots, x_n \quad s.t. \quad ||x_i - x_j|| = d_{ij}, (i,j) \in S, \tag{13.9}$$

where $||\cdot||$ denotes a norm in a real vector space and it is calculated as the Euclidean distance 2-norm in this chapter. The original problem (13.9) can be reformulated as a mathematical global optimization problem (GOP)

$$(P) \quad \min P(X) = \sum_{(i,j)\in S} w_{ij}(||x_i - x_j||^2 - d_{ij}^2)^2 \tag{13.10}$$

in the terms of finding the global minimum of the function $P(X)$, where w_{ij}, $(i,j) \in S$ are positive weights, $X = (x_1, x_2, \ldots, x_n)^T \in R^{n\times 3}$ [431] and usually S has many fewer than $n^2/2$ elements due to the error in the theoretical or experimental data [262, 721]. There may even not exist any solution $x_1, x_2, \ldots, x_n$ to satisfy the distance constraints in (13.9), for example when data for atoms $i, j, k \in S$ violate the triangle inequality; in this case, we may add a perturbation term $-\epsilon^T X$ to $P(X)$:

$$(P_\epsilon) \quad \min P_\epsilon(X) = \sum_{(i,j)\in S} w_{ij}(||x_i - x_j||^2 - d_{ij}^2)^2 - \epsilon^T X, \tag{13.11}$$

where $\epsilon \geq 0$. In some cases, instead exact values d_{ij}, $(i,j) \in S$ can be found, we can only specify lower and upper bounds on the distances: $l_{ij} \leq ||x_i - x_j|| \leq u_{ij}$, $(i,j) \in S$; in such cases we may penalize all the unsatisfied constraints into the objective function of (P_ϵ) by adding $\sum_{(i,j)\in S} \left(\max\left\{l_{i,j}^2 - ||x_i - x_j||^2, 0\right\}\right)^2 + \left(\max\left\{||x_i - x_j||^2 - u_{i,j}^2, 0\right\}\right)^2$ into $P_\epsilon(X)$ [262, 721], where we may let d_{ij} be the interatomic distance (less than 6 Å) for the pair in successive residues of a protein and set $l_{ij} = (1 - 0.05)d_{ij}$ and $u_{ij} = (1 + 0.05)d_{ij}$ [262]. In this chapter we will use the canonical duality approach introduced in Sect. 13.2 [221, 232, 233] to solve (13.9), (13.10), and (13.11). Because the canonical dual is a perfect dual with zero duality gap between prime and dual problems, we can get the accurate global optimal solutions of problems (13.9), (13.10), and(13.11). Thus by canonical dual approach we may successfully construct the molecular structure of prion AGAAAAGA amyloid fibrils as follows.

If we look at the prion AGAAAAGA molecular modeling problem as a MDGP with two anchors and two sensors, we can easily construct the prion AGAAAAGA amyloid fibril models. In fact we may let the coordinates of these two anchors being variable. But, these two anchors belong to one body of Chains A and B, and the two sensors belong to another body of Chains G and H. This is a simple two-body problem model of theoretical physics, i.e. Einstein's absolute relative theory. Hence, we may look the coordinates of two anchors being fixed. The constructions will be based on the most recently released experimental molecular structures of human M129 prion peptide 127–132 (PDB entry 3NHC released into Protein Data Bank (www.rcsb.org) on 04-AUG-2010) (in brief, this chapter will use the PrP structured region 127–132 to do homology modelling for the PrP unstructured region 113–120). The atomic-resolution structure of this peptide is a steric zipper, with strong vdW interactions between β-sheets and HBs to maintain the β-strands (Fig. 13.2, where the dashed lines denote the HBs). In Fig. 13.2 we see that G (H) chains (i.e. β-sheet 2) of 3NHC.pdb can be obtained from A (B) chains (i.e. β-sheet 1) by

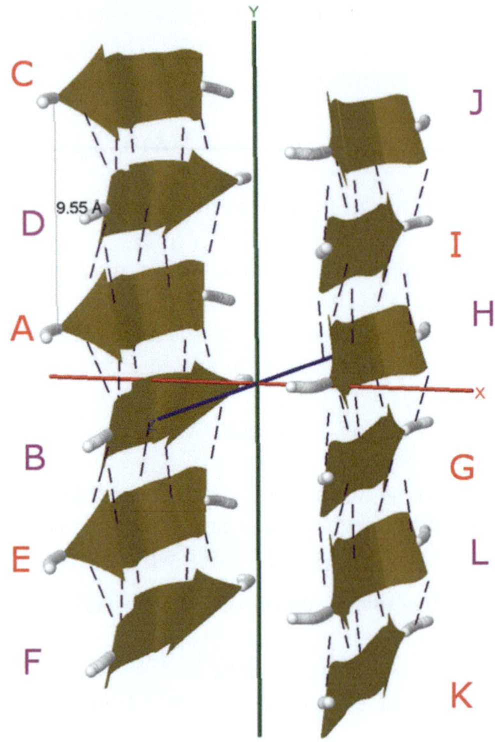

Fig. 13.2 Protein fibril structure of human M129 prion GYMLGS (127–132)

$$G(H) = \begin{pmatrix} 1 & 0 & 0 \\ 0 & -1 & 0 \\ 0 & 0 & -1 \end{pmatrix} A(B) + \begin{pmatrix} 9.07500 \\ 4.77650 \\ 0.00000 \end{pmatrix}, \qquad (13.12)$$

and other chains can be got by

$$I(J) = G(H) + \begin{pmatrix} 0 \\ 9.5530 \\ 0 \end{pmatrix}, K(L) = G(H) + \begin{pmatrix} 0 \\ -9.5530 \\ 0 \end{pmatrix}, \qquad (13.13)$$

$$C(D) = A(B) + \begin{pmatrix} 0 \\ 9.5530 \\ 0 \end{pmatrix}, E(F) = A(B) + \begin{pmatrix} 0 \\ -9.5530 \\ 0 \end{pmatrix}. \qquad (13.14)$$

Basing on the template 3NHC.pdb from Protein Data Bank, three prion AGAAAAGA palindrome amyloid fibril models – an AGAAAA model (Model-chapt13-1), a GAAAAG model (Model-chapt13-2), and an AAAAGA model (Model-chapt13-3) will be successfully constructed in this chapter. AB chains of Model-chapt13-1 to Model-chapt13-3 were respectively got from AB chains of 3NHC.pdb using the mutate module of the free package Swiss-PdbViewer (SPDBV

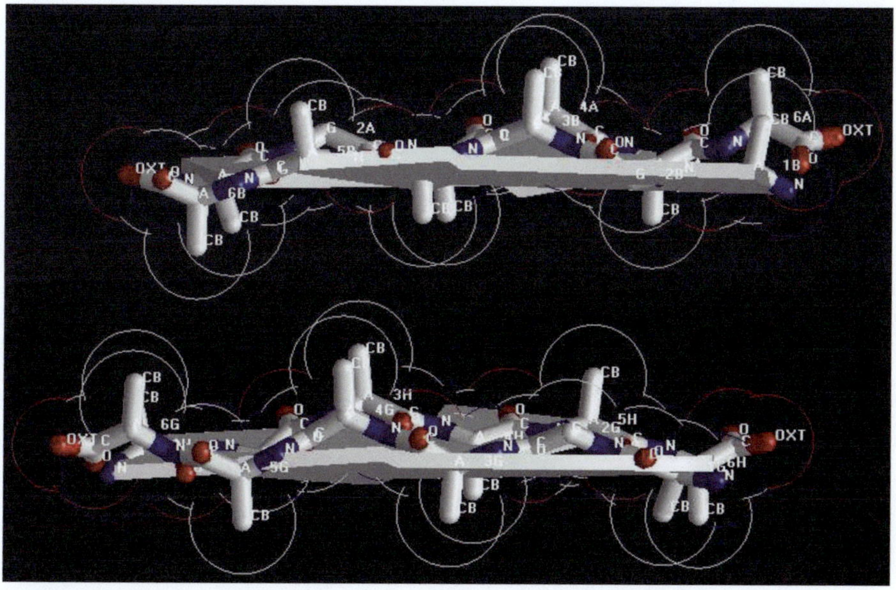

Fig. 13.3 Far vdW contacts of AG chains and BH chains of Model-chapt13-1

Version 4.01) (spdbv.vital-it.ch). It is pleasant to see that almost all the hydrogen bonds are still kept after the mutations, where for the donor O (oxygen) atom and the acceptor H (hydrogen) atom if the distance cutoff is less than 3.00 Å and the angle cutoff is less than 120.00 degrees then a hydrogen bond is kept; thus we just need to consider the vdW contacts only. Making mutations for GH chains of 3NHC.pdb, we can get the GH chains of Model-chapt13-1 to Model-chapt13-3. However, the vdW contacts between A chain and G chain, between B chain and H chain are too far at this moment (Figs. 13.3, 13.4, and 13.5) because the shortest distance of atoms between Chain A and Chain G, and between Chain B and Chain H, is still very larger than the double size of the vdW radius of CB carbon atom. Seeing Figs. 13.3, 13.4, and 13.5, we may know that for Model-chapt13-1 to Model-chapt13-3 at least two vdW interactions between A.ALA3.CB-G.ALA4.CB, B.ALA4.CB-H.ALA3.CB should be maintained. Fixing the coordinates of A.ALA3.CB and B.ALA4.CB (two anchors) ((6.014,5.917,0.065), (5.658,1.630,-0.797)), letting d equal to the twice of the vdW radius of Carbon atom (i.e. $d = 3.4$ Å), and letting the coordinates of G.ALA4.CB and H.ALA3.CB (two sensors) be variables, we may get a simple MDGP with 6 variables and its dual with 2 variables:

$$P(x_1, x_2) = \frac{1}{2}\left\{(x_{11} - 6.014)^2 + (x_{12} - 5.917)^2 + (x_{13} - 0.065)^2 - 3.4^2\right\}^2$$

$$+\frac{1}{2}\left\{(x_{21} - 5.658)^2 + (x_{22} - 1.630)^2 + (x_{23} + 0.797)^2 - 3.4^2\right\}^2,$$

$$P^d(\varsigma_1, \varsigma_2) = -11.56\varsigma_1 - \frac{1}{2}\varsigma_1^2 - 11.56\varsigma_2 - \frac{1}{2}\varsigma_2^2.$$

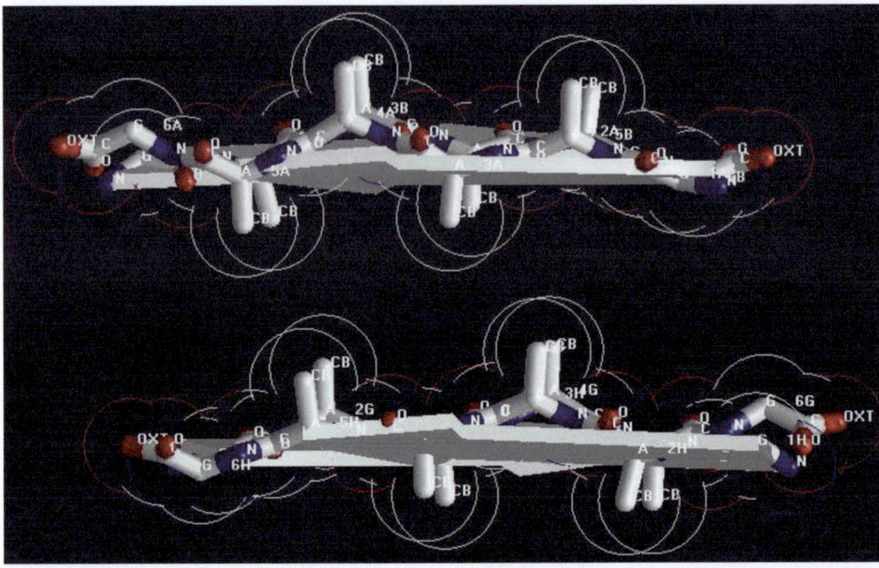

Fig. 13.4 Far vdW contacts of AG chains and BH chains of Model-chapt13-2

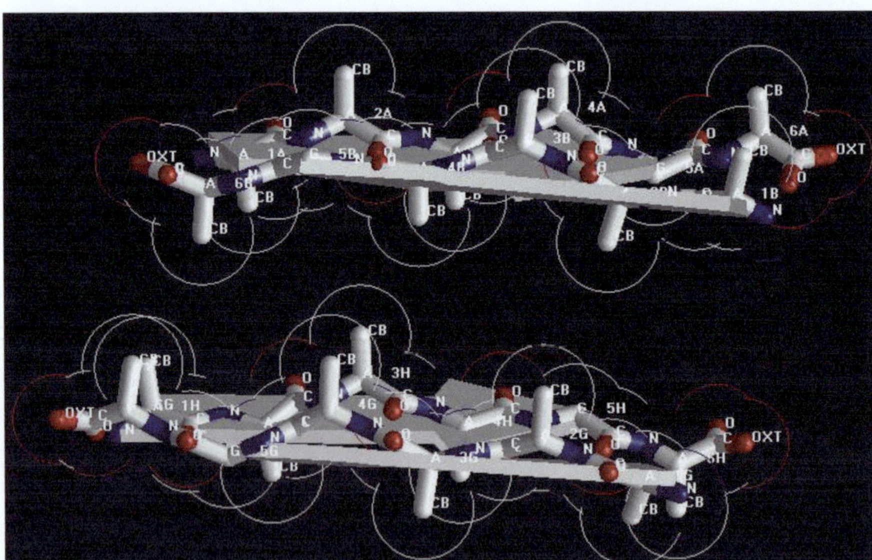

Fig. 13.5 Far vdW contacts of AG chains and BH chains of Model-chapt13-3

We can get a local maximal solution (−11.56,−11.56) for $P^d(\varsigma_1, \varsigma_2)$ and its corresponding local maximal solution to $P(x_1, x_2)$. But we need the global maximal solution of $P^d(\varsigma_1, \varsigma_2)$. Thus, by introducing perturbation parameters $\epsilon = 0.05$, we have to seek the global optimal solutions from the perturbed problems of $P(x_1, x_2)$ and $P^d(\varsigma_1, \varsigma_2)$:

$$\begin{aligned} P_\epsilon(x_1, x_2) &= \frac{1}{2}\{(x_{11} - 6.014)^2 + (x_{12} - 5.917)^2 + (x_{13} - 0.065)^2 - 3.4^2\}^2 \\ &\quad + \frac{1}{2}\{(x_{21} - 5.658)^2 + (x_{22} - 1.630)^2 + (x_{23} + 0.797)^2 - 3.4^2\}^2 \\ &\quad - 0.05x_{11} - 0.05x_{12} - 0.05x_{13} - 0.05x_{21} - 0.05x_{22} - 0.05x_{23}, \\ P_\epsilon^d(\varsigma_1, \varsigma_2) &= 59.6233\varsigma_1 - 0.5\varsigma_1^2 + 23.7451\varsigma_2 - 0.5\varsigma_2^2 \\ &\quad - \frac{1}{2}\left(\frac{(0.05 + 12.028\varsigma_1)^2}{2\varsigma_1} + \frac{(0.05 + 11.834\varsigma_1)^2}{2\varsigma_1} + \frac{(0.05 + 0.130\varsigma_1)^2}{2\varsigma_1}\right) \\ &\quad - \frac{1}{2}\left(\frac{(0.05 + 11.316\varsigma_2)^2}{2\varsigma_2} + \frac{(0.05 + 3.2600\varsigma_2)^2}{2\varsigma_2} + \frac{(0.05 - 1.594\varsigma_2)^2}{2\varsigma_2}\right). \end{aligned}$$

We can easily get the global maximal solution $(0.0127287, 0.0127287) \in \{\varsigma \in R^2 | \varsigma_i > 0, i = 1, 2\}$ for $P_\epsilon^d(\varsigma_1, \varsigma_2)$. Then, we get its corresponding solution for $P_\epsilon(x_1, x_2)$:

$$\bar{x} = (7.97807, 7.88107, 2.02907, 7.62207, 3.59407, 1.16707).$$

By Theorem 5.1, we know that $\bar{x}$ is a global minimal solution of $P_\epsilon(x_1, x_2)$. We set $\bar{x}$ as the coordinates of G.ALA4.CB and H.ALA3.CB and taking the average value we get

$$G(H) = \begin{pmatrix} 1 & 0 & 0 \\ 0 & -1 & 0 \\ 0 & 0 & -1 \end{pmatrix} A(B) + \begin{pmatrix} 1.96407 \\ 9.51107 \\ 1.23207 \end{pmatrix}. \tag{13.15}$$

By (13.15) we can get very close vdW contacts between A chain and G chain, between B chain and H chain (Figs. 13.6, 13.7, and 13.8). Thus, we successfully constructed Model-chapt13-1 to Model-chapt13-3, and through further refinements by the Amber 11 package [91] we at last get the optimal Models (Figs. 13.9, 13.10, and 13.11). We find the RMSD (root mean square deviation) between Figs. 13.6, 13.7, and 13.8 and Figs. 13.9, 13.10, and 13.11 is 0 Å; this implies that the Amber 11 refinements are not necessary and the CDT is good enough to get the optimal Model-chapt13-1 to Model-chapt13-3 as illuminated in Figs. 13.6, 13.7, and 13.8. The other CDIJ and EFKL chains can be got by parallelizing ABGH chains in the use of mathematical formulas (13.13)∼(13.14).

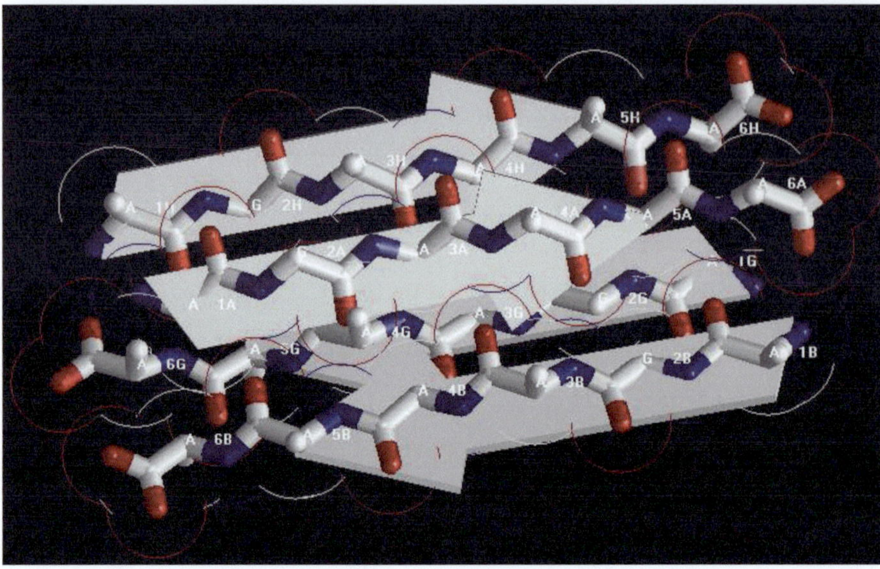

Fig. 13.6 Close vdW contacts of AG chains and BH chains of Model-chapt13-1

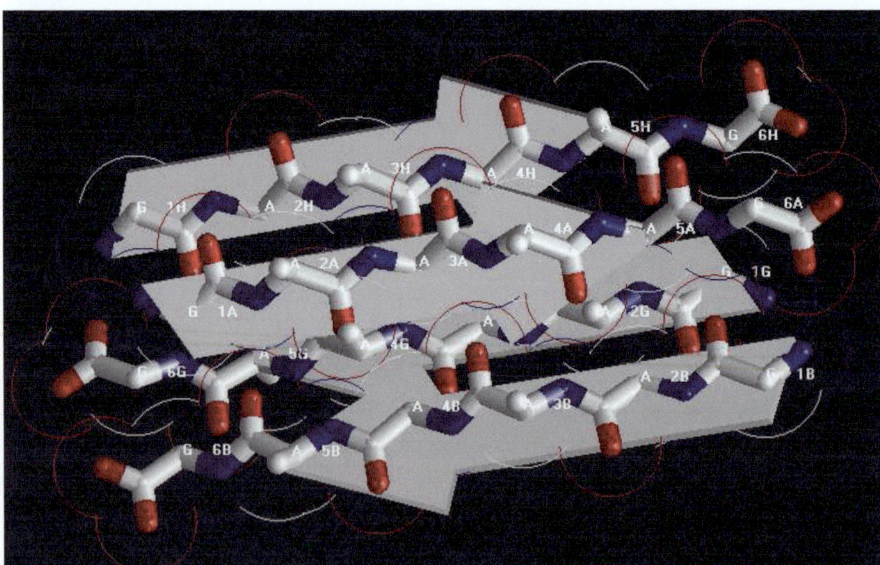

Fig. 13.7 Close vdW contacts of AG chains and BH chains of Model-chapt13-2

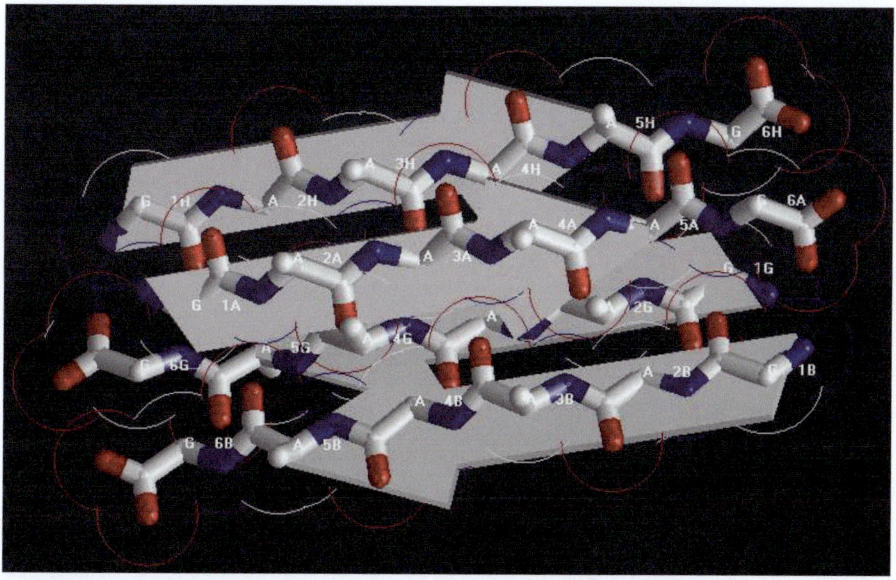

Fig. 13.8 Close vdW contacts of AG chains and BH chains of Model-chapt13-3

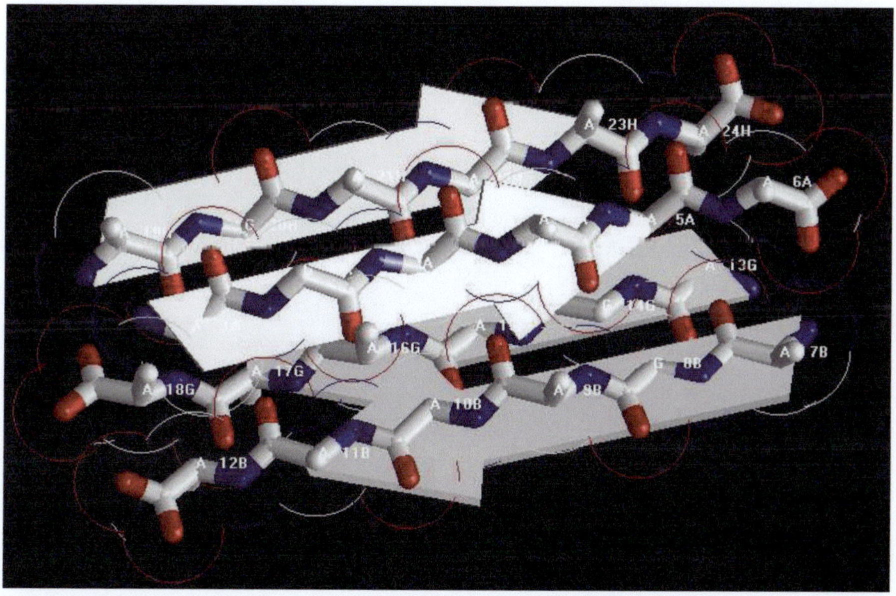

Fig. 13.9 Optimal structure of prion AGAAAAGA amyloid fibril Model-chapt13-1

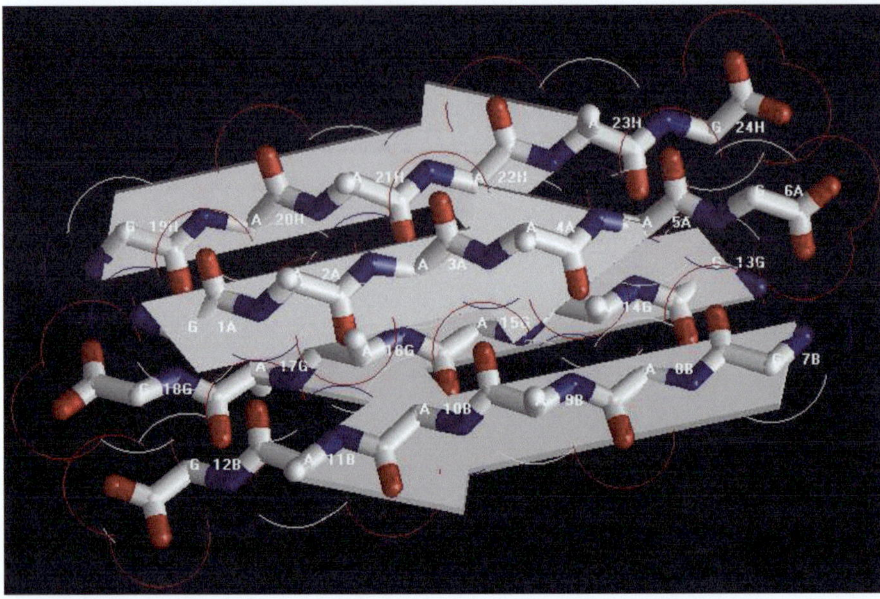

Fig. 13.10 Optimal structure of prion AGAAAAGA amyloid fibril Model-chapt13-2

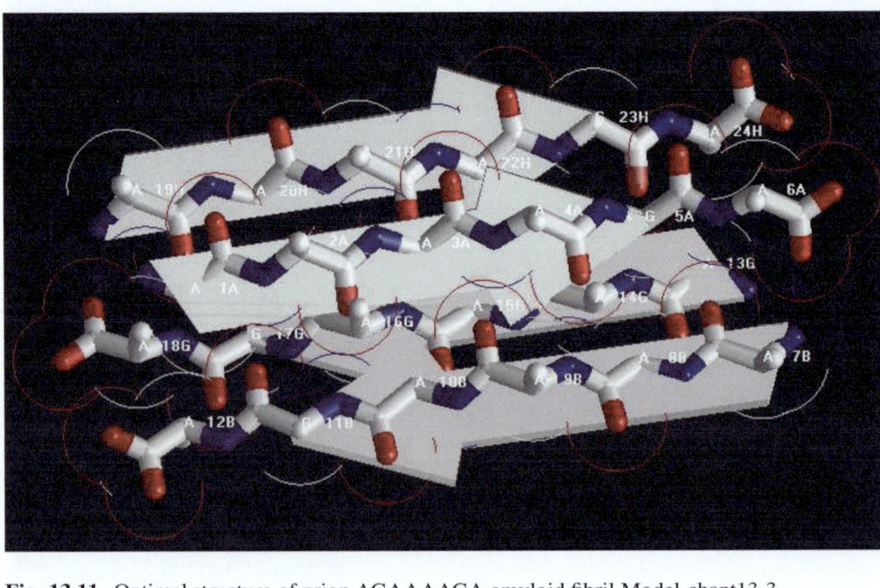

Fig. 13.11 Optimal structure of prion AGAAAAGA amyloid fibril Model-chapt13-3

As the end of this section, we give some remarks on the Model-chapt13-1 to Model-chapt13-3. (1) The canonical dual approach exactly makes the closest CB atoms between Chain A and Chain G, and between Chain B and Chain H, just being equal to the double size of the vdW radius of CB carbon atom (Figs. 13.6, 13.7, and 13.8) and this is the perfect structure of the Model-chapt13-1 to Model-chapt13-3. Figures 13.9, 13.10, and 13.11 were got by the further refinements through the SDCG optimization methods of Amber 11 package. The zero RMSD value between Figs. 13.6, 13.7, and 13.8 and Figs. 13.9, 13.10, and 13.11 implies to us that the canonical dual approach of this chapter works well. (2) The SDCG optimization methods of Amber 11 package automatically considered the bond angles and dihedral angles, and during the canonical dual molecular model building and optimization procedure, the perfect bond angles and dihedral angles automatically produced by the Swiss-PdbViewer package are still being kept. (3) The molecular modeling problem of this chapter is in fact a very simple two-body problem of theoretical physics, i.e. Einstein's absolute relative theory. In mathematics, it is a sensor network problem with two anchors and two sensors.

13.4 Concluding Remarks

This chapter presents an important method and provides useful information for treatments of prion diseases. X-ray crystallography is a powerful tool to determine the protein 3D structure. However, it is time-consuming and expensive, and not all proteins can be successfully crystallized, particularly for membrane proteins. Although NMR spectroscopy is indeed a very powerful tool in determining the 3D structures of membrane proteins, it is also time-consuming and costly. Due to the noncrystalline and insoluble nature of the neurodegenerative amyloid fibril or plaque, little structural data on the prion AGAAAAGA segment is available. Under these circumstances, the novel canonical dual computational approach introduced in this chapter showed its power in the molecular modeling of prion AGAAAAGA amyloid fibrils. This indicated that computational approaches or introducing novel mathematical formulations and physical concepts into molecular biology can significantly stimulate the development of biological and medical science [696]. The optimal atomic-resolution structures of prion AGAAAAGA amyloid fibils presented in this chapter are useful for the drive to find treatments for prion diseases in the field of medicinal chemistry.

13.5 Other Models

The simplified LJ potential minimization problem is

$$\min f(x) = 4\sum_{i=1}^{N}\sum_{j=1,j<i}^{N}\left(\frac{1}{\tau_{ij}^{6}} - \frac{1}{\tau_{ij}^{3}}\right) \quad s.t. \quad x \in R^{n},$$

where $\tau_{ij} = (x_{3i-2} - x_{3j-2})^2 + (x_{3i-1} - x_{3j-1})^2 + (x_{3i} - x_{3j})^2$, $(x_{3i-2}, x_{3i-1}, x_{3i})$ is the coordinates of atom i in R^3, $i, j = 1, 2, \ldots, N(\geq 2$ integer), $n = 3N$ and N is the whole number of atoms. The nonconvexity of the objective function and the huge number of local minima, which is growing exponentially with N, interest many mathematical optimization experts. In this section, the canonical dual theory elegantly tackles this problem illuminated by the amyloid fibril molecular model building.

13.5.1 Overview

Neutral atoms are subject to two distinct forces in the limit of large distance and short distance: a dispersion force (i.e. attractive vdW force) at long ranges, and a repulsion force, the result of overlapping electron orbitals. The LJ potential represents this behavior (en.wikipedia.org/wiki/LJ_potential, or [397] and references therein). The LJ potential is of the form

$$V(r) = 4\varepsilon\left[(\frac{\sigma}{r})^{12} - (\frac{\sigma}{r})^{6}\right], \tag{13.16}$$

where r is the distance between two atoms, ε is the depth of the potential well and σ is the atom diameter; these parameters can be fitted to reproduce experimental data or deduced from results of accurate quantum chemistry calculations. The $(\frac{\sigma}{r})^{12}$ term describes repulsion and the $(\frac{\sigma}{r})^{6}$ term describes attraction (Fig. 13.12). In Fig. 13.12 we may see two points: (I) $V(r) = 0$ (but the value of $V(r)$ is not the minimal value)

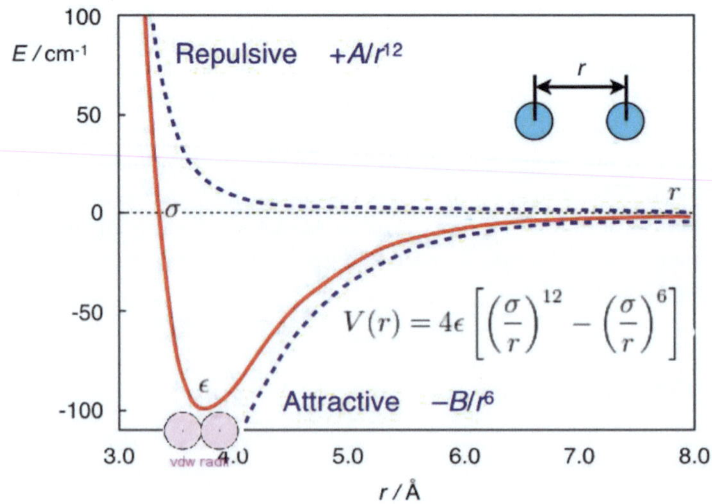

Fig. 13.12 The LJ Potential (formulas (13.16) and (13.22)) (see website: homepage.mac.com/swain/CMC/DDResources)

when $r = \sigma$ (i.e. the distance between two atoms equals to the sum of *atom radii* of the atoms); and (II) when $r = 2^{1/6}\sigma$ (i.e. the distance between two atoms equals to the sum of *vdW radii* of the atoms), the value of $V(r)$ reaches its minimal value $-\varepsilon$ (i.e. the bottom of the potential well; the force between the atoms is zero at this point). This section is written based on (II). If we introduce the coordinates of the atoms whose number is denoted by N and let $\varepsilon = \sigma = 1$ be the reduced units, the form (13.16) becomes

$$f(x) = 4\sum_{i=1}^{N} \sum_{j=1,j<i}^{N} \left(\frac{1}{\tau_{ij}^{6}} - \frac{1}{\tau_{ij}^{3}} \right), \tag{13.17}$$

where $\tau_{ij} = (x_{3i-2} - x_{3j-2})^2 + (x_{3i-1} - x_{3j-1})^2 + (x_{3i} - x_{3j})^2 = ||X_i - X_j||_2^2$ and $(x_{3i-2}, x_{3i-1}, x_{3i})$ is the coordinates of atom i, $i,j = 1,2,\ldots,N(\geq 2)$. The minimization of LJ potential $f(x)$ on R^n (where $n = 3N$) is an optimization problem:

$$\min_{s.t. x \in R^{3N}} f(x). \tag{13.18}$$

This optimization problem interests many optimization experts, for example, Pardalos [460], Xue [652–656], Huang et al. [295, 296].

For (13.18), when its global optimization solution is reached, the value r in (13.16) should be the sum of two *vdW radii* of the two atoms interacted. The three dimensional structure of a molecule with N atoms can be described by specifying the 3-Dimensional coordinate positions $X_1, X_2, \ldots, X_N \in R^3$ of all its atoms. Given bond lengths r_{ij} between a subset S of the atom pairs, the determination of the molecular structure is

$$(P_0) \quad \textit{to find} \quad X_1, X_2, \ldots, X_N \quad s.t. \quad ||X_i - X_j|| = r_{ij}, (i,j) \in S, \tag{13.19}$$

where $||\cdot||$ denotes a norm in a real vector space and it is calculated as the Euclidean distance 2-norm in this section. Equation (13.19) can be reformulated as a mathematical global optimization problem (GOP)

$$(P) \quad \min P(X) = \textstyle\sum_{(i,j)\in S} w_{ij}(||X_i - X_j||^2 - r_{ij}^2)^2 \tag{13.20}$$

in the terms of finding the global minimum of the function $P(X)$, where w_{ij}, $(i,j) \in S$ are positive weights, $X = (X_1, X_2, \ldots, X_N)^T \in R^{N\times 3}$ [431] and usually S has many fewer than $N^2/2$ elements due to the error in the theoretical or experimental data [262, 721]. There may even not exist any solution $X_1, X_2, \ldots, X_N$ to satisfy the distance constraints in (13.19), for example when data for atoms $i, j, k \in S$ violate the triangle inequality; in this case, we may add a perturbation term $-\epsilon^T X$ to $P(X)$:

$$(P_\epsilon) \quad \min P_\epsilon(X) = \textstyle\sum_{(i,j)\in S} w_{ij}(||X_i - X_j||^2 - r_{ij}^2)^2 - \epsilon^T X, \tag{13.21}$$

where $\epsilon \geq 0$. Thus, the L-J potential optimization problem (13.18) is rewritten into the optimization problem (13.21).

Problem (13.21) is just the minimization problem of sum of fourth-order polynomials, which can be elegantly solved by the canonical dual theory (CDT) in optimization [221, 232, 233]. We apply the above theory to an amyloid fibril molecular model building problem. In the next subsection, the molecular model building works of prion AGAAAAGA amyloid fibrils will be done by the CDT. We find that just very slight refinement is needed by the optimization programs of computational chemistry package Amber 11 to get the optimal molecular models. This implies to us the effectiveness of the CDT to solve our L-J potential minimization problem. Thus, when using the time-consuming and costly X-ray crystallography or NMR spectroscopy we still cannot determine the 3D structure of a protein, we may introduce computational approaches or novel mathematical formulations and physical concepts into molecular biology to study molecular structures. These concluding remarks will be made lastly.

13.5.2 Applications to a Lennard-Jones Potential Optimization Problem

In 2007, Sawaya et al. [517] got a breakthrough finding: the atomic structures of all amyloid fibrils revealed steric zippers, with strong vdW interactions (LJ) between β-sheets and HBs to maintain the β-strands. Similarly as (13.16), i.e. the potential energy for the vdW interactions (Fig. 13.12) between β-sheets:l

$$V_{LJ}(r) = \frac{A}{r^{12}} - \frac{B}{r^{6}}, \tag{13.22}$$

the potential energy for the HBs between the β-strands has a similar formula

$$V_{HB}(r) = \frac{C}{r^{12}} - \frac{D}{r^{10}}, \tag{13.23}$$

where A, B, C, D are constants given. Thus, the amyloid fibril molecular model building problem is reduced to well solve the optimization problem (13.18) or (13.21) (in this section we apply the CDT introduced in Sect. 13.2 to solve (13.21)).

In this section, we will use suitable templates 3nvf.pdb, 3nvg.pdb and 3nvh.pdb from the Protein Data Bank (www.rcsb.org/) to build some amyloid fibril models.

13.5.2.1 3NVF

Constructions of the AGAAAAGA amyloid fibril molecular structures of prion 113–120 region are based on the most recently released experimental molecular structures of IIHFGS segment 138–143 from human prion (PDB entry 3NVF

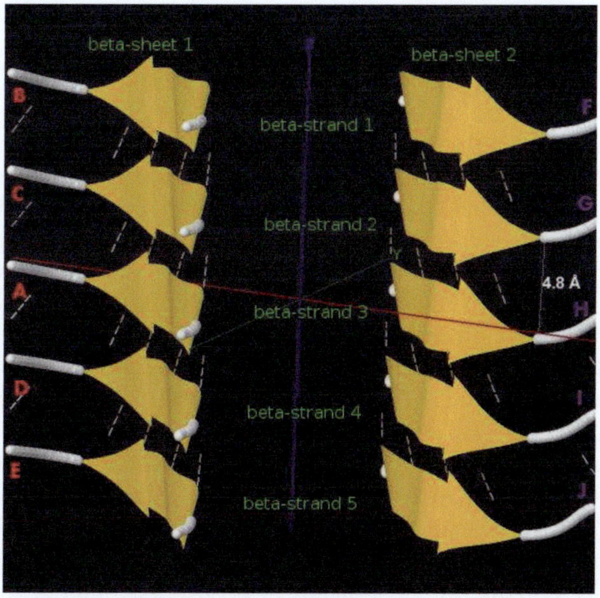

Fig. 13.13 Protein fibril structure of IIHFGS segment 138–143 from human prion. The *purple dashed lines* denote the hydrogen bonds. A, B, ..., I, J denote the 10 chains of the fibrils

released into Protein Data Bank (www.rcsb.org) on 2011-03-02) [18]. The atomic-resolution structure of this peptide is a steric zipper, with strong vdW interactions between β-sheets and HBs to maintain the β-strands (Fig. 13.13). In Fig. 13.13 we see that H chain (i.e. β-sheet 2) of 3NVF.pdb can be obtained from A chain (i.e. β-sheet 1) by

$$H = \begin{pmatrix} -1 & 0 & 0 \\ 0 & -1 & 0 \\ 0 & 0 & 1 \end{pmatrix} A + \begin{pmatrix} 27.546 \\ 0 \\ 0 \end{pmatrix}, \tag{13.24}$$

and other chains can be got by

$$C(G) = A(H) + \begin{pmatrix} 0 \\ 0 \\ 4.8 \end{pmatrix}, B(F) = A(H) + 2\begin{pmatrix} 0 \\ 0 \\ 4.8 \end{pmatrix}, \tag{13.25}$$

$$D(I) = A(H) - \begin{pmatrix} 0 \\ 0 \\ 4.8 \end{pmatrix}, E(J) = A(H) - 2\begin{pmatrix} 0 \\ 0 \\ 4.8 \end{pmatrix}. \tag{13.26}$$

Basing on the template 3NVF.pdb from Protein Data Bank, three prion AGAAAAGA palindrome amyloid fibril models – an AGAAAA model (3nvf-Model 1), a GAAAAG model (3nvf-Model 2), and an AAAAGA model (3nvf-Model 3) will be successfully constructed in this section. Chain A of 3nvf-Models 1–3 were got from A Chain of 3NVF.pdb using the mutate module of the free package Swiss-PdbViewer (SPDBV Version 4.01) (spdbv.vital-it.ch). It is pleasant to see that almost all the hydrogen bonds are still kept after the mutations; thus we just need to consider the vdW contacts only. Making mutations for H Chain of 3NVF.pdb, we can get H Chain of 3nvf-Models 1–3. However, we find that the vdW contacts between A Chain and H Chain are too far at this moment. We know that for 3nvf-Model 1 at least the vdW interaction between A.GLY2.CA-H.GLY2.CA, A.ALA4.CB-H.GLY2.CA should be maintained, for 3nvf-Model 2 at least three vdW interactions between A.ALA4.CB-H.ALA2.CB, A.ALA2.CB-H.ALA2.CB, A.ALA2.CB-H.ALA4.CB should be maintained, and for 3nvf-Model 3 at least three vdW interactions between A.ALA2.CB-H.ALA2.CB, A.ALA2.CB-H.ALA4.CB, A.ALA4.CB-H.ALA2.CB should be maintained. Fixing the coordinates of A.GLY2.CA and A.ALA4.CB (two anchors) ((–10.919,–3.862,–1.487), (6.357,1.461,–1.905)) for 3nvf-Model 1, fixing the coordinates of A.ALA2.CB and A.ALA4.CB (two anchors) ((11.959,–2.844,–1.977), (6.357,1.461,–1.905)) for 3nvf-Models 2–3, letting d equal to the twice of the vdW radius of Carbon atom (i.e. $d = 3.4$ Å), and letting the coordinates of H.GLY2.CA of 3nvf-Model 1 (two sensors) and the coordinates of H.ALA2.CB and H.ALA4.CB of 3nvf-Models 2–3 (two sensors) be variables, we may get a simple MDGP with 3/6 variables and its dual with 2/3 variables for 3nvf-Model 1:

$$\begin{aligned} P_\epsilon(x_1) = &\ \frac{1}{2}\left\{(x_{11}+10.919)^2+(x_{12}+3.862)^2+(x_{13}+1.487)^2-3.4^2\right\}^2 \\ &+\frac{1}{2}\left\{(x_{11}-6.357)^2+(x_{12}-1.461)^2+(x_{13}+1.905)^2-3.4^2\right\}^2 \\ &-(0.05x_{11}+0.05x_{12}+0.05x_{13}), \end{aligned}$$

$$\begin{aligned} P_\epsilon^d(\varsigma_1,\varsigma_2) = &\ 124.7908\varsigma_1-\frac{1}{2}\varsigma_1^2+34.615\varsigma_2-\frac{1}{2}\varsigma_2^2 \\ &-\frac{1}{2}\begin{pmatrix}0.05-21.838\varsigma_1+12.714\varsigma_2\\ 0.05-7.724\varsigma_1+2.922\varsigma_2\\ 0.05-2.974\varsigma_1-3.81\varsigma_2\end{pmatrix}^T\begin{pmatrix}\frac{1}{2\varsigma_1+2\varsigma_2} & 0 & 0\\ 0 & \frac{1}{2\varsigma_1+2\varsigma_2} & 0\\ 0 & 0 & \frac{1}{2\varsigma_1+2\varsigma_2}\end{pmatrix} \\ &\begin{pmatrix}0.05-21.838\varsigma_1+12.714\varsigma_2\\ 0.05-7.724\varsigma_1+2.922\varsigma_2\\ 0.05-2.974\varsigma_1-3.81\varsigma_2\end{pmatrix}. \end{aligned}$$

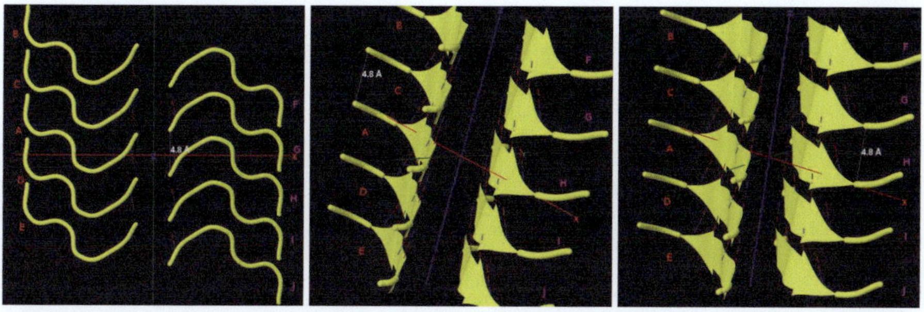

Fig. 13.14 Protein fibril structure of 3nvf-Models 1–3 (from *left* to *right* respectively) for prion AGAAAAGA segment 113–120. The *purple dashed lines* denote the hydrogen bonds. A, B, ..., I, J denote the 10 chains of the fibrils

We can get a global maximal solution (70.1836,70.1812) for $P_\epsilon^d(\varsigma_1, \varsigma_2)$ and its corresponding local maximal solution to $P_\epsilon(x_1)$:

$$\bar{x} = (-2.28097, -1.20037, -1.69582).$$

By Theorem 13.2.1 we know that $\bar{x}$ is a global minimal solution of $P_\epsilon(x_1)$. Thus we get

$$H = \begin{pmatrix} -1 & 0 & 0 \\ 0 & -1 & 0 \\ 0 & 0 & 1 \end{pmatrix} A + \begin{pmatrix} -4.5619 \\ -2.4009 \\ 0.0004 \end{pmatrix} \tag{13.27}$$

for 3nvf-Model 1, whose other chains can be got by (13.25)–(13.26) (Fig. 13.14). For 3nvf-Models 2–3, similarly we may get a simple MDGP with 6 variables and its dual with 3 variables:

$$\begin{aligned} P_\epsilon(x_1, x_2) = \; & \frac{1}{2}\left\{(x_{11} - 11.959)^2 + (x_{12} + 2.844)^2 + (x_{13} + 1.977)^2 - 3.4^2\right\}^2 \\ & + \frac{1}{2}\left\{(x_{21} - 11.959)^2 + (x_{22} + 2.844)^2 + (x_{23} + 1.977)^2 - 3.4^2\right\}^2 \\ & + \frac{1}{2}\left\{(x_{11} - 6.357)^2 + (x_{12} - 1.461)^2 + (x_{13} + 1.905)^2 - 3.4^2\right\}^2 \\ & - (0.05x_{11} + 0.05x_{12} + 0.05x_{13} + 0.05x_{21} + 0.05x_{22} + 0.05x_{23}), \end{aligned}$$

$$P_\epsilon^d(\varsigma_1,\varsigma_2,\varsigma_3) = 143.4545\varsigma_1 - \frac{1}{2}\varsigma_1^2 + 143.4545\varsigma_2 - \frac{1}{2}\varsigma_2^2 + 34.6150\varsigma_3 - \frac{1}{2}\varsigma_3^2$$

$$-\frac{1}{2}\begin{pmatrix} 0.05+23.9180\varsigma_1+12.7140\varsigma_3 \\ 0.05-5.6880\varsigma_1+2.9220\varsigma_3 \\ 0.05-3.9540\varsigma_1-3.8100\varsigma_3 \\ 0.05+23.9180\varsigma_2 \\ 0.05-5.6880\varsigma_2 \\ 0.05-3.9540\varsigma_2 \end{pmatrix}^T \begin{pmatrix} \frac{1}{2\varsigma_1+2\varsigma_3} & 0 & 0 & 0 & 0 & 0 \\ 0 & \frac{1}{2\varsigma_1+2\varsigma_3} & 0 & 0 & 0 & 0 \\ 0 & 0 & \frac{1}{2\varsigma_1+2\varsigma_3} & 0 & 0 & 0 \\ 0 & 0 & 0 & \frac{1}{2\varsigma_2} & 0 & 0 \\ 0 & 0 & 0 & 0 & \frac{1}{2\varsigma_2} & 0 \\ 0 & 0 & 0 & 0 & 0 & \frac{1}{2\varsigma_2} \end{pmatrix}$$

$$\begin{pmatrix} 0.05+23.9180\varsigma_1+12.7140\varsigma_3 \\ 0.05-5.6880\varsigma_1+2.9220\varsigma_3 \\ 0.05-3.9540\varsigma_1-3.8100\varsigma_3 \\ 0.05+23.9180\varsigma_2 \\ 0.05-5.6880\varsigma_2 \\ 0.05-3.9540\varsigma_2 \end{pmatrix}.$$

We can get a global maximal solution (0.920088,0.0127286,0.921273) for $P_\epsilon^d(\varsigma_1,\varsigma_2,\varsigma_3)$ and its corresponding local maximal solution to $P_\epsilon(x_1,x_2)$:

$$\bar{x} = (9.16977, -0.676538, -1.9274, 13.9231, -0.879925, -0.0129248).$$

By Theorem 13.2.1 we know that $\bar{x}$ is a global minimal solution of $P_\epsilon(x_1,x_2)$. Thus we get

$$H = \begin{pmatrix} -1 & 0 & 0 \\ 0 & -1 & 0 \\ 0 & 0 & 1 \end{pmatrix} A + \begin{pmatrix} 20.8459 \\ -2.1533 \\ 0.6638 \end{pmatrix}. \tag{13.28}$$

for 3nvg-Models 2–3, whose other chains can be got by (13.25)–(13.26) (Fig. 13.14).

We find 3nvf-Model 1 has some atoms with bad/close contacts, 3nvf-Model 2 has 6 bad/close contacts, and 3nvf-Model 3 has no bad/close contact. This means it not necessary at all to further refine 3nvf-Model 3. We remove these bad contacts by performing energy minimization using Amber 11 [91]. Even if there are no obvious bad contacts, it is still a good idea to run a short energy minimization to relax the structures a bit. We will perform the energy minimization in 2 stages. In the first stage, we'll only minimize the water molecules and hold the protein fixed for 500 steps of steepest descent method and then 500 steps of conjugate gradient method. Our goal is just to remove bad contacts, there is no need to go overboard with minimization. In the second stage we proceed directly to minimizing the entire system as a whole for 1500 steps of steepest descent method and then 1000 steps of conjugate gradient method. RMSD (root mean square deviation) is an indicator for structural changes in a protein. It is used to measure the scalar spatial distance between atoms of the same type (for example the C_α atoms) for two structures in different time. The RMSDs between the last snapshot after the refinement and the snapshot illuminated in Fig. 13.14 are 2.15796, 1.3089087, 1.045318 Å for these three 3nvf-Models respectively. The very small values of RMSD are very good measure of precision of CDT for our model building. This shows us that CDT performs well.

13.5.2.2 3NVG

In this subsection, the constructions of the AGAAAAGA amyloid fibril molecular structures of prion 113–120 region are based on the most recently released experimental molecular structures of MIHFGN segment 137–142 from mouse prion (PDB entry 3NVG released into Protein Data Bank (www.rcsb.org) on 2011-03-02) [18]. The atomic-resolution structure of this peptide is a steric zipper, with strong vdW interactions between β-sheets and HBs to maintain the β-strands (Fig. 13.15). In Fig. 13.15 we see that H Chain (i.e. β-sheet 2) of 3NVG.pdb can be obtained from A Chain (i.e. β-sheet 1) by

$$H = \begin{pmatrix} -1 & 0 & 0 \\ 0 & 1 & 0 \\ 0 & 0 & -1 \end{pmatrix} A + \begin{pmatrix} -27.28 \\ 2.385 \\ 15.738 \end{pmatrix}, \tag{13.29}$$

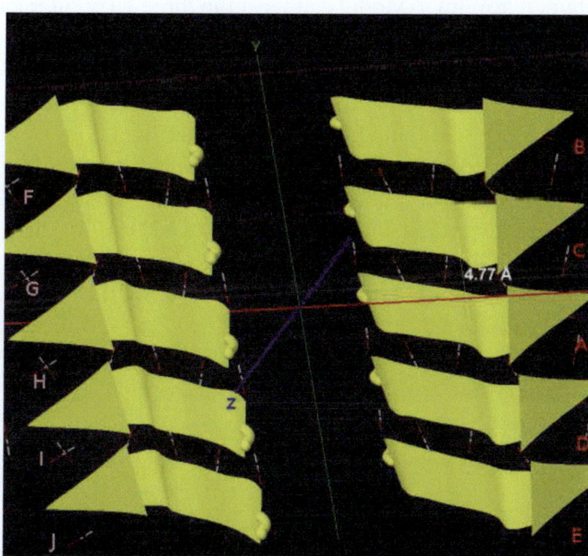

Fig. 13.15 Protein fibril structure of MIHFGN segment 137–142 from mouse prion. The *purple dashed lines* denote the hydrogen bonds. A, B, . . . , I, J denote the 10 chains of the fibril

and other chains can be got by

$$C(G) = A(H) + \begin{pmatrix} 0 \\ 4.77 \\ 0 \end{pmatrix}, B(F) = A(H) + 2 \begin{pmatrix} 0 \\ 4.77 \\ 0 \end{pmatrix}, \tag{13.30}$$

$$D(I) = A(H) - \begin{pmatrix} 0 \\ 4.77 \\ 0 \end{pmatrix}, E(J) = A(H) - 2 \begin{pmatrix} 0 \\ 4.77 \\ 0 \end{pmatrix}. \tag{13.31}$$

Basing on the template 3NVG.pdb from Protein Data Bank, three prion AGAAAAGA palindrome amyloid fibril models – an AGAAAA model (3nvg-Model 1), a GAAAAG model (3nvg-Model 2), and an AAAAGA model (3nvg-Model 3) will be successfully constructed in this section. Chain A of 3nvg-Models 1–3 were got from A Chain of 3NVG.pdb using the mutate module of the free package Swiss-PdbViewer (SPDBV Version 4.01) (spdbv.vital-it.ch). It is pleasant to see that almost all the hydrogen bonds are still kept after the mutations; thus we just need to consider the vdW contacts only. Making mutations for H Chain of 3NVG.pdb, we can get the H Chains of 3nvg-Models 1–3. However, the vdW contacts between A Chain and H Chain are too far at this moment. We know that for 3nvg-Model 1 at least the three vdW interaction between A.GLY2.CA-H.GLY2.CA, A.GLY2.CA-H.ALA4.CB, A.ALA4.CB-H.GLY2.CA should be maintained, for 3nvg-Model 2 at least the three vdW interactions between A.ALA2.CB-H.ALA2.CB, A.ALA2.CB-H.ALA4.CB, A.ALA4.CB-H.ALA2.CB should be maintained, and for 3nvg-Model 3 at least the three vdW interactions between A.ALA2.CB-H.ALA2.CB, A.ALA2.CB-H.ALA4.CB, A.ALA4.CB-H.ALA2.CB should be maintained. Fixing the coordinates of A.GLY2.CA and A.ALA4.CB (two anchors) ((−11.159,−2.241,4.126), (−5.865,−2.618,8.696)) for 3nvg-Model 1, fixing the coordinates of A.ALA2.CB and A.ALA4.CB (two anchors) ((−12.040,−2.675,5.307), (−5.865,−2.618,8.696)) for 3nvg-Models 1–2, letting d equal to the twice of the vdW radius of Carbon atom (i.e. $d = 3.4$ Å), and letting the coordinates of H.GLY2.CA and H.ALA4.CB of 3nvg-Model 1 (two sensors) and the coordinates of H.ALA2.CB and H.ALA4.CB of 3nvg-Models 2–3 (two sensors) be variables, we may get a simple MDGP with 6 variables and its dual with 3 variables for 3nvg-Model 1:

$$\begin{aligned} P_\epsilon(x_1, x_2) = &\ \frac{1}{2}\left\{(x_{11} + 11.159)^2 + (x_{12} + 2.241)^2 + (x_{13} - 4.126)^2 - 3.4^2\right\}^2 \\ &+ \frac{1}{2}\left\{(x_{21} + 11.159)^2 + (x_{22} + 2.241)^2 + (x_{23} - 4.126)^2 - 3.4^2\right\}^2 \\ &+ \frac{1}{2}\left\{(x_{11} + 5.865)^2 + (x_{12} + 2.618)^2 + (x_{13} - 8.696)^2 - 3.4^2\right\}^2 \\ &-(0.05x_{11} + 0.05x_{12} + 0.05x_{13} + 0.05x_{21} + 0.05x_{22} + 0.05x_{23}), \end{aligned}$$

$$P^d_\epsilon(\varsigma_1,\varsigma_2,\varsigma_3)=135.009238\varsigma_1-\frac{1}{2}\varsigma_1^2+135.009238\varsigma_2-\frac{1}{2}\varsigma_2^2+105.3125\varsigma_3-\frac{1}{2}\varsigma_3^2$$

$$-\frac{1}{2}\begin{pmatrix} 0.05-22.318\varsigma_1-11.73\varsigma_3 \\ 0.05-4.482\varsigma_1-5.236\varsigma_3 \\ 0.05+8.252\varsigma_1+17.392\varsigma_3 \\ 0.05-22.318\varsigma_2 \\ 0.05-4.482\varsigma_2 \\ 0.05+8.252\varsigma_2 \end{pmatrix}^T \begin{pmatrix} \frac{1}{2\varsigma_1+2\varsigma_3} & 0 & 0 & 0 & 0 & 0 \\ 0 & \frac{1}{2\varsigma_1+2\varsigma_3} & 0 & 0 & 0 & 0 \\ 0 & 0 & \frac{1}{2\varsigma_1+2\varsigma_3} & 0 & 0 & 0 \\ 0 & 0 & 0 & \frac{1}{2\varsigma_2} & 0 & 0 \\ 0 & 0 & 0 & 0 & \frac{1}{2\varsigma_2} & 0 \\ 0 & 0 & 0 & 0 & 0 & \frac{1}{2\varsigma_2} \end{pmatrix}$$

$$\begin{pmatrix} 0.05-22.318\varsigma_1-11.73\varsigma_3 \\ 0.05-4.482\varsigma_1-5.236\varsigma_3 \\ 0.05+8.252\varsigma_1+17.392\varsigma_3 \\ 0.05-22.318\varsigma_2 \\ 0.05-4.482\varsigma_2 \\ 0.05+8.252\varsigma_2 \end{pmatrix}.$$

We can get a global maximal solution (0.708403,0.0127287,0.699001) for $P^d_\epsilon(\varsigma_1,\varsigma_2,\varsigma_3)$ and its corresponding local maximal solution to $P_\epsilon(x_1,x_2)$:

$$\bar{x}=(-8.51192,-2.41048,6.4135,-9.19493,-0.276929,6.09007).$$

By Theorem 13.2.1 we know that $\bar{x}$ is a global minimal solution of $P_\epsilon(x_1,x_2)$. Thus we get

$$H=\begin{pmatrix} -1 & 0 & 0 \\ 0 & 1 & 0 \\ 0 & 0 & -1 \end{pmatrix}A+\begin{pmatrix} -18.133923 \\ 0.6673703 \\ 11.955023 \end{pmatrix} \tag{13.32}$$

for 3nvg-Model 1, whose other chains can be got by (13.30)–(13.31) (Fig. 13.16). For 3nvg-Models 2–3, similarly we may get a simple MDGP with 6 variables and its dual with 3 variables:

$$\begin{aligned} P_\epsilon(x_1,x_2) = {} & \frac{1}{2}\left\{(x_{11}+12.040)^2+(x_{12}+2.675)^2+(x_{13}-5.307)^2-3.4^2\right\}^2 \\ & +\frac{1}{2}\left\{(x_{21}+12.040)^2+(x_{22}+2.675)^2+(x_{23}-5.307)^2-3.4^2\right\}^2 \\ & +\frac{1}{2}\left\{(x_{11}+5.865)^2+(x_{12}+2.618)^2+(x_{13}-8.696)^2-3.4^2\right\}^2 \\ & -(0.05x_{11}+0.05x_{12}+0.05x_{13}+0.05x_{21}+0.05x_{22}+0.05x_{23}), \end{aligned}$$

$$P^d_\epsilon(\varsigma_1,\varsigma_2,\varsigma_3)=168.721474\varsigma_1-\frac{1}{2}\varsigma_1^2+168.721474\varsigma_2-\frac{1}{2}\varsigma_2^2+105.312565\varsigma_3-\frac{1}{2}\varsigma_3^2$$

$$-\frac{1}{2}\begin{pmatrix} 0.05-24.080\varsigma_1-11.73\varsigma_3 \\ 0.05-5.35\varsigma_1-5.236\varsigma_3 \\ 0.05+10.614\varsigma_1+17.392\varsigma_3 \\ 0.05-24.080\varsigma_2 \\ 0.05-5.35\varsigma_2 \\ 0.05+10.614\varsigma_2 \end{pmatrix}^T \begin{pmatrix} \frac{1}{2\varsigma_1+2\varsigma_3} & 0 & 0 & 0 & 0 & 0 \\ 0 & \frac{1}{2\varsigma_1+2\varsigma_3} & 0 & 0 & 0 & 0 \\ 0 & 0 & \frac{1}{2\varsigma_1+2\varsigma_3} & 0 & 0 & 0 \\ 0 & 0 & 0 & \frac{1}{2\varsigma_2} & 0 & 0 \\ 0 & 0 & 0 & 0 & \frac{1}{2\varsigma_2} & 0 \\ 0 & 0 & 0 & 0 & 0 & \frac{1}{2\varsigma_2} \end{pmatrix}$$

$$\begin{pmatrix} 0.05-24.080\varsigma_1-11.73\varsigma_3 \\ 0.05-5.35\varsigma_1-5.236\varsigma_3 \\ 0.05+10.614\varsigma_1+17.392\varsigma_3 \\ 0.05-24.080\varsigma_2 \\ 0.05-5.35\varsigma_2 \\ 0.05+10.614\varsigma_2 \end{pmatrix}.$$

We can get a global maximal solution (0.849735,0.0127287,0.84036) for $P^d_\epsilon(\varsigma_1,\varsigma_2,\varsigma_3)$ and its corresponding local maximal solution to $P_\epsilon(x_1,x_2)$:

$$\bar{x}=(-8.95484,-2.63187,7.00689,-10.0759,-0.710929,7.27107).$$

By Theorem 13.2.1 we know that $\bar{x}$ is a global minimal solution of $P_\epsilon(x_1,x_2)$. Thus we get

$$H=\begin{pmatrix} -1 & 0 & 0 \\ 0 & 1 & 0 \\ 0 & 0 & -1 \end{pmatrix}A+\begin{pmatrix} -19.3102 \\ 0.6644 \\ 13.5316 \end{pmatrix}. \tag{13.33}$$

for 3nvg-Models 2–3, whose other chains can be got by (13.30)–(13.31) (Fig. 13.16).

We did same refinements for 3nvg-Models 1–3 as for 3nvf-Models 1–3. The RMSDs between the last snapshot after the refinement and the snapshot illuminated in Fig. 13.16 are 1.572438, 1.404648, 1.464767 Å for these three 3nvg-Models respectively. The very small values of RMSD show us that CDT performs well and precisely for 3nvg-Model building.

13.5.2.3 3NVH

Similar as the above two subsections, this subsection constructs the AGAAAAGA amyloid fibril molecular structures of prion 113–120 region basing on the most recently released experimental molecular structures of MIHFGND segment 137–143 from mouse prion (PDB entry 3NVH released into Protein Data Bank

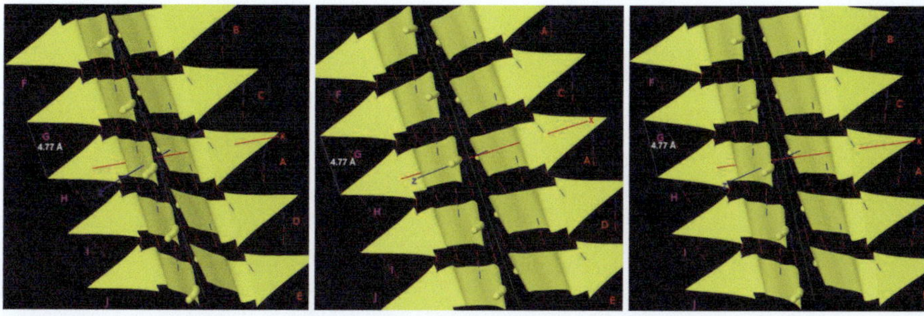

Fig. 13.16 Protein fibril structure of 3nvg-Models 1–3 (from *left* to *right* respectively) for prion AGAAAAGA segment 113–120. The *purple dashed lines* denote the hydrogen bonds. $A, B, \ldots, I, J$ denote the 10 chains of the fibrils

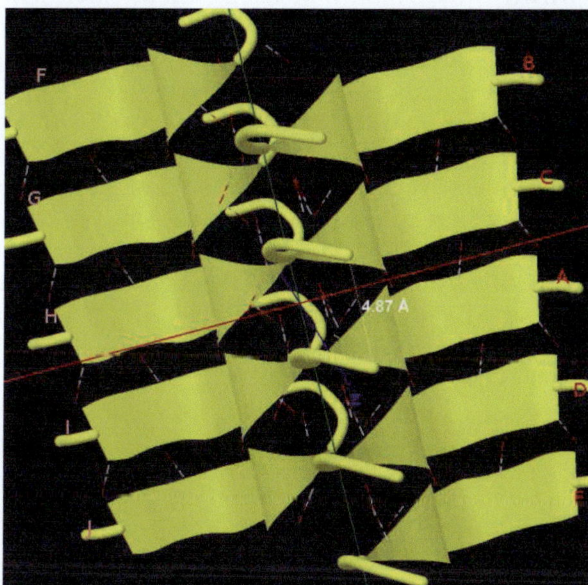

Fig. 13.17 Protein fibril structure of MIHFGND segment 137–143 from mouse prion. The *purple dashed lines* denote the hydrogen bonds. $A, B, \ldots, I, J$ denote the 10 chains of the fibril

(www.rcsb.org) on 2011-03-02) [18]. The atomic-resolution structure of this peptide is a steric zipper, with strong vdW interactions between β-sheets and HBs to maintain the β-strands (Fig. 13.17). In Fig. 13.17 we see that H chain (i.e. β-sheet 2) of 3NVH.pdb can be obtained from A chain (i.e. β-sheet 1) by

$$H = \begin{pmatrix} -1 & 0 & 0 \\ 0 & 1 & 0 \\ 0 & 0 & -1 \end{pmatrix} A + \begin{pmatrix} 0 \\ 2.437 \\ -15.553 \end{pmatrix}, \tag{13.34}$$

and other chains can be got by

$$C(G) = A(H) + \begin{pmatrix} 0 \\ 4.87 \\ 0 \end{pmatrix}, B(F) = A(H) + 2\begin{pmatrix} 0 \\ 4.87 \\ 0 \end{pmatrix}, \tag{13.35}$$

$$D(I) = A(H) - \begin{pmatrix} 0 \\ 4.87 \\ 0 \end{pmatrix}, E(J) = A(H) - 2\begin{pmatrix} 0 \\ 4.87 \\ 0 \end{pmatrix}. \tag{13.36}$$

Basing on the template 3NVH.pdb from Protein Data Bank, three prion AGAAAAGA palindrome amyloid fibril models – an AGAAAAG model (3nvh-Model 1), a GAAAAGA model (3nvh-Model 2) will be successfully constructed in this section. A chain of 3nvh-Models 1–2 were got from A chain of 3NVH.pdb using the mutate module of the free package Swiss-PdbViewer (SPDBV Version 4.01) (spdbv.vital-it.ch). It is pleasant to see that almost all the hydrogen bonds are still kept after the mutations; thus we just need to consider the vdW contacts only. Making mutations for H chain of 3NVH.pdb, we can get the H chains of 3nvh-Models 1–2. However, the vdW contacts between A chain and H chain are too far at this moment (≥4.25 Å). we may know that for 3nvh-Models 1–2 at least one vdW interaction between A.ALA4.CB-H.ALA4.CB should be maintained. Fixing the coordinates of A.ALA4.CB (the anchor) ((1.731,–1.514,–7.980)), letting d equal to the twice of the vdW radius of Carbon atom (i.e. $d = 3.4$ Å), and letting the coordinate of H.ALA4.CB (one sensor) be variables, we may get a simple MDGP with 3 variables and its dual with 1 variable:

$$\begin{aligned} P_\epsilon(x_1) = & \frac{1}{2}\{(x_{11} - 1.731)^2 + (x_{12} + 1.514)^2 + (x_{13} + 7.980)^2 - 3.4^2\}^2 \\ & -0.05x_{11} - 0.05x_{12} - 0.05x_{13}, \\ P_\epsilon^d(\varsigma_1) = & 57.409\varsigma_1 - \frac{1}{2}\varsigma_1^2 \\ & \frac{(0.05 + 3.462\varsigma_1)^2 + (0.05 - 3.028\varsigma_1)^2 + (0.05 - 15.96\varsigma_1)^2}{4\varsigma_1}. \end{aligned}$$

We can easily get the global maximal solution $0.0127287 \in \{\varsigma \in R^1 | \varsigma_i > 0, i = 1\}$ for $P_\epsilon^d(\varsigma_1)$. Then, we get its corresponding solution for $P_\epsilon(x_1)$:

$$\bar{x} = (3.69507, 0.450071, -6.01593).$$

By Theorem 13.2.1 we know that $\bar{x}$ is a global minimal solution of $P_\epsilon(x_1)$, i.e. for H.ALA4.CB. Thus we get

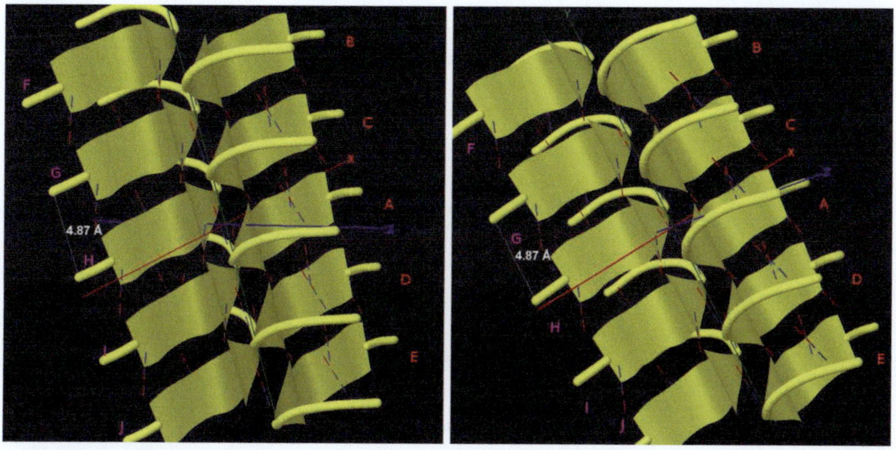

Fig. 13.18 Protein fibril structure of 3nvh-Models 1–2 (from *left* to *right* respectively) for prion AGAAAAGA segment 113–120. The *purple dashed lines* denote the hydrogen bonds. *A*, *B*, ..., *I*, *J*, *K*, *L* denote the 12 chains of the fibrils

$$H = \begin{pmatrix} -1 & 0 & 0 \\ 0 & 1 & 0 \\ 0 & 0 & -1 \end{pmatrix} A + \begin{pmatrix} 5.42607 \\ 1.964071 \\ -13.99593 \end{pmatrix} \quad (13.37)$$

for 3nvh-Models 1–2, whose other chains can be got by (13.35)–(13.36) (Fig. 13.18).

We carried on the same refinements for 3nvh-Models 1–2 as for 3nvf-Models 1–3 and 3nvg-Models 1–3. The RMSDs between the last snapshot after the refinement and the snapshot illuminated in Fig. 13.18 are 1.534417, 1.572836 Å for the two 3nvh-Models respectively. The very small values of RMSD again show to us that CDT performs well and precisely for 3nvg-Model building.

13.5.2.4 Refined 3nvf-Models 1–3, 3nvg-Models 1–3, 3nvh-Models 1–2

The amyloid fibril models of prion AGAAAAGA segment refined by Amber 11 are illuminated in Figs. 13.19, 13.20, and 13.21. All these models are without any bad contact now (checked by package Swiss-PdbViewer), and the vdW interactions between the two β-sheets are in a very perfect way now. All the initial structures before dealt by CDT approach have very far vdW contacts between the two β-sheets. CDT easily made the vdW contacts come closer and reach a state with the lowest potential energy, which has perfect vdW contacts as shown in Figs. 13.19, 13.20, and 13.21.

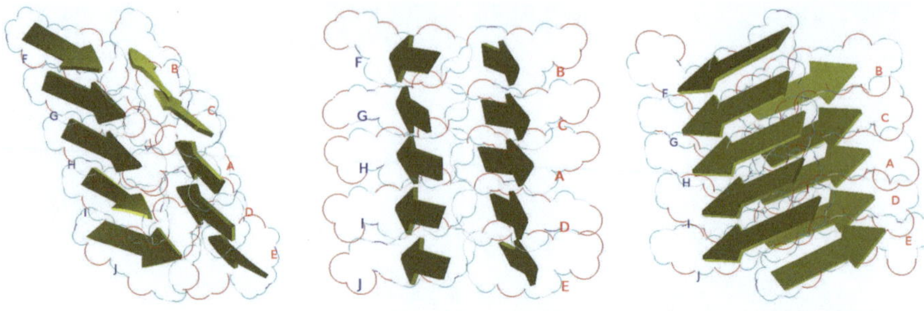

Fig. 13.19 Perfect 3nvf-Models 1–3 (from left to right respectively) for prion AGAAAAGA segment 113–120. The *purple dashed lines* denote the hydrogen bonds. *A*, *B*, ..., *I*, *J* denote the 10 chains of the fibrils

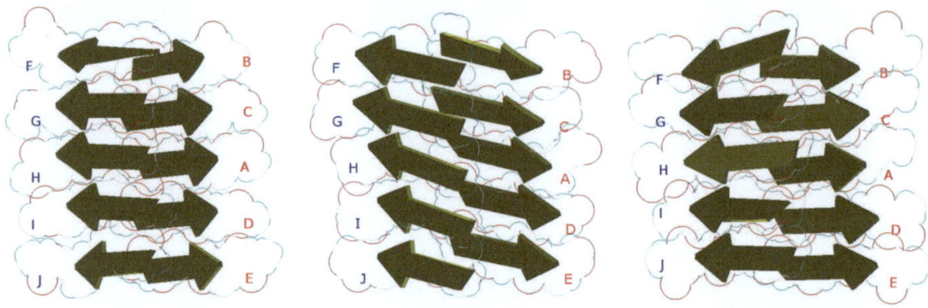

Fig. 13.20 Perfect 3nvg-Models 1–3 (from left to right respectively) for prion AGAAAAGA segment 113–120. The *purple dashed lines* denote the hydrogen bonds. *A*, *B*, ..., *I*, *J* denote the 10 chains of the fibrils

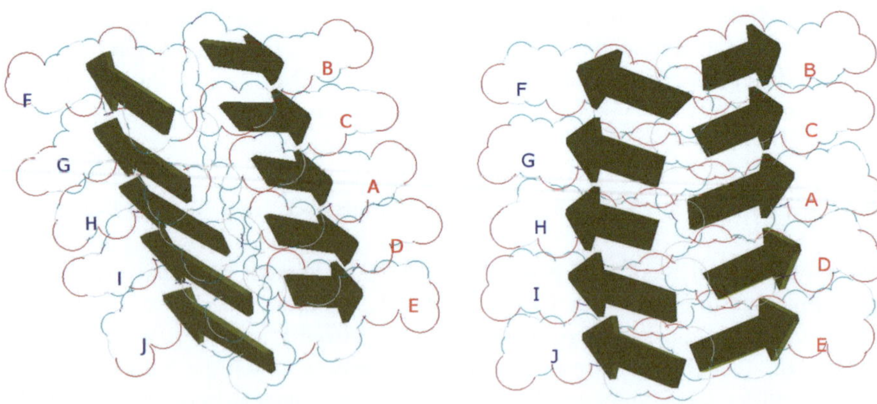

Fig. 13.21 Perfect 3nvh-Models 1–2 (from left to right respectively) for prion AGAAAAGA segment 113–120. The *purple dashed lines* denote the hydrogen bonds. *A*, *B*, ..., *I*, *J*, *K*, *L* denote the 12 chains of the fibrils

13.5.3 Concluding Remarks

Global optimization of LJ clusters is a challenging problem for researchers in the field of biology, physics, chemistry, computer science, materials science, and especially for experts in mathematical optimization research field because of the nonconvexity of the L-J potential energy function and enormous local minima on the potential energy surface. In March 2008, American Mathematical Programming Society specially produced one whole issue, No. 76, to discuss this problem. In this section through clever use of global optimization techniques of Gao's canonical dual theory (CDT), we successfully tackle this challenging problem illuminated by the amyloid fibril molecular model building. Clearly, this section shows to readers that CDT is very useful and powerful to tackle challenging problems in optimization area and many other areas.

13.5.4 3MD4/5 Models

Constructions of the AGAAAAGA amyloid fibril molecular structures of prion 113–120 region are based on the most recently released experimental molecular structures of GYMLGS/GYVLGS segment 127–132 from prion membrane protein (PDB entry 3MD4/5 released into Protein Data Bank (www.rcsb.org) on 2011-05-25) [375]. The atomic-resolution structure of this peptide is a steric zipper, with strong vdW interactions between β-sheets and HBs to maintain the β-strands (Figs. 13.22 and 13.23).

Basing on the template 3MD4.pdb from Protein Data Bank, three prion AGAAAAGA palindrome amyloid fibril models – an AGAAAAG model (3md4-Model 1), a GAAAAGA model (3md4-Model 2) and a AAAAGA model (3md4-Model 3) will be successfully constructed in this subsection. AB Chains of 3md4-Models 1–3 (Figs. 13.24, 13.25, and 13.26) were got from AB Chains of 3md4.pdb using the mutate module of the free package Swiss-PdbViewer (SPDBV Version 4.01) (spdbv.vital-it.ch). It is pleasant to see that the hydrogen bonds are kept after the mutations.

Thus, we just need to refine/relax the Models by the Steepest Decent-Conjugate Gradient optimization program of the Amber 11 (Figs. 13.27, 13.28, and 13.29).

Basing on the template 3MD5.pdb from Protein Data Bank, three prion AGAAAAGA palindrome amyloid fibril models – an AGAAAAG model (3md5-Model 1), a GAAAAGA model (3md5-Model 2) and a AAAAGA model (3md5-Model 3) will be successfully constructed in this subsection. AB Chains of 3md5-Models 1–3 (Figs. 13.30, 13.31, and 13.32) were got from AB Chains of 3md5.pdb using the mutate module of the free package Swiss-PdbViewer (SPDBV Version 4.01) (spdbv.vital-it.ch). However, the vdW contacts between A chain and B chain are too far at this moment (Figs. 13.30, 13.31, and 13.32).

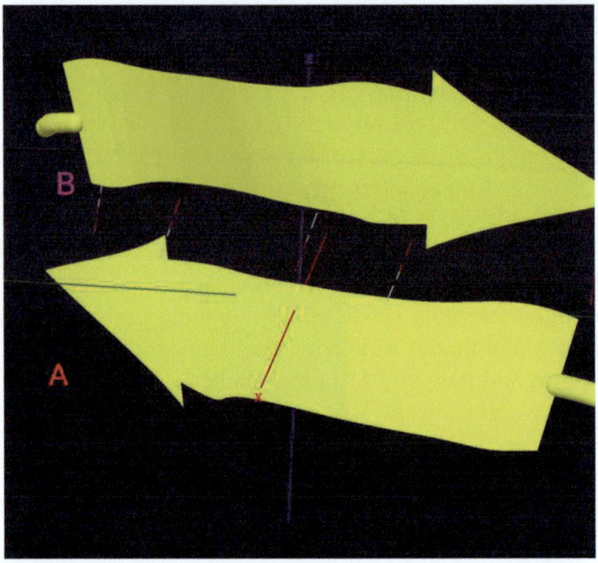

Fig. 13.22 Protein fibril structure of GYMLGS segment 127–132 from prion membrane protein. The *purple dashed lines* denote the hydrogen bonds. *A*, *B* denote the 2 chains of the fibril

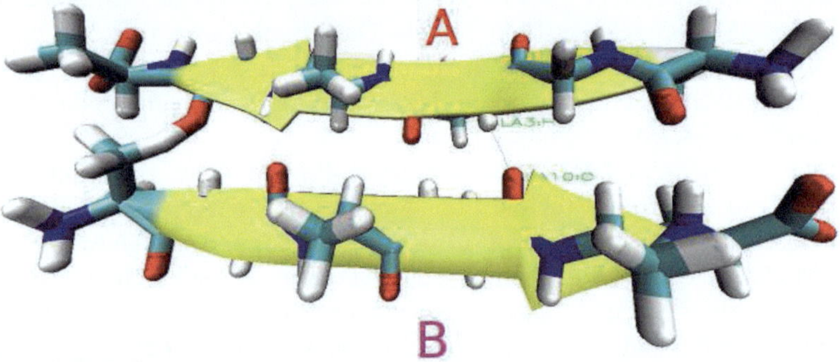

Fig. 13.23 Protein fibril structure of GYVLGS segment 127–132 from prion membrane protein. The *purple dashed lines* denote the hydrogen bonds. **A**, **B** denote the 2 chains of the fibril

Seeing Fig. 13.30, we may know that for 3mg5-Model 1 at least two vdW interactions between B.ALA130.CB-A.ALA130.CB and B.ALA130.CB-A.ALA132.CB should be maintained. Fixing the coordinate of B.ALA130.CB (the anchor) ((11.796, 7.063, 8.213)), letting d equal to the twice of the vdW radius of Carbon atom (i.e. $d = 3.4\,\text{Å}$), and letting the coordinates of A.ALA130.CB (a sensor) or A.ALA132.CB (a sensor) be variables, we may get a simple MDGP with 3 variables and its dual with 1 variable:

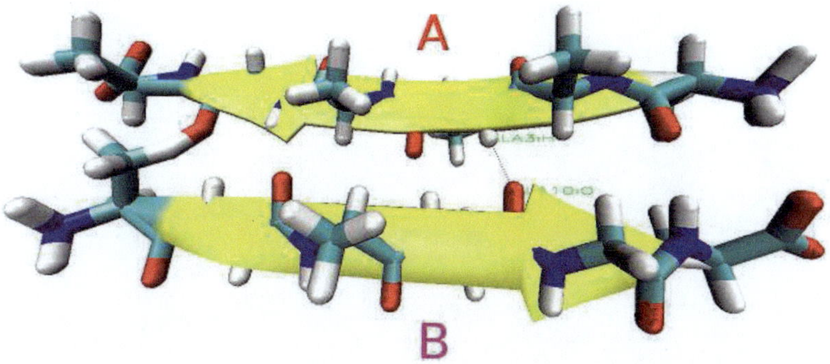

Fig. 13.24 vdW and HB contacts of AB chains of 3md4-Model 1

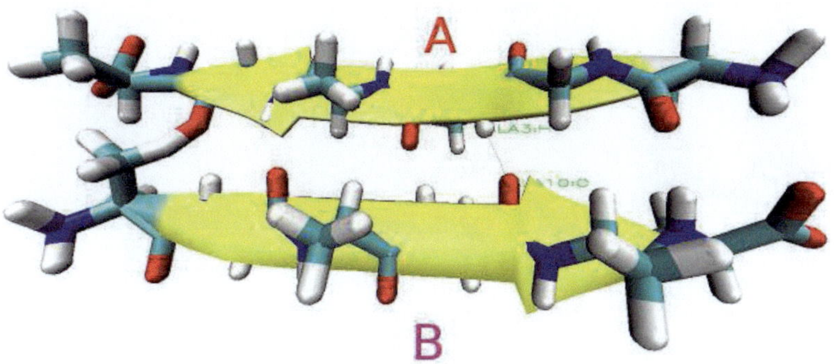

Fig. 13.25 vdW and HB contacts of AB chains of 3md4-Model 2

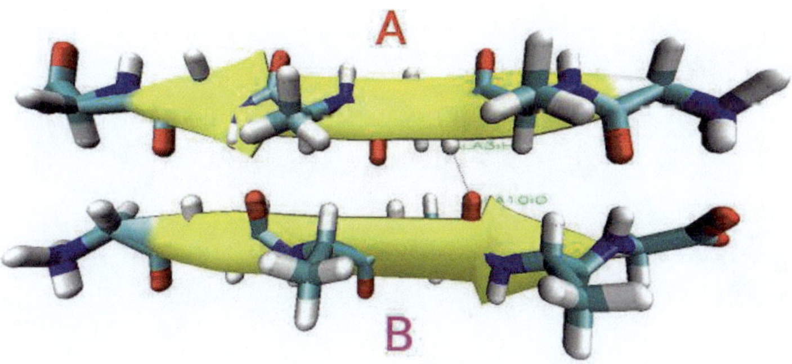

Fig. 13.26 vdW and HB contacts of AB chains of 3md4-Model 3

$$
\begin{aligned}
P_\epsilon(x_1) = \; & \frac{1}{2}\left\{(x_{11}-11.796)^2+(x_{12}-7.063)^2+(x_{13}-8.213)^2-3.4^2\right\}^2 \\
& -0.05x_{11}-0.05x_{12}-0.05x_{13},
\end{aligned}
$$

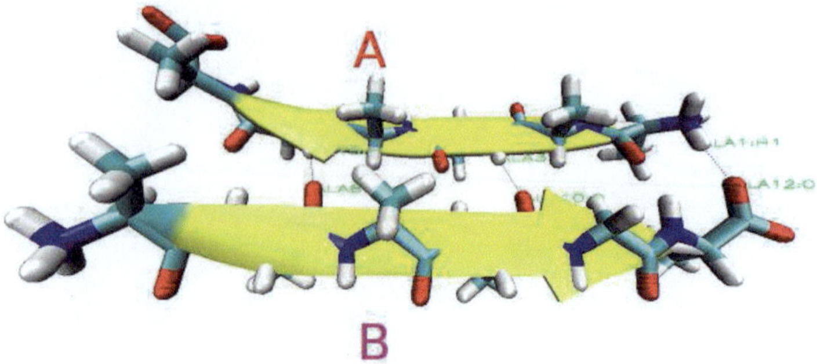

Fig. 13.27 Refined AB chains of 3md4-Model 1

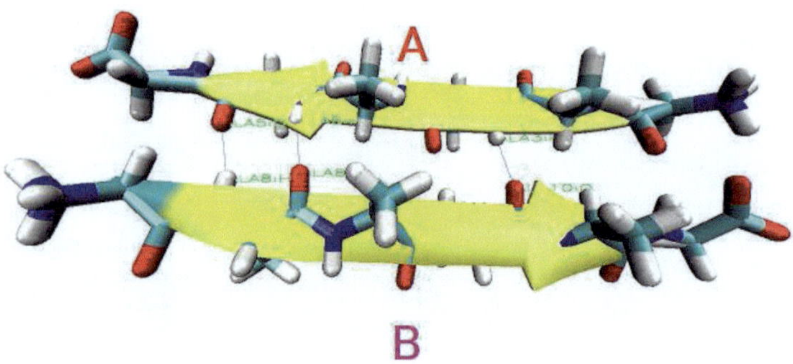

Fig. 13.28 Refined AB chains of 3md4-Model 2

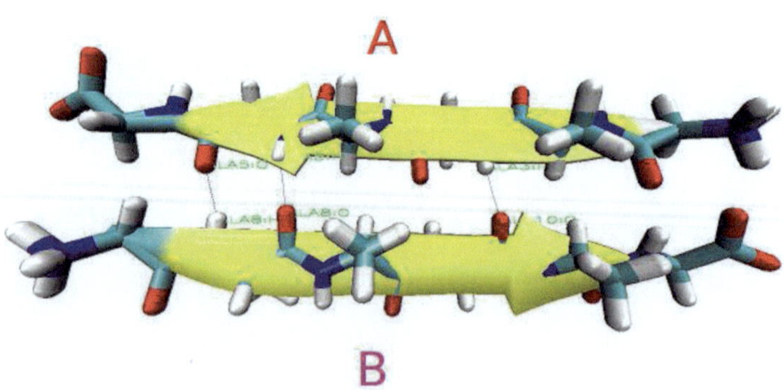

Fig. 13.29 Refined AB chains of 3md4-Model 3

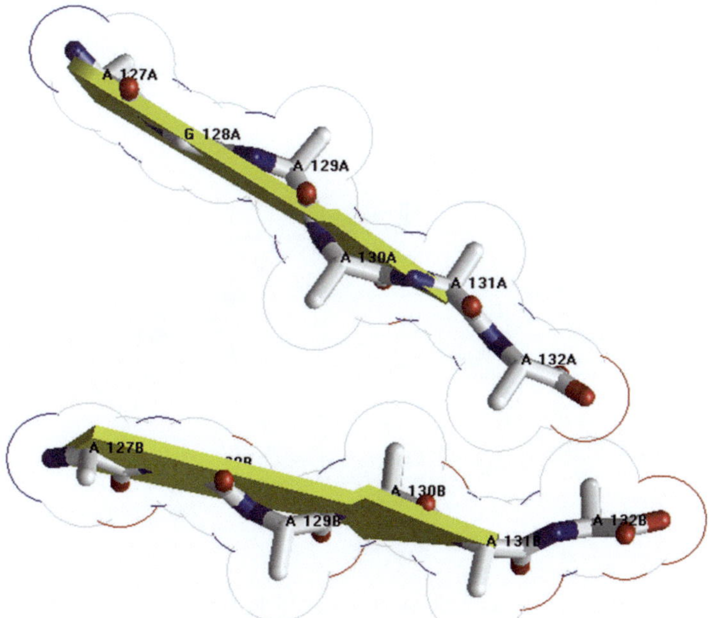

Fig. 13.30 Far vdW and HB contacts of AB chains of 3md5-Model 1

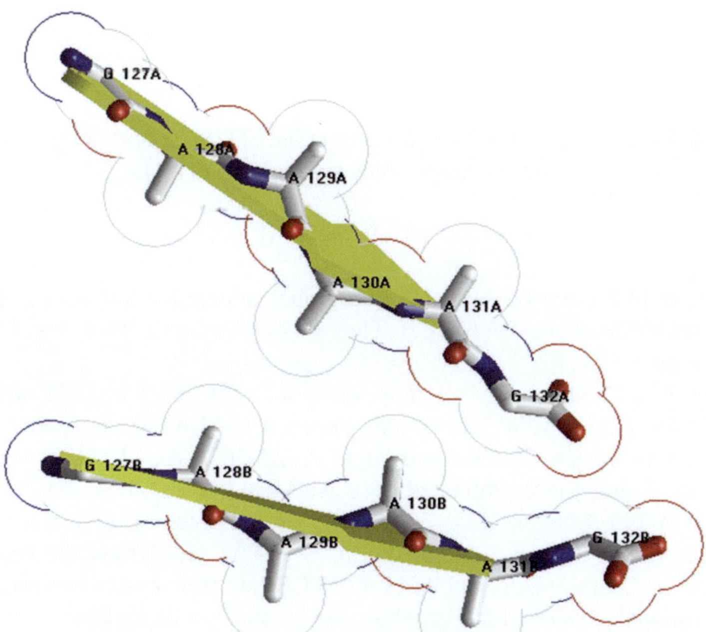

Fig. 13.31 Far vdW and HB contacts of AB chains of 3md5-Model 2

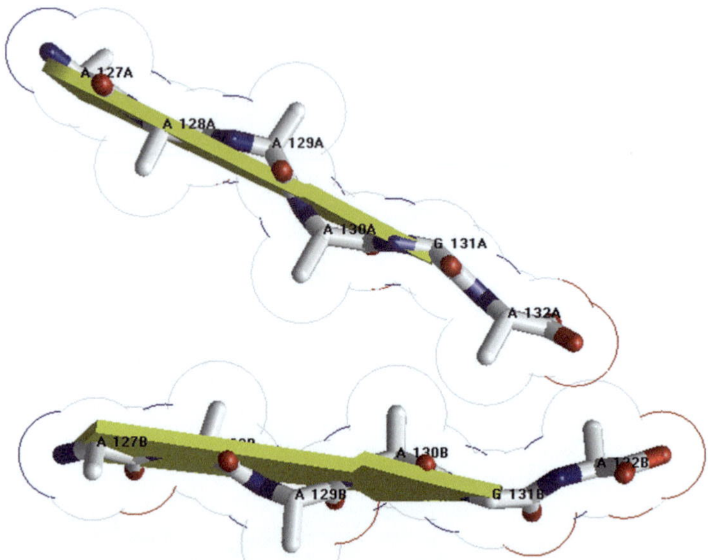

Fig. 13.32 Far vdW and HB contacts of AB chains of 3md5-Model 3

$$P^d_\epsilon(\varsigma_1) = 244.925\varsigma_1 - \frac{1}{2}\varsigma_1^2$$
$$\frac{(0.05 + 23.592\varsigma_1)^2 + (0.05 + 14.126\varsigma_1)^2 + (0.05 + 16.426\varsigma_1)^2}{4\varsigma_1}.$$

We can easily get the global maximal solution $0.0127287 \in \{\varsigma \in R^1 | \varsigma_i > 0, i = 1\}$ for $P^d_\epsilon(\varsigma_1)$. Then, we get its corresponding solution for $P_\epsilon(x_1)$:

$$\bar{x} = (13.7601, 9.02707, 10.1771).$$

By Theorem 13.2.1 we know that $\bar{x}$ is a global minimal solution of $P_\epsilon(x_1)$, i.e. for A.ALA130.CB or A.ALA132.CB. Thus we get B = A+1.9641 (Fig. 13.33) for 3md5-Model1.

Seeing Figs. 13.31∼13.32, we may know that for 3mg5-Model2 and 3mg5-Model3 at least three vdW interactions between B.ALA130.CB-A.ALA130.CB, B.ALA128.CB-A.ALA130.CB, and B.ALA128.CB-A.ALA128.CB should be maintained. Fixing the coordinates of B.ALA130.CB and B.ALA128.CB (two anchors) ((11.796, 7.063, 8.213), (5.849, 6.703, 6.531)), letting d equal to the twice of the vdW radius of Carbon atom (i.e. $d = 3.4$ Å), and letting the coordinates of A.ALA130.CB (a sensor) and A.ALA128.CB (another sensor) be variables, we may get a simple MDGP with 6 variables and its dual with 3 variable:

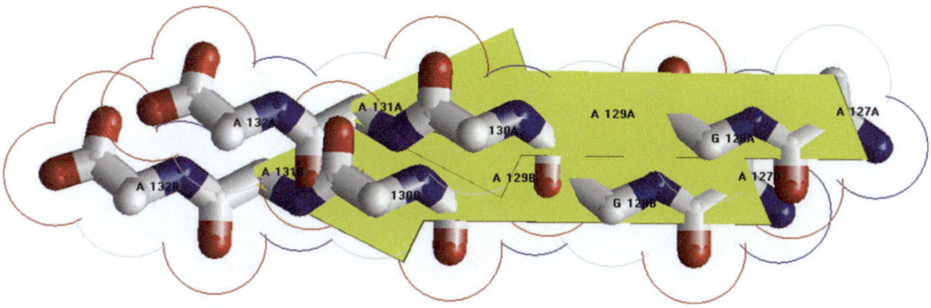

Fig. 13.33 Close vdW and HB contacts of AB chains of 3md5-Model 1

$$
\begin{aligned}
P_\epsilon(x_1,x_2) = & \frac{1}{2}\left\{(x_{11}-11.796)^2+(x_{12}-7.063)^2+(x_{13}-8.213)^2-3.4^2\right\}^2 \\
& +\frac{1}{2}\left\{(x_{11}-5.849)^2+(x_{12}-6.703)^2+(x_{13}-6.531)^2-3.4^2\right\}^2 \\
& +\frac{1}{2}\left\{(x_{21}-5.849)^2+(x_{22}-6.703)^2+(x_{23}-6.531)^2-3.4^2\right\}^2 \\
& -0.05x_{11}-0.05x_{12}-0.05x_{13}-0.05x_{21}-0.05x_{22}-0.05x_{23},
\end{aligned}
$$

$$
\begin{aligned}
P_\epsilon^d(\varsigma_1,\varsigma_2,\varsigma_3) = & \ 244.925\varsigma_1-\frac{1}{2}\varsigma_1^2+110.235\varsigma_2-\frac{1}{2}\varsigma_2^2+110.235\varsigma_3-\frac{1}{2}\varsigma_3^2 \\
& -\frac{1}{2}\begin{pmatrix} 0.05+23.592\varsigma_1+11.698\varsigma_2 \\ 0.05+14.126\varsigma_1+13.406\varsigma_2 \\ 0.05+16.426\varsigma_1+13.062\varsigma_2 \\ 0.05+11.698\varsigma_3 \\ 0.05+13.406\varsigma_3 \\ 0.05+13.062\varsigma_3 \end{pmatrix}^T
\begin{pmatrix} \frac{1}{2\varsigma_1+2\varsigma_2} & 0 & 0 & 0 & 0 & 0 \\ 0 & \frac{1}{2\varsigma_1+2\varsigma_2} & 0 & 0 & 0 & 0 \\ 0 & 0 & \frac{1}{2\varsigma_1+2\varsigma_2} & 0 & 0 & 0 \\ 0 & 0 & 0 & \frac{1}{2\varsigma_3} & 0 & 0 \\ 0 & 0 & 0 & 0 & \frac{1}{2\varsigma_3} & 0 \\ 0 & 0 & 0 & 0 & 0 & \frac{1}{2\varsigma_3} \end{pmatrix} \\
& \begin{pmatrix} 0.05+23.592\varsigma_1+11.698\varsigma_2 \\ 0.05+14.126\varsigma_1+13.406\varsigma_2 \\ 0.05+16.426\varsigma_1+13.062\varsigma_2 \\ 0.05+11.698\varsigma_3 \\ 0.05+13.406\varsigma_3 \\ 0.05+13.062\varsigma_3 \end{pmatrix}.
\end{aligned}
$$

We can easily get a global maximal solution $(0.00503155, 0.0154485, 0.0127287) \in \{\varsigma \in R^3 | \varsigma_i > 0, i = 1,2,3\}$ for $P_\epsilon^d(\varsigma_1,\varsigma_2,\varsigma_3)$. Take $\bar{\varsigma}_3 = 0.0127287$ and get its corresponding prime problem solution:

$$\bar{x}_2 = (7.81307, 8.66707, 8.49507).$$

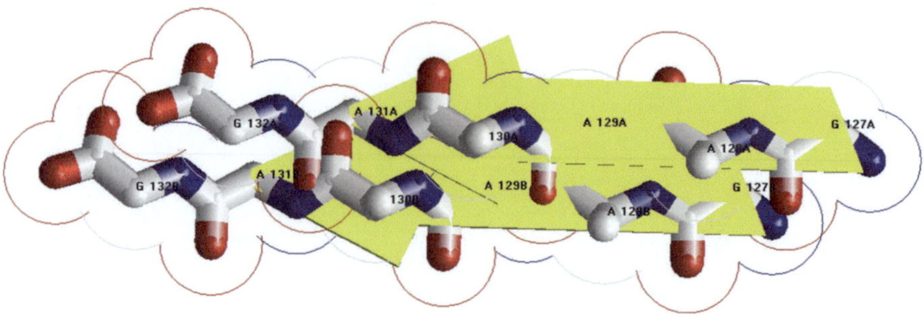

Fig. 13.34 Close vdW and HB contacts of AB chains of 3md5-Model 2

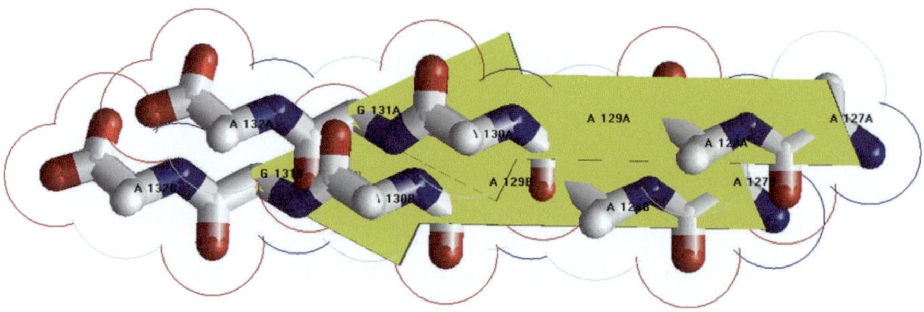

Fig. 13.35 Close vdW and HB contacts of AB chains of 3md5-Model 3

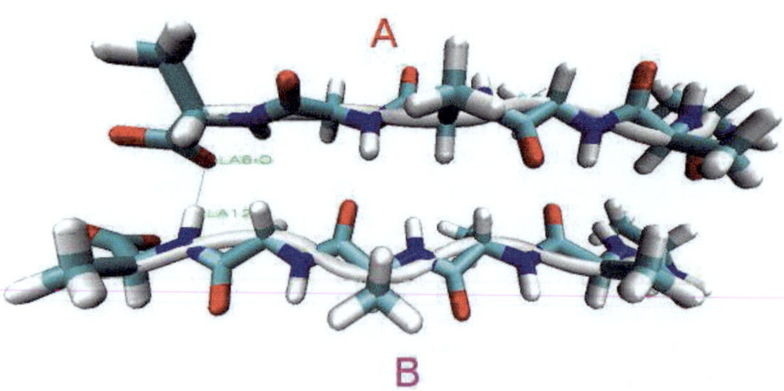

Fig. 13.36 Refined AB chains of 3md5-Model 1

By Theorem 13.2.1 we know that $\bar{x} = (8.53076, 8.01214, 8.16493, 7.81307, 8.66707, 8.49507)$ is a global minimal solution of $P_\epsilon(x_1, x_2)$, i.e. for A.ALA130.CB and A.ALA128.CB. Thus we get B = A+1.9641 (Figs. 13.34∼13.35) for 3md5-Model2 and 3md5-Model3.

We just need to refine/relax the Models by the Steepest Decent-Conjugate Gradient optimization program of the Amber 11 (Figs. 13.36, 13.37, and 13.38).

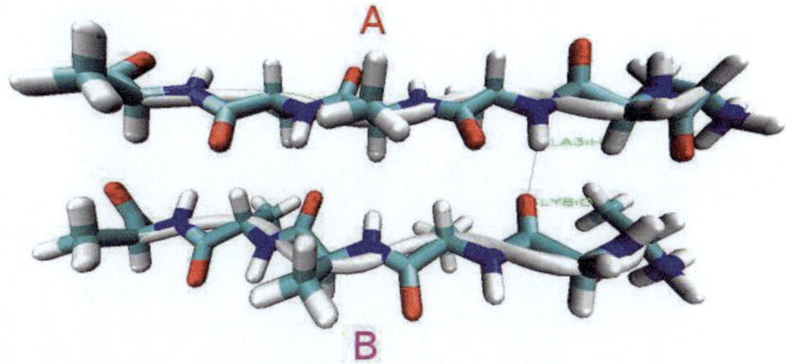

Fig. 13.37 Refined AB chains of 3md5-Model 2

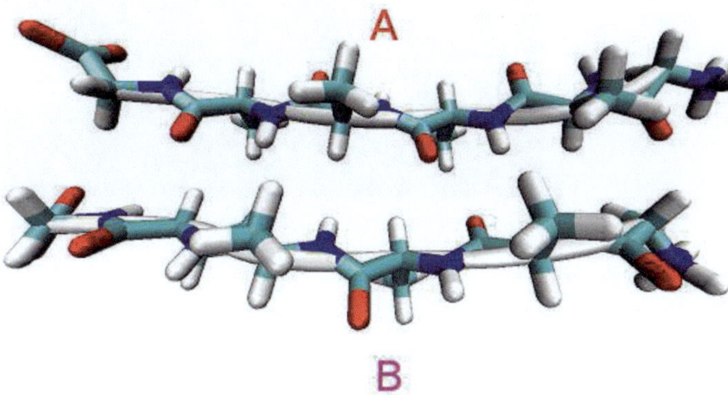

Fig. 13.38 Refined AB chains of 3md5-Model 3

13.5.5 3NVE Models

Constructions of the AGAAAAGA amyloid fibril molecular structures of prion 113–120 region are based on the most recently released experimental molecular structures of MMHFGN segment 138–143 from Syrian Hamster prion (PDB entry 3NVE released into Protein Data Bank (www.rcsb.org) on 2011-03-02) [18]. The atomic-resolution structure of this peptide is a steric zipper, with strong vdW interactions between β-sheets and HBs to maintain the β-strands (Fig. 13.39).

In Fig. 13.39 we see that G (H) chains (i.e. β-sheet 2) of 3NVE.pdb can be obtained from A (B) chains (i.e. β-sheet 1) by

$$G(H) = A(B) + \begin{pmatrix} 0 \\ 11.784 \\ 0 \end{pmatrix}, \tag{13.38}$$

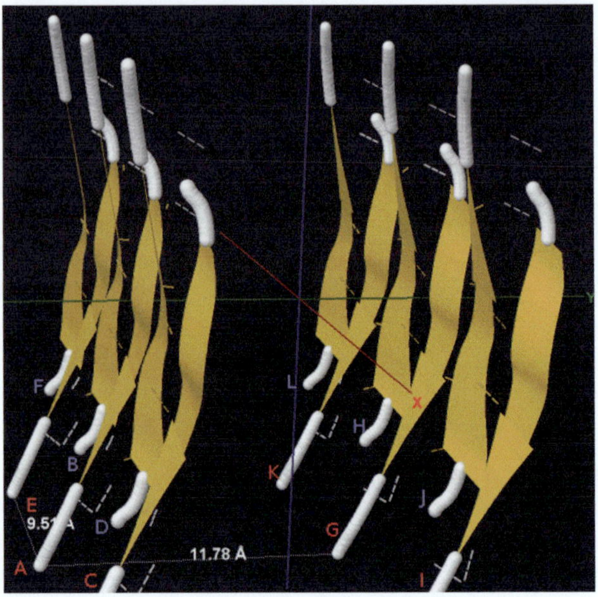

Fig. 13.39 Protein fibril structure of MMHFGN segment 138–143 from Syrian Hamster prion. The *purple dashed lines* denote the hydrogen bonds. *A*, *B*, ..., *K*, *L* denote the 12 chains of the fibril

and other chains can be got by

$$C(D) = A(B) + \begin{pmatrix} 9.513 \\ 0 \\ 0 \end{pmatrix}, E(F) = A(B) + \begin{pmatrix} -9.513 \\ 0 \\ 0 \end{pmatrix}, \tag{13.39}$$

$$J(I) = H(G) + \begin{pmatrix} 9.513 \\ 0 \\ 0 \end{pmatrix}, L(K) = H(G) + \begin{pmatrix} -9.513 \\ 0 \\ 0 \end{pmatrix}. \tag{13.40}$$

Basing on the template 3NVE.pdb from Protein Data Bank, three prion AGAAAAGA palindrome amyloid fibril models – an AGAAAA model (3nve-Model 1), a GAAAAG model (3nve-Model 2), and an AAAAGA model (3nve-Model 3) will be successfully constructed in this subsection. AB chains of 3nve-Models 1–3 were respectively got from AB chains of 3NVE.pdb using the mutate module of the free package Swiss-PdbViewer (SPDBV Version 4.01) (spdbv.vital-it.ch). It is pleasant to see that almost all the hydrogen bonds are still kept after the mutations; thus we just need to consider the vdW contacts only. Making mutations for GH chains of 3NVE.pdb, we can get the GH chains of 3nve-Models 1–3. However, the vdW contacts between A chain and G chain, between B chain and H chain are too far at this moment (Figs. 13.40, 13.41, and 13.42).

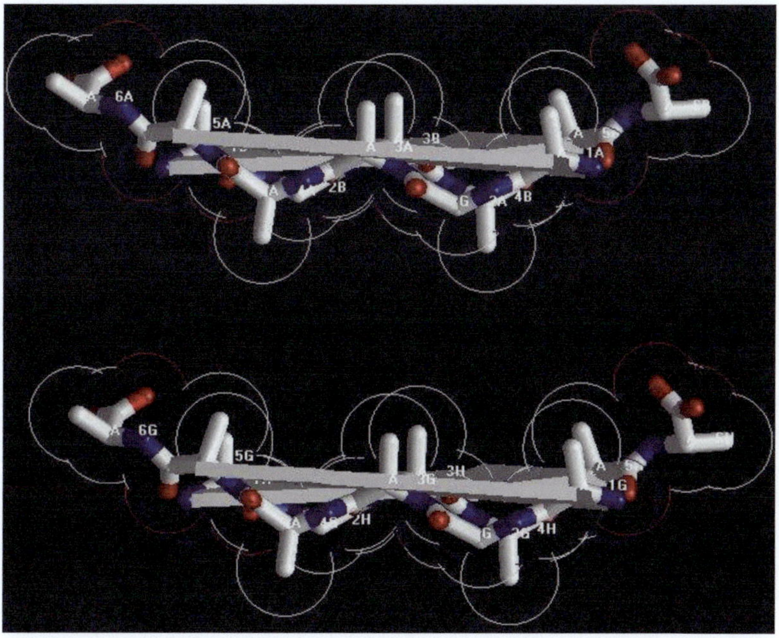

Fig. 13.40 Far vdW contacts of AG chains and BH chains of 3nve-Model 1

Seeing Fig. 13.40, we may know that for 3nve-Model 1 at least two vdW interactions between A.ALA4.CB-G.ALA3.CB, B.ALA4.CB-G.ALA3.CB should be maintained. Seeing Figs. 13.41∼13.42, we may know that for 3nve-Models 2∼3 at least the two vdW interactions between A.ALA4.CB-G.ALA3.CB, A.ALA2.CB-G.ALA3.CB should be maintained. Fixing the coordinates of A.ALA4.CB and B.ALA4.CB (two anchors) ((5.150,−1.295,5.372), (−0.586, 0.444,12.766)) for 3nve-Model 1, fixing the coordinates of A.ALA4.CB, A.ALA2.CB (two anchors) ((5.150,−1.295,5.372),(4.624,−0.327,12.132)) for 3nve-Models 1∼2, letting d equal to the twice of the vdW radius of Carbon atom (i.e. $d = 3.4$ Å), and letting the coordinates of G.ALA3.CB of 3nve-Models 1–3 (one sensor) be variables, we may get a simple MDGP with 3 variables and its dual with 2 variables for 3nve-Model 1:

$$P(x_1) = \frac{1}{2}\left\{(x_{11} - 5.150)^2 + (x_{12} + 1.295)^2 + (x_{13} - 5.372)^2 - 3.4^2\right\}^2$$
$$+\frac{1}{2}\left\{(x_{11} + 0.586)^2 + (x_{12} + 0.444)^2 + (x_{13} - 12.766)^2 - 3.4^2\right\}^2,$$

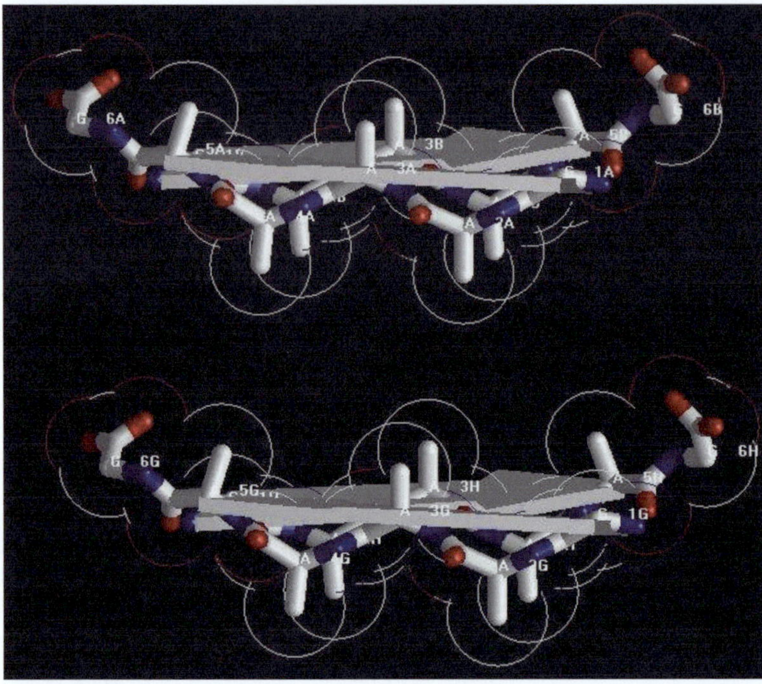

Fig. 13.41 Far vdW contacts of AG chains and BH chains of 3nve-Model 2

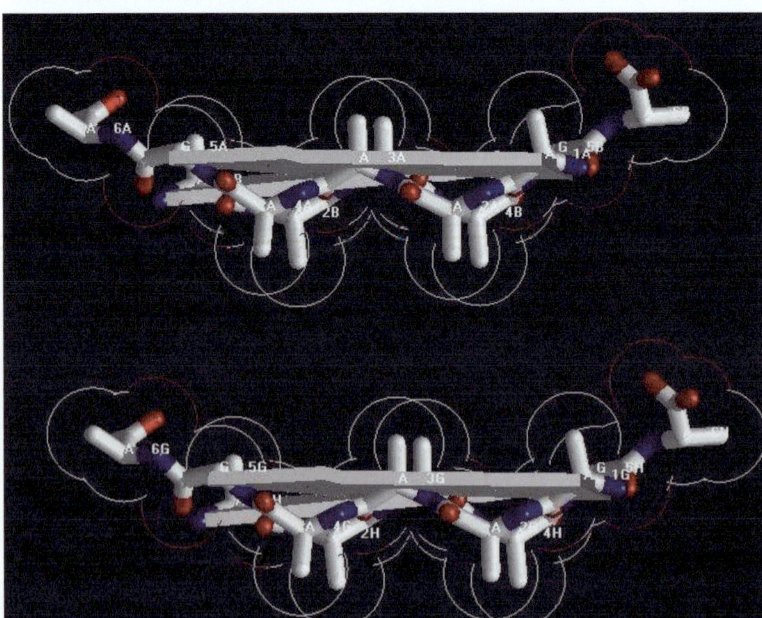

Fig. 13.42 Far vdW contacts of AG chains and BH chains of 3nve-Model 3

$$P^d(\varsigma_1,\varsigma_2) = 45.4979\varsigma_1 - \frac{1}{2}\varsigma_1^2 + 151.9513\varsigma_2 - \frac{1}{2}\varsigma_2^2$$
$$-\frac{1}{2}\begin{pmatrix} 10.300\varsigma_1 - 1.172\varsigma_2 \\ -2.590\varsigma_1 - 0.888\varsigma_2 \\ 10.744\varsigma_1 + 25.532\varsigma_2 \end{pmatrix}^T \begin{pmatrix} \frac{1}{2\varsigma_1+2\varsigma_2} & 0 & 0 \\ 0 & \frac{1}{2\varsigma_1+2\varsigma_2} & 0 \\ 0 & 0 & \frac{1}{2\varsigma_1+2\varsigma_2} \end{pmatrix}$$
$$\begin{pmatrix} 10.300\varsigma_1 - 1.172\varsigma_2 \\ -2.590\varsigma_1 - 0.888\varsigma_2 \\ 10.744\varsigma_1 + 25.532\varsigma_2 \end{pmatrix}.$$

We can get a global maximal solution (10.5143,10.5143) for $P^d(\varsigma_1,\varsigma_2)$ and its corresponding local maximal solution to $P(x_1)$:

$$\bar{x} = (2.282, -0.8695, 9.069).$$

By Theorem 13.2.1 we know that $\bar{x}$ is a global minimal solution of $P(x_1)$. Thus we get

$$G(H) = A(B) + \begin{pmatrix} -2.868 \\ 0.4255 \\ 3.6970 \end{pmatrix} \tag{13.41}$$

for 3nve-Model 1, whose other chains can be got by (13.39)–(13.40).

For 3nve-Models 2∼3, similarly we may get a simple MDGP with 3 variables and its dual with 2 variables:

$$P_\epsilon(x_1) = \frac{1}{2}\left\{(x_{11}-5.150)^2 + (x_{12}+1.295)^2 + (x_{13}-5.372)^2 - 3.4^2\right\}^2$$
$$+\frac{1}{2}\left\{(x_{11}-4.624)^2 + (x_{12}+0.327)^2 + (x_{13}-12.132)^2 - 3.4^2\right\}^2$$
$$-(0.05x_{11} + 0.05x_{12} + 0.05x_{13}),$$

$$P^d_\epsilon(\varsigma_1,\varsigma_2) = 45.4979\varsigma_1 - \frac{1}{2}\varsigma_1^2 + 157.1137\varsigma_2 - \frac{1}{2}\varsigma_2^2$$
$$-\frac{1}{2}\begin{pmatrix} 0.05 + 10.3\varsigma_1 + 9.248\varsigma_2 \\ 0.05 - 2.59\varsigma_1 - 0.654\varsigma_2 \\ 0.05 + 10.744\varsigma_1 + 24.264\varsigma_2 \end{pmatrix}^T \begin{pmatrix} \frac{1}{2\varsigma_1+2\varsigma_2} & 0 & 0 \\ 0 & \frac{1}{2\varsigma_1+2\varsigma_2} & 0 \\ 0 & 0 & \frac{1}{2\varsigma_1+2\varsigma_2} \end{pmatrix}$$

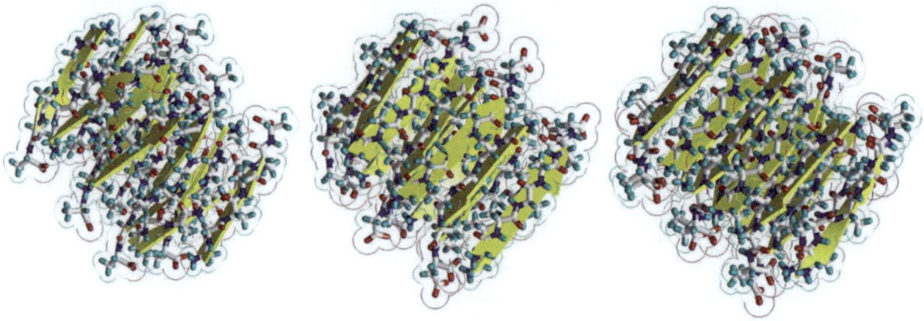

Fig. 13.43 The optimal 3NVE AGAAAAGA amyloid fibril models

$$\begin{pmatrix} 0.05 + 10.3\varsigma_1 + 9.248\varsigma_2 \\ 0.05 - 2.59\varsigma_1 - 0.654\varsigma_2 \\ 0.05 + 10.744\varsigma_1 + 24.264\varsigma_2 \end{pmatrix}.$$

We can get a global maximal solution (0.0112166,0.0000329004) for $P_\epsilon^d(\varsigma_1, \varsigma_2)$ and its corresponding local maximal solution to $P_\epsilon(x_1)$:

$$\bar{x} = (7.37079, -0.0252351, 7.6141).$$

By Theorem 13.2.1 we know that $\bar{x}$ is a global minimal solution of $P_\epsilon(x_1)$. Thus we get

$$G(H) = A(B) + \begin{pmatrix} 2.4938 \\ 0.7858 \\ -1.1379 \end{pmatrix} \tag{13.42}$$

for 3nve-Models 2∼3, whose other chains can be got by (13.39)–(13.40). Then the Models are refined by the SDCG optimization program of Amber 10 and the perfect 3NVE prion AGAAAAGA amyloid fibril models were got (Fig. 13.43).

Chapter 14
The Hybrid Method of Evolutionary Computations with Simulated Annealing

X-ray crystallography and NMR spectroscopy are two powerful tools to determine the protein 3D structure. However, not all proteins can be successfully crystallized, particularly for membrane proteins. Although NMR spectroscopy is indeed very powerful in determining the 3D structures of membrane proteins, same as X-ray crystallography, it is still very time-consuming and expensive. Under many circumstances, due to the noncrystalline and insoluble nature of some proteins, X-ray and NMR cannot be used at all. Computational approaches, however, allow us to obtain a description of the protein 3D structure at a submicroscopic level.

14.1 Overview

Prion diseases are invariably fatal and highly infectious neurodegenerative diseases affecting humans and animals. The neurodegenerative diseases such as CJD, vCJDs, iCJD, fCJD, sCJD, GSS, FFI, Kuru in humans, scrapie in sheep, BSE or 'mad-cow' disease in cattle, CWD in white-tailed deer, elk, mule deer, moose, TME in mink, FSE in cat, EUE in nyala, oryx, greater kudu, and SE in ostrich etc. belong to prion diseases. By now there have not been some effective therapeutic approaches or medications to treat all these prion diseases.

Prion diseases are amyloid fibril diseases. The normal cellular prion protein (PrP^C) is rich in α-helices but the infectious prions (PrP^{Sc}) are rich in β-sheets amyloid fibrils. The conversion of PrP^C to PrP^{Sc} is believed to involve a conformational change from a predominantly α-helical protein (about 42 % α-helix and 3 % β-sheet) to a protein rich in β-sheets (about 30 % α-helix and 43 % β-sheet) [257].

Many experimental studies such as [67, 68, 70, 288, 315, 316, 357, 445, 619] have shown two important ideas: (i1) the hydrophobic region (113–120) AGAAAAGA of prion proteins is critical in the conversion from a soluble PrP^C into an insoluble PrP^{Sc} fibrillar form; and (i2) normal AGAAAAGA is an inhibitor/blocker of prion

J. Zhang, *Molecular Structures and Structural Dynamics of Prion Proteins and Prions*, Focus on Structural Biology 9, DOI 10.1007/978-94-017-7318-8_14

Fig. 14.1 Protein fibril structure of human M129 prion GYMLGS (127–132) (PDB ID: 3NHC). The *dashed lines* denote the hydrogen bonds. *A*, *B*, ..., *K*, *L* denote the 12 chains of the fibril

diseases. PrP lacking the palindrome could not convert to PrP^{Sc} and also did not generate proteinase K resistance. The presence of residues 119 and 120 (the two last residues within the motif AGAAAAGA) seems to be crucial for this inhibitory effect. The replacement of glycine at residues 114 and 119 by alanine led to the inability of the peptide to build fibrils but it nevertheless increased. The A117V variant is linked to the GSS disease. Furthermore, we computationally clarified by ourselves that prion AGAAAAGA segment indeed has an amyloid fibril forming property: Fig. 14.1. By now the 3D structure of the AGAAAGA palindrome peptide has not been known. The physiological conditions such as pH [85] and temperature [594] will affect the propensity to form fibrils in this region. However, laboratory experiences have shown that using traditional experimental methods is very difficult to obtain atomic-resolution structures of AGAAAAGA due to the noncrystalline and insoluble nature of the amyloid fibril [571, 710] and its unstable nature. By introducing novel mathematical canonical dual formulations and computational approaches, in this chapter we may construct atomic-resolution molecular structures for prion (113–120) AGAAAAGA amyloid fibrils.

The 3D atomic resolution structure of PrP (106–126), i.e. TNVKHVAGAAAA-GAVVGGLGG, can be looked as the structure of a control peptide [102, 373]. Ma and Nussinov [408] established homology structure of AGAAAAGA and its molecular dynamics simulation studies [408]. Recently, Wagoner et al. studied the structure of GAVAAAAVAG of mouse prion protein [594, 595]. All these structures (including the structures of this current chapter) of amyloid cross-β spines can be reduced to the ones of 8 classes of steric zippers reported in [517]. Thus, we might be able to say that the molecular modeling strategy reported in this study should be reliable.

Many studies have indicated that computational approaches or introducing novel mathematical formulations and physical concepts into molecular biology can significantly stimulate the development of biological and medical science. Various computer computational approaches were used to address the problems related to "amyloid fibril" [86, 118, 120, 130, 609, 620]. Here, we would like to use the simulated annealing evolutionary computations to build the optimal atomic-resolution amyloid fibril models in hopes to be used for controlling prion diseases. The templates from the Protein Data Bank (pdb) (www.rcsb.org) and the modeling computational algorithms are different among this current study and previous studies in Chaps. 11 and 12. Chapter 11 used all the pdb files of [517] and the Insight II package to establish the models. Chapter 12 used 3FVA.pdb (the NNQNTF (173–178) segment from elk prion protein) and hybrid SADG method to build the models. This current study uses 3NHC.pdb (the GYMLGS (127–132) segment from human M129 prion protein) as the modelling template and simulated annealing evolutionary computations (SAEC) to build the models, where SAECs were got from the hybrid algorithms of [4] by simply replacing the DG method by the SA algorithm of [24] and numerical computational results show that SAECs can successfully pass the test of more than 40 well-known benchmark global optimization problems:

Algorithm 14.1 (SA-SAES($\mu + \lambda$)).
Step 0. Randomly generate μ parents, where each parent $z_k = (\mathbf{x}_k, \boldsymbol{\sigma}_k)$.
Step 1. Apply SA on each parent $\mathbf{x}_k$.
Step 2. Set $\tau = \left(\sqrt{\left(2\sqrt{(n)}\right)}\right)^{-1}$ and $\tau' = \left(\sqrt{(2n)}\right)^{-1}$.
Step 3. Until λ children are generated, do
Step 4. Select two parents $z_k = (\mathbf{x}_k, \boldsymbol{\sigma}_k)$ and $z_l = (\mathbf{x}_l, \boldsymbol{\sigma}_l)$ at random to generate child $\mathbf{y}_j = (\mathbf{x}_j, \boldsymbol{\sigma}_j)$.
Step 5. Discrete recombination: for each variable x_{ji} and step size σ_{ji} in $\mathbf{y}_j$, do ($x_{ji} = x_{ki}$ and $\sigma_{ji} = \sigma_{ki}$) or ($x_{ji} = x_{li}$ and $\sigma_{ji} = \sigma_{li}$)
Step 6. Mutation: For each x_{ji} and step size σ_{ji} in $\mathbf{y}_j$

$$x'_{ji} = x_{ji} + \sigma_{ji} N_j(0, 1) \tag{14.1}$$

$$\sigma'_{ji} = \sigma_{ji} \exp(\tau' N(0, 1) + \tau N_j(0, 1)) \tag{14.2}$$

Step 7. If the number of children is less than λ, go to Step 4.
Step 8. Select the best μ individuals among all the $\mu + \lambda$ parents and children.
Step 9. Apply SA on the best individual among the selected μ individuals.
Step 10. If the stopping criteria are satisfied, stop, else go to step 2.

Algorithm 14.2 (SA-SACEP).
Step 0. Randomly generate μ parents and evaluate them, where each parent $z_k = (\mathbf{x}_k, \boldsymbol{\sigma}_k)$.

Step 1. Apply SA on each parent $\mathbf{x}_k$.
Step 2. Set $\tau = \left(\sqrt{\left(2\sqrt{(n)}\right)}\right)^{-1}$ and $\tau' = \left(\sqrt{(2n)}\right)^{-1}$.
Step 3. For each parent, generate a child as follows

$$x'_{ji} = x_{ji} + \sigma_{ji} N_j(0, 1)$$

$$\sigma'_{ji} = \sigma_{ji} \exp(\tau' N(0, 1) + \tau N_j(0, 1))$$

Step 4. Evaluate all children
Step 5. Undertake a tournament y for each parent and child as follows: select ζ individuals with replacement from the joint set of parents and children. For each individual z of the ζ individuals, if y is better than z, add 1 to the fitness of y.
Step 6. Select the best μ individuals among all parents and children with the highest fitness.
Step 7. Apply SA on the best individual among the selected μ individuals.
Step 8. If the stopping criteria are satisfied, stop, else go to step 1.

Numerical results show us Self-Adaptive Evolutionary Strategy $\mu + \lambda$ (SAES($\mu + \lambda$)) method and Self-Adaptive Classical Evolutionary Programming (SACEP) method can successfully work for all our test problems. The simulated annealing algorithm is a sequential computing algorithm and evolutionary algorithms are parallel computing algorithms. So, in this section, using Simulated Annealing (SA) method, we improve them. We use SA as a search operator once for SAES($\mu + \lambda$) method, and once for SACEP method. Both the algorithms designed in this section simply work by applying the SA on all individuals in the population of the initial generation. In subsequent generations, SA is applied only for the best solutions found so far [686].

The atomic structures of all amyloid fibrils revealed steric zippers, with strong vdW interactions between β-sheets and hydrogen bonds (HBs) to maintain the β-strands [517]. The vdW contacts of atoms are described by the Lennard-Jones (LJ) potential energy:

$$V_{LJ}(r) = 4\varepsilon \left[(\frac{\sigma}{r})^{12} - (\frac{\sigma}{r})^{6} \right], \tag{14.3}$$

where ε is the depth of the potential well and σ is the atom diameter; these parameters can be fitted to reproduce experimental data or deduced from results of accurate quantum chemistry calculations. The $(\frac{\sigma}{r})^{12}$ term describes repulsion and the $(\frac{\sigma}{r})^{6}$ term describes attraction. If we introduce the coordinates of the atoms whose number is denoted by N and let $\varepsilon = \sigma = 1$ be the reduced units, the form (14.3) becomes

$$f(x) = 4\sum_{i=1}^{N}\sum_{j=1,j<i}^{N}\left(\frac{1}{\tau_{ij}^{6}} - \frac{1}{\tau_{ij}^{3}}\right), \tag{14.4}$$

where $\tau_{ij} = (x_{3i-2} - x_{3j-2})^2 + (x_{3i-1} - x_{3j-1})^2 + (x_{3i} - x_{3j})^2$, $(x_{3i-2}, x_{3i-1}, x_{3i})$ is the coordinates of atom i, $N \geq 2$. The minimization of LJ potential $f(x)$ on R^n (where $n = 3N$) is an optimization problem:

$$\min f(x) \quad subject \quad to \quad x \in R^{3N}. \tag{14.5}$$

Similarly as (14.3) – the potential energy for the vdW interactions between β-sheets:

$$V_{LJ}(r) = \frac{A}{r^{12}} - \frac{B}{r^{6}}, \tag{14.6}$$

the potential energy for the HBs between the β-strands has the formula

$$V_{HB}(r) = \frac{C}{r^{12}} - \frac{D}{r^{10}}, \tag{14.7}$$

where A, B, C, D are given constants. Thus, the amyloid fibril molecular modeling problem is deduced into well solve the mathematical optimization problem (14.5).

This chapter is organized as follows. In Sect. 14.2, we first describe how to build the prion AGAAAAGA amyloid fibril molecular models, and then explain how the models with only six variables are built and can be solved by any optimization algorithm. At the end of Sect. 14.2 the models are done a little refinement by Amber 11 [91]. At last, we conclude that when using the time-consuming and costly X-ray crystallography or NMR spectroscopy we still cannot determine the protein 3D structure, we may introduce computational approaches or novel mathematical formulations and physical concepts into molecular biology to study molecular structures. This concluding remark will be made in the last section.

14.2 Prion AGAAAAGA Amyloid Fibril Models' Molecular Modeling and Optimizing

Constructions of the AGAAAAGA amyloid fibril molecular structures of prion 113–120 region are based on the most recently released experimental molecular structures of human M129 prion peptide 127–132 (PDB entry 3NHC released into Protein Data Bank on 04-AUG-2010). The atomic-resolution structure of this peptide is a steric zipper, with strong vdW interactions between β-sheets and HBs to maintain the β-strands (Fig. 14.1).

In Fig. 14.1 we see that G (H) chains (i.e. β-sheet 2) of 3NHC.pdb can be obtained from A (B) chains (i.e. β-sheet 1) by

$$G(H) = \begin{pmatrix} 1 & 0 & 0 \\ 0 & -1 & 0 \\ 0 & 0 & -1 \end{pmatrix} A(B) + \begin{pmatrix} 9.07500 \\ 4.77650 \\ 0.00000 \end{pmatrix}, \tag{14.8}$$

and other chains can be got by

$$I(J) = G(H) + \begin{pmatrix} 0 \\ 9.5530 \\ 0 \end{pmatrix}, K(L) = G(H) + \begin{pmatrix} 0 \\ -9.5530 \\ 0 \end{pmatrix}, \tag{14.9}$$

$$C(D) = A(B) + \begin{pmatrix} 0 \\ 9.5530 \\ 0 \end{pmatrix}, E(F) = A(B) + \begin{pmatrix} 0 \\ -9.5530 \\ 0 \end{pmatrix}. \tag{14.10}$$

Basing on template 3NHC.pdb from Protein Data Bank, three prion AGAAAAGA palindrome amyloid fibril models – an AAAAGA model (Model-chapt14-1), a GAAAAG model (Model-chapt14-2), and an AAAAGA model (Model-chapt14-3) will be successfully constructed in this chapter. AB chains of Model-chapt14-1 to Model-chapt14-3 were respectively got from AB chains of 3NHC.pdb using the mutate module of the free package Swiss-PdbViewer (SPDBV Version 4.01) (spdbv.vital-it.ch). It is pleasant to see that almost all the hydrogen bonds are still kept after the mutations; thus we just need to consider the vdW contacts only. Making mutations for GH chains of 3NHC.pdb, we can get the GH chains of Model-chapt14-1 to Model-chapt14-3. However, the vdW contacts between A chain and G chain, between B chain and H chain are too far at this moment (Fig. 14.2). Seeing Fig. 14.2, we may know that for Model-chapt14-1 to Model-chapt14-3 at least two vdW interactions between A.ALA3.CB-G.ALA4.CB, B.ALA4.CB-H.ALA3.CB should be maintained. Fixing the coordinates of A.ALA3.CB and B.ALA4.CB, letting the coordinates of G.ALA4.CB and H.ALA3.CB be variables, we may get a simple LJ potential energy minimization problem (14.5) just with six variables. For solving this six variable optimization problem, *any* optimization computational algorithm can be used to solve this low-dimensional mathematical optimization problem; for example, in this chapter we may use the SAEC algorithm.

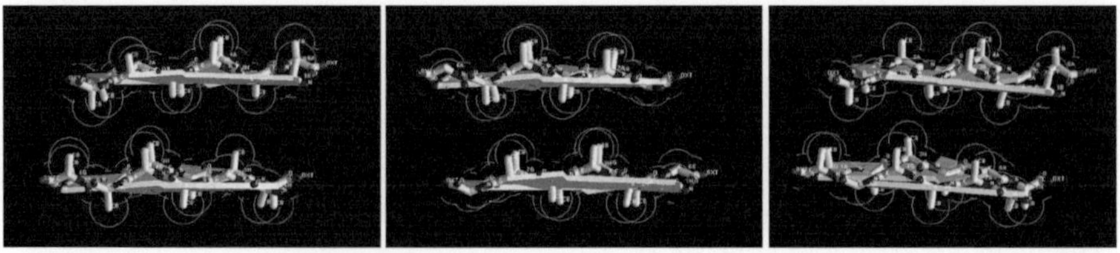

Fig. 14.2 At initial state, the vdW contacts between AG chains (β-sheet 1) and BH chains (β-sheet 2) of Model-chapt14-1 to Model-chapt14-3 are very far

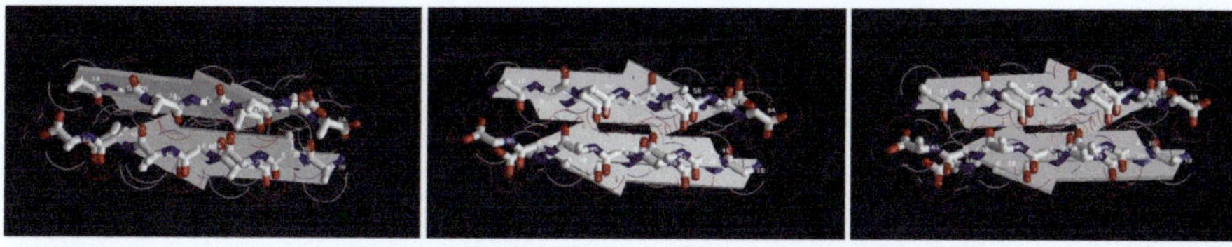

Fig. 14.3 After LJ potential energy minimization, the vdW contacts between AG chains (β-sheet 1) and BH chains (β-sheet 2) of Model-chapt14-1 to Model-chapt14-3 become very closer

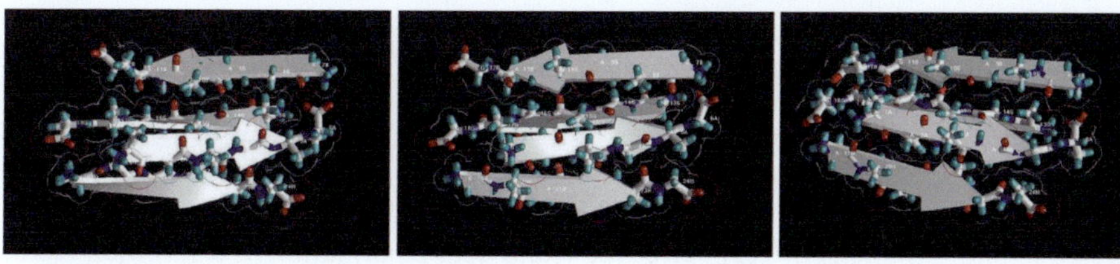

Fig. 14.4 After Amber 11 refinement of the whole potential energy minimization, the optimal structure of prion AAAAGA amyloid fibril Model-chapt14-1 to Model-chapt14-3

Setting the coordinates of G.ALA4.CB and H.ALA3.CB as initial solutions, running a SAEC algorithm, for Model-chapt14-1 to Model-chapt14-3 we get

$$G(H) = \begin{pmatrix} 1 & 0 & 0 \\ 0 & -1 & 0 \\ 0 & 0 & -1 \end{pmatrix} A(B) + \begin{pmatrix} -0.703968 \\ 7.43502 \\ -0.33248 \end{pmatrix}. \qquad (14.11)$$

By (14.11) we can get close vdW contacts between A chain and G chain, between B chain and H chain (Fig. 14.3). Furthermore, we may employ the Amber 11 package [91] to slightly optimize Model-chapt14-1 to Model-chapt14-3 and at last get Model-chapt14-1 to Model-chapt14-3 with stable total potential energies (Fig. 14.4). The other CDIJ and EFKL chains can be got by parallelizing ABGH chains in the use of mathematical formulas (14.7)–(14.8).

14.3 Concluding Remarks

X-ray crystallography is a powerful tool to determine the protein 3D structure. However, it is time-consuming and expensive, and not all proteins can be successfully crystallized, particularly for membrane proteins. Although NMR spectroscopy is indeed a very powerful tool in determining the 3D structures of membrane proteins, it is also time-consuming and costly. Due to the noncrystalline and insoluble nature of the amyloid fibril, little structural data on the prion AGAAAAGA segment is available. Under these circumstances, the novel simple strategy introduced in this chapter can well do the molecular modeling of prion AGAAAAGA amyloid fibrils. This indicated that computational approaches or introducing novel mathematical formulations and physical concepts into molecular biology can significantly stimulate the development of biological and medical science. The optimal atomic-resolution structures of prion AGAAAAGA amyloid fibils presented in this chapter are useful for the drive to find treatments for prion diseases in the field of medicinal chemistry.

14.4 Another Application

14.4.1 Introduction

X-ray crystallography, NMR spectroscopy, and dual polarization interferometry, etc. are indeed very powerful tools to determine the 3D structure of a protein (including the membrane protein), though they are time-consuming and costly. However, for some proteins, due to their unstable, noncrystalline and insoluble nature, these tools cannot work. Under this condition, mathematical and physical theoretical methods and computational approaches allow us to obtain a description of the protein 3D structure at a submicroscopic level. This chapter presents some practical and useful mathematical optimization computational approaches to produce 3D structures of the Prion AGAAAAGA Amyloid Fibrils, from an energy minimization point of view.

X-ray crystallography finds the X-ray final structure of a protein, which usually need refinements in order to produce a better structure. The computational methods presented in this section can be also acted as a tool for the refinements.

All neurodegenerative diseases including Parkinson's, Alzheimer's, Huntington's, and Prion's have a similarity, which is they all featured amyloid fibrils (en.wikipedia.org/wiki/Amyloid and references [238, 245, 434, 439, 462, 517, 552, 636] and references therein). A prion is a misshapen protein that acts like an infectious agent (hence the name, which comes from the words protein and infection). Prions cause a number of fatal diseases such as 'mad cow' disease in cattle, scrapie in sheep and kuru and CJD in humans. Prion diseases (being rich in β-sheets, about 43 % β-sheet) [85, 95, 169, 257, 447, 457, 491] belong to neurodegenerative diseases. Many experimental studies such as [67, 68, 70, 85, 237, 267, 274, 288, 315, 316, 322, 351, 352, 357, 361, 445, 514, 619, 674] have shown that the normal hydrophobic region (113–120) AGAAAAGA of prion proteins is an inhibitor/blocker of prion diseases. PrP lacking this palindrome could not convert to prion diseases. The presence of residues 119 and 120 (the two last residues within the motif AGAAAAGA) seems to be crucial for this inhibitory effect. The replacement of Glycine at residues 114 and 119 by Alanine led to the inability of the peptide to build fibrils but it nevertheless increased. The A117V variant is linked to the GSS disease. The physiological conditions such as pH [85] and temperature

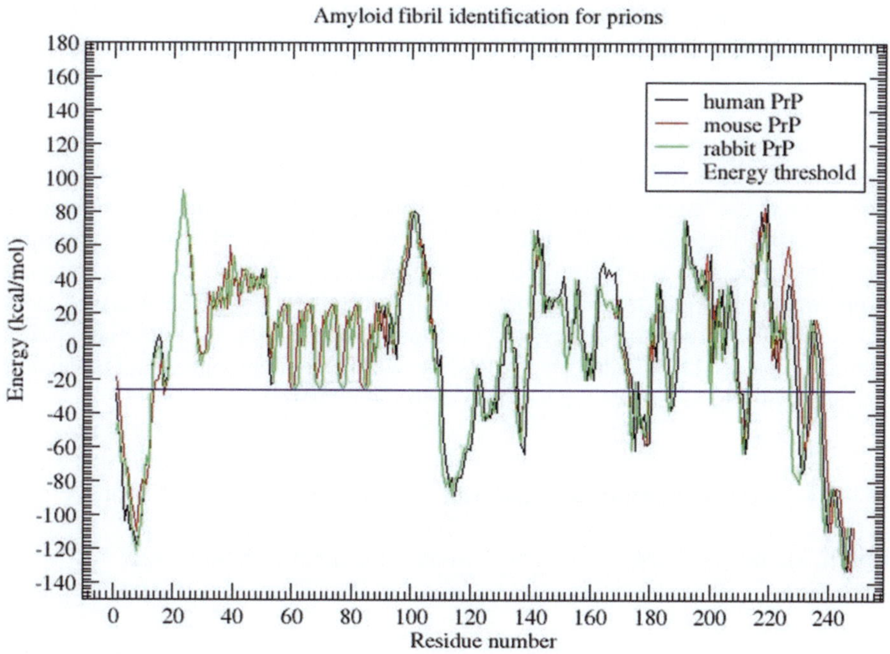

Fig. 14.5 Prion AGAAAAGA (113-120) is surely and clearly identified as the amyloid fibril formation region, because its energy is less than the amyloid fibril formation threshold energy of −26 kcal/mol [694]

[594] will affect the propensity to form fibrils in this region. The 3D atomic resolution structure of PrP (106–126), i.e. TNVKHVAGAAAAGAVVGGLGG, can be looked as the structure of a control peptide [102, 373]. Ma and Nussinov [408] established homology structure of AGAAAAGA and did its molecular dynamics simulation studies [408]. Recently, Wagoner et al. computer simulation studied the structure of GAVAAAAVAG of mouse prion protein [594, 595]. Furthermore, the author computationally clarified that prion AGAAAAGA segment indeed has an amyloid fibril forming property (Fig. 14.5). However, to the best of the author's knowledge, there is little X-ray or NMR structural data available to date on AGAAAAGA (which falls just within the N-terminal unstructured region (1–123) of prion proteins) due to its unstable, noncrystalline and insoluble nature. This section will computationally study the MM structures of this region of prions.

14.4.2 Molecular Structures of AGAAAAGA Amyloid Fibrils

"Amyloid is characterized by a cross-β sheet quaternary structure" and "recent X-ray diffraction studies of microcrystals revealed atomistic details of core region of amyloid" (en.wikipedia.org/wiki/Amyloid and references [238, 245, 434, 439, 462, 517, 552, 636] and references therein). All the quaternary structures of amyloid cross-β spines can be reduced to the one of eight classes of steric zippers of [517], with strong vdW interactions between β-sheets and HBs to maintain the β-strands.

A new era in the structural analysis of amyloids started from the 'steric zipper'-β-sheets [439]. As the two sheets zip up, HYDs (& vdWs) have been formed. The extension of the 'steric zipper' above and below (i.e. the β-strands) is maintained by HBs (but there is no HB between the two β-sheets). This is the common structure associated with some 20 neurodegenerative amyloid diseases, ranging from Alzheimer's and type-II diabetes to prion diseases. For prion AGAAAAGA amyloid fibril structure, basing on the common property of potential energy minimization of HYDs, vdWs, and HBs, we will present computational molecular structures of prion AGAAAAGA amyloid fibrils.

14.4.2.1 Review on Materials and Methods, and Results of MM Models

Hybrid Method of Steepest Descent Conjugate Gradient with Simulated Annealing

X-ray crystallography finds the X-ray final structure of a protein, which usually need refinements using a simulated annealing protocol in order to produce a better structure. Thus, it is very amenable to use simulated annealing (SA) to format the models constructed. Zhang [680, 683] presents a hybrid method of global search SA with local steepest descent (SD), conjugate gradient (CG) searches. The hybrid method is executed with the following three procedures. (1) Firstly the SD method and then the CG method are executed. These two local search methods are traditional optimization methods. The former has nice convergence but is slow when close to minimums. The latter is efficient but its gradient RMS and GMAX gradient [91] do not have a good convergence. (2) When models cannot be optimized further, we employ standard SA global search procedure. (3) Lastly, the SD and CG methods are used to refine the models. The PDB [47] templates used in [680, 683] are 2OKZ.pdb, 2ONW.pdb, 2OLX.pdb, 2OMQ.pdb, 2ON9.pdb, 2ONV.pdb, 2ONA.pdb, 1XYO.pdb, 2OL9.pdb, 2OMN.pdb, 2ONX.pdb, 2OMP.pdb, 1YJP.pdb of [517], but only the 2OMP and 1YJP template-based three MM-Models (Figs. 6a$\sim$6c in [680]) are successfully passed through the SDCG-SA-SDCG computational procedures.

Hybrid Method of Discrete Gradient with Simulated Annealing

Zhang [680, 683] used 3FVA.pdb as the pdb template to build two MM-Models (Figs. 11$\sim$12 in [680]). The Models were built using a hybrid SA Discrete Gradient (DG [22, 42]) method. Then the Models were optimized using SDCGSA-SDCG methods as in [680].

Computational Method of Canonical Dual Global Optimization Theory

Zhang [681, 682] used 3NHC.pdb, 3NVF/G/H/E.pdb templates to build several MM-Models (Figs. 9∼11 in [681], and Figs. 5∼8 in [682]). These Models were built in the use of canonical dual global optimization theory [221, 232, 233] and then refined by SDCG-SA-SDCG methods as in [680].

14.4.2.2 New Material and Method, and New MM-Models

New Material

This section uses a suitable pdb file template 3NHD.pdb (the GYVLGS segment 127–132 from human prion with V129 [17]) from the Protein Data Bank to build MM-models of AGAAAAGA amyloid fibrils for prions.

New Computational Method – Computational Method of Simulated Annealing Evolutionary Computations

The computational methods used to build the new MM-Models will be simulated annealing evolutionary computations (SAECs), where SAECs were got from the hybrid algorithms of [4] by simply replacing the DG method by the SA algorithm of [24] and numerical computational results show that SAECs can successfully pass the test of more than 40 well-known benchmark global optimization problems [684].

New MM-Models

The atomic-resolution X-ray structure of 3NHD.pdb is a steric zipper, with strong vdW interactions between β-sheets and HBs to maintain the β-strands (Fig. 14.6). By observations of the 3rd column of coordinates of 3NHD.pdb and Fig. 14.6, G(H) chains (i.e. β-sheet 2) of 3NHD.pdb can be calculated from A(B) chains (i.e. β-sheet 1) by Eq. 14.12 and other chains can be calculated by Eqs. 13.12, 13.13, and 13.14:

$$G(H) = \begin{pmatrix} -1 & 0 & 0 \\ 0 & 1 & 0 \\ 0 & 0 & -1 \end{pmatrix} A(B) + \begin{pmatrix} -20.5865 \\ 9.48 \\ 0 \end{pmatrix}, \tag{14.12}$$

$$K(L) = G(H) + \begin{pmatrix} 0 \\ 0 \\ 9.59 \end{pmatrix}, I(J) = G(H) + \begin{pmatrix} 0 \\ 0 \\ -9.59 \end{pmatrix}, \tag{14.13}$$

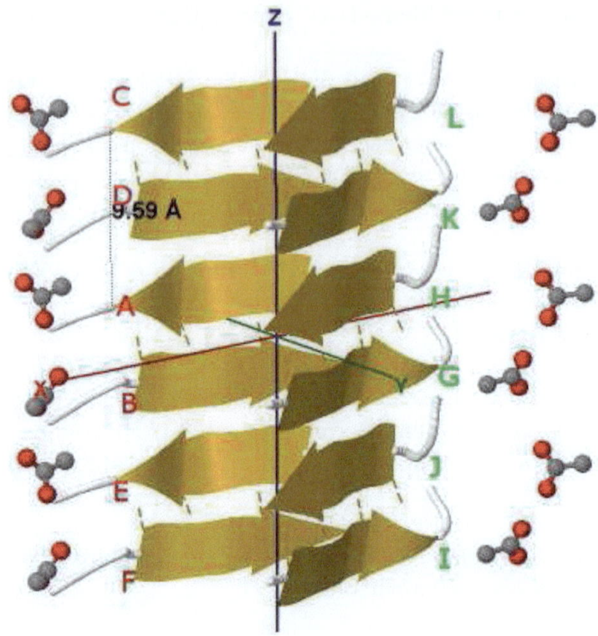

Fig. 14.6 Protein fibril structure of human V129 prion GYVLGS (127–132) (PDB ID: 3NHD). The *dashed lines* denote the hydrogen bonds. $A, B \ldots K, L$ denote the chains of the fibril

$$C(D) = A(B) + \begin{pmatrix} 0 \\ 0 \\ 9.59 \end{pmatrix}, E(F) = A(B) + \begin{pmatrix} 0 \\ 0 \\ -9.59 \end{pmatrix}. \qquad (14.14)$$

Basing on template 3NHD.pdb from Protein Data Bank, three prion AGAAAAGA palindrome amyloid fibril models – an AGAAAA model (Model-chapt14-4), a GAAAAG model (Model Model-chapt14-5), and an AAAAGA model (Model Model-chapt14-6) – will be successfully constructed in this section. Because the template is a segment of six residues, the three shorter prion fragments are selected. This chapter does not perform calculations on the full AGAAAAGA. Chains AB of the three Models were respectively got from AB chains of 3NHD.pdb using the mutate module of the free package Swiss-PdbViewer (SPDBV Version 4.01) (http://spdbv.vital-it.ch). It is pleasant to see that almost all the hydrogen bonds are still kept after the mutations; thus we just need to consider the vdW contacts only. Making mutations for GH chains of 3NHD.pdb, we can get the GH chains of Models Model-chapt14-4∼Model-chapt14-6. However, the vdW contacts between Chain A and Chain G, between B chain and H chain are too far at this moment (Figs. 14.7, 14.8, and 14.9). Seeing Figs. 14.7, 14.8, and 14.9, we may know that for Model-chapt14-4 at least 3 vdW interactions B6.ALA.CB-H3.ALA.CB-B4.ALA.CB-H5.ALA.CB should be maintained (their distances in Fig. 14.7 are 7.82, 8.36, 9.04 Å respectively), for Model-chapt14-5 at least 3 vdW interactions G4.ALA.CB-A3.ALA.CB-G2.ALA.CB-A5.ALA.CB should be maintained (their distances in

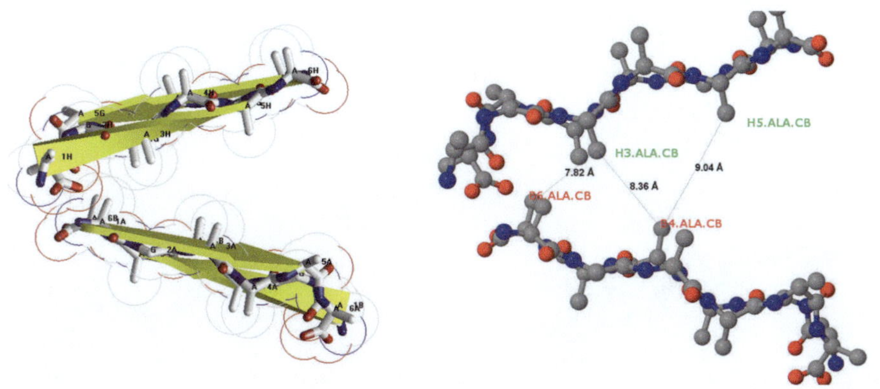

Fig. 14.7 At initial state, the vdW contacts between AB chains (β-sheet 1) and GH chains (β-sheet 2) of Model Model-chapt14-4 are very far

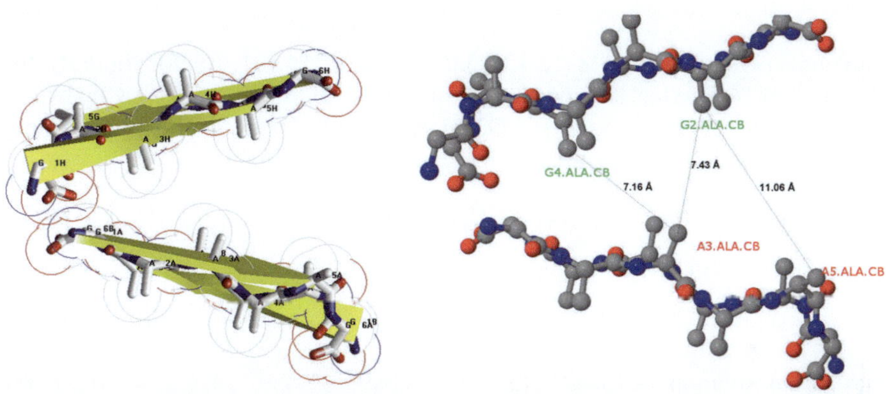

Fig. 14.8 At initial state, the vdW contacts between AB chains (β sheet 1) and GH chains (β-sheet 2) of Model Model-chapt14-5 are very far

Fig. 14.8 are 7.16, 7.43, 9.31 Å respectively), and for Model-chapt14-6 at least 3 vdW interactions A1.ALA.CB-G4.ALA.CB-A3.ALA.CB-G2.ALA.CB should be maintained (their distances in Fig. 14.9 are 3.45, 7.16, 7.43 Å respectively). For Model-chapt14-4, fixing the coordinates of B6.ALA.CB and B4.ALA.CB, letting the coordinates of H3.ALA.CB and H5.ALA.CB be variables, we may get a simple Lennard-Jones (LJ) potential energy minimization problem just with six variables:

$$
\begin{aligned}
\min f(x) = \; & 1/2\{(x_{11} + 16.359)^2 + (x_{12} - 9.934)^2 + (x_{13} + 3.526)^2 - 3.4^2\}^2 \\
& +1/2\{(x_{21} + 9.726)^2 + (x_{22} - 8.530)^2 + (x_{23} + 3.613)^2 - 3.4^2\}^2 \\
& +1/2\{(x_{11} + 9.726)^2 + (x_{12} - 8.530)^2 + (x_{13} + 3.613)^2 - 3.4^2\}^2
\end{aligned}
$$

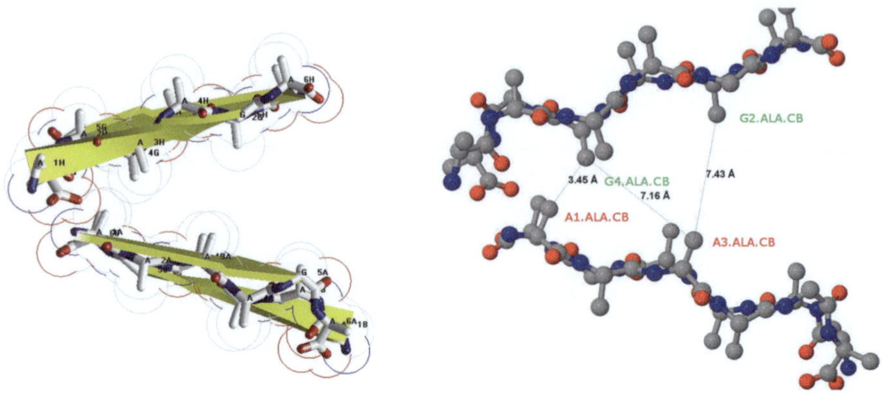

Fig. 14.9 At initial state, the vdW contacts between AB chains (β-sheet 1) and GH chains (β-sheet 2) of Model Model-chapt14-6 are very far

with initial solution (−12.928, 12.454, 3.034; −6.635, 14.301, 2.628). Similarly, for Model-chapt14-5 fixing the coordinates of A3.ALA.CB and A5.ALA.CB, letting the coordinates of G4.ALA.CB and G2.ALA.CB be variables, we may get a simple LJ potential energy minimization problem just with six variables:

$$\begin{aligned}\min f(x) = {} & 1/2\{(x_{11} + 8.655)^2 + (x_{12} - 8.153)^2 + (x_{13} - 1.770)^2 - 3.4^2\}^2 \\ & +1/2\{(x_{21} + 8.655)^2 + (x_{22} - 8.153)^2 + (x_{23} - 1.770)^2 - 3.4^2\}^2 \\ & +1/2\{(x_{21} + 2.257)^2 + (x_{22} - 6.095)^2 + (x_{23} - 3.078)^2 - 3.4^2\}^2\end{aligned}$$

with initial solution (−13.909, 12.227, −0.889; −7.439, 14.419, −2.033); for Model-chapt14-6, fixing the coordinates of A1.ALA.CB and A3.ALA.CB, letting the coordinates of G4.ALA.CB and G2.ALA.CB be variables, we may get a simple LJ potential energy minimization problem with six variables:

$$\begin{aligned}\min f(x) = {} & 1/2\{(x_{11} + 15.632)^2 + (x_{12} - 9.694)^2 + (x_{13} - 0.687)^2 - 3.4^2\}^2 \\ & +1/2\{(x_{11} + 8.655)^2 + (x_{12} - 8.153)^2 + (x_{13} - 1.770)^2 - 3.4^2\}^2 \\ & +1/2\{(x_{21} + 8.655)^2 + (x_{22} - 8.153)^2 + (x_{23} - 1.770)^2 - 3.4^2\}^2\end{aligned}$$

with initial solution (−13.909, 12.227, −0.889; −7.439, 14.419, −2.033).

We may use any optimization algorithms or packages to easily solve problems the above three minimization problems and get their respective global optimal solutions (−13.062, 9.126, −3.336; −12.344, 6.695, −2.457), (−11.275, 6.606, 3.288; −5.461, 7.124, 2.424), (−12.149, 8.924, 1.229; -9.256, 11.007, 3.517), which were got by the SAEC algorithms in this chapter. Input these global optimal solutions into Eq. 14.12, take average and tests then we get Eq. 14.15:

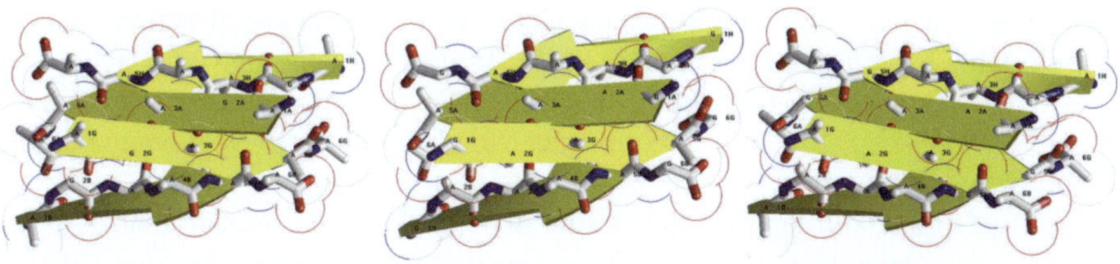

Fig. 14.10 After LJ potential energy minimization, the vdW contacts of Model-chapt14-4 ~ Model-chapt14-6 (from left to right) become very closer (the distances are illuminated by the overlap of border of CB atoms' surface)

$$G(H) = \begin{pmatrix} -1 & 0 & 0 \\ 0 & 1 & 0 \\ 0 & 0 & -1 \end{pmatrix} A(B) + \begin{pmatrix} -20.2788 \\ -0.0821 \\ 0.5609 \end{pmatrix}. \tag{14.15}$$

By Eq. 14.15, we can get close vdW contacts (Fig. 14.10).

From Figs. 14.7, 14.8, and 14.10, we may see that the Optimization algorithm works and the computational experiences show us we had better at least define two sensors and two anchors in order to form a zipper between the two β-sheets. Next, in order to remove very close bad contacts, we relax Fig. 14.10 by a slight SDCG-Optimization in the use of Amber 11 [91] and we get the optimized MM-Models of Model-chapt14-4~Model-chapt14-6. The other CDEF and LKJI chains can be got by parallelizing ABGH chains in the use of mathematical Eqs. 13.12~13.13. The new amyloid fibril models are useful for the drive to find treatments for prion diseases in the field of medicinal chemistry. The computational algorithms presented in this section and their references therein are useful in materials science, drug design, etc.

Because the three minimization problems here are the optimization problem with six variables only and these optimization problems are to minimize fourth-order polynomials, the proposed SAEC method of this chapter and other computational methods can easily get the same optimal solutions to optimize the above three models, Model-chapt14-4~Model-chapt14-6.

The Practical LBFGS Quasi-Newtonian Method

Energy minimization (EM), with the images at the endpoints fixed in space, of the total system energy provides a minimum energy path. EM can be done using SD, CG, and LBFGS (Limited-memory Broyden-Fletcher-Goldfarb-Shanno). SD is robust and easy to implement but it is not most efficient especially when closer to minimum. CG is slower than SD in the early stages but more efficient when closer to minimum. The hybrid of SD-CG will make SD more efficient than SD or CG alone. However, CG cannot be used to find the minimization energy path, for example, when "forces are truncated according to the tangent direction, making it impossible to define a Lagrangian" [149]. In this case, the powerful and faster quasi-Newtonian method (e.g. the LBFGS quasi-Newtonian minimizer) can be used [73, 149, 394, 443, 718]. We briefly introduce the LBFGS quasi-Newtonian method as follows.

Newton's method in optimization explicitly calculates the Hessian matrix of the second order derivatives of the objective function and the reverse of the Hessian matrix [178]. The convergence of this method is quadratic, so it is faster than SD or CG. In high dimensions, finding the inverse of the Hessian is very expensive. In some cases, the Hessian is a non-invertible matrix, and furthermore in some cases, the Hessian is symmetric indefinite. Qusi-Newton methods thus appear to overcome all these shortcomings.

Quasi-Newton methods (a special case of variable metric methods) are to approximate the Hessian. Currently, the most common quasi-Newton algorithms are the SR1 formula, the BHHH method, the widespread BFGS method and its limited /low-memory extension LBFGS, and Broyden's methods. In Amber [91] and Gromacs [577], LBFGS is used, and the hybrid of LBFGS with CG – a Truncated Newton linear CG method with optional LBFGS Preconditioning [443] – is used in Amber [91].

14.4.2.3 A New Thinking About Protein 3D-Structure Constructions

If a NMR or X-ray structure of a protein has not been determined and stored in PDB bank yet, we still can easily get the 3D-structural frame of the protein. For example, before 2005 when we did not know the NMR structure of rabbit prion protein, we could get its homology model structure using the NMR structure of the human prion protein (PDB id: 1QLX) as the template [700]. We may use the homology structure to determine the 3D-structural frame of a protein when its NMR or X-ray structure has not been determined yet. The determination is an optimization problem described as follows.

"Very often in a structural analysis, we want to approximate a secondary structural element with a single straight line" ([72]: page 212). For example, Fig. 14.11 uses two straight lines that act as the longitudinal axis of β-Strand A (i.e. A chain), β-Strand B (i.e. B chain) respectively. Each straight line should be positioned among the Ca atoms so that it is closest to all these Ca atoms in a least-squares sense, which is to minimize the sum of the squares of the perpendicular distances (d_i) from the Ca atoms to the strand/helix axis:

$$S^* = \min S = \sum_{i=1}^{N} \|d_i\|^2. \tag{14.16}$$

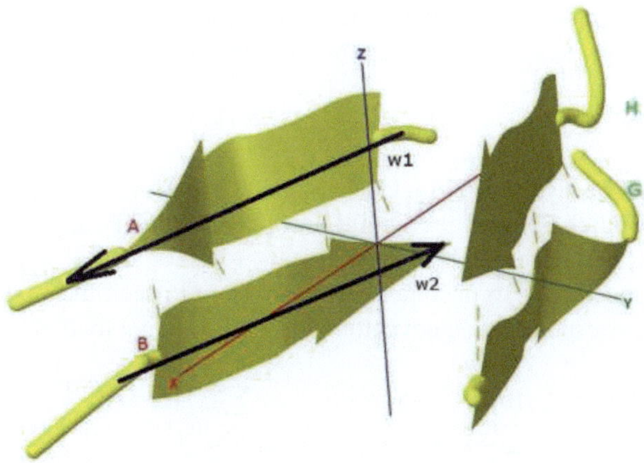

Fig. 14.11 The 3D-structural frame of AB chains of Model-chapt14-4 in Fig. 14.7 with two β-Strands

Define the vector $w = (w_x, w_y, w_z)^T$ for the axis. Then d_i represents the perpendicular vector going from Ca atom a to the axis:

$$\|d_i\|^2 = \|a^{(i)}\|^2 \sin^2\theta_i = \|a^{(i)}\|^2(1-\cos^2\theta_i) = \|a^{(i)}\|^2\left(1 - \frac{(a^{(i)T}w)^2}{\|a^{(i)}\|^2\|w\|^2}\right) \tag{14.17}$$

$$= \left(a_x^{(i)2} + a_y^{(i)2} + a_z^{(i)2}\right)\left(1 - \frac{\left(a_x^{(i)}w_x + a_y^{(i)}w_y + a_z^{(i)}w_z\right)^2}{\left(a_x^{(i)2} + a_y^{(i)2} + a_z^{(i)2}\right)\left(w_x^2 + w_y^2 + w_z^2\right)}\right). \tag{14.18}$$

According to Eqs. 14.16~13.18, for the β-Strand A – β-Strand B of AB chains, we get the following two optimization problems for Model-chapt14-4 respectively:

$$\min S_A =$$
$$((-16.196)^2 + 8.315^2 + 1.061^2)\{1 - (-16.196w_x + 8.315w_y + 1.061w_z)^2/[((-16.196)^2 + 8.315^2 + 1.061^2)(w_x^2 + w_y^2 + w_z^2)]\}$$
$$+((-12.977)^2 + 6.460^2 + 1.908^2)\{1 - (-12.977w_x + 6.460w_y + 1.908w_z)^2/[((-12.977)^2 + 6.460^2 + 1.908^2)(w_x^2 + w_y^2 + w_z^2)]\}$$
$$+((-9.178)^2 + 6.745^2 + 1.448^2)\{1 - (-9.178w_x + 6.745w_y + 1.448w_z)^2/[((-9.178)^2 + 6.745^2 + 1.448^2)(w_x^2 + w_y^2 + w_z^2)]\}$$
$$+((-6.455)^2 + 4.112^2 + 1.558^2)\{1 - (-6.455w_x + 4.112w_y + 1.558w_z)^2/[((-6.455)^2 + 4.112^2 + 1.558^2)(w_x^2 + w_y^2 + w_z^2)]\}$$
$$+((-3.006)^2 + 5.750^2 + 1.782^2)\{1 - (-3.006w_x + 5.750w_y + 1.782w_z)^2/[((-3.006)^2 + 5.750^2 + 1.782^2)(w_x^2 + w_y^2 + w_z^2)]\}$$
$$+((-1.226)^2 + 2.750^2 + 0.233^2)\{1 - (-1.226w_x + 2.750w_y + 0.233w_z)^2/[((-1.226)^2 + 2.750^2 + 0.233^2)(w_x^2 + w_y^2 + w_z^2)]\}.$$

$$\min S_B =$$
$$((-0.959)^2 + 2.950^2 + (-4.817)^2)\{1 - (-0.959w_x + 2.950w_y - 4.817w_z)^2/[((-0.959)^2 + 2.950^2 + (-4.817)^2)(w_x^2 + w_y^2 + w_z^2)]\}$$
$$+((-3.465)^2 + 4.999^2 + (-2.846)^2)\{1 - (-3.465w_x + 4.999w_y - 2.846w_z)^2/[((-3.465)^2 + 4.999^2 + (-2.846)^2)(w_x^2 + w_y^2 + w_z^2)]\}$$
$$+((-7.213)^2 + 4.412^2 + (-3.340)^2)\{1 - (-7.213w_x + 4.412w_y - 3.340w_z)^2/[((-7.213)^2 + 4.412^2 + (-3.340)^2)(w_x^2 + w_y^2 + w_z^2)]\}$$
$$+((-9.954)^2 + 7.078^2 + (-3.168)^2)\{1 - (-9.954w_x + 7.078w_y - 3.168w_z)^2/[((-9.954)^2 + 7.078^2 + (-3.168)^2)(w_x^2 + w_y^2 + w_z^2)]\}$$
$$+((-13.660)^2 + 6.241^2 + (-3.137)^2)\{1 - (-13.660w_x + 6.241w_y - 3.137w_z)^2/[((-13.660)^2 + 6.241^2 + (-3.137)^2)(w_x^2 + w_y^2 + w_z^2)]\}$$
$$+((-16.702)^2 + 8.507^2 + (-3.074)^2)\{1 - (-16.702w_x + 8.507w_y - 3.074w_z)^2/[((-16.702)^2 + 8.507^2 + (-3.074)^2)(w_x^2 + w_y^2 + w_z^2)]\}.$$

We solve $\min S_A$ and $\min S_B$ (taking the average of the coordinates of Ca atoms as initial solutions), getting their optimal solutions

$$w_1 = (-10.751, 6.428, 1.411)^T, w_2 = (-7.960, 4.579, -2.256)^T$$

respectively (Fig. 14.11). We may use the vectors w_1, w_2 and Eq. 14.15 to construct Chains GH and then build an optimal Model-chapt14-4 [3, 19, 210, 451, 665]. In ([72]: pages 213–216), $w_x^2 + w_y^2 + w_z^2 = 1$ (i.e. w is a unit vector) is restrained and Eq. 14.16 becomes into a problem to seek the smallest eigenvalue (S^*) and its corresponding eigenvector w of the following matrix:

$$\begin{pmatrix} \sum_{i=1}^N (ay^{(i)})^2 + (az^{(i)})^2 & -\sum_{i=1}^N ax^{(i)} ay^{(i)} & -\sum_{i=1}^N az^{(i)} ax^{(i)} \\ -\sum_{i=1}^N ax^{(i)} ay^{(i)} & \sum_{i=1}^N (az^{(i)})^2 + (ax^{(i)})^2 & -\sum_{i=1}^N ay^{(i)} az^{(i)} \\ -\sum_{i=1}^N az^{(i)} ax^{(i)} & -\sum_{i=1}^N ay^{(i)} az^{(i)} & \sum_{i=1}^N (ax^{(i)})^2 + (ay^{(i)})^2 \end{pmatrix}.$$

This matrix is symmetric and positive definite, and its eigenvectors form an orthogonal basis for the set of atoms under consideration. In physics, it is called the inertial tensor involving studies of rotational inertia and its eigenvectors are called the principle axes of inertia. Furthermore, we may also notice that Eq. 14.16 can be rewritten as

$$\min(\sum_{i=1}^N \|d_i\|^2)^2 \quad s.t. \quad w^T w = 1, \tag{14.19}$$

where $\|d_i\|^2 = (a^{(i)}x^2 + a^{(i)}y^2 + a^{(i)}y^2)(w_x^2 + w_y^2 + w_z^2) - (a_x^{(i)} w_x + a_y^{(i)} w_y + a_z^{(i)} w_z)^2$. Thus, Eq. 14.19 can be easily solved by the canonical dual global optimization theory [221, 232, 233], by the ways of solving the canonical dual of Eq. 14.19 or solving the quadratic differential equations of the prime-dual Gao-Strang complementary function [221, 232, 233] through some ordinary or partial differential equation computational strategies.

14.4.3 Conclusions

To date the hydrophobic region AGAAAAGA palindrome (113–120) of the unstructured N-terminal region (1–123) of prions has little existing experimental structural data available. This section successfully constructs three molecular structure models for AGAAAAGA palindrome (113–120) by using some suitable template 3NHD.pdb from Protein Data Bank and refinement of the Models with several optimization techniques within AMBER 11 [690]. These models should be very helpful for the experimental studies of the hydrophobic region AGAAAAGA palindrome of prion proteins (113–120) when the NMR or X-ray molecular structure of prion AGAAAAGA peptide has not been easily determined yet. These constructed Model-chapt14-4∼Model-chapt14-6 for amyloid fibrils may be useful for the goals of medicinal chemistry.

This section also introduces numerous practical computational approaches to construct the molecular models when it is difficult to obtain atomic-resolution structures of proteins with traditional experimental methods of X-ray and NMR etc., due to the unstable, noncrystalline and insoluble nature of these proteins. Known structures can be perfectly reproduced by these computational methods, which can be compared with contemporary methods. As we all know, X-ray crystallography finds the X-ray final structure of a protein, which usually need refinements using a SA protocol in order to produce a better structure. SA is a global search procedure and usually it is better to hybrid with local search procedures. Thus, the computational methods introduced in this section should be better than SA along to refine X-ray final structures.

In molecular biology of prion neuroscience, the AGAAAAGA peptide has rich in Alanine and every two amino acids have one Alanine. As we all know, the high Glycine (and, to a less extent, alanine) content can allow for tight packing of the sheets, which contributes to amyloids' rigid structure and tensile strength. Recently, the crystal structures of polymorphic prion protein $\beta 1$ peptides were determined and revealed variable steric zipper conformations too. The structures of (I) 4TUT.pdb (Class 7), 4UBY/Z.pdb (Class 8), (II) 4W5M/P/L.pdb (Class 8), (III) 4WBU/V.pdb (Class 8), (3NHC/D.pdb (Class 8)) and (IV) 4W5Y/67/71.pdb (Class 6) in the PDB Bank can be used to contruct the amyloid structures at prion AGAAAAGA region.

Chapter 15
Simulated Annealing Refined Replica Exchange Global Search Algorithm

In the Metropolis-Hastings-based MCMC (MHMCMC) [16, 25, 299, 502, 628], a Markov stochastic process is built to sample a target probability $p(x) = C^{-1}e^{-f(x)}$ for a current state x, where $f(x)$ is the objective function (which is the likelihood in this chapter) and C denotes the normalization constant. For SA, a variable temperature *Temp* is introduced into the objective function of the target distribution, i.e. $p(x) = C^{-1}e^{-f(x)/Temp}$. For RE [196, 380, 382, 559, 568], a new state y is generated from the current state x of the Markov process by drawing y with a transition probability $q(x, y)$, and the new state y is accepted with the probability $\min(1, [p(y)q(y, x)]/[p(x)q(x, y)])$. In the RE implementation, for a replica j at each iteration step l, the local Markov chain move from the conformation state x_j^l to the new state x_j^{l+1} accepted with the probability $\min(1, e^{-1/Temp(j)(f(x_j^{l+1})-f(x_j^l))})$, and the replica transition from j to $j+1$ at two neighbouring temperatures $Temp(j)$ and $Temp(j+1)$ with the accepted probability $\min(1, e^{-1/Temp(j)f(x_{j+1}^{l+1})-1/Temp(j+1)f(x_j^{l+1})+1/Temp(j+1)f(x_{j+1}^{l+1})+1/Temp(j)f(x_j^{l+1})})$. The temperatures are monotonically decreasing in order for the convenience of simulated annealing in use and the transition step size of RE is larger for higher temperature and smaller for the lower temperature. The negative binomial distribution is used to calculate the maximum likelihood, where the likelihood is formed by assuming Gaussian errors around the deterministic model. For the simulated annealing refinement for RE, the neignbourhood scheme of [24] is still in use here. The pseudo-code of the simulated annealing refined REMCMC algorithm is presented by the author as follows:

J. Zhang, *Molecular Structures and Structural Dynamics of Prion Proteins and Prions*, Focus on Structural Biology 9, DOI 10.1007/978-94-017-7318-8_15

Algorithm 15.1.

```
Define random and constant parameters
Initialization:
offset =0
for (i=1 to replicas) do

   Set up MCMC random and constant Parameters for each replica
   Temperature scheme to produce Temp(i)
   MCMC Initialized at Temp(i) to get MSEIRSinit

endfor
Do MCMC with RE:
for (i_set=1 to Total_set) do

   Set(i) of Total_set:
   for (j=1 to replicas) do
      Input Parameters and MSEIRSinit
      Call MCMC to get new Parameters and each likelihood, accepting new MC move
      with Metropolis criterion for each replica
      Output new Parameters, and the Likelihood_last(j), Likelihood_last_cold(j)
   endfor
   j = offset +1

   while (j+1 ≤ replicas) do

      k=j+1
      Δ = (1/Temp(k) − 1/Temp(j))*(Likelihood_last_cold(j)−Likelihood_last_cold(k))
      if (Δ ≤ 0) then

         Output Likelihood_last(j)
         swap Parameters and Labels of j and k

      else

         Generate a random number Rand of [0,1]
         if (Rand ≤ e^(−Δ)) then

              Output Likelihood_last(j)
              swap Parameters and Labels of j and k

         endif

      endif
      j=j+2

   endwhile
   offset = 1- offset

endfor
Adjust each of Temps by SA & repeat the algorithm until reach Temp_target.
```

The author used the mortality data from five English towns (Blackburn, Leicester, Manchester, Newcastle and Wigan) collected during the 1918–1919 influenza pandemic [63, 428] to test the above SARE global search algorithm. In Figs. 15.1, 15.2, 15.3, 15.4 and 15.5, we may see that SARE global search algorithm performs better than all other local search MH, MM algorithms.

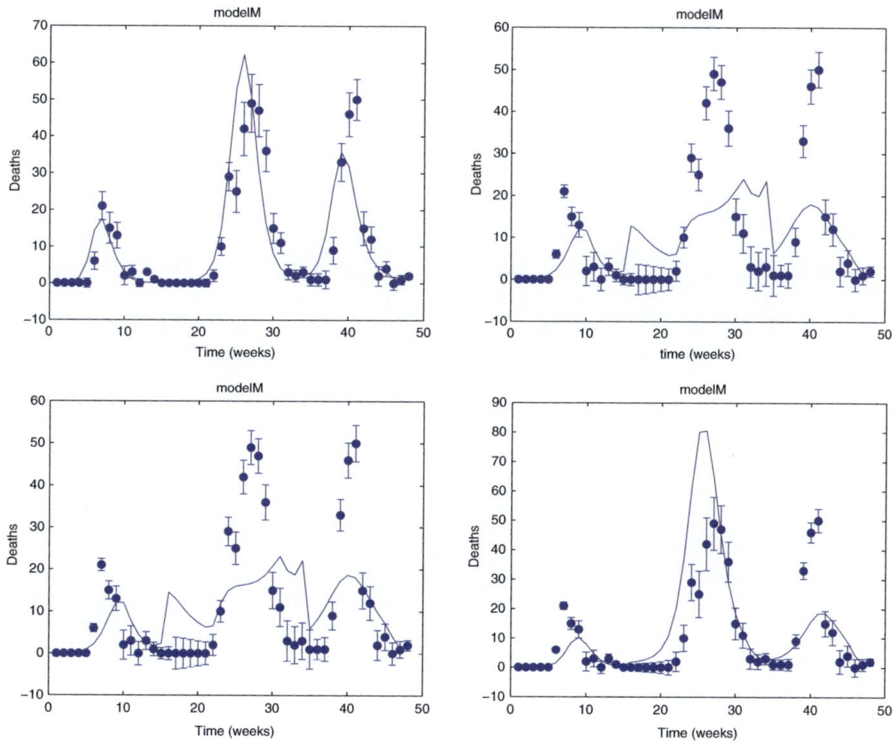

Fig. 15.1 Death data and its fitted model mortality curve for the Blackburn population (from left to right: SARE (10,000 iterations), MII (10,000 iterations), MH (50,000 iterations), MM (14,999 iterations))

REMD (replica-exchange molecular dynamics [91, 550]) has also been used by the author to study the prion AGAAAAGA amyloid fibrils. The system is built by Amber 10 [89] via Leap and run minimization via Sander. Here is a brief description of the system and the procedure used to generate the topology and coordinate files: we will build the peptide AGAAAAGA in Leap. We will build the linear sequence, and we will use the ff99SB force field and the "mbondi2" radii that are appropriate for the igb = 5 option in sander. Here is the input file for leap:
leap.in

```
source leaprc.ff99SB
set default PBradii mbondi2
m = sequence { ACE ALA GLY ALA ALA ALA ALA GLY ALA NHE }
saveamberparm m alagly8.prmtop alagly8.inpcrd
quit
```

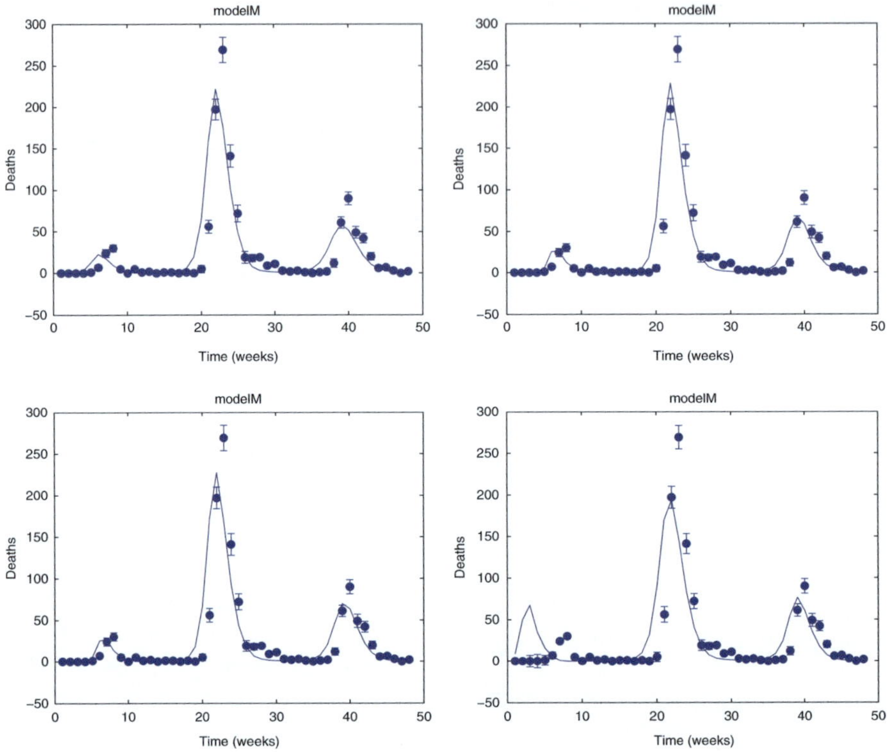

Fig. 15.2 Death data and its fitted model mortality curve for the Leicester population (from left to right: SARE (10,000 iterations), MH (10,000 iterations), MH (50,000 iterations), MM (14,999 iterations))

We have tleap read this file with the following command: "tleap -f leap.in". This should produce alagly8.prmtop and alagly8.inpcrd. Make sure to check the leap output and log file for any errors. Thus, the MM for AGAAAAGA peptide was built. Then, we can do REMD simulations as *ambermd.org/tutorials/advanced/tutorial7/*.

New techniques such as reverse Monte Carlo, sequence Monte Carlo, particle Markov chain Monte Carlo, etc. must be able to furthermore improve the above Algorithm 15.1.

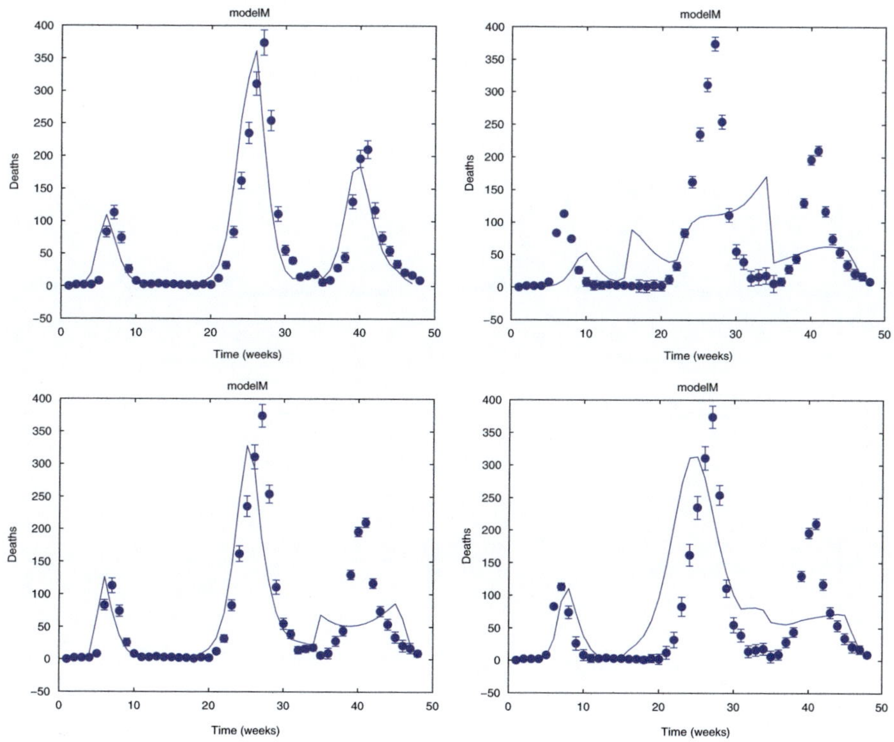

Fig. 15.3 Death data and its fitted model mortality curve for the Manchester population (from left to right: SARE (10,000 iterations), MH (10,000 iterations), MH (50,000 iterations), MM (14,999 iterations))

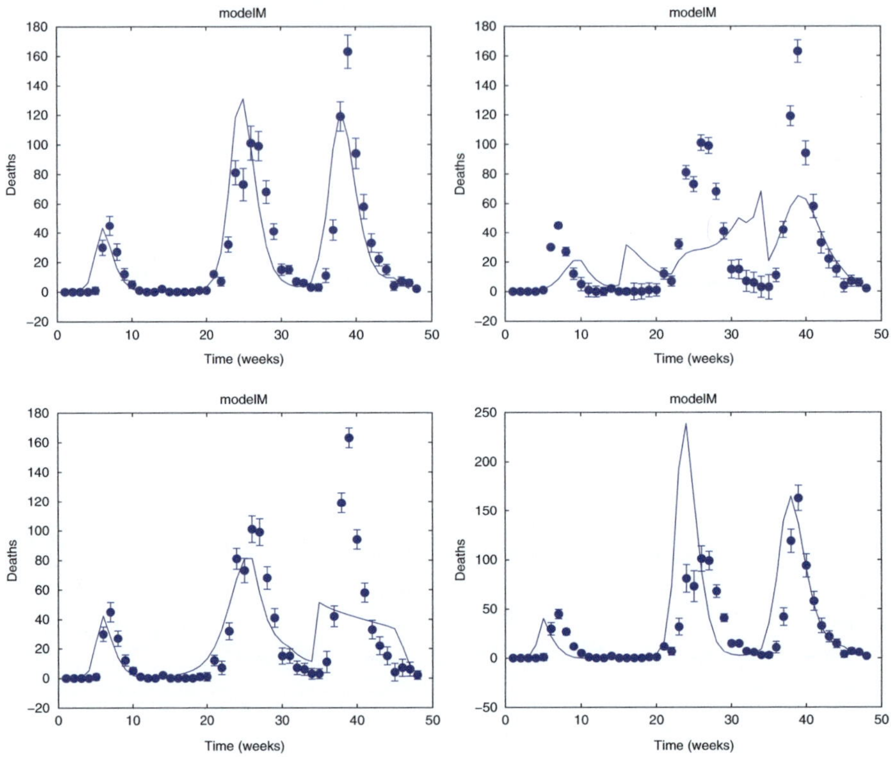

Fig. 15.4 Death data and its fitted model mortality curve for the Newcastle population (from left to right: SARE (10,000 iterations), MH (10,000 iterations), MH (50,000 iterations), MM (14,999 iterations))

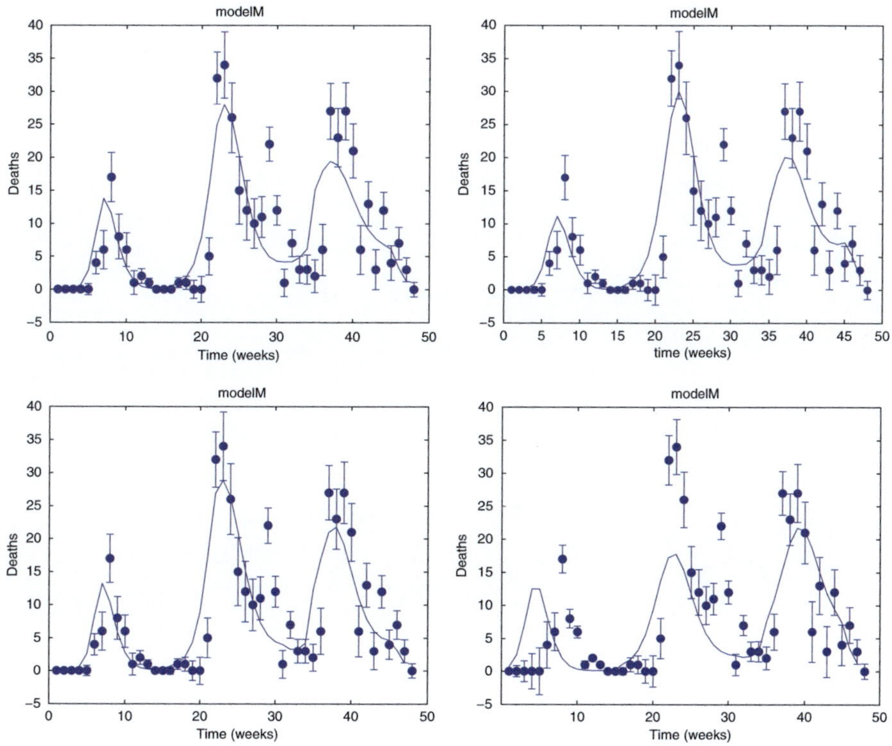

Fig. 15.5 Death data and its fitted model mortality curve for the Wigan population (from left to right: SARE (10,000 iterations), MH (10,000 iterations), MH (50,000 iterations), MM (14,999 iterations))

Chapter 16
LBFGS Quasi-Newtonian Methods for Molecular Modeling Prion AGAAAAGA Amyloid Fibrils

16.1 Introduction

Neurodegenerative diseases including Parkinson's, Alzheimer's, Huntington's, and Prion's were found they all featured amyloid fibrils [103, 207, 308, 438, 573, 627]. Amyloid is characterized by a cross-β sheet quaternary structure and recent X-ray diffraction studies of microcrystals revealed atomistic details of core region of amyloid [439, 517]. All the quaternary structures of amyloid cross-β spines can be reduced to one of the eight classes of steric zippers of [517], with strong vdW interactions between β-sheets and HBs to maintain the β-strands. A new era in the structural analysis of amyloids started from the 'steric zipper' – β-sheets [439]. As the two β-sheets zip up, HYDs & vdWs have been formed. The extension of the 'steric zipper' above and below (i.e. the β-strands) is maintained by HBs (but usually there is no HB between the two β-sheets). This is the common structure associated with some 20 neurodegenerative amyloid diseases. We first do some mathematical analysis for the common structure. Let r be the distance between two atoms, the vdW contacts of the two atoms are described by the LJ potential energy:

$$V_{LJ}(r) = 4\epsilon \left((\frac{\sigma}{r})^{12} - (\frac{\sigma}{r})^{6} \right), \tag{16.1}$$

where ϵ is the depth of the potential well and σ is the atom diameter; these parameters can be fitted to reproduce experimental data or deduced from results of accurate quantum chemistry calculations. The $(\frac{\sigma}{r})^{12}$ item describes repulsion and the $-(\frac{\sigma}{r})^{6}$ item describes attraction (Fig. 16.1). If we introduce the coordinates of the atoms whose number is denoted by N and let $\epsilon = \sigma = 1$ be the reduced units, then, for N atoms, Eq. 16.1 is

J. Zhang, *Molecular Structures and Structural Dynamics of Prion Proteins and Prions*, Focus on Structural Biology 9, DOI 10.1007/978-94-017-7318-8_16

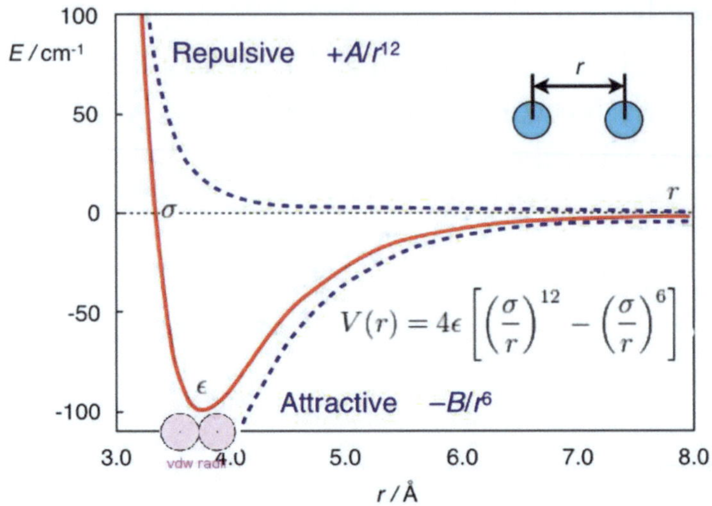

Fig. 16.1 The LJ potential energy (Eqs. 16.1 and 16.4) (can be seen in Fig. 1 of [687])

$$f(x) = 4\sum_{i=1}^{N}\sum_{j=1,j<i}^{N}\left(\frac{1}{\tau_{ij}^{6}} - \frac{1}{\tau_{ij}^{3}}\right), \tag{16.2}$$

where $\tau_{ij} = (x_{3i-2} - x_{3j-2})^2 + (x_{3i-1} - x_{3j-1})^2 + (x_{3i} - x_{3j})^2$, and $(x_{3i-2}, x_{3i-1}, x_{3i})$ is the coordinates of atom i, $N \geq 2$. The minimization of LJ potential energy $f(x)$ on R^n (where $n = 3N$) is an optimization problem

$$\min_{x \in R^n} f(x), \tag{16.3}$$

which is a well-known and challenging test problem for global optimization (see http://www-wales.ch.cam.ac.uk/CCD.html and its recent references such as [347, 530, 549, 661, 698]). It is very hard for global optimization to directly solve Eq. 16.3 even without a large number of atoms. Similarly as Eq. 16.1 – i.e.

$$V_{LJ}(r) = \frac{A}{r^{12}} - \frac{B}{r^{6}}, \tag{16.4}$$

the potential energy for the HBs between β-strands is

$$V_{HB}(r) = \frac{C}{r^{12}} - \frac{D}{r^{10}}, \tag{16.5}$$

where A, B, C, D are given constants and usually most of the HBs are still kept during the phase of molecular modeling. Thus, the amyloid fibril molecular modeling problem can be reduced to solve the optimization problem Eq. 16.3 though it is not easy to accurately solve Eq. 16.3 for a large molecule.

Alternatively, we have found another way to solve Eq. 16.3 [681]. Seeing Fig. 16.1, we may know that the optimization problem Eq. 16.3 reaches its optimal value at the bottom of the LJ potential well, where the distance between two atoms equals to the sum of vdW radii of the two atoms. Hence, the amyloid fibril molecular modeling problem can be looked as a MDGP (Molecular Distance Geometry Problem) [262]. As an example to explain MDGP, the problem of locating sensors in telecommunication networks is a DGP. In such a case, the positions of some sensors are known (which are called anchors) and some of the distances between sensors (which can or cannot be anchors) are known. The DGP is to locate the positions of all the sensors. The MDGP looks sensors as atoms and their telecommunication network as a molecule. In mathematics, the following Eqs. 16.6, 16.7 and 16.8 can express the MDGP for Eq. 16.3. The 3D structure of a molecule with N atoms can be described by specifying the 3D coordinate positions $x_1, x_2, \ldots, x_N \in R^3$ of all its atoms. Given bond lengths d_{ij} between a subset S of the atom pairs, the determination of the molecular 3D structure is

$$(\mathcal{P}_0) \quad to \quad find \quad x_1, x_2, \ldots, x_N \quad such \quad that \quad ||x_i - x_j|| = d_{ij}, (i,j) \in S, \tag{16.6}$$

where $|| \cdot ||$ denotes a norm in a real vector space and in this chapter it is calculated as the Euclidean distance 2-norm. Equation 16.6 can be reformulated as a mathematical global optimization problem

$$(\mathcal{P}) \quad \min P(X) = \sum_{(i,j) \in S} w_{ij}(||x_i - x_j||^2 - d_{ij}^2)^2 \tag{16.7}$$

in the terms of finding the global minimum of the function $P(X)$, where $w_{ij}, (i,j) \in S$ are positive weights, $X = (x_1, x_2, \ldots, x_N)^T \in R^n$ [431] and usually S has fewer elements than $N^2/2$ due to the error in the theoretical or experimental data [262, 721]. Even there may not exist any solution $x_1, x_2, \ldots, x_N$ to satisfy the distance constraints in Eq. 16.6, for example when data for atoms $i, j, k \in S$ violate the triangle inequality; in this case, we may add a perturbation item $-\varepsilon^T X$ to $P(X)$:

$$(\mathcal{P}_\varepsilon) \quad \min P_\varepsilon(X) = \sum_{(i,j) \in S} w_{ij}(||x_i - x_j||^2 - d_{ij}^2)^2 - \varepsilon^T X, \tag{16.8}$$

where $\varepsilon \geq 0$. In some cases, instead exact values $d_{ij}, (i,j) \in S$ can be found, we can only specify lower and upper bounds on the distances: $l_{ij} \leq ||x_i - x_j|| \leq u_{ij}, (i,j) \in S$; in such cases we may penalize all the unsatisfied constraints into the objective function of $(\mathcal{P}_\varepsilon)$ by adding $\sum_{(i,j)\in S} \left(\max\left\{l_{i,j}^2 - ||x_i - x_j||^2, 0\right\}\right)^2 + \left(\max\left\{||x_i - x_j||^2 - u_{i,j}^2, 0\right\}\right)^2$ into $P_\varepsilon(X)$ [262, 721], where we may let d_{ij} be the interatomic distance (less than 6 Å) for the pair in successive residues of a protein and set $l_{ij} = (1 - 0.05)d_{ij}$ and $u_{ij} = (1 + 0.05)d_{ij}$ [262]. In this chapter, we aim to solve Eq. 16.8 (or Eq. 16.3) for modeling amyloid fibril molecular 3D structures.

Neurodegeneration is the progressive loss of structure or function of neurons, including death of neurons. A prion is a misshapen protein that acts like an infectious agent (but not requiring either DNA, RNA, or both) to cause a number of fatal diseases. Prion diseases are rich in β-sheets (compared with the normal prion protein PrP^C in rich of α-helices) and are so-called "protein structural conformational" diseases. The normal hydrophobic region 113–120 AGAAAAGA peptide of prion proteins is an inhibitor/blocker of prion diseases. PrP lacking this palindrome could not convert to prion diseases. Brown et al. pointed out that the AGAAAAGA peptide was found to be necessary (though not sufficient) for blocking the toxicity and amyloidogenicity of PrP 106–126, and the peptide AGAA does not form fibrils [67]. The minimum sequence necessary for fibril formation should be AGAAA, AGAAAA, AGAAAAG, AGAAAAGA and GAAAAGA, but the molecular structures of these fibrils have not known yet. This chapter addresses an important problem on modeling the 3D molecular structures of prion AGAAAAGA amyloid fibrils of neurodegenerative diseases. The rest of this chapter is arranged as follows. In the next section, i.e. Sect. 16.2, an improved LBFGS Quasi-Newtonian method is presented for solving Eq. 16.8. Section 16.3 implements this Quasi-Newtonian method by constructing an 3D molecular structure of prion AGAAAAGA amyloid fibrils of neurodegenerative prion diseases. Numerical results of computations show that the method designed in Sect. 16.2 is very effective and successful. This concluding remark will be made in the last section, i.e. Sect. 16.4.

16.2 Methods

In a (macro)molecular system, if it is very far from equilibrium, then the forces may be excessively large, a robust EM (energy minimization) is required; another reason to perform an EM is the removal of all kinetic energy from the system: EM reduces the thermal noise in the structures and potential energies [577]. EM, with the images at the endpoints fixed in space, of the total system energy provides a minimum energy path. EM can be done using SD (steepest descent), CG (conjugate gradient), and Limited-memory Broyden-Fletcher-Goldfarb-Shanno (LBFGS) methods.

Three kinds of possible EM methods are: (1) derivative-free methods – that require only function evaluations, e.g. the simplex method and its variants; (2) derivative information methods – the partial derivatives of the potential energy with respect to all coordinates are known and the forces are minimized, e.g. SD, CG methods; and (3) second derivative information methods, e.g. LBFGS method. "SD is based on the observation that if the real-valued function $f(x)$ is defined and differentiable in a neighborhood of a point x_0 then $f(x)$ decreases fastest if one goes from x_0 in the direction of the negative gradient of $f(x)$ at x_0" and SD local search method converges fast [538]. SD is robust and easy to implement but it is not most efficient especially when closer to minimum; at this moment, we may use the efficient CG. CG is slower than SD in the early stages but more efficient when closer to minimum. CG algorithm adds an orthogonal vector to the current direction of the search, and then moves them in another direction nearly perpendicular to this vector. The hybrid of SD-CG will make SD or CG more efficient than SD or CG alone. However, CG cannot be used to find the EM path, for example, when "forces are truncated according to the tangent direction, making it impossible to define a Lagrangian" [91, 149]. In this case, the powerful and faster quasi-Newtonian method (e.g. the LBFGS quasi-Newtonian minimizer) can be used [73, 77, 149, 394, 443, 718]. We briefly introduce the LBFGS quasi-Newtonian method as follows.

Newton's method in optimization explicitly calculates the Hessian matrix of the second-order derivatives of the objective function and the reverse of the Hessian matrix [178]. The convergence of this method is quadratic, so it is faster than SD or CG. In high dimensions, finding the inverse of the Hessian is very expensive. In some cases, the Hessian is a non-invertible matrix, and furthermore in some cases, the Hessian is symmetric indefinite. Quasi-Newton methods thus appear to overcome all these shortcomings.

Quasi-Newton methods (a special case of variable metric methods) are to approximate the Hessian. Currently, the most common quasi-Newton algorithms are the SR1 formula, the BHHH method, the widespread BFGS method and its limited/low-memory extension LBFGS, DFP, MS, and Broyden's methods [48, 401, 444, 614]. In Amber [91] and Gromacs [577], LBFGS is used, and the hybrid of LBFGS with CG – a Truncated Newton linear CG method with optional LBFGS Preconditioning [443] – is used in Amber [91].

For BFGS method, whether it converges at all on nonconvex problems is still an open problem. In fact, Powell [477] gave a counter-example that shows that BFGS with an inexact line-search search may fail to converge [163, 417, 670]. Li and Fukushima [378] proposed a modified BFGS method for nonconvex objective function. Basing on [378, 394, 443, 649, 660], in this chapter we present an improved LBFGS method described as follows [291] – which presents the nonmonotone line search technique [258, 269] for the Wolfe-type search. The improved LBFGS method presented in this chapter is much better than the standard BFGS method in view of the CPU time (see Fig. 16.2) tested through more than 30 nonlinear programming problems (where each selected problem is regular, that is, its first and second derivatives exist and are continuous everywhere, and each problem is with different dimensions, i.e., 100, 500, 1000 and 10,000 dimensions) and its mathematical theory to support this algorithm can be seen from the Supplementary Materials [291] listed at the end of this chapter. This chapter implements the Wolfe-type search by the approximation technique of piecewise linear/quadratic function [22].

Algorithm 16.1 (An Improved LBFGS Method for minimizing nonconvex function [291]).

Step 0: Choose an initial point $x_0 \in R^n$, an initial positive definite matrix H_0, and choose constants σ_1,σ_2 such that $0 < \sigma_1 < \sigma_2 < 1$, and choose an positive integer m_1. Let $k = 0$.

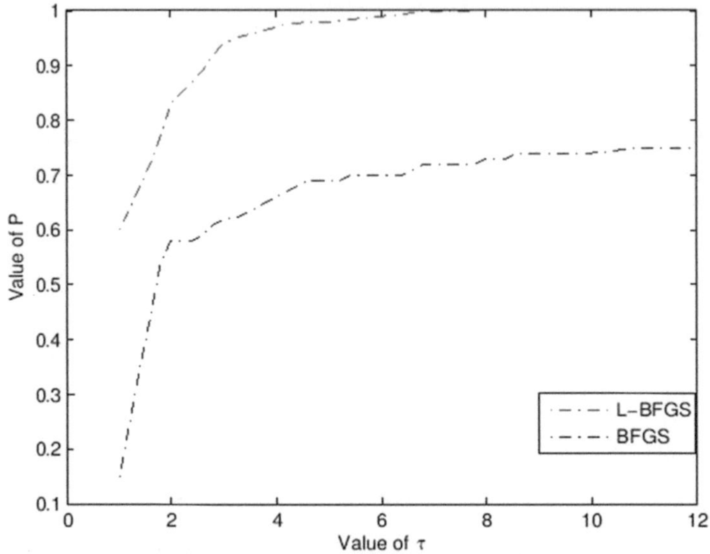

Fig. 16.2 Performance based on CPU time, where $\tau \in R^1$ is a factor of the best possible ratio for CPU performance ratio of solvers L-BFGS and BFGS, and P is the probability distribution of the CPU performance ratio. Detailed mathematical formula for P can be seen in [186]

Step 1: If $\|g_k\| = 0$, then output x_k and stop; otherwise, go to *Step 2*.
Step 2: Solve the following linear equation to get d_k:

$$d_k = -H_k g_k,$$

Step 3: Find a step-size $\lambda_k > 0$ satisfying the Wolfe-type line search conditions:

$$f(x_k + \lambda_k d_k) \leq f(x_k) + \sigma_1 \lambda_k g_k^T d_k, \tag{16.9}$$

$$g(x_k + \lambda_k d_k)^T d_k \geq \sigma_2 g_k^T d_k. \tag{16.10}$$

Step 4: Let the next iterate by $x_{k+1} = x_k + \lambda_k d_k$. Calculate g_{k+1} and $\|g_{k+1}\|$.
Step 5: Let $s_k = x_{k+1} - x_k = \lambda_k d_k, y_k = g_{k+1} - g_k$, $\gamma_k = \|g_k\|$, then $y_k^* = y_k + \gamma_k s_k$.
Step 6: Let $\bar{m} = \min\{k+1, m_1\}$. Update H_k following the formula

$$\begin{aligned} H_{k+1} = &(V_k^{*T} \cdots V_{k-\bar{m}+1}^{*T}) H_{k-\bar{m}+1} (V_{k-\bar{m}+1}^{*} \cdots V_k^{*}) \\ &+ \omega_{k-\bar{m}+1}^{*} (V_{k-1}^{*T} \cdots V_{k-\bar{m}+2}^{*T}) s_{k-\bar{m}+1} s_{k-\bar{m}+1}^T (V_{k-\bar{m}+2}^{*} \cdots V_{k-1}^{*}) \\ &+ \cdots + \omega_k^{*} s_k s_k^T, \end{aligned} \tag{16.11}$$

where $\omega_k^* = \frac{1}{s_k^T y_k^*}$, $V_k^* = I - \omega_k^* y_k^* s_k^T$ (when $k = 0$, $H_0 = \frac{y_0^T s_0}{\|y_0\|^2}$, $V_0^* = I - \frac{y_0^* s_0^T}{s_0^T y_0^*}$, $H_1 = (I - \frac{y_0^* s_0^T}{s_0^T y_0^*})^T H_0 (I - \frac{y_0^* s_0^T}{s_0^T y_0^*}) + \frac{1}{s_0^T y_0^*} s_k s_k^T$).
Step 7: Let $k = k + 1$ and go to *Step 1*.

In Algorithm 16.1, g_k denotes the gradient of $f(x)$ at x_k, and the convergence and the R-linear convergent rate of Algorithm 16.1 are guaranteed by the following assumptions:

(A1) the objective function $f : R^n \to R$ is twice continuously differentiable, i.e. $f \in C^2$;
(A2) f has Lipschitz continuous gradients and Hessians, i.e., there exist constants $L, M_2 > 0$ such that

$$\|g(x) - g(y)\| \le L\|x - y\|, \quad \forall x, y \in R^n,$$
$$\|G(x) - G(x^*)\| \le M_2\|x - x^*\|,$$

where G denotes the Hessian and x^* is a stationary point (i.e. x_k converges to x^* where $g(x^*) = 0$);
(A3) $G(x^*)$ is positive definite; and
(A4) the level set of f

$$\Omega = \{x \in R^n \mid f(x) \le f(x_0)\},$$

is bounded, where $x_0 \in R^n$ is an initial point.

Next we give the detailed mathematical proof of the convergence and R-linear convergent rate of Algorithm 16.1.

In order to establish the convergence for Algorithm 16.1, we represent Eq. 16.11 as

$$B_k^{l+1} = B_k^l - \frac{B_k^l s_l s_l^T B_k^l}{s_l^T B_k^l s_l} + \frac{y_l^* y_l^{*T}}{y_k^{*T} s_l}, \quad l = k - \bar{m} + 1, \cdots, k, \tag{16.12}$$

where $B_k = H_k^{-1}$, $s_l = x_{l+1} - x_l$, $y_l^* = y_l + \gamma_l s_l$, and $B_k^{k-\bar{m}+1} = B_0$ for all k.

Theorem 16.2.1 *Let Assumptions (A1)~(A4) hold, and* $\{x_k\}$ *denotes the sequence generated by Algorithm 16.1. Then* $\lim_{k\to\infty} \inf \|g_k\| = 0$*; moreover, there exits a constant* $t \in [0, 1)$ *such that*

$$f(x_k) - f(x^*) \le t^k [f(x_0) - f(x^*)],$$

which is to say that x_k *converges to the minimum R-linearly.*

Proof. Denote by M_1 an upper bound of $\|g_k\|$ on Ω and by $y_k^* = y_k + \gamma_k s_k$, we can prove that

$$\frac{\|y_k^{*T}\|^2}{y_k^{*T} s_k} \leq \frac{(L + M_1)^2}{\gamma},$$

where γ is a sufficiently small positive number. By this formula and taking the trace operation in both sides of Eq. 16.12, we get

$$\begin{aligned} tr(B_{k+1}) &= tr(B_k^{k-\bar{m}+1}) - \sum_{l=k-\bar{m}+1}^{k} \frac{\|B_l s_l\|^2}{s_l^T B_l s_l} + \sum_{l=k-\bar{m}+1}^{k} \frac{\|y_l^*\|^2}{y_l^{*T} s_l} \\ &= \cdots \\ &\leq tr(B_0) - \sum_{l=0}^{k} \frac{\|B_l s_l\|^2}{s_l^T B_l s_l} + \sum_{l=0}^{k} \frac{\|y_l^*\|^2}{y_l^{*T} s_l} \\ &\leq tr(B_0) - \sum_{l=0}^{k} \frac{\|B_l s_l\|^2}{s_l^T B_l s_l} + \frac{(L+M_1)^2}{\gamma}(k+1) \\ &\leq M_3, \end{aligned} \tag{16.13}$$

where M_3 is a positive constant. We also take the determinant and by Eq. 16.13, we have

$$\begin{aligned} det(B_{k+1}) &= det(B_k^{k-\bar{m}+1}) \prod_{l=k-\bar{m}+1}^{k} \frac{y_l^{*T}}{s_l^T B_l s_l} \\ &= \cdots = det(B_0) \prod_{l=0}^{k} \frac{y_l^{*T}}{s_l^T B_l s_l} \geq det(B_0)(\tfrac{\gamma}{M_3})^{k+1} \\ &\geq M_4, \end{aligned} \tag{16.14}$$

where M_4 is a positive constant. Combining Eqs. 16.13 and 16.14, we obtain a constant $\delta > 0$ such that $\cos(\theta_k) \geq \delta$, where $\cos(\theta_k) = \frac{s_k^T B_k s_k}{\|s_k\| \|B_k s_k\|}$.

Because $g(x)$ is Lipschitz continuous and by the Wolfe-type line search condition Eq. 16.10, we have

$$(\sigma_2 - 1) g_k^T d_k \leq (g_{k+1} - g_k)^T d_k \leq \lambda_k L \|d_k\|^2,$$

which implies

$$\lambda_k \geq \frac{(\sigma_2 - 1)}{L} \frac{g_k^T d_k}{\|d_k\|^2}. \tag{16.15}$$

Using the Wolfe-type line search condition Eqs. 16.9 and 16.15, we have

$$\begin{aligned} f_{k+1} &\leq f_k + \sigma_1 \lambda_k g_k^T d_k \leq f_k + \sigma_1 \tfrac{(\sigma_2-1)}{L} \tfrac{(g_k^T d_k)^2}{\|d_k\|^2} \\ &= f_k + \sigma_1 \tfrac{(\sigma_2-1)}{L} \|g_k\|^2 \cos^2 \theta_k. \end{aligned}$$

Then using the monotonicity and the bound of $\{f_k\}$ on Ω, we obtain

$$\sum_{k=1}^{\infty} \|g_k\|^2 \cos^2 \theta_k < \infty.$$

Combining $\cos \theta_k \geq \delta$, we get

$$\lim_{k \to \infty} \inf \|g_k\| = 0.$$

Under the Assumptions (A2)$\sim$(A3), there is a neighbourhood $U(x^*)$ of x^* such that for all $x \in U(x^*)$,

$$\|g(x)\| \geq \|g(x) - g(x^*)\| \geq m\|x - x^*\|, \tag{16.16}$$

and for all $d \in R^n$,

$$d^T G(x) d \geq m\|d\|^2, \tag{16.17}$$

where m is a positive constant got according to the Assumptions. Hence, Eqs. 16.16$\sim$16.17 hold with $x = x_k$ for all k sufficiently large.

Since

$$\begin{aligned} y_k^{*T} s_k &= (y_k + \gamma_k s_k)^T s_k \geq y_k^T s_k \\ &\geq -(1 - \sigma_2) g_k^T s_k \\ &= (1 - \sigma_2)\|g_k\|\,\|s_k\| \cos \theta_k, \end{aligned}$$

and $y_k^{*T} s_k \leq (L + M_1)^2 \|s_k\|^2$, if denoting $\alpha_1 = \frac{1-\sigma_1}{(L+M_1)^2}$, we can obtain

$$\|s_k\| \geq \frac{1 - \sigma_1}{(L + M_1)^2} \|g_k\| \cos \theta_k = \alpha_1 \|g_k\| \cos \theta_k. \tag{16.18}$$

Combining Eq. 16.18 and the Wolfe-type line search condition Eq. 16.9, we get

$$\begin{aligned} f_{k+1} - f^* &\leq f_k - f^* + \sigma_1 g_k^T s_k \\ &= f_k - f^* - \sigma_1 \|g_k\|\,\|s_k\| \cos \theta_k \\ &\leq f_k - f^* - \sigma_1 \alpha_1 \|g_k\|^2 \cos^2 \theta_k. \end{aligned}$$

Using Eq. 16.16, then Eq. 16.18 implies for all k sufficiently large,

$$\begin{aligned} f_{k+1} - f^* &\leq f_k - f^* - \sigma_1 \alpha_1 \|g_k\|^2 \cos^2 \theta_k \\ &\leq f_k - f^* - \sigma_1 \alpha_1 m^2 \|x_k - x^*\|^2 \cos^2 \theta_k. \end{aligned} \tag{16.19}$$

By Taylor's expansion, then there exists a positive number $M_5 > 0$ such that

$$f_k - f^* \leq M_5 \|x_k - x^*\|^2$$

for all k sufficiently large.

Let $\alpha_2 = \frac{\sigma_1 \alpha_1 m^2}{M_5}$, indeed Eq. 16.19 implies that

$$\begin{aligned} f_{k+1} - f^* &\leq f_k - f^* - \sigma_1 \alpha_1 m^2 \|x_k - x^*\|^2 \cos^2 \theta_k \\ &\leq f_k - f^* - \alpha_2 (f_k - f^*) \cos^2 \theta_k \\ &= (1 - \alpha_2 \cos^2 \theta_k)(f_k - f^*) \end{aligned}$$

for all sufficiently large k.

Hence, there is a constant $t = 1 - \alpha_2 \cos^2 \theta_k \in [0, 1)$ such that

$$f(x_k) - f(x^*) \leq t^k [f(x_0) - f(x^*)]. \tag{16.20}$$

By Eq. 16.17, we have

$$f_k - f^* \geq \frac{1}{2} m \|x_k - x^*\|^2,$$

which, together with Eq. 16.20, implies

$$\|x_k - x^*\| \leq t^{k/2} [\frac{2(f(x_0) - f(x^*))}{m}]^{1/2}.$$

This is to say the sequence $\{x_k\}$ is R-linearly convergent. The proof of this theorem is complete. □

In [291], the following nonmonotone line search technique for stepsize λ_k is used:

$$f(x_k + \lambda_k d_k) \leq \max_{0 \leq j \leq M_0} f(x_{k-j}) + \sigma_1 \lambda_k g_k^T d_k,$$

$$g(x_k + \lambda_k d_k)^T d_k \geq \max\{\sigma_2, 1 - (\lambda_k ||d_k||)^p\} g_k^T d_k,$$

where $p \in (-\infty, 1)$, and M_0 is a nonnegative integer. This is a difference between the algorithm of [291] and this chapter. All in all, it is well known that quasi-Newton method is an efficient solution method for unconstrained and continuously differentiable minimization problem [74, 76, 177]. However, it needs computing and storage of the updated matrix which is an approximation to the Hessian matrix at each iteration of the method. Hence, its efficiency may decrease when it is applied to large scale optimization problem. To overcome the drawback, limited memory quasi-Newton method is proposed [442]. The main ideal of this method is nearly identical to that of the standard BFGS method, and the only difference is that the inverse Hessian approximation is not formed explicitly, but defined by a small number, say $\bar{m}$, of BFGS updates. This technique received much attention in recent years and numerical experiments show that it is very competitive [242, 394], and its global convergence and R-linear convergence rate with Wolfe line search are established for the uniformly convex case [77, 394]. Since the limited memory BFGS method may suffer from ill-conditions for small value of $\bar{m}$, Al-Baali [12] made some modifications to the method and establish its global convergence based on the same assumptions, and Byrd et al. [75] derives new representation of limited memory quasi-Newton matrices for the benefit of computing the updated matrix. Recently, a non-monotone line search is introduced, see e.g., [258, 269]. Then it is showed to be more competitive and practical for solving nonlinear optimization problems, and [671] established the global convergence of this line search applied to limited memory BFGS method based on the uniformly convex assumption. Motivated by the above observation, it turn out that in two respects the limited memory BFGS method is much less effective. First, we note that the convergence analysis of these method are focused on the uniformly convex assumption and little is known for nonconvex case. Second, numerical experiments have suggested the main weakness of limited memory method is that it may converge very slowly in terms of number of iterations for ill-conditioned problems. The purpose of the above Algorithm 16.1 is to reduce these defects and Fig. 16.2 shows the effectiveness of the proposed algorithm. We will apply it into the molecular modeling of prion AGAAAAGA amyloid fibrils in the next section.

16.3 Results and Discussion

From their research of prion, scientists found that the cross-β structure of peptides is with the nature of self-aggregation, the self-aggregating to form fibers. This provides us a new research idea for nanomaterials. HBs can be formed between peptide β-strands, and one peptide monomer connects together with another in accordance with the specific structure to form fibers. Many laboratories in the world are synthesizing peptides that can self-aggregate to form fibers, and want to be able to control the growth of the fiber to find out new functional materials [14, 257]. The studies of this chapter not only benefit nanometerials research, but also benefit the research on neurodegenerative amyloid fibril diseases. Prion AGAAAAGA peptide has been reported to own an amyloid fibril forming property (initially described in 1992 by Gasset et al. of Prusiner's Group) [67, 68, 70, 85, 95, 102, 237, 267, 274, 288, 315, 316, 322, 351, 352, 357, 361, 373, 408, 445, 514, 594, 595, 619, 674], but there has not been experimental structural bioinformatics for this segment yet due to the unstable, noncrystalline and insoluble nature of this region. Furthermore, Zhang [680] did accurate calculations to confirm the amyloid fibril property at this region (Fig. 16.3).

16.3.1 Material for the Molecular Modeling

This chapter uses a suitable pdb file template 2OMP.pdb (the LYQLEN peptide derived from human insulin residues 13–18 [517]) from the Protein Data Bank to build an 8-chain AGAAAAGA prion amyloid fibril molecular model to illuminate Algorithm 16.1 works very well. To choose 2OMP.pdb (Fig. 16.4) as the modeling template is due to it can pass all the long procedures of SDCG-SA (equilibrations & productions)-SDCG of [680]. By observations of Fig. 16.4 and the second column

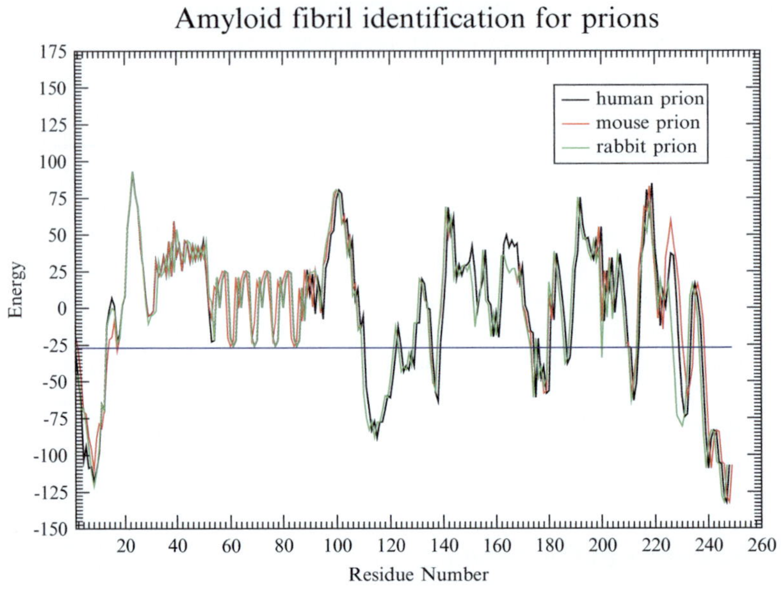

Fig. 16.3 Prion AGAAAAGA (113–120) segment is clearly and surely identified as the amyloid fibril formation region, because its energy is less than the amyloid fibril formation threshold energy −26 kcal/mol [695]

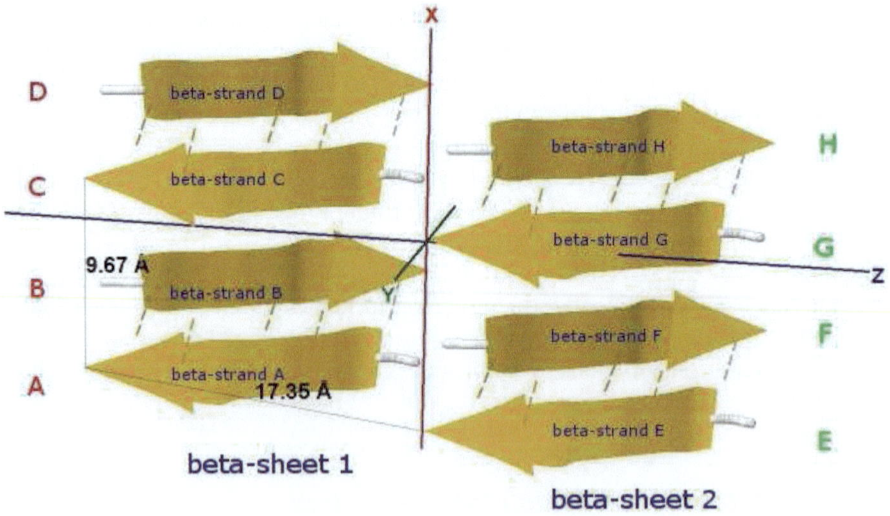

Fig. 16.4 Protein fibril structure of human insulin LYQLEN (13–18) (PDB id: 2OMP). The *dashed lines* denote the HBs between the pairs of β-strands. *A, B, C, D, E, F, G, H* denote the chains of the fibril. The pair of β-sheets 1 & 2 forms a completely dry interface by vdWs, and between many pairs of β-sheets wet interfaces are formed with water molecules

of coordinates of 2OMP.pdb, we know that E(F) chains can be calculated on the XZ-plane from A(B) chains by Eq. 16.21 and other chains can be got by a parallel up (or down) along the X-axis by Eqs. 16.22∼16.23:

$$E(F) = A(B) + (-1.885, 0, 17.243), \tag{16.21}$$

$$C(D) = A(B) + (9.666, 0, 0), \tag{16.22}$$

$$G(H) = E(F) + (9.666, 0, 0). \tag{16.23}$$

16.3.2 New Molecular Modeling Homology Model

Basing on the template 2OMP.pdb from the Protein Data Bank (Fig. 16.4), the AGAAAAGA palindrome amyloid fibril model of prions (denoted as Model-chapt16-1) will be constructed. Chains AB of Model-chapt16-1 will be got from AB Chains of 2OMP.pdb using the mutate module of the free package Swiss-PdbViewer (SPDBV Version 4.01) (http://spdbv.vital-it.ch). It is pleasant to see that some HBs are still kept after the mutations; thus we just need to consider the vdWs only. Making mutations for EF Chains of 2OMP.pdb, we can get the EF Chains of Model-chapt16-1. Then we add GLY and ALA residues by XLEaP module of Amber 11. However, the vdWs between Chain A and Chain E, between B Chain and F Chain are too far at this moment (Fig. 16.5, where the twice of the vdW radius of CB atom is 3.4 Å).

In [680] the commercial package InsightII (http://accelrys.com/) is used to build models. Instead of InsightII, because this package is not available by the authors, this chapter uses Algorithm 16.1 to build and optimize Model 1. In "Zipper 1", fixing the coordinates of A.ALA3.CB, A.ALA1.CB and letting the coordinates of E.ALA6.CB, E.ALA4.CB be variables, we may get an optimization problem:

$$\begin{aligned} \min 4 & \left\{ \frac{1}{\left[(x_{11}-1.071)^2+(x_{12}-2.986)^2+(x_{13}-1.888)^2\right]^6} - \frac{1}{\left[(x_{11}-1.071)^2+(x_{12}-2.986)^2+(x_{13}-1.888)^2\right]^3} \right\} \\ +4 & \left\{ \frac{1}{\left[(x_{21}-1.071)^2+(x_{22}-2.986)^2+(x_{23}-1.888)^2\right]^6} - \frac{1}{\left[(x_{21}-1.071)^2+(x_{22}-2.986)^2+(x_{23}-1.888)^2\right]^3} \right\} \\ +4 & \left\{ \frac{1}{\left[(x_{21}-1.135)^2+(x_{22}+0.763)^2+(x_{23}-7.209)^2\right]^6} - \frac{1}{\left[(x_{21}-1.135)^2+(x_{22}+0.763)^2+(x_{23}-7.209)^2\right]^3} \right\} \end{aligned} \tag{16.24}$$

or

$$\begin{aligned} \min \tfrac{1}{2} & \left\{(x_{11}-1.071)^2 + (x_{12}-2.986)^2 + (x_{13}-1.888)^2 - 3.4^2\right\}^2 \\ +\tfrac{1}{2} & \left\{(x_{21}-1.071)^2 + (x_{22}-2.986)^2 + (x_{23}-1.888)^2 - 3.4^2\right\}^2 \\ +\tfrac{1}{2} & \left\{(x_{21}-1.135)^2 + (x_{22}+0.763)^2 + (x_{23}-7.209)^2 - 3.4^2\right\}^2 \\ & -0.05\left\{x_{11}+x_{12}+x_{13}+x_{21}+x_{22}+x_{23}\right\} \end{aligned} \tag{16.25}$$

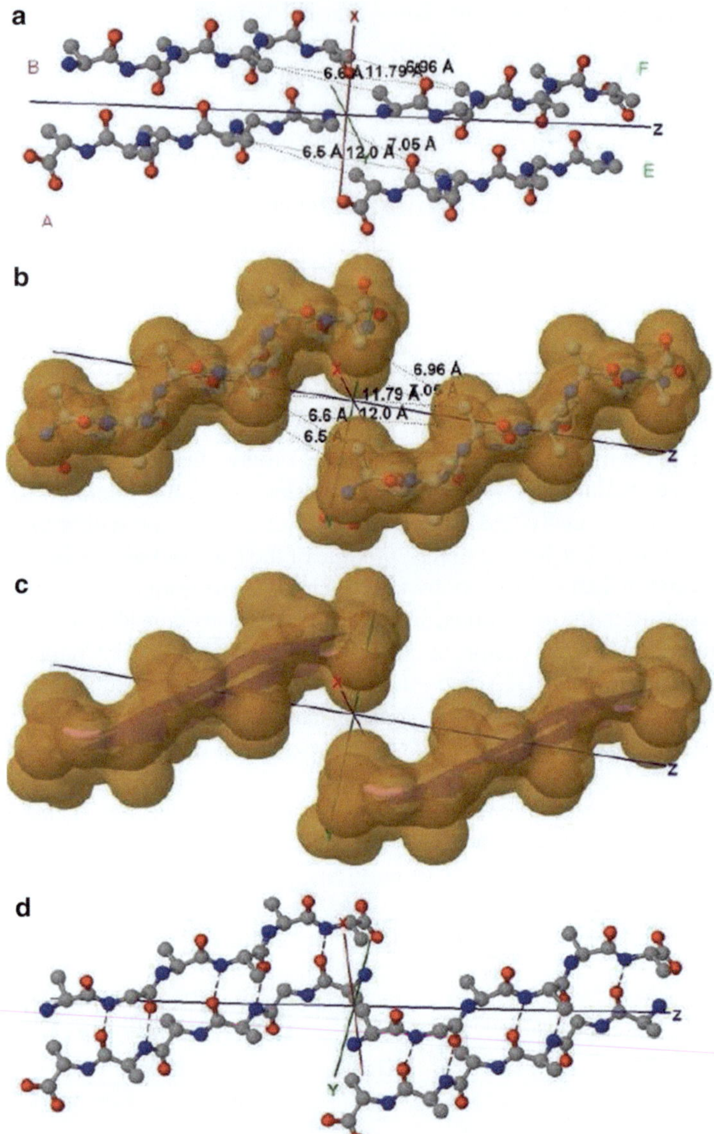

Fig. 16.5 (**a**) Shows the distances of "Zipper 1"-E.ALA6.CB-A.ALA3.CB-E.ALA4.CB-A.ALA1.CB are 6.5, 12.0 and 7.05 Å respectively, and the distances of "Zipper 2"-F.ALA1.CB-B.ALA4.CB-F.ALA3.CB-B.ALA6.CB are 6.6, 11.79, 6.96 Å respectively. (**b**) Shows the far vdW surface. (**c**) Shows the *violet colored* ABEF Chains of Fig. 16.3. (**d**) Shows HBs: A/E.ALA5.O-B/F.GLY2.N, A/E.ALA5.N-B/F.GLY2.O, A/E.ALA3.O-B/F.ALA4.N, A/E.ALA3.N-B/F.ALA4.O, A/E.ALA1.O-B/F.ALA6.N

with an initial solution (−0.067, 5.274, 7.860; −1.119, 1.311, 13.564). Similarly, in "Zipper 2", fixing the coordinates of B.ALA4.CB, B.ALA6.CB and letting the coordinates of F.ALA1.CB, F.ALA3.CB be variables, we get another optimization problem:

$$\begin{aligned}\min 4 &\left\{ \frac{1}{\left[(x_{11}-5.446)^2+(x_{12}-2.796)^2+(x_{13}-2.662)^2\right]^6} - \frac{1}{\left[(x_{11}-5.446)^2+(x_{12}-2.796)^2+(x_{13}-2.662)^2\right]^3} \right\}\\ +4 &\left\{ \frac{1}{\left[(x_{21}-5.446)^2+(x_{22}-2.796)^2+(x_{23}-2.662)^2\right]^6} - \frac{1}{\left[(x_{21}-5.446)^2+(x_{22}-2.796)^2+(x_{23}-2.662)^2\right]^3} \right\}\\ +4 &\left\{ \frac{1}{\left[(x_{21}-5.201)^2+(x_{22}+1.125)^2+(x_{23}-7.873)^2\right]^6} - \frac{1}{\left[(x_{21}-5.201)^2+(x_{22}+1.125)^2+(x_{23}-7.873)^2\right]^3} \right\}\end{aligned} \tag{16.26}$$

or

$$\begin{aligned}\min \tfrac{1}{2} &\left\{(x_{11}-5.446)^2+(x_{12}-2.796)^2+(x_{13}-2.662)^2-3.4^2\right\}^2\\ +\tfrac{1}{2} &\left\{(x_{21}-5.446)^2+(x_{22}-2.796)^2+(x_{23}-2.662)^2-3.4^2\right\}^2\\ +\tfrac{1}{2} &\left\{(x_{21}-5.201)^2+(x_{22}+1.125)^2+(x_{23}-7.873)^2-3.4^2\right\}^2\\ &-0.05\left\{x_{11}+x_{12}+x_{13}+x_{21}+x_{22}+x_{23}\right\}\end{aligned} \tag{16.27}$$

with an initial solution (4.714, 4.878, 8.881; 4.170, 1.360, 14.292). Next, we solve Eqs. 16.25 and 16.27 by Algorithm 16.1.

We first solve Eq. 16.25 in the use of Algorithm 16.1. We set $\sigma_1 = 10^{-4}$, $\sigma_2 = 0.1 \sim 0.9$, $m_1 = 3 \sim 7$, take the initial solution $x_0 = (-0.067, 5.274, 7.860; -1.119, 1.311, 13.564)$ and calculate its gradient $g_0 = (-69.7747, 140.135, 365.852, -752.075, -285.005, 3576.69)$ and its Hessian matrix $H_0 =$

$$\begin{pmatrix} 66.4497 & -10.415 & -27.1845 & 0 & 0 & 0\\ -10.415 & 82.2093 & 54.6557 & 0 & 0 & 0\\ -27.1845 & 54.6557 & 203.929 & 0 & 0 & 0\\ 0 & 0 & 0 & 380.664 & -4.02618 & -159.578\\ 0 & 0 & 0 & -4.02618 & 369.586 & -25.5081\\ 0 & 0 & 0 & -159.578 & -25.5081 & 1048.02 \end{pmatrix}, \tag{16.28}$$

which is a positive definite matrix with eigenvalues (1085.02, 372.093, 341.157, 230.049, 61.2695, 61.2695). Then Algorithm 16.1 hybridized with simulated annealing global optimal search (in order to bring local optimal solutions to jump out of local traps, replacing the discrete gradient local optimal search method in Algorithm 16.1 of [698] by the Algorithm 16.1 of this chapter) is executed and the optimal solution (3.027, 4.954, 3.856; 1.679, 1.777, 5.011) for Eq. 16.25 is got.

Similarly, for Eq. 16.27, we take the initial solution $x_0 = (4.714, 4.878, 8.881; 4.170, 1.360, 14.292)$ and calculate its gradient $g_0 = (-46.8782, 133.142, 397.798, -401.192, -182.604, 3436.46)$ and its Hessian matrix $H_0 =$

$$\begin{pmatrix} 66.1163 & -6.0961 & -18.2092 & 0 & 0 & 0 \\ -6.0961 & 81.3119 & 51.7918 & 0 & 0 & 0 \\ -18.2092 & 51.7918 & 218.677 & 0 & 0 & 0 \\ 0 & 0 & 0 & 339.302 & -2.9188 & -85.8315 \\ 0 & 0 & 0 & -2.9188 & 361.487 & -2.99786 \\ 0 & 0 & 0 & -85.8315 & -2.99786 & 1034.38 \end{pmatrix}, \tag{16.29}$$

which is a positive definite matrix with eigenvalues (1044.83, 361.8, 328.538, 238.159, 63.973, 63.973). The optimal solution (7.412, 4.760, 4.624; 5.887, 1.451, 5.757) for Eq. 16.27 is got.

We set (3.027, 4.954, 3.856; 1.679, 1.777, 5.011) as the coordinates of E.ALA6.CB and E.ALA4.CB, (7.412, 4.760, 4.624; 5.887, 1.451, 5.757) as the coordinates of F.ALA1.CB and F.ALA3.CB, and taking the average value we get

$$E(F) = A(B) + (1.0335, 1.0823, 0.9723). \tag{16.30}$$

By Eq. 16.30 we can get very close vdW contacts between A (B) chains and E(F) chains (Fig. 16.5b), and other chains of Model-chapt16-1 can be got by Eqs. 16.22∼16.23 and Eq. 16.31:

$$I(J) = A(E) - (9.666, 0, 0). \tag{16.31}$$

The initial structure of Model-chapt16-1 illuminated in Fig. 16.5a,b – is not the optimal structure with the lowest total potential energy. The initial structure also has no hydrogen atoms (so no hydrogen bonds existed) and water molecules added. For each Chain, the C-terminal and N-terminal atoms also have problems. Clearly there are a lot of close/bad contacts between β-strand atoms as illuminated in Fig. 16.5a,b. We used the ff03 force field of AMBER 11,in a neutral pH environment. The amyloid fibrils were surrounded with a 8 Å layer of TIP3PBOX water molecules using the XLEaP module of AMBER 11. 1944 waters and 408 hydrogen atoms were added for Model-chapt16-1 by the XLEaP module. The solvated amyloid fibril was minimized by the method.

The LJ potential energy of atoms' vdW interactions is just a part of the total potential energy of a protein, and by observations from Model-chapt16-1 computed by Eqs. 16.30, 16.22∼16.23 and 16.31 we can see that the contacts between β-strand atoms and β-sheet atoms are too close/bad. Thus, we need to relax Model-chapt16-1 computed. The relaxation is done in the use of local search LBFGS Quasi-Newton method (lbfgs_memory_depth = 3) within AMBER 11 [91]. The relaxed/optimized Model-chapt16-1 is illuminated in Fig. 16.6c. Seeing Fig. 16.6d compared with Fig. 16.5b, we may know the vdW interactions between the two β-sheets are very relaxed/optimized now. Figure 16.6c shows the Model-chapt16-1 of optimal molecular structure for prion AGAAAAGA amyloid fibrils.

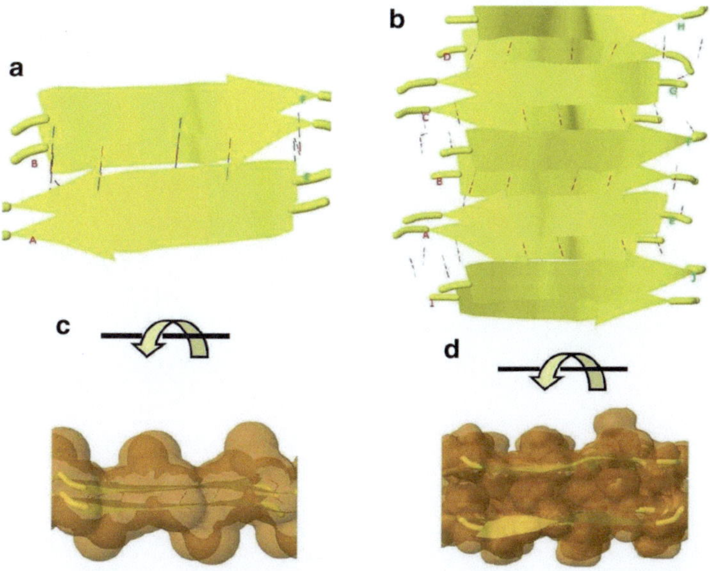

Fig. 16.6 *Left*: The close vdW surface contacts between Chains A(B) and E(F) of Model-chapt16-1 after solving Eqs. 16.25 and 16.27; *Right*: The constructed Model 1 with 10 chains. The *dashed lines* denote HBs

16.4 Conclusion

In a (macro)molecular system, a robust energy minimization is very necessarily required. Mathematical optimization minimization methods find a place to apply in these systems. Because in physics the (macro)molecular system usually is not a simple two-body problem of system, local search optimization methods are very useful in the applications to the (macro)molecular system. On anther sense, when a protein is unstable, noncrystalline or insoluble and very difficult to detect its 3D structure by the expensive and costly NMR and X-ray, theoretical mathematical or physical computational method can be used to produce the 3D structure of the protein. Moreover, even the X-ray crystallography finds the X-ray final structure of a protein, we still need refinements using theoretical protocols in order to produce a better structure. The theoretical computational method – an improved LBFGS Quasi-Newtonian mathematical optimization method – presented in this chapter and other mathematical optimization methods mentioned in this chapter should be very useful in the protein molecular modeling research field [693].

This chapter also shows the effectiveness of the improved LBFGS mathematical optimization method presented. Prion AGAAAAGA amyloid fibrils have not much structural information. This chapter presents some bioinformatics on the molecular structures of prion AGAAAAGA amyloid fibrils in the sense of theoretical emphasis. The structures may be helpful in the advance in the biochemical knowledge of prion protein misfolding or instability and in the future applications for therapeutic agent design.

Chapter 17
Particle Swarm Global Optimization Search Algorithm

17.1 A Brief Introduction to Particle-Swarm-Type Algorithms

Two optimization techniques employed in CALYPSO (www.calypso.cn) [265, 379, 393, 403, 405, 604, 605, 709, 719] are listed as follows. (1) As other successful heuristic algorithms, PSO [650] makes no or very few assumptions on the initial solutions(/particles) and the solution space usually is very large. (2) PSO can conquer energy barriers of local minimal solutions by PSO's swarm intelligence; locally searched PSO algorithm can easily deal with complex problems with many local energy barriers. A special case of PSO is the fireworks algorithm (FWA) for optimization. FWA consists of an explosive operator, a mutation operation, a mapping rule, and a selection strategy; it has good properties of explosion, instantaneity, simplicity, locality, emergent properties, distributed parallelism, diversity, scalability, and adaptability.

List the swarm in a tabu will produce the successful heuristic Tabu search (TS) algorithm. A TS is to impose a tabu list that stores a number of previously visited solutions to enable a search process to negotiate otherwise difficult regions. A practical TS procedure of is described as follows.

Step 1. Make an initial solution as the current solution.
Step 2. Evaluate the current solution. If it is the best so far, record it as best solution.
Step 3. Update the tabu list

  (a) If Tabu list not full, add current solution to the list.
  (b) Else, replace the oldest members with current solution.

Step 4. Generate and evaluation the number of (e.g. 1000) possible moves from the current solution.
Step 5. Rank all the moves in an order of the energy.

J. Zhang, *Molecular Structures and Structural Dynamics of Prion Proteins and Prions*, Focus on Structural Biology 9, DOI 10.1007/978-94-017-7318-8_17

Step 6. Examine all the moves in the rank order of *Step 5.*

(a) If move owns the lowest energy than the best solution so far, accept it and go to *Step 7.*
(b) If move is not in the tabu list, accept it and goto *Step 7.*
(c) If no acceptable move, exit.

Step 7. If the maximum number of iterations (e.g. 10000) is reached, exist with the best solution found; else goto *Step 2.* If the best solution has not changed for the maximum number of iterations, TS algorithm failed and change to other algorithms (e.g. Borwein reflection methods, Rubinov cutting angle global search methods, etc.).

TS algorithm has been used in MM, docking, quantitative structure-activity relationship (QSAR) etc. In protein folding, ant colony optimization (ACO) algorithms [667] have also been applied. The ACO algorithm is a member of PSO algorithms, based on the behavior of ants seeking a path between their colony and a source of food; ACO can be reduced to finding good paths through graphs of graph theory. Being similar to ACO, bee colony optimization (BCO) algorithm is an algorithm based on the intelligent foraging behavior of honey bee swarm; the colony consists of three groups of bees: employed bees, onlookers and scouts.

We also noticed the Cuckoo search, firefly algorithm, and bat algorithm etc. developed by Xin-she Yang (Cambridge University) et al. However, all the above heuristic algorithms have no strict mathematical proofs [10, 64, 78, 79, 155, 157, 249, 376, 416, 429, 475, 476, 539, 546, 668] of their algorithmic convergence.

Chapter 18
A Summary of the Research Works on AGAAAAGA

The highly conserved hydrophobic palindrome AGAAAAGA PrP 113–120 has been considered essential to PrP conformational conversion. First we give a survey of the research works on *prion AGAAAAGA* listed in the PubMed database (*http://www.ncbi.nlm.nih.gov/pubmed/*).

- Abskharon et al. [7] reported the full-length $HuPrP^C$ can adopt a more elaborate $\beta 0$-$\beta 1$-$\alpha 1$-$\beta 2$-$\alpha 2$-$\alpha 3$ structural organization than the canonical $\beta 1$-$\alpha 1$-$\beta 2$-$\alpha 2$-$\alpha 3$ prion-like fold. From this structure, it appears that the palindromic motif AGAAAAGA mediates β-enrichment in the PrP^C monomer as one of the early events in the conversion of PrP^C into PrP^{Sc} [7].
- The C-terminally-truncated human prion protein variant Y145Stop (i.e. HuPrP 23–144) is a valuable model for understanding the fundamental properties of amyloid formation. To examine the role of AGAAAAGA segment in fibrillization of PrP 23–144, a deletion variant Δ 113–120 PrP 23–144 (in which the palindrome sequence is missing) is used [322]. The deletion results in an altered amyloid β-core without affecting amyloidogenicity or seeding specificity; this concludes that the core of some amyloids contains "essential" (nucleation-determining) and "nonessential" regions, with the latter being "movable" in amino acid sequence space [322].
- The amyloid fibrils formed by the polypeptides of PrP 113–127, AGAAAAGAVVGGLGG, are taken as the model compound to investigate the biophysical principles governing the steric zipper amyloid fibril formation [102, 373, 517]. The target fibrils adopt the structural motif of Class 7 steric zipper, which is formed by stacking of antiparallel β-sheet layers with residue $117 + k$ forming backbone hydrogen bonds to residue $120 - k$ [102, 373, 517].
- Computer simulations of amyloid fibril formation by the Syrian hamster prion protein (SHaPrP) residues AGAAAAGA, the mouse prion protein (MoPrP) residues VAGAAAAGAV, and their variations GA_6G (a longer uninterrupted Alanine stretch flanked by Glycine), $(AG)_4$ (a complete disruption

J. Zhang, *Molecular Structures and Structural Dynamics of Prion Proteins and Prions*, Focus on Structural Biology 9, DOI 10.1007/978-94-017-7318-8_18

of hydrophobic residues), A_8, GAAAGAAA (a mimic of Aβ 29–36), A_{10}, V_{10}, GAVAAAAVAG (uninterrupted hydrophobic sequence), VAVAAAAVAV (less flexible than MoPrP 111–120) are studied in [594]. The first two peptides are thought to act as the velcro that holds the parent prion proteins together in amyloid structures and can form fibrils themselves [594].

- AGAAAAGA of HuPrP 105–210 is reported being involved into the oligomerization and engaged in intra- and/or inter-molecular interactions [514]. The SHaPrP 109–122 peptide near the AGAAAAGA region is observed to form steric zipper fibrils by the data of crystal solid-state NMR and MD [373].
- The cellular isoform of the prion protein PrP^C is located at the cell membrane. Studies have shown that exposure of cells to copper (Cu) causes internalisation of PrP^C in vitro, and deletion mutation studies have shown that the palindromic region, amino acids 113–120 with the sequence AGAAAAGA is essential for copper-induced internalisation to occur [267]. Kourie et al. [352] studied the copper modulation of ion channels of PrP 106–126 mutant prion peptide fragments and found that the hydrophobic core AGAAAAGA is not a Cu^{2+}-binding site (but Met 109 and His 111 can form ion channels) [352].
- The AGAAAAGA palindrome in PrP is required not only for the attainment of the PrP^{Sc} conformation but also to generate a productive PrP^{Sc}-PrP^C complex that leads to the propagation of PrP^{Sc} [445]. In contrast to wild type (wt) PrP, PrP lacking the palindrome (PrP Δ 112–119) neither converted to PrP^{Sc} nor generated proteinase K-resistant PrP. In [445], we also know that synthetic peptides corresponding to the so-called "toxic peptide" PrP 106–126 segment form fibrils in solution with β-sheet structure, suggesting that PrP 106–126 segment may feature in PrP^{Sc}-PrP^C associations, and larger peptides of PrP 107–142 antagonize the in vitro conversion of PrP^C to the protease-resistant state in a cell-free conversion model.
- Zanuy et al. [674] reported that, for AGAAAAGA, the antiparallel strand orientation is preferred within the sheets but the parallel orientation is preferred between sheets [408, 674]. AGAAAAGA is one of the most highly amyloidogenic peptides and oligomers of AGAAAAGA were found to be stable when the size is 6–8 (hexamer to octamer) [408]. Here the AGAAAAGA model of [408] used for their MD is a homology structure.
- In [619], the following bioinformatics on AGAAAAGA is known. AGAAAAGA displays the highest tendency to form amyloid. Peptides containing AGAAAAGA are toxic to neurons in culture, whereby the sequence AGAAAAGA was found to be necessary but not sufficient for the neurotoxic effect [67]. The synthetic peptides derived from the central part of PrP^C 106–141 have an inhibitory effect on this conversion of PrP^C to PK-resistant PrP^{Sc}. The presence of residues 119 and 120 (the two last residues within the motif AGAAAAGA) seems to be crucial for this inhibitory effect. Mutant PrP molecules carrying deletions of amino acids 108–121 or 114–121 are not convertible to PrP^{Sc}; therefore, the central hydrophobic region, spanning all or most of the sequence AGAAAAGA, plays an important role in the

PrP^{Sc}-PrP^{C} conversion process. Wegner et al. [619] assessed the effect of mutations at and around the AGAAAAGA hydrophobic sequence on protease resistance and found that mutations in the central AGAAAAGA hydrophobic region lead to immediate alterations in PrP structure and processing [619].

- The amyloidogenic and hydrophobic core AGAAAAGA has been implicated in modulation of neurotoxicity and the secondary structure of PrP 106–126 [316, 351], which is dependent on the formation of aggregated fibril structures regulated by the AGAAAAGA core [351].
- Brown [67] reported that AGAAAAGA blocks the toxicity of PrP 106–126, suggesting that this sequence is necessary (but insufficient) for the interaction of PrP 106–126 with neurons and targeting or use of the AGAAAAGA peptide may represent a therapeutic opportunity for controlling prion disease [67].
- HuPrP 106–126 has been shown to be highly fibrillogenic and toxic to neurons in vitro. Jobling et al. [316] found that the AGAAAAGA 113–120 hydrophobic core sequence is important for PrP 106–126 toxicity probably by influencing its assembly into a neurotoxic structure and the hydrophobic sequence may similarly affect aggregation and toxicity observed in prion diseases [316].
- Chabry et al. [95] reported that peptides from the central part of the hamster PrP 106–141 (where residues in the vicinity of positions 106–141 of PrP^{Sc} and/or PrP^{C} are critically involved in the intermolecular interactions that lead to PrP^{Sc} formation) could completely inhibit the conversion induced by preformed PrP^{Sc} and the presence of residues 119 and 120 from the highly hydrophobic sequence AGAAAAGA (PrP 113–120) was crucial for an efficient inhibitory effect [95].
- Holscher et al. [288] reported that the presence of AGAAAAGA of mouse PrP^{C} plays an important role in the conversion process of PrP^{C} into PrP^{Sc} and that a deletion mutant lacking these codons indeed behaves as a dominant-negative mutant with respect to PrP^{Sc} accumulation [288].
- PrP AGAAAAGA is the most highly amyloidogenic peptide and is conserved across all species [237]. Gasset et al. [237] reported there are similarities between the PrP sequence AGAAAAGA and that of silkworm fibroin, and the homology between PrP sequence AGAAAAGAVVGGLGG and that of spider fibroin [237]. Thus, we may say that the hydrophobic region of PrP 109–136 should be a region to form β-sheets and amyloid polymers, instead of α-helices of the Garnier-Robson analysis [234].

Seeing the above survey, we know that prion AGAAAAGA peptide has been reported to own an amyloid fibril property (initially described in 1992 by Gasset et al. of Prusiner's Group). However, there has not been traditional X-ray or NMR experimental structural bioinformatics for this octapeptide yet, due to the unstable, noncrystalline and insoluble nature of this region, which just falls within the N-terminal unstructured region of prion proteins. Studies on atomic-resolution structures of the AGAAAAGA peptide will prove useful in future experimental studies on this region, aspects of the structure or the dynamics of this region should play a role in the aggregation process, and knowledge of these may be useful for the goals of medicinal chemistry for controlling prion diseases. Zhang [680] successfully constructed three amyloid fibril models (denoted as Models 1–3) for the PrP 113–120 AGAAAAGA region.

Model 1 and Models 2–3 belong to Class 7 and 1 of [517] respectively, i.e. Model 1 is β-strand antiparallel, face = back, up-up (Fig. 18.1) (where numerical results show the agreement with [102, 408]), and Models 2–3 are β-strand parallel, face-to-face, up-up (Figs. 18.2 and 18.3). In all these models, there is about 5 Å between the two closest adjacent β-sheets, maintained by HYDs, and about 4.5 Å between the two closest adjacent β-strands, which are linked by HBs. Computational approaches of global optimization, local search EM, SA and structural bioinformatics etc. or introducing novel mathematical formulations and physical concepts into molecular biology may allow us to obtain a description of the protein 3D structure at a submicroscopic level for prion AGAAAAGA amyloid fibrils [680, 681, 683, 685, 689, 698, 705].

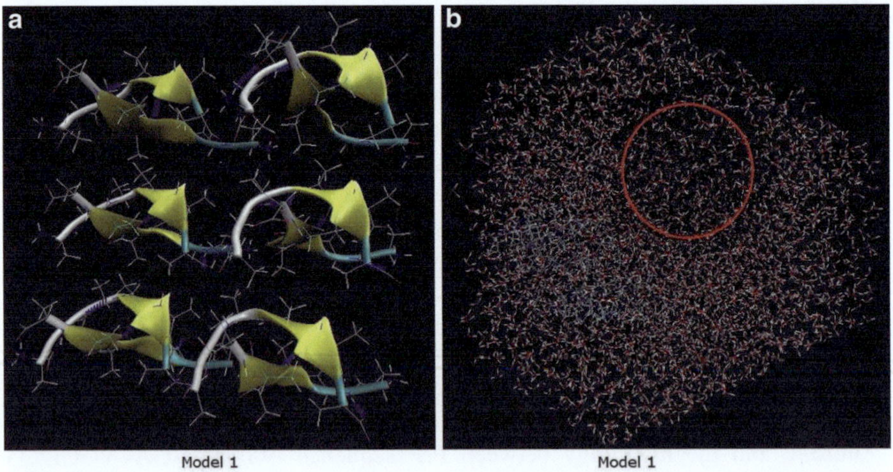

Fig. 18.1 Model 1. Model 1(**a**): *Blue dashed lines* denote the HBs between the pairs of β-strands. Model 1(**b**): The *red circle* denotes there are very few water molecules in the truncated octahedral box of TIP3P waters

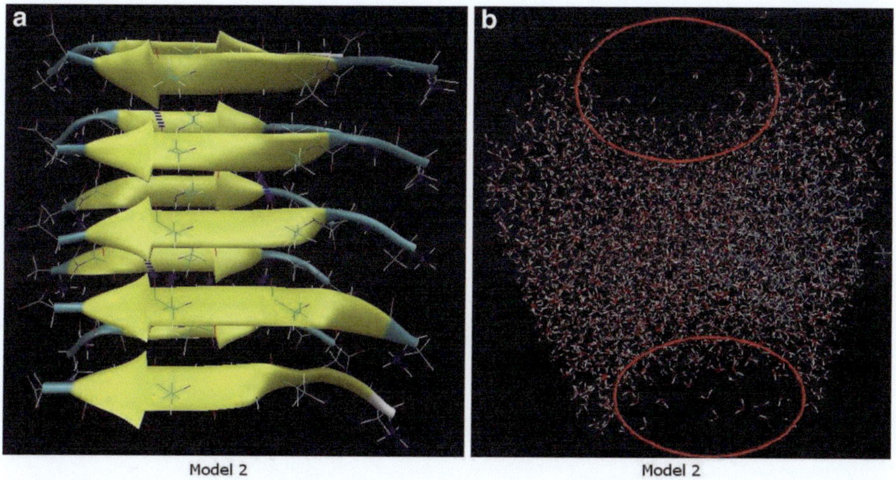

Fig. 18.2 Model 2. Model 2(**a**): *Blue dashed lines* denote the HBs between the pairs of β-strands. Model 2(**b**): The *red circles* denote there are very few water molecules in the truncated octahedral box of TIP3P waters

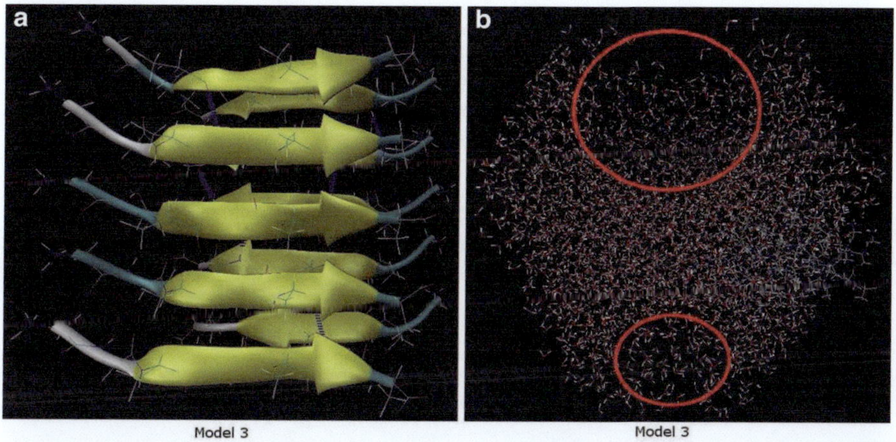

Fig. 18.3 Model 3. Model 3(**a**): *Blue dashed lines* denote the HBs between the pairs of β-strands. Model 3(**b**): The *red circles* denote there are very few water molecules in the truncated octahedral box of TIP3P waters

References

1. Aarts E, Korst J (1989) Simulated annealing and Boltzmann machines. Wiley, Chichester/New York. ISBN:978-0-471-92146-2
2. Aarts EHL, Laarhoven PJMV (1985) Statistical cooling: a general approach to combinatorial optimization problems. Philips J Res 40(4):193–226
3. Abagyan RA, Maiorov VN (1988) A simple quantitative representation of polypeptide chain folds: comparison of protein tertiary structures. J Biomol Struct Dyn 5(6):1267–1279
4. Abbass HA, Bagirov AM, Zhang JP (2003) The discrete gradient evolutionary strategy method for global optimization. In: The IEEE congress on evolutionary computation (CEC2003), Canberra, vol 1. IEEE-Press, pp 435–442
5. Abbass HA, Newton CS, Sarker R (2002) Heuristic and optimization for knowledge discovery. Idea Group Publishing, Hershey
6. Abraham MJ, van der Spoel D, Lindahl E, Hess B, GROMACS development team (2014) GROMACS User manual version 5.0.4. www.gromacs.org
7. Abskharon RN, Giachin G, Wohlkonig A, Soror SH, Pardon E, Legname G, Steyaert J (2014) Probing the N-terminal β-sheet conversion in the crystal structure of the human prion protein bound to a nanobody. J Am Chem Soc 136(3):937–944
8. Aguzzi A, Heikenwalder M (2006) Pathogenesis of prion diseases: current status and future outlook. Nat Rev Microbiol 4(10):765–775
9. Ahn I, Son HS (2007) Comparative bioinformatics analysis of prion proteins isolated from reptile, rodent, ruminant, and human species. Exp Mol Med 39(6):769–777
10. Ai YS, Liu PC, Zheng TY (1998) Adaptive hybrid global inversion algorithm. Sci China (Series D) 41(2):137–143
11. Al-Baali M (1985) Descent property and global convergence of the Fletcher-Reeves method with inexact line search. IMA J Numer Anal 5(1):121–124
12. Al-Baali M (1999) Improved Hessian approximations for the limited memory BFGS method. Numer Algorithms 22(1):99–112
13. Alonso DO, DeArmond SJ, Cohen FE, Daggett V (2001) Mapping the early steps in the pH-induced conformational conversion of the prion protein. Proc Natl Acad Sci U S A 98(6):2985–2989
14. Alper T, Cramp W, Haig D, Clarke M (1967) Does the agent of scrapie replicate without nucleic acid? Nature 214(5090):764–766
15. Andraos J (2008) Kinetic plasticity and the determination of product ratios for kinetic schemes leading to multiple products without rate laws: new methods based on directed graphs. Can J Chem 86(4):342–357

J. Zhang, *Molecular Structures and Structural Dynamics of Prion Proteins and Prions*, Focus on Structural Biology 9, DOI 10.1007/978-94-017-7318-8

16. Andrieu C, De Freitas N, Doucet A, Jordan MI (2003) An Introduction to MCMC for machine learning. Mach Learn 50(1–2):5–43
17. Apostol MI, Sawaya MR, Cascio D, Eisenberg D (2010) Crystallographic studies of prion protein (PrP) segments suggest how structural changes encoded by polymorphism at residue 129 modulate susceptibility to human prion disease. J Biol Chem 285(39):29671–29675
18. Apostol MI, Wiltzius JJW, Sawaya MR, Cascio D, Eisenberg D (2011) Atomic structures suggest determinants of transmission barriers in mammalian prion disease. Biochem 50(13):2456–2463
19. Aqvist J (1986) A simple way to calculate the axes of an α-helix. Comput Chem 10(2):97–99
20. Arfken G (1985) The method of steepest descents. §7.4 in mathematical methods for physicists, 3rd edn. Academic, Orlando, pp 428–436
21. Bagirov AM (2003) Continuous subdifferential approximations and their applications. J Math Sci 15(5):2567–2609
22. Bagirov AM, Karasözen B, Sezer M (2008) Discrete gradient method: derivative-free method for nonsmooth optimization. J Optim Theory Appl 137(2):317–334
23. Bagirov A, Karmitsa N, Mäkelä MM (2014) Introduction to nonsmooth optimization – theory, practice and software. Springer, London/New York. ISBN:978-3-319-08113-7
24. Bagirov AM, Zhang JP (2003) Comparative analysis of the cutting angle and simulated annealing methods in global optimization. Optimization 52(4–5):363–378
25. Baldi P, Brunak S (2001) Bioinformatics: the machine learning approach, 2nd edn. MIT, Cambridge. Chapter 4.5
26. Baral PK, Swayampakula M, Rout MK, Kav NN, Spyracopoulos L, Aguzzi A, James MN (2014) Structural basis of prion inhibition by phenothiazine compounds. Structure 22(2):291–303
27. Barducci A, Chelli R, Procacci P, Schettino V (2005) Misfolding pathways of the prion protein probed by molecular dynamics simulations. Biophys J 88(2):1334–1343
28. Barducci A, Chelli R, Procacci P, Schettino V, Gervasio FL, Parrinello M (2006) Metadynamics simulation of prion protein: β-structure stability and the early stages of misfolding. J Am Chem Soc 128(8):2705–2710
29. Barlow RM, Rennie JC (1976) The fate of ME7 scrapie infection in rats, guinea-pigs and rabbits. Res Vet Sci 21(1):110–111
30. Barnham KJ, Cappai R, Beyreuther K, Masters CL, Hill AF (2006) Delineating common molecular mechanisms in Alzheimer's and prion diseases. Trends Biochem Sci 31(8):465–472
31. Baron H, Baron-Van Evercooren A, Brucher JM (1988) Antiserum to scrapie-associated fibril protein reacts with amyloid plaques in familial transmissible dementia. J Neuropathol Exp Neurol 47(2):158–165
32. Barron C, Gomez S, Romero D, Saavedra A (1999) A genetic algorithm for Lennard-Jones atomic clusters. Appl Math Lett 12(7):85–90
33. Barron RM, Thomson V, Jamieson E, Melton DW, Ironside J, Will R, Manson JC (2001) Changing a single amino acid in the N-terminus of murine PrP alters TSE incubation time across three species barriers. EMBO J 20(18):5070–5078
34. Barron RM, Thomson V, King D, Shaw J, Melton DW, Manson JC (2003) Transmission of murine scrapie to P101L transgenic mice. J Gen Virol 84(Pt 11):3165–3172
35. Barron-Romero C (2005) Minimum search space and efficient methods for structural cluster optimization. arXiv0504030v5
36. Barry RA, Kent SB, McKinley MP, Meyer RK, DeArmond SJ, Hood LE, Prusiner SB (1986) Scrapie and cellular prion proteins share polypeptide epitopes. J Infect Dis 153(5):848–854
37. Barry RA, McKinley MP, Bendheim PE, Lewis GK, DeArmond SJ, Prusiner SB (1985) Antibodies to the scrapie protein decorate prion rods. J Immunol 135(1):603–613
38. Barry RA, Vincent MT, Kent SB, Hood LE, Prusiner SB (1988) Characterization of prion proteins with monospecific antisera to synthetic peptides. J Immunol 140(4):1188–1193
39. Bastian FO, Sanders DE, Forbes WA, Hagius SD, Walker JV, Henk WG, Enright FM, Elzer PH (2007) Spiroplasma spp. from transmissible spongiform encephalopathy brains or ticks induce spongiform encephalopathy in ruminants. J Med Microbiol 56(Pt 9):1235–1242

40. Baylis M, Goldmann W (2004) The genetics of scrapie in sheep and goats. Curr Mol Med 4(4):385–396
41. Bazan JF, Fletterick RJ, McKinley MP, Prusiner SB (1987) Predicted secondary structure and membrane topology of the scrapie prion protein. Protein Eng 1(2):125–135
42. Beliakov G, Monsalve Tobon JE, Bagirov AM (2003) Parallelization of the discrete gradient method of non-smooth optimization and its applications. In: Computational science – ICCS 2003. Lecture notes in computer science, vol 2659/2003. Springer, Berlin/New York, pp 592–601
43. Bencsik A, Lezmi S, Hunsmann G, Baron T (2001) Close vicinity of PrP expressing cells (FDC) with noradrenergic fibers in healthy sheep spleen. Dev Immunol 8(3–4):235–241
44. Bendheim PE, Barry RA, DeArmond SJ, Stites DP, Prusiner SB (1984) Antibodies to a scrapie prion protein. Nature 310(5976):418–421
45. Berg JM, Tymoczko J, Stryer L (2012) Biochemistry, 7th edn. W H Freeman Publisher, New York
46. Beringue V, Vilotte JL, Laude H (2008) Prion agent diversity and species barrier. Vet Res 39(4):47–76
47. Berman HM, Westbrook J, Feng Z, Gilliland G, Bhat TN, Weissig H, Shindyalov IN, Bourne PE (2000) The protein data bank. Nucl Acids Res 28(1):235–242
48. Berndt EK, Hall BH, Hall RE, Hausman J (1974) Estimation and inference in nonlinear structural models. Ann Econ Soc Meas 3(4):653–665
49. Berti F, Gaggelli E, Guerrini R, Janicka A, Kozlowski H, Legowska A, Miecznikowska H, Migliorini C, Pogni R, Remelli M, Rolka K, Valensin D, Valensin G (2007) Structural and dynamic characterization of copper(II) binding of the human prion protein outside the octarepeat region. Chemistry 13(7):1991–2001
50. Bett C, Fernández-Borges N, Kurt TD, Lucero M, Nilsson KP, Castilla J, Sigurdson CJ (2012) Structure of the $\beta 2$-$\alpha 2$ loop and interspecies prion transmission. FASEB J 26(7):2868–2876
51. Bhandarkar M, Bhatele A, Bohm E, Brunner R et al (2012) NAMD user's guide version 2.9, University of Illinois and Beckman Institute, Urbana
52. Biasini E, Tapella L, Restelli E, Pozzoli M, Massignan T, Chiesa R (2010) The hydrophobic core region governs mutant prion protein aggregation and intracellular retention. Biochem J 430(3):477–486
53. Bilbro GL, Snyder WE (1991) Optimization of functions with many minima. IEEE Trans Syst Man Cyb 21(4):840–849
54. Biljan I, Giachin G, Ilc G, Zhukov I, Plavec J, Legname G (2012) Structural basis for the protective effect of the human prion protein carrying the dominant-negative E219K polymorphism. Biochem J 446(2):243–251
55. Biljan I, Ilc G, Giachin G, Legname G, Plavec J (2012) Structural rearrangements at physiological pH: nuclear magnetic resonance insights from the V210I human prion protein mutant. Biochemistry 51(38):7465–7474
56. Biljan I, Ilc G, Giachin G, Raspadori A, Zhukov I, Plavec J, Legname G (2011) Toward the molecular basis of inherited prion diseases: NMR structure of the human prion protein with V210I mutation. J Mol Biol 412(4):660–673
57. Biswas S, Langeveld JP, Tipper D, Lu S (2006) Intracellular accumulation of a 46 kDa species of mouse prion protein as a result of loss of glycosylation in cultured mammalian cells. Biochem Biophys Res Commun 349(1):153–161
58. Bitel CL, Feng Y, Souayah N, Frederikse PH (2010) Increased expression and local accumulation of the prion protein, Alzheimer ASS peptides, superoxide dismutase 1, and nitric oxide synthases 1 & 2 in muscle in a rabbit model of diabetes. BMC Physiol 10:18. doi:10.1186/1472-6793-10-18
59. Bjorndahl TC, Zhou GP, Liu XH, Pineiro RP, Semenchenko V, Saleem F, Acharya S, Bujold A, Sobsey CA, Wishart DS (2011) Detailed biophysical characterization of the acid-induced PrP^{C} to PrP^{Sc} conversion process. Biochemistry 50(7):1162–1173
60. Bockman JM, Prusiner SB, Tateishi J, Kingsbury DT (1987) Immunoblotting of Creutzfeldt-Jakob disease prion proteins: host species-specific epitopes. Ann Neurol 21(6):589–595

61. Bode L, Pocchiari M, Gelderblom H, Diringer H (1985) Characterization of antisera against scrapie-associated fibrils (SAF) from affected hamster and cross-reactivity with SAF from scrapie-affected mice and from patients with Creutzfeldt-Jakob disease. J Gen Virol 66(Pt 11):2471–2478
62. Bohachevsky IO, Johnson ME, Stein ML (1986) Generalized simulated annealing for function optimization. Technometrics 28(3):209–217
63. Bootsma MCJ, Ferguson NM (2007) The effect of public health measures on the 1918 influenza pandemic in U.S. cities. Proc Natl Acad Sci U S A 104(18):7588–7593
64. Branch MA, Coleman TF, Li Y (1999) A subspace, interior, and conjugate gradient method for large-scale bound-constrained minimization problems. SIAM J Sci Comput 21(1):1–23
65. Branden C, Tooze J (1999) Introduction to protein structure, 2nd edn. Garland Science Publishing, New York
66. Brooks DG, Verdini WA (1988) Computational experience with generalized simulated annealing over continuous variables. Am J Math Manag Sci 8(3–4):425–449
67. Brown DR (2000) Prion protein peptides: optimal toxicity and peptide blockade of toxicity. Mol Cell Neurosci 15(1):66–78
68. Brown DR (2001) Microglia and prion disease. Microsc Res Tech 54(2):71–80
69. Brown D (2001) The 'recipe for disaster' that killed 80 and left a £5bn bill. The Daily Telegraph (London). Retrieved May 2015. http://www.telegraph.co.uk/news/uknews/1371964/The-recipe-for-disaster-that-killed-80-and-left-a-5bn-bill.html
70. Brown DR, Herms J, Kretzschmar HA (1994) Mouse cortical cells lacking cellular PrP survive in culture with a neurotoxic PrP fragment. Neuroreport 5(16):2057–2060
71. Brun A, Castilla J, Ramírez MA, Prager K, Parra B, Salguero FJ, Shiveral D, Sánchez C, Sánchez-Vizcaíno JM, Douglas A, Torres JM (2004) Proteinase K enhanced immunoreactivity of the prion protein-specific monoclonal antibody 2A11. Neurosci Res 48(1):75–83
72. Burkowski FJ (2009) Structural bioinformatics: an algorithmic approach. Chapman and Hall/CRC, Boca Raton. ISBN:978-1-5848-8683-9 (Hardcover)
73. Byrd RH, Lu P, Nocedal J (1995) A limited memory algorithm for bound constrained optimization. SIAM J Sci Stat Comput 16(5):1190–1208
74. Byrd RH, Nocedal J (1989) A tool for the analysis of quasi-Newton methods with application to unconstrained minimization. SIAM J Numer Anal 26(3):727–739
75. Byrd RH, Nocedal J, Schnabel RB (1994) Representations of quasi-Newton matrices and their use in limited memory methods. Math Program 63(1–3):129–156
76. Byrd RH, Nocedal J, Yuan YX (1987) Global convergence of a class of quasi-Newton methods on convex problems. SIAM J Numer Anal 24(5):1171–1189
77. Byrd RH, Nocedal J, Zhu C (1995) Towards a discrete Newton method with memory for large-scale optimization. Optimization Technology Center Report OTC-95-1, EEC Department, North-western University
78. Byrd RH, Schnabel RB, Schultz GA (1987) A trust region algorithm for nonlinearly constrained optimization. SIAM J Numer Anal 24(5):1152–1170
79. Byrd RH, Schnabel RB, Shultz GA (1988) Approximate solution of the trust region problem by minimization over two-dimensional subspaces. Math Program 40(1–3):247–263
80. Cai L, Wang Y, Wang JF, Chou KC (2011) Identification of proteins interacting with human SP110 during the process of viral infections. Med Chem 7(2):121–126
81. Call ME, Schnell JR, Xu C, Lutz RA, Chou JJ, Wucherpfennig KW (2006) The structure of the zetazeta transmembrane dimer reveals features essential for its assembly with the T cell receptor. Cell 127(2):355–368
82. Call ME, Wucherpfennig KW, Chou JJ (2010) The structural basis for intramembrane assembly of an activating immunoreceptor complex. Nat Immunol 11(11):1023–1029
83. Calzolai L, Lysek DA, Guntert P, Von Schroetter C, Zahn R, Riek R, Wuthrich K (2000) NMR structures of three single-residue variants of the human prion protein. Proc Natl Acad Sci U S A 97(15):8340–8345

84. Calvo F, Doye JPK, Wales DJ (2001) Quantum partition functions from classical distributions. Application to rare-gas clusters. J Chem Phys 114(17):7312–7329
85. Cappai R, Collins SJ (2004) Structural biology of prions. In: Rabenau HF, Cinatl J, Doerr HW (eds) Prions – a challenge for science, medicine and the public health system. Contributions to microbiology, vol 11. Karger, Basel, pp 14–32
86. Carter DB, Chou KC (1998) A model for structure dependent binding of Congo Red to Alzeheimer beta-amyloid fibrils. Neurobiol Aging 19(1):37–40
87. Case DA, Babin V, Berryman JT, Betz RM, Cai Q, Cerutti DS, Cheatham TE, Darden III TA, Duke RE, Gohlke H, Goetz AW, Gusarov S, Homeyer N, Janowski P, Kaus J, Kolossváry I, Kovalenko A, Lee TS, LeGrand S, Luchko T, Luo R, Madej B, Merz KM, Paesani F, Roe DR, Roitberg A, Sagui C, Salomon-Ferrer R, Seabra G, Simmerling CL, Smith W, Swails J, Walker RC, Wang J, Wolf RM, Wu X, Kollman PA (2014) AMBER 14, University of California, San Francisco
88. Case DA, Darden TA, Caldwell JW, Cheatham III TE, Simmerling CL, Wang J, Duke RE, Luo R, Merz KM, Pearlman DA, Crowley M, Walker RC, Zhang W, Wang B, Hayik S, Roitberg A, Seabra G, Wong KF, Paesani F, Wu X, Brozell S, Tsui V, Gohlke H, Yang L, Tan C, Mongan J, Hornak V, Cui G, Beroza P, Mathews DH, Schafmeister C, Ross WS, Kollman PA (2006) AMBER 9, University of California, San Francisco
89. Case DA, Darden TA, Cheatham III TE, Simmerling CL, Wang J, Duke RE, Luo R, Crowley M, Walker RC, Zhang W, Merz KM, Wang B, Hayik S, Roitberg A, Seabra G, Kolossváry I, Wong KF, Paesani F, Vanicek J, Wu X, Brozell SR, Steinbrecher T, Gohlke H, Yang L, Tan C, Mongan J, Hornak V, Cui G, Mathews DH, Seetin MG, Sagui C, Babin V, Kollman PA (2008) AMBER 10, University of California, San Francisco
90. Case DA, Darden TA, Cheatham TE, Simmerling III CL, Wang J, Duke RE, Luo R, Merz KM, Wang B, Pearlman DA, Crowley M, Brozell S, Tsui V, Gohlke H, Mongan J, Hornak V, Cui G, Beroza P, Schafmeister C, Caldwell JW, Ross WS, Kollman PA (2004) AMBER 8, University of California, San Francisco
91. Case DA, Darden TA, Cheatham TE, Simmerling III CL, Wang J, Duke RE, Luo R, Walker RC, Zhang W, Merz KM, Roberts BP, Wang B, Hayik S, Roitberg A, Seabra G, Kolossváry I, Wong KF, Paesani F, Vanicek J, Liu J, Wu X, Brozell SR, Steinbrecher T, Gohlke H, Cai Q, Ye X, Wang J, Hsieh MJ, Cui G, Roe DR, Mathews DH, Seetin MG, Sagui C, Babin V, Luchko T, Gusarov S, Kovalenko A, Kollman PA (2010) AMBER 11, University of California, San Francisco
92. Case DA, Pearlman DA, Caldwell JW, Cheatham III TE, Wang J, Ross WS, Simmerling CL, Darden TA, Merz KM, Stanton RV, Cheng AL, Vincent JJ, Crowley M, Tsui V, Gohlke H, Radmer RJ, Duan Y, Pitera J, Massova I, Seibel GL, Singh UC, Weiner PK, Kollman PA (2002) AMBER 7, University of California, San Francisco
93. Caughey B, Race R, Vogel M, Buchmeier M, Chesebro B (1988) In vitro expression of cloned PrP cDNA derived from scrapie-infected mouse brain: lack of transmission of scrapie infectivity. Ciba Found Symp 135:197–208 (chapter 13)
94. Cerný V (1985) Thermodynamical approach to the traveling salesman problem: an efficient simulation algorithm. J Optim Theor Appl 45(1):41–51
95. Chabry J, Caughey B, Chesebro B (1998) Specific inhibition of in vitro formation of protease-resistant prion protein by synthetic peptides. J Biol Chem 273(21):13203–13207
96. Chan CH, Yu TH, Wong KB (2011) Stabilizing salt-bridge enhances protein thermostability by reducing the heat capacity change of unfolding. PLoS ONE 6(6):e21624
97. Chen HF (2009) Aggregation mechanism investigation of the GIFQINS cross-β amyloid fibril. Comput Biol Chem 33(1):41–45
98. Chen PY, Lin CC, Chang YT, Lin SC, Chan SI (2002) One o-linked sugar can effect the coil-to-β structural transition of the prion peptide. Proc Natl Acad Sci U S A 99(20):12633–12638
99. Chen C, Chen L, Zou X, Cai P (2009) Prediction of protein secondary structure content by using the concept of Chou's pseudo amino acid composition and support vector machine. Protein Peptide Lett 16(1):27–31

100. Chen W, van der Kamp MW, Daggett V (2010) Diverse effects on the native β-sheet of the human prion protein due to disease-associated mutations. Biochem 49(45):9874–9881
101. Cheng CJ, Daggett V (2014) Molecular dynamics simulations capture the misfolding of the bovine prion protein at acidic pH. Biomolecules 4(1):181–201
102. Cheng HM, Tsai TWT, Huang WYC, Lee HK, Lian HY, Chou FC, Mou Y, Chu J, Chan JC (2011) Steric zipper formed by hydrophobic peptide fragment of Syrian hamster prion protein. Biochem 50(32):6815–6823
103. Chiang PK, Lam MA, Luo Y (2008) The many faces of amyloid beta in Alzheimer's disease. Curr Mol Med 8(6):580–584
104. Christen B, Hornemann S, Damberger FF, Wüthrich K (2009) Prion protein NMR structure from tammar wallaby (Macropus eugenii) shows that the $\beta 2$-$\alpha 2$ loop is modulated by long-range sequence effects. J Mol Biol 389(5):833–845
105. Christen B, Hornemann S, Damberger FF, Wüthrich K (2012) Prion protein mPrP [F175A](121–231): structure and stability in solution. J Mol Biol 423(4):496–502
106. Chianini F, Fernández-Borges N, Vidal E, Gibbard L, Pintado B, de Castro J, Priola SA, Hamilton S, Eatona LS, Finlayson J, Pang Y, Steele P, Reid HW, Dagleish MP, Castilla J(2012) Rabbits are not resistant to prion infection. Proc Natl Acad Sci U S A 109(13):5080–5085
107. Cho HJ (1986) Antibody to scrapie-associated fibril protein identifies a cellular antigen. J Gen Virol 67(Pt 2):243–253
108. Choi JK, Park SJ, Jun YC, Oh JM, Jeong BH, Lee HP, Park SN, Carp RI, Kim YS (2006) Generation of monoclonal antibody recognized by the GxxxG motif (Glycine zipper) of prion protein. Hybridoma (Larchmt) 25(5):271–277
109. Chou KC (1985) Low-frequency motions in protein molecules: beta-sheet and beta-barrel. Biophys J 48(2):289–297
110. Chou KC (1988) Review: low-frequency collective motion in biomacromolecules and its biological functions. Biophys Chem 30(1):3–48
111. Chou KC (1989) Low-frequency resonance and cooperativity of hemoglobin. Trends Biochem Sci 14(6):212–213
112. Chou KC (1992) Energy-optimized structure of antifreeze protein and its binding mechanism. J Mol Biol 223(2):509–517
113. Chou KC (1989) Graphic rules in steady and non-steady enzyme kinetics. J Biol Chem 264(20):12074–12079
114. Chou KC (1990) Review: applications of graph theory to enzyme kinetics and protein folding kinetics. Steady and non-steady state systems. Biophys Chem 35(1):1–24
115. Chou KC (1995) A novel approach to predicting protein structural classes in a (20–1)-D amino acid composition space. PROTEINS: Struct, Funct Gene 21(4):319–344
116. Chou KC (2001) Prediction of protein cellular attributes using pseudo-amino acid composition. PROTEINS: Struct, Funct, Gene (Erratum: ibid., 2001, Vol 44, 60) 43(3):246–255
117. Chou KC (2004) Molecular therapeutic target for type-2 diabetes. J Proteome Res 3(6):1284–1288
118. Chou KC (2004) Insights from modelling the tertiary structure of BACE2. J Proteome Res 3(5):1069–1072
119. Chou KC (2004) Modelling extracellular domains of GABA-A receptors: subtypes 1, 2, 3, and 5. Biochem Biophys Res Commun 316(3):636–642
120. Chou KC (2004) Review: structural bioinformatics and its impact to biomedical science. Curr Med Chem 11(16):2105–2134
121. Chou KC (2004) Insights from modelling the 3D structure of the extracellular domain of alpha7 nicotinic acetylcholine receptor. Biochem Biophy Res Co 319(2):433–438
122. Chou KC (2005) Modeling the tertiary structure of human cathepsin-E. Biochem Biophys Res Commun 331(1):56–60
123. Chou KC (2005) Coupling interaction between thromboxane A2 receptor and alpha-13 subunit of guanine nucleotide-binding protein. J Proteome Res 4(5):1681–1686
124. Chou KC (2010) Graphic rule for drug metabolism systems. Curr Drug Metab 11(4):369–378

125. Chou KC (2011) Some remarks on protein attribute prediction and pseudo amino acid composition (50th anniversary year review). J Theor Biol 273(1):236–247
126. Chou KC, Carlacci L (1991) Simulated annealing approach to the study of protein structures. Protein Eng 4(6):661–667
127. Chou KC, Carlacci L (1991) Energetic approach to the folding of alpha/beta barrels. Proteins: Struct, Funct, Genet 9(4):280–295
128. Chou KC, Carlacci L, Maggiora GM (1990) Conformational and geometrical properties of idealized beta-barrels in proteins. J Mol Biol 213(2):315–326
129. Chou KC, Chen NY (1977) The biological functions of low-frequency phonons. Sci Sin 20:447–457
130. Chou KC, Howe WJ (2002) Prediction of the tertiary structure of the beta-secretase zymogen. Biochem Biophys Res Commun 292(3):702–708
131. Chou KC, Maggiora GM, Scheraga HA (1992) The role of loop-helix interactions in stabilizing four-helix bundle proteins. Proc Natl Acad Sci U S A 89(16):7315–7319
132. Chou KC, Nemethy G, Rumsey S, Tuttle RW, Scheraga HA (1986) Interactions between two beta-sheets: energetics of beta/beta packing in proteins. J Mol Biol 188(4):641–649
133. Chou KC, Nemethy G, Scheraga HA (1983) Effects of amino acid composition on the twist and the relative stability of parallel and antiparallel beta-sheets. Biochemistry 22(26):6213–6221
134. Chou KC, Nemethy G, Scheraga, HA (1983) Role of interchain interactions in the stabilization of right-handed twist of β-sheets. J Mol Biol 168(2):389–407
135. Chou KC, Nemethy G, Scheraga HA (1990) Review: energetics of interactions of regular structural elements in proteins. Acc Chem Res 23(5):134–141
136. Chou KC, Pottle M, Nemethy G, Ueda Y, Scheraga HA (1982) Structure of beta-sheets: origin of the right-handed twist and of the increased stability of antiparallel over parallel sheets. J Mol Biol 162(1):89–112
137. Chou KC, Scheraga HA (1982) Origin of the right-handed twist of beta-sheets of poly-L-valine chains. Proc Natl Acad Sci U S A 79(22):7047–7051
138. Chou KC, Shen HB (2009) Review: recent advances in developing web-servers for predicting protein attributes. Nat Sci 1(2):63–92
139. Chou KC, Wei DQ, Zhong WZ (2003) Binding mechanism of coronavirus main proteinase with ligands and its implication to drug design against SARS (Erratum: ibid., 2003, Vol 310, 675). Biochem Biophys Res Comm 308(1):148–151
140. Chou KC, Wu ZC, Xiao X (2011) iLoc-Euk: a multi-label classifier for predicting the subcellular localization of singleplex and multiplex eukaryotic proteins. PLoS ONE 6(3):e18258
141. Chou KC, Zhang CT (1995) Review: prediction of protein structural classes. Crit Rev Biochem Mol Biol 30(4):275–349
142. Chou KC, Zhang CT, Maggiora GM (1994) Solitary wave dynamics as a mechanism for explaining the internal motion during microtubule growth. Biopolymers 34(1):143–153
143. Chou KC, Zhou GP (1982) Role of the protein outside active site on the diffusion-controlled reaction of enzyme. J Am Chem Soc 104(5):1409–1413
144. Christen B, Damberger FF, Pérez DR, Hornemann S, Wüthrich K (2013) Structural plasticity of the cellular prion protein and implications in health and disease. Proc Natl Acad Sci U S A 110(21):8549–8554
145. Christen B, Hornemann S, Damberger FF, Wüthrich K (2009) Prion protein NMR structure from tammar wallaby (Macropus eugenii) shows that the $\beta 2$-$\alpha 2$ loop is modulated by long-range sequence effects. J Mol Biol 389(5):833–845
146. Christen B, Hornemann S, Damberger FF, Wüthrich K (2012) Prion protein mPrP [F175A](121–231): structure and stability in solution. J Mol Biol 423(4):496–502
147. Christen B, Pérez DR, Hornemann S, Wüthrich K (2008) NMR structure of the bank vole prion protein at 20 degrees C contains a structured loop of residues 165–171. J Mol Biol 383(2):306–312

148. Chu NK, Shabbir W, Bove-Fenderson E, Araman C, Lemmens-Gruber R, Harris DA, Becker CF (2014) A C-terminal membrane anchor affects the interactions of prion proteins with lipid membranes. J Biol Chem 289(43):30144–30160
149. Chu J, Trout BL, Brooks BR (2003) A superlinear minimization scheme for the nudged elastic band method. J Chem Phys 119(24):12708–12717
150. Ciccotosto GD, Cappai R, White AR (2008) Neurotoxicity of prion peptides on cultured cerebellar neurons. In: Hill AF (ed) Prion protein protocols methods. Methods in molecular biology, vol 459. Humana Press, Totowa, pp 83–96 (chapter 6)
151. Clarke FH (1983) Optimization and nonsmooth analysis. Wiley, New York. ISBN:978-0-898712-56-8
152. Cohn H, Fielding M (1999) Simulated annealing: searching for an optimal temperatur schedule. SIAM J Optim 9(3):779–802
153. Coleman BM, Harrison CF, Guo B, Masters CL, Barnham KJ, Lawson VA, Hill AF (2014) Pathogenic mutations within the hydrophobic domain of the prion protein lead to the formation of protease sensitive prion species with increased lethality. J Virol 88(5):2690–2703
154. Coleman T, Shalloway D, Wu Z (1994) A parallel build-up algorithm for global energy minimizations of molecular clusters using effective energy simulated annealing. J Glob Optim 4(2):171–185
155. Coleman TF, Verma A (2001) A preconditioned conjugate gradient approach to linear equality constrained minimization. Comput Optim Appl 20(1):61–72
156. Cong X, Bongarzone S, Giachin G, Rossetti G, Carloni P, Legname G (2013) Dominantnegative effects in prion diseases: insights from molecular dynamics simulations on mouse prion protein chimeras. J Biomol Struct Dyn 31(8):829–840
157. Conn AR, Gould NIM, Toint PL (2000) Trust-region methods (MPS-SIAM series on optimization). SIAM, Philadelphia. ISBN:978-0-898714-60-9
158. Corana A, Marchesi M, Martini C, Ridella S (1987) Minimizing multimodal functions of continuous variables with the 'Simulated Annealing' algorithm. ACM Trans Math Softw 13(3):262–280
159. Cosentino U, Pitea D, Moro G, Saracino GA, Caria P, Varì RM, Colombo L, Forloni G, Tagliavini F, Salmona M (2008) The anti-fibrillogenic activity of tetracyclines on PrP 106–126: a 3D-QSAR study. J Mol Model 14(10):987–994
160. Courageot MP, Daude N, Nonno R, Paquet S, Di Bari MA, Le Dur A, Chapuis J, Hill AF, Agrimi U, Laude H, Vilette D (2008) A cell line infectible by prion strains from different species. J Gen Virol 89(Pt 1):341–347
161. Crichton RR, Ward RJ (2006) Metal-based neurodegeneration, from molecular mechanisms to therapeutic strategies. Wiley, Chichester. ISBN:978-1-119-97714-8
162. Cui T, Daniels M, Wong BS, Li R, Sy MS, Sassoon J, Brown DR (2003) Mapping the functional domain of the prion protein. Eur J Biochem 270(16):3368–3376
163. Dai YH (2002) Convergence properties of the BFGS algorithm. SIAM J Optim 13(3):693–701
164. Dai YH, Yuan YX (1996) Convergence properties of the conjugate descent method. Adv in Math 25:552–562
165. Daidone I, Nola AD, Smith JC (2011) Molecular origin of Gerstmann-Straussler-Scheinker syndrome: insight from computer simulation of an amyloidogenic prion peptide. Biophys J 100(12):3000–3007
166. Damberger FF, Christen B, Pérez DR, Hornemann S, Wüthrich K (2011) Cellular prion protein conformation and function. Proc Natl Acad Sci U S A 108(42):17308–17313
167. Das A, Chakrabarti BK (eds) (2005) Quantum annealing and related optimization methods. Lecture note in physics, vol 679. Springer, Heidelberg. ISBN:3-540-27987-3
168. Daskalov A, Gantner M, Wälti MA, Schmidlin T, Chi CN, Wasmer C, Schütz A, Ceschin J, Clavé C, Cescau S, Meier B, Riek R, Saupe SJ (2014) Contribution of specific residues of the β-solenoid fold to HET-s prion function, amyloid structure and stability. PLoS Pathog 10(6):e1004158
169. Daude N (2004) Prion diseases and the spleen. Viral Immunol 17(3):334–349

170. Daude N, Ng V, Watts JC, Genovesi S, Glaves JP, Wohlgemuth S, Schmitt-Ulms G, Young H, McLaurin J, Fraser PE, Westaway D (2010) Wild-type Shadoo proteins convert to amyloid-like forms under native conditions. J Neurochem 113(1):92–104
171. Deaven DM, Ho KM (1995) Molecular geometry optimization with a genetic algorithm. Phys Rev Lett 75(2):288–291
172. Deaven DM, Tit N, Morris JR, Ho KM (1996) Structural optimization of Lennard-Jones clusters by a genetic algorithm. Chem Phys Lett 256(1–2):195–200
173. De Fea KA, Nakahara DH, Calayag MC, Yost CS, Mirels LF, Prusiner SB, Lingappa VR (1994) Determinants of carboxyl-terminal domain translocation during prion protein biogenesis. J Biol Chem 269(24):16810–16820
174. Dekkers A, Aarts E (1991) Global optimization and simulated annealing. Math Prog 50(1–3):367–393
175. De Marco ML, Daggett V (2004) From conversion to aggregation: protofibril formation of the prion protein. Proc Natl Acad Sci U S A 101(8):2293–2298
176. Demyanov VF, Rubinov AM (2000) Quasidifferentiability and related topics. Kluwer Academic, Dordrecht/Boston. ISBN:0-7923-6284-5
177. Dennis JE, More JJ (1977) Quasi-Newton method, motivation and theory. SIAM Rev 19(1):46–89
178. Dennis JE, Schnabel RB (1996) Numerical methods for unconstrained optimization and nonlinear equations. SIAM, Philadelphia. ISBN:978-0-898713-64-0
179. Desai R, Patil R (1996) SALO: combining simulated annealing and local optimization for efficient global optimization. In: Proceedings of the 9th Florida AI research symposium, Key West, pp 233–237
180. De Simone A, Dodson GG, Verma CS, Zagari A, Fraternali F (2005) Prion and water; tight and dynamical hydration sites have a key role in structural stability. Proc Natl Acad Sci U S A 102(21):7535–7540
181. De Vicente J, Lanchares J, Hermida R (2003) Placement by thermodynamic simulated annealing. Phys Lett A 317(5–6):415–423
182. Dima RI, Thirumalai D (2002) Exploring the propensities of helices in PrP^C to form β sheet using NMR structures and sequence alignments. Biophys J 83(3):1268–1280
183. Dima RI, Thirumalai D (2004) Probing the instabilities in the dynamics of helical fragments from mouse PrP^C. Proc Natl Acad Sci U S A 101(43):15335–15340
184. Di Martino A, Bigon E, Corona G, Callegaro L (1991) Production and characterization of antibodies to mouse scrapie-amyloid protein elicited by non-carrier linked synthetic peptide immunogens. J Mol Recognit 4(2–3):85–91
185. Ding H, Luo L, Lin H (2009) Prediction of cell wall lytic enzymes using Chou's amphiphilic pseudo amino acid composition. Protein Pept Lett 16(4):351–355
186. Dolan ED, Moré JJ (2002) Benchmarking optimization software with performance profiles. Math Program Ser A 91(2):201–213
187. Dong CF, Wang XF, An R, Chen JM, Shan B, Han L, Lei YJ, Han J, Dong XP (2007) Interaction analysis between various PrP fusion proteins and the tubulin in vitro. Bing Du Xue Bao 23(1):28–32
188. Doss CG, Rajith B, Rajasekaran R, Srajan J, Nagasundaram N, Debajyoti C (2013) In silico analysis of prion protein mutants: a comparative study by molecular dynamics approach. Cell Biochem Biophys 67(3):1307–1318
189. Doye J (1999) Global optimization and multiple-funnel landscapes: Lennard-Jones clusters. In: International workshop on global optimization, Firenze, 28 Sept–2 Oct 1999
190. Doye JPK, Wales DJ (1995) Magic numbers and growth sequences of small face-centred-cubic and decahedral clusters. Chem Phys Lett 247(4–6):339–347
191. Doye JPK, Wales DJ, Berry RS (1995) The effect of the range of the potential on the structures of clusters. J Chem Phys 103(10):4234–4249
192. Dragani B, Cocco R, Principe DR, Paludi D, Aceto A (2001) Conformational properties of five peptides corresponding to the entire sequence of glutathione transferase domain II. Arch Biochem Biophys 389(1):15–21

193. Du QS, Huang RB, Wang SQ, Chou KC (2010) Designing inhibitors of M2 proton channel against H1N1 swine influenza virus. PLoS ONE 5(2):e9388
194. Du QS, Sun H, Chou KC (2007) Inhibitor design for SARS coronavirus main protease based on "distorted key theory". Med Chem 3(1):1–6
195. Dupiereux I, Zorzi W, Rachidi W, Zorzi D, Pierard O, Lhereux B, Heinen E, Elmoualij B (2006) Study on the toxic mechanism of prion protein peptide 106-126 in neuronal and non neuronal cells. J Neurosci Res 84(3):637–646
196. Earl DJ, Deem MW (2005) Parallel tempering: theory, applications, and new perspectives. Phys Chem Chem Phys 7:3910–3916
197. Eghiaian F, Grosclaude J, Lesceu S, Debey P, Doublet B, Tréguer E, Rezaei H, Knossow M (2004) Insight into the PrPC–>PrPSc conversion from the structures of antibody-bound ovine prion scrapie-susceptibility variants. Proc Natl Acad Sci U S A 101(28):10254–10259
198. El-Bastawissy E, Knaggs MH, Gilbert IH (2001) Molecular dynamics simulations of wild-type and point mutation human prion protein at normal and elevated temperature. J Mol Graph Model 20(2):145–154
199. Fang SC, Gao DY, Sheu RL, Wu SY (2008) Canonical dual approach to solving 0-1 quadratic programming problems. J Ind Manag Optim 4(1):125–142
200. Farges J, de Feraudy MF, Raoult B, Torchet G (1985) Cluster models made of double icosahedron units. Surf Sci 156(Part 1):370–378
201. Farquhar CF, Somerville RA, Ritchie LA (1989) Post-mortem immunodiagnosis of scrapie and bovine spongiform encephalopathy. J Virol Methods 24(1-2):215–221
202. Fei L, Perrett S (2009) Disulfide bond formation significantly accelerates the assembly of Ure2p fibrils because of the proximity of a potential amyloid stretch. J Biol Chem 284(17):11134–11141
203. Fernandez-Borges N, Chianini F, Erana H, Vidal E, Eaton SL, Pintado B, Finlayson J, Dagleish MP, Castilla J (2012) Naturally prion resistant mammals: a utopia? Prion 6(5):425–429
204. Fernandez-Funez P, Casas-Tinto S, Zhang Y, Gomez-Velazquez M, Morales-Garza MA, Cepeda-Nieto AC, Castilla J, Soto C, Rincon-Limas DE (2009) In vivo generation of neurotoxic prion protein: role for hsp70 in accumulation of misfolded isoforms. PLoS Genet 5(6):e1000507
205. Fernandez-Funez P, Zhang Y, Casas-Tinto S, Xiao X, Zou WQ, Rincon-Limas DE (2010) Sequence-dependent prion protein misfolding and neurotoxicity. J Biol Chem 285(47):36897–36908
206. Fernandez-Funez P, Zhang Y, Sanchez-Garcia J, Jensen K, Zou WQ, Rincon-Limas DE (2011) Pulling rabbits to reveal the secrets of the prion protein. Commun Integr Biol 4(3):262–266
207. Ferreira ST, Vieira MN, De Felice FG (2007) Soluble protein oligomers as emerging toxins in Alzheimer's and other amyloid diseases. IUBMB life 59(4–5):332–345
208. Fielding M (2000) Simulated annealing with an optimal fixed temperature. SIAM J Optim 11(2):289–307
209. Fletcher R (1987) Practical method of optimization, vol I: unconstrained optimization, 2nd edn. Wiley, New York
210. Foote J, Raman A (2000) A relation between the principal axes of inertia and ligand binding. Proc Natl Acad Sci U S A 97(3):978–983
211. Forrest S (1993) Genetic algorithms: principles of natural selection applied to computation. Science 261(5123):872–878
212. Fox BL (1995) Simulated annealing: folklore, facts, and directions. In: Monte Carlo and Quasi-Monte carlo methods in scientific computing. Springer, New York, pp 17–48
213. Freeman DL, Doll JD (1985) Quantum Monte Carlo study of the thermodynamic properties of argon clusters: the homogeneous nucleation of argon in argon vapor and "magic number" distributions in argon vapor. J Chem Phys 82(1):462–471
214. Frenkel D, Frenkel D, Smit B (2001) Understanding molecular simulation (computational science) – from algorithms to applications, 2nd edn. (Hardcover). Academic Press, San Diego. ISBN:978-0-12-267351-1

215. Friedman-Levi Y, Mizrahi M, Frid K, Binyamin O, Gabizon R (2013) PrPST, a soluble, protease resistant and truncated PrP form features in the pathogenesis of a genetic prion disease. PLoS ONE 8(7):e69583
216. Gabizon R, McKinley MP, Groth D, Prusiner SB (1988) Immunoaffinity purification and neutralization of scrapie prion infectivity. Proc Natl Acad Sci U S A 85(18):6617–6621
217. Gabizon R, McKinley MP, Groth D, Westaway D, DeArmond SJ, Carlson GA, Prusiner SB (1989) Immunoaffinity purification and neutralization of scrapie prions. Prog Clin Biol Res 317:583–600
218. Gaggelli E, Bernardi F, Molteni E, Pogni R, Valensin D, Valensin G, Remelli M, Luczkowski M, Kozlowski H (2005) Interaction of the human prion PrP (106–126) sequence with copper(II), manganese(II), and zinc(II): NMR and EPR studies. J Am Chem Soc 127(3):996–1006
219. Gao DY (1997) Dual extremum principles in finite deformation theory with applications to post-buckling analysis of extended nonlinear beam theory. Appl Mech Rev 50(11):S64–S71
220. Gao DY (1999) General analytic solutions and complementary variational principles for large deformation nonsmooth mechanics. Meccanica 34(3):169–198
221. Gao DY (2000) Duality principles in nonconvex systems: theory, methods and applications. Kluwer Academic, Dordrecht/Boston/London. ISBN:0-7923-6145-8
222. Gao DY (2000) Analytic solution and triality theory for nonconvex and nonsmooth variational problems with applications. Nonlinear Anal 42(7):1161–1193
223. Gao DY (2000) Canonical dual transformation method and generalized triality theory in nonsmooth global optimization. J Glob Optim 17(1):127–160
224. Gao DY (2003) Perfect duality theory and complete set of solutions to a class of global optimization. Optim 52(4–5):467–493
225. Gao DY, Strang G (1989) Geometric nonlinearity: potential energy, complementary energy, and the gap function. Quart Appl Math XLVII(3):487–504
226. Gao DY (2004) Canonical duality theory and solutions to constrained nonconvex quadratic programming. J Glob Optim 29(4):377–399
227. Gao DY (2005) Sufficient conditions and perfect duality in nonconvexs minimization with inequality. J Ind Manag Optim 1(1):53–63
228. Gao DY (2006) Complete solutions and extremality criteria to polynomial optimization problems. J Glob Opt 35(1):131–143
229. Gao DY (2007) Solutions and optimality criteria to box constrained nonconvex minimization problems. J Ind Manag Optim 3(2):293–304
230. Gao DY (2008) Advances in canonical duality theory with applications to global optimization. In: Proceedings foundations of computer-aided process operations (FOCAPO 2008), Cambridge, 29 June–2 July 2008
231. Gao C, Lei YJ, Han J, Shi Q, Chen L, Guo Y, Gao YJ, Chen JM, Jiang HY, Zhou W, Dong XP (2006) Recombinant neural protein PrP can bind with both recombinant and native apolipoprotein E in vitro. Acta Biochim Biophys Sin (Shanghai) 38(9):593–601
232. Gao DY, Ruan N, Pardalos PM (2012) Canonical dual solutions to sum of fourth-order polynomials minimization problems with applications to sensor network localization. In: Boginski VL, Commander CW, Pardalos PM, Ye YY (eds) Sensors: theory, algorithms, and applications. Springer optimization and its applications, vol 61, chapter 3, pp 37–54. ISBN:978-0-387-88618-3
233. Gao DY, Wu CZ (2012) On the triality theory in global optimization (I) unconstrained problems. arXiv:1104.2970v2 (http://arxiv.org/abs/1104.2970v2)
234. Garnier J, Osguthorpe DJ, Robson B (1978) Analysis of the accuracy and implications of simple methods for predicting the secondary structure of globular proteins. J Mol Biol 120(1):97–120
235. Garrec J, Tavernelli I, Rothlisberger U (2013) Two misfolding routes for the prion protein around pH 4.5. PLoS Comput Biol 9(5):e1003057
236. Garssen GJ, Van Keulen LJ, Farquhar CF, Smits MA, Jacobs JG, Bossers A, Meloen RH, Langeveld JP (2000) Applicability of three anti-PrP peptide sera including staining of tonsils and brainstem of sheep with scrapie. Microsc Res Tech 50(1):32–39

237. Gasset M, Baldwin MA, Lloyd DH, Gabriel JM, Holtzman DM, Cohen F, Fletterick R, Prusiner SB (1992) Predicted α-helical regions of the prion protein when synthesized as peptides form amyloid. Proc Natl Acad Sci U S A 89(22):10940–10944
238. Gazit E (2002) A possible role for pi-stacking in the self-assembly of amyloid fibrils. FASEB J 16(1):77–83
239. Geman S, Geman D (1984) Stochastic relaxation, Gibbs distribution and the Bayesian restoration in images. IEEE Trans Pattern Anal Mach Intell 6(6):721–741
240. Gerber R, Tahiri-Alaoui A, Hore PJ, James W (2008) Conformational pH dependence of intermediate states during oligomerization of the human prion protein. Protein Sci 17(3):537–544
241. Gershenfeld N (1999) The nature of mathematical modeling. Cambridge University Press, Cambridge, Chapter 13, Section 3
242. Gilbert JC, Lemarichal C (1989) Some numerical experiments with variable storage quasi-Newton algorithms. Math Program 45(1–3):407–435
243. Gilbert JC, Nocedal J (1992) Global convergence properties of conjugate gradient methods for optimization. SIAM J Optim 2(1):21–42
244. Gilch S, Wopfner F, Renner-Müller I, Kremmer E, Bauer C, Wolf E, Brem G, Groschup MH, Schätzl HM (2003) Polyclonal anti-PrP auto-antibodies induced with dimeric PrP interfere efficiently with PrP^{Sc} propagation in prion-infected cells. J Biol Chem 278(20):18524–18531
245. Gilead S, Gazit E (2004) Inhibition of amyloid fibril formation by peptide analogues modified with alpha-aminoisobutyric acid. Angew Chem Int Ed Engl 43(31):4041–4044
246. Glockshuber R, Hornemann S, Billeter M, Riek R, Wider G, Wuthrich K (1998) Prion protein structural features indicate possible relations to signal peptidases. FEBS Lett 426(3):291–296
247. Goffe WL, Ferrier GD, Rogers J (1994) Global optimization of statistical functions with simulated annealing. J Econom 60(1–2):65–99
248. Golanska E, Hulas-Bigoszewska K, Sikorska B, Liberski PP (2005) Analyses of 14–3–3 protein in the cerebrospinal fluid in Creutzfeldt-Jakob disease. Preliminary report. Neurol Neurochir Pol 39(5):358–365
249. Goldfeld SM, Quandt RE, Trotter HF (1966) Maximization by quadratic hill-climbing. Econometrica 34(3):541–551
250. Gong K, Li L, Wang JF, Cheng F, Wei DQ, Chou KC (2009) Binding mechanism of H5N1 influenza virus neuraminidase with ligands and its implication for drug design. Med Chem 5(3):242–249
251. Gorfe AA, Caflisch A (2007) Ser170 controls the conformational multiplicity of the loop 166–175 in prion proteins: implication for conversion and species barrier. FASEB J 21(12):3279–3287
252. Gossert AD, Bonjour S, Lysek DA, Fiorito F, Wüthrich K (2005) Prion protein NMR structures of elk and of mouse/elk hybrids. Proc Natl Acad Sci U S A 102(3):646–650
253. Götz AW, Williamson MJ, Xu D, Poole D, Grand SL, Walker RC (2012) Routine microsecond molecular dynamics simulations with AMBER on GPUs. 1. generalized born. J Chem Theory Comput 8(5):1542–1555
254. Govaerts C, Wille H, Prusiner SB, Cohen EE (2004) Evidence for assembly of prions with left-handed β-helices into trimers. Proc Natl Acad Sci U S A 101(22):8342–8347
255. Granville V, Krivanek M, Rasson JP (1994) Simulated annealing: a proof of convergence. IEEE T Pattern Anal Mach Intell 16(6):652–656
256. Greenes JW, Supowit KJ (1986) Simulated annealing without rejected moves. IEEE Trans Comput-Aided Des 5(1):221–228
257. Griffith JS (1967) Self-replication and scrapie. Nature 215(5105):1043–1044
258. Grippo L, Lamparello F, Lucidi S (1986) A nonmonotone line search technique for Newton's method. SIAM J Numer Anal 23(4):707–716
259. Groschup MH, Harmeyer S, Pfaff E (1997) Antigenic features of prion proteins of sheep and of other mammalian species. J Immunol Methods 207(1):89–101
260. Groschup MH, Langeveld J, Pfaff E (1994) The major species specific epitope in prion proteins of ruminants. Arch Virol 136(3–4):423–431

261. Groschup MH, Pfaff E (1993) Studies on a species-specific epitope in murine, ovine and bovine prion protein. J Gen Virol 74(Pt 7):1451–1456
262. Grosso A, Locatelli M, Schoen F (2009) Solving molecular distance geometry problems by global optimization algorithms. Comput Optim Appl 43(1):23–37
263. Guest WC, Cashman NR, Plotkin SS (2010) Electrostatics in the stability and misfolding of the prion protein: salt bridges, self-energy, and salvation. Biochem Cell Biol 88(2):371–381
264. Guest WC, Cashman NR, Plotkin SS (2011) A theory for the anisotropic and inhomogeneous dielectric properties of proteins. Phys Chem Chem Phys 13(13):6286–6295
265. Guillaume CL, Gregoryanz E, Degtyareva O, McMahon MI, Hanfland M, Evans S, Guthrie M, Sinogeikin SV, Mao HK (2011) Cold melting and solid structures of dense lithium. Nature Phys 7(3):211–214
266. Guiroy DC, Marsh RF, Yanagihara R, Gajdusek DC (1993) Immunolocalization of scrapie amyloid in non-congophilic, non-birefringent deposits in golden Syrian hamsters with experimental transmissible mink encephalopathy. Neurosci Lett 155(1):112–115
267. Haigh CL, Edwards K, Brown DR (2005) Copper binding is the governing determinant of prion protein turnover. Mol Cell Neurosci 30(2):186–196
268. Haire LF, Whyte SM, Vasisht N, Gill AC, Verma C, Dodson EJ, Dodson GG, Bayley PM (2004) The crystal structure of the globular domain of sheep prion protein. J Mol Biol 336(5):1175–1183
269. Han JY, Liu GH (1997) Global convergence analysis of a new nonmonotone BFGS algorithm on convex objective functions. Comput Optim Appl 7(3):277–289
270. Handisurya A, Gilch S, Winter D, Shafti-Keramat S, Maurer D, Schätzl HM, Kirnbauer R (2007) Vaccination with prion peptide-displaying papillomavirus-like particles induces autoantibodies to normal prion protein that interfere with pathologic prion protein production in infected cells. FEBS J 274(7):1747–1758
271. Hanoux V, Wijkhuisen A, Alexandrenne C, Créminon C, Boquet D, Couraud JY (2009) Polyclonal anti-idiotypic antibodies which mimic an epitope of the human prion protein. Mol Immunol 46(6):1076–1083
272. Harris DA, Huber MT, van Dijken P, Shyng SL, Chait BT, Wang R (1993) Processing of a cellular prion protein: identification of N- and C-terminal cleavage sites. Biochemistry 32(4):1009–1016
273. Harrison CF, Barnham KJ, Hill AF (2007) Neurotoxic species in prion disease: a role for PrP isoforms? J Neurochem 103(5):1709–1720
274. Harrison CF, Lawson VA, Coleman BM, Kim YS, Masters CL, Cappai R, Barnham KJ, Hill AF (2010) Conservation of a glycine-rich region in the prion protein is required for uptake of prion infectivity. J Biol Chem 285(26):20213–20223
275. Hashimoto K, Mannen T, Nukina N (1992) Immunohistochemical study of kuru plaques using antibodies against synthetic prion protein peptides. Acta Neuropathol 83(6):613–617
276. Hasnain SS, Murphy LM, Strange RW, Grossmann JG, Clarke AR, Jackson GS, Collinge J (2001) XAFS study of the high-affinity copper-binding site of HuPrP(91–231) and its low-resolution structure in solution. J Mol Biol 311(3):467–473
277. Hay B, Prusiner SB, Lingappa VR (1987) Evidence for a secretory form of the cellular prion protein. Biochemistry 26(25):8110–8115
278. Hayat M, Khan A (2011) Predicting membrane protein types by fusing composite protein sequence features into pseudo amino acid composition. J Theor Biol 271(1):10–17
279. Hedar AR, Fukushima M (2002) Hybrid simulated annealing and direct search method for nonlinear unconstrained global optimization. Optim Methods Softw 17(5):891–912
280. Hedar AR, Fukushima M (2003) Simplex coding genetic algorithm for the global optimization of nonlinear functions. In: Tanino T, Tanaka T, Inuiguchi M (eds) Multi-objective programming and goal programming. Springer, Berlin/New York, pp 135–140
281. Henriques ST, Pattenden LK, Aguilar MI, Castanho MA (2008) PrP (106–126) does not interact with membranes under physiological conditions. Biophys J 95(4):1877–1889
282. Herrmann T, Güntert P, Wüthrich K (2002) Protein NMR structure determination with automated NOE assignment using the new software CANDID and the torsion angle dynamics algorithm DYANA. J Mol Biol 319(1):209–227

283. Hestenes MR, Stiefel E (1952) Method of conjugate gradient for solving linear system. J Res Nat Bur Stand 49(6):409–436
284. Hoare MR, Pal P (1971) Statics and stability of small cluster neclei. Nature (Phys Sci) 230(9):5–8
285. Hoare MR, Pal P (1971) Physical cluster mechanic: statics and energy surfaces for monatomic systems. Adv Chem Phys. 20(84):161–196
286. Hoare MR, Pal P (1972) Geometry and stability of "spherical" f.c.c. microcrystallites. Nature (Phys Sci) 236(64):35–37
287. Holscher C, Bach UC, Dobberstein B (2001) Prion protein contains a second endoplasmic reticulum targeting signal sequence located at its C terminus. J Biol Chem 276(16):13388–13394
288. Holscher C, Delius H, Burkle A (1998) Overexpression of nonconvertible PrP^C delta114–121 in scrapie-infected mouse neuroblastoma cells leads to trans-dominant inhibition of wild-type PrP^{Sc} accumulation. J Virol 72(2):1153–1159
289. Hornemann S, Glockshuber R (1998) A scrapie-like unfolding intermediate of the prion protein domain PrP(121–231) induced by acidic pH. Proc Natl Acad Sci U S A 95(11):6010–6014
290. Horowitz E, Sahni S (1990) Fundamentals of Data structures in Pascal, 3rd edn. Computer Science Press, New York
291. Hou YT, Wang YJ (2012) The modified limited memory BFGS method for large-scale optimization, M.Sc. degree thesis, Qufu Normal University, Qufu
292. Hu Y, Storey C (1991) Global convergence result for conjugate gradient methods. J Optim Theor Appl 71(2):399–405
293. Huang D, Caflisch A (2015) Evolutionary conserved Tyr169 stabilizes the $\beta 2$-$\alpha 2$ loop of prion protein. J Am Chem Soc 137(8):2948–2957
294. Huang P, Lian F, Wen Y, Guo C, Lin D (2013) Prion protein oligomer and its neurotoxicity. Acta Biochim Biophys Sin (Shanghai) 45(6):442–451
295. Huang HX, Pardalos PM (2002) Multivariable partition approach for optimization problems. Cybern Syst Anal 2:134–147
296. Huang HX, Pardalos PM, Shen ZJ (2002) Equivalent formulations and necessary optimality conditions for the Lennard-Jones problem. J Glob Optim 22(1–4):97–118
297. Humphrey W, Dalke A, Schulten K (1996) VMD – visual molecular dynamics. J Mol Graph 14(1):33–38
298. Iannuzzi L, Palomba R, Di Meo GP, Perucatti A, Ferrara L (1998) Comparative FISH-mapping of the prion protein gene (PRNP) on cattle, river buffalo, sheep and goat chromosomes. Cytogenet Cell Genet 81(3–4):202–204
299. Iba Y (2001) Extended ensemble Monte Carlo. Int J Mod Phys C 12(5):623–656
300. Ikegami Y, Ito M, Isomura H, Momotani E, Sasaki K, Muramatsu Y, Ishiguro N, Shinagawa M (1991) Pre-clinical and clinical diagnosis of scrapie by detection of PrP protein in tissues of sheep. Vet Rec 128(12):271–275
301. Ilc G, Giachin G, Jaremko M, Jaremko L, Benetti F, Plavec J, Zhukov I, Legname G (2010) NMR structure of the human prion protein with the pathological Q212P mutation reveals unique structural features. PLoS ONE 5(7):e11715
302. Imran M, Mahmood S, Babar ME, Hussain R, Yousaf MZ, Abid NB, Lone KP (2012) PRNP gene variation in Pakistani cattle and buffaloes. Gene 505(1):180–185
303. Ingber L (1989) Very fast simulated re-annealing. Math Comput Model 12(8):967–973
304. Ingber L (1992) Genetic algorithms and very fast simulated re-annealing: a comparison. Math Comput Model 16(11):87–100
305. Ingber L (1993) Simulated annealing: practice versus theory. Math Comput Model 18(11):29–57
306. Ingber L (1996) Adaptive simulated annealing (ASA): lessons learned. Contr Cyber 25(1):33–54
307. Inouye H, Kirschner DA (1996) Refined fibril structures: the hydrophobic core in Alzheimer's amyloid β-protein and prion as revealed by X-ray diffraction. Ciba Found Symp 199:22–35, (chapter 3); discussion 35–39

308. Irvine GB, El-Agnaf OM, Shankar GM, Walsh DM (2008) Protein aggregation in the brain: the molecular basis for Alzheimer's and Parkinson's diseases. Mol Med Camb Mass 14(7–8):451–464
309. Ishiguro N, Inoshima Y, Sassa Y, Takahashi T (2009) Molecular characterization of chicken prion proteins by C-terminal-specific monoclonal antibodies. Vet Immunol Immunopathol 128(4):402–406
310. Ishikawa T, Kuwata K (2010) Interaction analysis of the native structure of prion protein with quantum chemical calculations. J Chem Theor Comput 6(2):538–547
311. Izard T, Lawrence MC, Malby RL, Lilley GG, Colman PM (1994) The three-dimensional structure of N-acetylneuraminate lyase from Escherichia coli. Structure 2(5):361–369. http://www.cell.com/structure/retrieve/pii/S0969212600000381
312. Jackman R, Schmerr MJ (2003) Analysis of the performance of antibody capture methods using fluorescent peptides with capillary zone electrophoresis with laser-induced fluorescence. Electrophoresis 24(5):892–896
313. James TL, Liu H, Ulyanov NB, Farr-Jones S, Zhang H, Donne DG, Kaneko K, Groth D, Mehlhorn I, Prusiner SB, Cohen FE (1997) Solution structure of a 142-residue recombinant prion protein corresponding to the infectious fragment of the scrapie isoform. Proc Natl Acad Sci U S A 94(19):10086–10091
314. Jeong BH, Jin HT, Choi EK, Carp RI, Kim YS (2012) Lack of association between 14-3-3 β gene (YWHAB) polymorphisms and sporadic Creutzfeldt-Jakob disease (CJD). Mol Biol Rep 39(12):10647–10653
315. Jobling MF, Huang X, Stewart LR, Barnham KJ, Curtain C, Volitakis I, Perugini M, White AR, Cherny RA, Masters CL, Barrow CJ, Collins SJ, Bush AI, Cappai R (2001) Copper and Zinc binding modulates the aggregation and neurotoxic properties of the prion peptide PrP 106–126. Biochem 40(27):8073–8084
316. Jobling MF, Stewart LR, White AR, McLean C, Friedhuber A, Maher F, Beyreuther K, Masters CL, Barrow CJ, Collins SJ, Cappai R (1999) The hydrophobic core sequence modulates the neurotoxic and secondary structure properties of the prion peptide 106–126. J Neurochem 73(4):1557–1565
317. Johnson DS, Aragon CR, McGeoch LA, Schevon C (1987) Optimization by simulated annealing: an experimental evaluation; Part I. AT&T Bell Laboratories, Murray Hill
318. Johnson DS, Aragon CR, McGeoch LA, Schevon C (1989) Optimization by simulated annealing: an experimental evaluation; Part I, graph partitioning. Oper Res 37(6):865–892
319. Johnson DS, Aragon CR, McGeoch LA, Schevon C (1990) Optimization by simulated annealing: an experimental evaluation; Part II, graph colouring and number partitioning. Oper Res 39(3):378–406
320. Jones AEW, Forbes GW (1995) An adaptive simulated annealing algorithm for global optimization over continuous variables. J Glob Optim 6(1):1–37
321. Jones CE, Klewpatinond M, Abdelraheim SR, Brown DR, Viles JH (2005) Probing copper2+ binding to the prion protein using diamagnetic nickel2+ and 1H NMR: the unstructured N terminus facilitates the coordination of six copper2+ ions at physiological concentrations. J Mol Biol 346(5):1393–1407
322. Jones EM, Wu B, Surewicz K, Nadaud PS, Helmus JJ, Chen S, Jaroniec CP, Surewicz WK (2011) Structural polymorphism in amyloids: new insights from studies with Y145Stop prion protein fibrils. J Biol Chem 286(49):42777–42784
323. Joszai V, Turi I, Kallay C, Pappalardo G, Di Natale G, Rizzarelli E, Sovago I (2012) Mixed metal copper(II)-nickel(II) and copper(II)-zinc(II) complexes of multihistidine peptide fragments of human prion protein. J Inorg Biochem 112:17–24. doi:10.1016/j.jinorgbio.2012.02.014
324. Julien O, Chatterjee S, Bjorndahl TC, Sweeting B, Acharya S, Semenchenko V, Chakrabartty A, Pai EF, Wishart DS, Sykes BD, Cashman NR (2011) Relative and regional stabilities of the hamster, mouse, rabbit, and bovine prion proteins toward urea unfolding assessed by nuclear magnetic resonance and circular dichroism spectroscopies. Biochemistry 50(35):7536–7545

325. Julien O, Chatterjee S, Thiessen A, Graether SP, Sykes BD (2009) Differential stability of the bovine prion protein upon urea unfolding. Protein Sci 18(10):2172–2182
326. Kabsch W, Sander C (1983) Dictionary of protein secondary structure: pattern recognition of hydrogen-bonded and geometrical features. Biopolymers 22(12):2577–2637
327. Kajava A, Squire JM, Parry DAD (2006) Fibrous proteins: amyloids, prions and beta proteins, advances in protein chemistry, vol 73. Elsevier Academic Press, Amsterdam/Boston/San Diego. ISBN:978-0-12-034273-0
328. Kandaswamy KK, Chou KC, Martinetz T, Moller S, Suganthan PN, Sridharan S, Pugalenthi G (2011) AFP-Pred: a random forest approach for predicting antifreeze proteins from sequence-derived properties. J Theor Biol 270(1):56–62
329. Kaneko K, Zulianello L, Scott M, Cooper CM, Wallace AC, James TL, Cohen FE, Prusiner SB (1997) Evidence for protein X binding to a discontinuous epitope on the cellular prion protein during scrapie prion propagation. Proc Natl Acad Sci U S A 94(19):10069–10074
330. Kanyo ZF, Pan KM, Williamson RA, Burton DR, Prusiner SB, Fletterick RJ, Cohen FE (1999) Antibody binding defines a structure for an epitope that participates in the PrP^C to PrP^{Sc} conformational change. J Mol Biol 293(4):855–863
331. Kascsak RJ, Fersko R, Pulgiano D, Rubenstein R, Carp RI (1997) Immunodiagnosis of prion disease. Immunol Invest 26(1–2):259–268
332. Kascsak RJ, Rubenstein R, Merz PA, Carp RI, Robakis NK, Wisniewski HM, Diringer H (1986) Immunological comparison of scrapie-associated fibrils isolated from animals infected with four different scrapie strains. J Virol 59(3):676–683
333. Kascsak RJ, Rubenstein R, Merz PA, Tonna-DeMasi M, Fersko R, Carp RI, Wisniewski HM, Diringer H (1987) Mouse polyclonal and monoclonal antibody to scrapie-associated fibril proteins. J Virol 61(12):3688–3693
334. Kelker M, Kim C, Chueh PJ, Guimont R, Morré DM, Morré DJ (2001) Cancer isoform of a tumor-associated cell surface NADH oxidase (tNOX) has properties of a prion. Biochemistry 40(25):7351–7354
335. Khan MQ (2010) Misfolding of particular PrP species and susceptibility to prion infection. Master degree thesis, University of Toronto
336. Khan MQ, Sweeting B, Mulligan VK, Arslan PE, Cashman NR, Pai EF, Chakrabartty A (2010) Prion disease susceptibility is affected by β-structure folding propensity and local side-chain interactions in PrP. Proc Natl Acad Sci U S A 107(46):19808–19813
337. Kim HJ, Tark DS, Lee YH, Kim MJ, Lee WY, Cho IS, Sohn HJ, Yokoyama T (2012) Establishment of a cell line persistently infected with chronic wasting disease prions. J Vet Med Sci 74(10):1377–1380
338. Kirby L, Agarwal S, Graham JF, Goldmann W, Gill AC (2010) Inverse correlation of thermal lability and conversion efficiency for five prion protein polymorphic variants. Biochemistry 49(7):1448–1459
339. Kirkpatrick S, Gelatt CD, Vecchi MP (1983) Optimization by simulated annealing. Science 220(4598):671–680
340. Kirkwood JK, Wells GA, Cunningham AA, Jackson SI, Scott AC, Dawson M, Wilesmith JW (1992) Scrapie-like encephalopathy in a greater kudu (Tragelaphus strepsiceros) which had not been fed ruminant-derived protein. Vet Rec 130(17):365–367
341. Kitamoto T, Tateishi J, Sawa H, Doh-Ura K (1989) Positive transmission of Creutzfeldt-Jakob disease verified by murine kuru plaques. Lab Invest 60(4):507–512
342. Klemm HM, Welton JM, Masters CL, Klug GM, Boyd A, Hill AF, Collins SJ, Lawson VA (2012) The prion protein preference of sporadic Creutzfeldt-Jakob disease subtypes. J Biol Chem 287(43):36465–36472
343. Klewpatinond M, Viles JH (2007) Fragment length influences affinity for Cu2+ and Ni2+ binding to His96 or His111 of the prion protein and spectroscopic evidence for a multiple histidine binding only at low pH. Biochem J 404(3):393–402
344. Knaus KJ, Morillas M, Swietnicki W, Malone M, Surewicz WK, Yee VC (2001) Crystal structure of the human prion protein reveals a mechanism for oligomerization. Nat Struct Biol 8(9):770–774

345. Knolmurodov K, Hirano Y, Ebisuzaki T (2003) MD simulations on the influence of disease-related amino acid mutations in the human prion protein. Chem-Bio Inf J 3(2):86–95
346. Kocisko DA, Caughey B (2006) Searching for anti-prion compounds: cell-based high-throughput in vitro assays and animal testing strategies. Methods Enzymol 412:223–234
347. Kolossváry I, Bowers KJ (2010) Global optimization of additive potential energy functions: predicting binary Lennard-Jones clusters. Phys Rev E Stat Nonlin Soft Matter Phys 82(5 Pt 2):056711
348. Komar AA, Lesnik T, Cullin C, Guillemet E, Ehrlich R, Reiss C (1997) Differential resistance to proteinase K digestion of the yeast prion-like (Ure2p) protein synthesized in vitro in wheat germ extract and rabbit reticulocyte lysate cell-free translation systems. FEBS Lett 415(1):6–10
349. Kong Q, Mills JL, Kundu B, Li X, Qing L, Surewicz K, Cali I, Huang S, Zheng M, Swietnicki W, Sönnichsen FD, Gambetti P, Surewicz WK (2013) Thermodynamic stabilization of the folded domain of prion protein inhibits prion infection in vivo. Cell Rep 4(2):248–254
350. Korth C, Stierli B, Streit P, Moser M, Schaller O, Fischer R, Schulz-Schaeffer W, Kretzschmar H, Raeber A, Braun U, Ehrensperger F, Hornemann S, Glockshuber R, Riek R, Billeter M, Wthrich K, Oesch B (1997) Prion (PrP^{Sc})-specific epitope defined by a monoclonal antibody. Nature 390(6655):74–77
351. Kourie JI (2001) Mechanisms of prion-induced modifications in membrane transport properties: implications for signal transduction and neurotoxicity. Chem Biol Interact 138(1):1–26
352. Kourie JI, Kenna BL, Tew D, Jobling MF, Curtain CC, Masters CL, Barnham KJ, Cappai R (2003) Copper modulation of ion channels of PrP[106–126] mutant prion peptide fragments. J Membr Biol 193(1):35–45
353. Kumar S, Nussinov R (2002) Close-range electrostatic interactions in proteins. ChemBioChem 3(7):604–617
354. Kunze S, Lemke K, Metze J, Bloukas G, Kotta K, Panagiotidis CH, Sklaviadis T, Bodemer W (2008) Atomic force microscopy to characterize the molecular size of prion protein. J Microsc 230(Pt 2):224–232
355. Kurt TD, Bett C, Fernández-Borges N, Joshi-Barr S, Hornemann S, Rülicke T, Castilla J, Wüthrich K, Aguzzi A, Sigurdson CJ (2014) Prion transmission prevented by modifying the $\beta 2$-α loop structure of host PrP^{C}. J Neurosci 34(3):1022–1027
356. Kurt TD, Jiang L, Bett C, Eisenberg D, Sigurdson CJ (2014) A proposed mechanism for the promotion of prion conversion involving a strictly conserved tyrosine residue in the $\beta 2$-$\alpha 2$ loop of PrP^{C}. J Biol Chem 289(15):10660–10667
357. Kuwata K, Matumoto T, Cheng H, Nagayama K, James TL, Roder H (2003) NMR-detected hydrogen exchange and molecular dynamics simulations provide structural insight into fibril formation of prion protein fragment 106–126. Proc Natl Acad Sci U S A 100(25):14790–14795
358. Kuwata K, Nishida N, Matsumoto T, Kamatari YO, Hosokawa-Muto J, Kodama K, Nakamura HK, Kimura K, Kawasaki M, Takakura Y, Shirabe S, Takata J, Kataoka Y, Katamine S (2007) Hot spots in prion protein for pathogenic conversion. Proc Natl Acad Sci U S A 104(29):11921–11926
359. Kuznetsov IB, Rackovsky S (2004) Comparative computational analysis of prion proteins reveals two fragments with unusual structural properties and a pattern of increase in hydrophobicity associated with disease-promoting mutations. Protein Sci 13(12):3230–3244
360. Laarhoven PJM, Aarts EHL (1987) Simulated annealing: theory and applications. Springer, Dordrecht/Boston
361. Laganowsky A, Liu C, Sawaya MR, Whitelegge JP, Park J, Zhao M, Pensalfini A, Soriaga AB, Landau M, Teng PK, Cascio D, Glabe C, Eisenberg D (2012) Atomic view of a toxic amyloid small oligomer. Science 335(6073):1228–1231
362. Lai XJ, Xu RC, Huang WQ (2011) Prediction of the lowest energy configuration for Lennard-Jones clusters. Sci China Chem 54(6):985–991
363. Lai XJ, Xu RC, Huang WQ (2011) Geometry optimization of bimetallic clusters using an efficient heuristic method. J Chem Phys 135(16):164109

364. Lange A, Gattin Z, Van Melckebeke H, Wasmer C, Soragni A, van Gunsteren WF, Meier BH (2009) A combined solid-state NMR and MD characterization of the stability and dynamics of the HET-s(218–289) prion in its amyloid conformation. Chembiochem 10(10):1657–1665
365. Larkin MA, Blackshields G, Brown NP, Chenna R, McGettigan PA, McWilliam H, Valentin F, Wallace IM, Wilm A, Lopez R, Thompson JD, Gibson TJ, Higgins DG (2007) Clustal W and Clustal X version 2.0. Bioinformatics 23(21):2947–2948
366. Lau A, Mays CE, Genovesi S, Westaway D (2012) RGG repeats of PrP-like Shadoo protein bind nucleic acids. Biochemistry 51(45):9029–9031
367. Laude H, Vilette D, Le Dur A, Archer F, Soulier S, Besnard N, Essalmani R, Vilotte JL (2002) New in vivo and ex vivo models for the experimental study of sheep scrapie: development and perspectives. C R Biol 325(1):49–57
368. Laws DD, Bitter HM, Liu K, Ball HL, Kaneko K, Wille H, Cohen FE, Prusiner SB, Pines A, Wemmer DE (2001) Solid-state NMR studies of the secondary structure of a mutant prion protein fragment of 55 residues that induces neurodegeneration. Proc Natl Acad Sci U S A 98(20):11686–11690
369. Lawson VA, Vella LJ, Stewart JD, Sharples RA, Klemm H, Machalek DM, Masters CL, Cappai R, Collins SJ, Hill AF (2008) Mouse-adapted sporadic human Creutzfeldt-Jakob disease prions propagate in cell culture. Int J Biochem Cell Biol 40(12):2793–2801
370. Leary RH (1997) Global optima of Lennard-Jones clusters. J Glob Optim 11(1):35–53
371. Lee S, Antony L, Hartmann R, Knaus KJ, Surewicz K, Surewicz WK, Yee VC (2010) Conformational diversity in prion protein variants influences intermolecular β-sheet formation. EMBO J 29(1):251–262
372. Lee S, Eisenberg D (2003) Seeded conversion of recombinant prion protein to a disulfide-bonded oligomer by a reduction-oxidation process. Nat Struct Biol 10(9):725–730
373. Lee SW, Mou Y, Lin SY, Chou FC, Tseng WH, Chen C, Lu CYD, Yu SSF, Chan JCC (2008) Steric zipper of the amyloid fibrils formed by residues 109 to 122 of the Syrian hamster prion protein. J Mol Biol 378(5):1142–1154
374. Lee DC, Sakudo A, Kim CK, Nishimura T, Saeki K, Matsumoto Y, Yokoyama T, Chen SG, Itohara S, Onodera T (2006) Fusion of Doppel to octapeptide repeat and N-terminal half of hydrophobic region of prion protein confers resistance to serum deprivation. Microbiol Immunol 50(3):203–209
375. Lee S, Yee VC (2011) Diversity in the cross-β spine structure of prion peptides. Journal:To be Published: http://www.rcsb.org/pdb/explore/explore.do?structureId=3MD4. doi:10.2210/pdb3md4/pdb and doi:10.2210/pdb3md5/pdb
376. Levenberg K (1944) A method for the solution of certain nonlinear problems in least squares. Quart Appl Math 2:164–166
377. Li X, Chen XD (2005) Global convergence of shortest-residual family of conjugate gradient methods without line search. Asia-Pac J Oper Res 22(4):529–538
378. Li DH, Fukushima M (2001) A modified BFGS method and its global convergence in nonconvex minimization. J Comput Math Appl 129(1–2):15–35
379. Li PF, Gao GY, Wang YC, Ma YM (2010) Crystal structures and exotic behavior of magnesium under pressure. J Phys Chem C 114(49):21745–21749
380. Li Y, Mascagni M, Gorin A (2009) A decentralized parallel implementation for parallel tempering algorithm. Parallel Comput 35(5):269–283
381. Li J, Mei FH, Xiao GF, Guo CY, Lin DH (2007) 1^H, 13^C and 15^N resonance assignments of rabbit prion protein 91–228. J Biomol NMR 38(2):181
382. Li Y, Protopopescub VA, Arnoldc N, Zhang X, Gorine A (2009) Hybrid parallel tempering and simulated annealing method. App Math Comput 212(1):216–228
383. Li YM, Tian B, Zhang BY, Dong XP (2001) Diagnosis of bovine spongiform encephalopathy and scrapie by Western blot. Sheng Wu Gong Cheng Xue Bao 17(5):494–497
384. Li L, Wei DQ, Wang JF, Chou KC (2007) Computational studies of the binding mechanism of calmodulin with chrysin. Biochem Biophys Res Commun 358(4):1102–1107
385. Liang J, Kong Q (2012) α-Cleavage of cellular prion protein. Prion 6(5):453–460

386. Liao QH, Gao QZ, Wei J, Chou KC (2011) Docking and molecular dynamics study on the inhibitory activity of novel inhibitors on epidermal growth factor receptor (EGFR). Med Chem 7(1):24–31
387. Liao YC, Tokes Z, Lim E, Lackey A, Woo CH, Button JD, Clawson GA (1987) Cloning of rat "prion-related protein" cDNA. Lab Invest 57(4):370–374
388. Liemann S, Glockshuber R (1999) Influence of amino acid substitutions related to inherited human prion diseases on the thermodynamic stability of the cellular prion protein. Biochemistry 38(11):3258–3267
389. Lin H, Ding H (2011) Predicting ion channels and their types by the dipeptide mode of pseudo amino acid composition. J Theor Biol 269(1):64–69
390. Lin DH, Wen Y (2011) Progresses on prion proteins. Scientia Sinica Chimica 41(4):683–698 (in Chinese)
391. Liu H, Farr-Jones S, Ulyanov NB, Llinas M, Marqusee S, Groth D, Cohen FE, Prusiner SB, James TL (1999) Solution structure of Syrian hamster prion protein rPrP(90–231). Biochemistry 38(17):5362–5377
392. Liu J, Han J, Yin H (1993) Global convergence of the Fletcher-Reeves algorithm with inexact line search. Manuscript, Institute of Applied Mathematics, Chinese Academy of Science, Beijing
393. Liu HY, Li Q, Zhu L, Ma YM (2011) Superhard and superconductive polymorphs of diamond-like BC3. Phys Lett A 375(3):771–774
394. Liu DC, Nocedal J (1989) On the limited memory method for large scale optimization. Math Program B 45:503–528
395. Liu A, Riek R, Zahn R, Hornemann S, Glockshuber R, Wüthrich K (1999) Peptides and proteins in neurodegenerative disease: helix propensity of a polypeptide containing helix 1 of the mouse prion protein studied by NMR and CD spectroscopy. Biopolymers 51(2):145–152
396. Locatelli M (2000) Simulated annealing algorithms for continuous global optimization: convergence conditions. J Optim Theory Appl 104(1):121–133
397. Locatelli M, Schoen F (2008) Structure prediction and global optimization. Optima Math Program Soc Newsl U S A 76:1–8
398. Loftus B, Rogers M (1997) Characterization of a prion protein (PrP) gene from rabbit, a species with apparent resistance to infection by prions. Gene 184(2):215–219
399. Lopez CD, Yost CS, Prusiner SB, Myers RM, Lingappa VR (1990) Unusual topogenic sequence directs prion protein biogenesis. Science 248(4952):226–229
400. Lucidi S, Piccioni M (1989) Random tunneling by means of acceptance-rejection sampling for global optimization. J Optim Theory Appl 62(2):255–277
401. Luenberger DG, Ye YY (2008) Linear and nonlinear programming, 3rd edn. International series in operations research & management science, vol 116. New York, Springer. ISBN:978-0-387-74502-2
402. Lührs T, Riek R, Güntert P, Wüthrich K (2003) NMR structure of the human doppel protein. J Mol Biol 326(5):1549–1557
403. Luo XY, Yang JH, Liu HY, Wu XJ, Wang YC, Ma YM, Wei SH, Gong XG, Xiang HJ (2011) Predicting two-dimensional boron-carbon compounds by the global optimization method. J Am Chem Soc 133(40):16285–16290
404. Lutz J, Brabeck C, Niemann HH, Kloz U, Korth C, Lingappa VR, Bürkle A (2010) Microdeletions within the hydrophobic core region of cellular prion protein alter its topology and metabolism. Biochem Biophys Res Commun 393(3):439–444
405. Lv J, Wang YC, Zhu L, Ma YM (2011) Predicted Novel high-pressure phases of lithium. Phys Rev Lett 106:15503–15506
406. Lysek DA, Schorn C, Nivon LG, Esteve-Moya V, Christen B, Calzolai L, von Schroetter C, Fiorito F, Herrmann T, Guntert P (2005) Prion protein NMR structures of cats, dogs, pigs, and sheep. Proc Natl Acad Sci U S A 102(3):640–645
407. Ma Q, Fan JB, Zhou Z, Zhou BR, Meng SR, Hu JY, Chen J, Liang Y (2012) The contrasting effect of macromolecular crowding on amyloid fibril formation. PLoS ONE 7(4):e36288

408. Ma BY, Nussinov R (2002) Molecular dynamics simulations of alanine rich β-sheet oligomers: insight into amyloid formation. Protein Sci 11(10):2335–2350
409. Mackay AL (1962) A dense non-crystallographic packing of equal spheres. Acta Cryst 15:916–918
410. McCoy AJ, Kortt AA (1997) The 1.8 $\mathring{A}$ crystal structure of winged bean albumin, the major albumin from Psophocarpus tetragonolobus (L) DC. J Mol Biol 269(5):881–891. www.sciencedirect.com/science/article/pii/S0022283697910676
411. Madec JY, Vanier A, Dorier A, Bernillon J, Belli P, Baron T (1997) Biochemical properties of protease resistant prion protein PrP^{Sc} in natural sheep scrapie. Arch Virol 142(8):1603–1612
412. Maiti NR, Surewicz WK (2001) The role of disulfide bridge in the folding and stability of the recombinant human prion protein. J Biol Chem 276(4):2427–2431
413. Mak V (2001) On the Asymmetric traveling salesman problem with replenishment arcs. Ph.D. thesis, Department of Mathematics and Statistics, The University of Melbourne
414. Mangels C, Kellner R, Einsiedel J, Weiglmeier PR, Rosch P, Gmeiner P, Schwarzinger S (2010) The therapeutically anti-prion active antibody-fragment scFv-W226: paramagnetic relaxation-enhanced NMR spectroscopy aided structure elucidation of the paratope-epitope interface. J Biomol Struct Dyn 28(1):13–22
415. Margittai M, Langen R (2008) Fibrils with parallel in-register structure constitute a major class of amyloid fibrils: molecular insights from electron paramagnetic resonance spectroscopy. Q Rev Biophys 41(3–4):265–297
416. Marquardt DW (1963) An algorithm for least-squares estimation of nonlinear parameters. SIAM J Appl Math 11(2):431–441
417. Mascarenhas WF (2004) The BFGS method with exact line searches fails for non-convex objective functions. Math Program Ser A 99(1):49–61
418. Mastrianni JA, Curtis MT, Oberholtzer JC, Da Costa MM, DeArmond S, Prusiner SB, Garbern JY (1995) Prion disease (PrP-A117V) presenting with ataxia instead of dementia. Neurology 45(11):2042–2050
419. Mays CE, Coomaraswamy J, Watts JC, Yang J, Ko KW, Strome B, Mercer RC, Wohlgemuth SL, Schmitt-Ulms G, Westaway D (2014) Endoproteolytic processing of the mammalian prion glycoprotein family. FEBS J 281(3):862–876
420. Mays CE, Yeom J, Kang HE, Bian J, Khaychuk V, Kim Y, Bartz JC, Telling GC, Ryou C (2011) In vitro amplification of misfolded prion protein using lysate of cultured cells. PLoS ONE 6(3):e18047
421. Meli M, Gasset M, Colombo G (2011) Dynamic diagnosis of familial prion diseases supports the $\beta 2$-$\alpha 2$ loop as a universal interference target. PLoS ONE 6(4):e19093
422. Merlot LTG, Boland N, Hughes BD, Stuckey PJ (2002) A hybrid algorithm for the examination timetabling problem. Lect Notes Comput Sci 2740:207–231
423. Merz PA, Kascsak RJ, Rubenstein R, Carp RI, Wisniewski HM (1987) Antisera to scrapie-associated fibril protein and prion protein decorate scrapie-associated fibrils. J Virol 61(1):42–49
424. Metropolis N, Rosenbluth AW, Rosenbluth MN, Teller AH, Teller E (1953) Equation of state calculations by fast computing machines. J Chem Phys 21(6):1087
425. Miki M, Hiroyasu T, Kasai M, Ikeuchi M (1999) Temperature parallel simulated annealing with adaptive neighbourhood for continuous optimization problem. In: Proceedings of the IASTED international conference parallel and distributed computing and systems, Boston, pp 302–316
426. Miller JM, Jenny AL, Taylor WD, Marsh RF, Rubenstein R, Race RE (1993) Immunohistochemical detection of prion protein in sheep with scrapie. J Vet Diagn Invest 5(3):309–316
427. Mills NL, Surewicz K, Surewicz WK, Sonnichsen FD (2009) Residue 129 polymorphism and conformational dynamics of familial prion diseases associated with the human prion protein variant D178N. doi:10.2210/pdb2k1d/pdb. To be published: http://www.rcsb.org/pdb/explore/explore.do?structureId=2K1D

428. Ministry of Health (1920) Reports on the pandemic of influenza 1918–1919, Reports on public health and medical subjects No. 4, Ministry of Health, London, England: His Majesty's Stationery Office
429. Mohabatkar H (2010) Prediction of cyclin proteins using Chou's pseudo amino acid composition. Protein Peptide Lett 17(10):1207–1214
430. Moré JJ, Sorensen DC (1983) Computing a trust region step. SIAM J Sci Stat Comput 4(3):553–572
431. More JJ, Wu ZJ (1997) Global continuation for distance geometry problems. SIAM J Optim 7(3):814–836
432. Moore RA, Vorberg I, Priola SA (2005) Species barriers in prion diseases – brief review. Arch Virol Suppl 19:187–202
433. Morales R, Abid K, Soto C (2007) The prion strain phenomenon: molecular basis and unprecedented features. Biochim Biophys Acta 1772(6):681–691
434. Morley JF, Brignull HR, Weyers JJ, Morimoto RI (2002) The threshold for polyglutamine-expansion protein aggregation and cellular toxicity is dynamic and influenced by aging in Caenorhabditis elegans. Proc Natl Acad Sci U S A 99(16):10417–10422
435. Morse PM, Feshbach H (1953) Asymptotic series; method of steepest descent. §4.6 in methods of theoretical physics, Part I. McGraw-Hill, New York, pp 434–443
436. Northby JA (1987) Structure and binding of the Lennard-Jones clusters: 13 <= N<=147. J Chem Phys 87(10):6166–6177
437. Muramoto T, Kitamoto T, Tateishi J, Goto I (1992) Successful transmission of Creutzfeldt-Jakob disease from human to mouse verified by prion protein accumulation in mouse brains. Brain Res 599(2):309–316
438. Nature Editorial (2001) More than just mad cow disease. Nature Struct Biol 8(4):281. doi:10.1038/86132
439. Nelson R, Sawaya MR, Balbirnie M, Madsen A, Riekel C, Grothe R, Eisenberg D (2005) Structure of the cross-beta spine of amyloid-like fibrils. Nature 435(7043):773–778
440. Ning L, Wang Q, Zheng Y, Liu H, Yao X (2015) Effects of the A117V mutation on the folding and aggregation of palindromic sequences (PrP113–120) in prion: insights from replica exchange molecular dynamics simulations. Mol Biosyst 11(2):647–655
441. Nisbet RM, Harrison CF, Lawson VA, Masters CL, Cappai R, Hill AF (2010) Residues surrounding the glycosylphosphatidylinositol anchor attachment site of PrP modulate prion infection: insight from the resistance of rabbits to prion disease. J Virol 84(13):6678–6686
442. Nocedal J (1980) Updating quasi-Newton matrices with limited storage. Math Comput 35(151):773–782
443. Nocedal J, Morales J (2000) Automatic preconditioning by limited memory quasi-Newton updating. SIAM J Optim 10(4):1079–1096
444. Nocedal J, Wright SJ (2006) Numerical optimization, 2nd edn. Springer, Berlin/New York. ISBN:978-0-387-30303-1
445. Norstrom EM, Mastrianni JA (2005) The AGAAAAGA palindrome in PrP is required to generate a productive PrP^{Sc}-PrP^{C} complex that leads to prion propagation. J Biol Chem 280(29):27236–27243
446. Oboznaya MB, Gilch S, Titova MA, Koroev DO, Volkova TD, Volpina OM, Schätzl HM (2007) Antibodies to a nonconjugated prion protein peptide 95–123 interfere with PrP^{Sc} propagation in prion-infected cells. Cell Mol Neurobiol 27(3):271–284
447. Ogayar A, Sánchez-Pérez M (1998) Prions: an evolutionary perspective. Int Microbiol 1(3):183–190
448. Okimoto N, Yamanaka K, Suenaga A, Hata M, Hoshimo T (2002) Molecular dynamics simulations of prion proteins – effect of Ala117$\rightarrow$Val mutation. Biophys J 82(5):2746–2757
449. Oliveira-Martins JB, Yusa S, Calella AM, Bridel C, Baumann F, Dametto P, Aguzzi A (2010) Unexpected tolerance of α-cleavage of the prion protein to sequence variations. PLoS ONE 5(2):e9107

450. Onodera T, Ikeda T, Muramatsu Y, Shinagawa M (1993) Isolation of scrapie agent from the placenta of sheep with natural scrapie in Japan. Microbiol Immunol 37(4):311–316
451. Orengo CA, Brown NP, Taylor WR (1992) Fast structure alignment for protein databank searching. Proteins: Struct Funct Genet 14(2):139–167
452. Ott CM, Akhavan A, Lingappa VR (2007) Specific features of the prion protein transmembrane domain regulate nascent chain orientation. J Biol Chem 282(15):11163–11171
453. Ott CM, Lingappa VR (2004) Signal sequences influence membrane integration of the prion protein. Biochemistry 43(38):11973–11982
454. Owen JP, Maddison BC, Whitelam GC, Gough KC (2007) Use of thermolysin in the diagnosis of prion diseases. Mol Biotechnol 35(2):161–170
455. Oxenoid K, Chou JJ (2005) The structure of phospholamban pentamer reveals a channel-like architecture in membranes. Proc Natl Acad Sci USA 102(31):10870–10875
456. Oztabak K, Ozkan E, Soysal I, Paya I, Un C (2009) Detection of prion gene promoter and intron1 indel polymorphisms in Anatolian water buffalo (Bubalus bubalis). J Anim Breed Genet 126(6):463–467
457. Pan KM, Baldwin M, Nguyen J (1993) Conversion of α-helices into β-sheets features in the formation of the scrapie prion proteins. Proc Natl Acad Sci U S A 90(23):10962–10966
458. Paquet E, Viktor HL (2015) Molecular dynamics, Monte Carlo simulations, and Langevin dynamics: a computational review. Biomed Res Int. doi:10.1155/2015/183918
459. Paramithiotis E, Pinard M, Lawton T, LaBoissiere S, Leathers VL, Zou WQ, Estey LA, Lamontagne J, Lehto MT, Kondejewski LH, Francoeur GP, Papadopoulos M, Haghighat A, Spatz SJ, Head M, Will R, Ironside J, O'Rourke K, Tonelli Q, Ledebur HC, Chakrabartty A, Cashman NR (2003) A prion protein epitope selective for the pathologically misfolded conformation. Nat Med 9(7):893–899
460. Pardalos PM, Shalloway D, Xue GL (1994) Optimization methods for computing global minima of nonconvex potential energy functions. J Glob Optim 4(2):117–133
461. Patel A, Vasiljevic S, Jones IM (2013) Unique structural properties associated with mouse prion triangle 105–125 protein. Prion 7(3):235–243
462. Pawar AP, Dubay KF, Zurdo J, Chiti F, Vendruscolo M, Dobson CM (2005) Prediction of "aggregation-prone" and "aggregation-susceptible" regions in proteins associated with neurodegenerative diseases. J Mol Biol 350(2):379–392
463. Perez DR, Damberger FF, Wuthrich K (2010) Horse prion protein NMR structure and comparisons with related variants of the mouse prion protein. J Mol Biol 400(2):121–128
464. Peretz D, Williamson RA, Kaneko K, Vergara J, Leclerc E, Schmitt-Ulms G, Mehlhorn IR, Legname G, Wormald MR, Rudd PM, Dwek RA, Burton DR, Prusiner SB (2001) Antibodies inhibit prion propagation and clear cell cultures of prion infectivity. Nature 412(6848):739–743
465. Peretz D, Williamson RA, Matsunaga Y, Serban H, Pinilla C, Bastidas RB, Rozenshteyn R, James TL, Houghten RA, Cohen FE, Prusiner SB, Burton DR (1997) A conformational transition at the N-terminus of the prion protein features in formation of the scrape isoform. J Mol Biol 273(3):614–622
466. Perez DR, Wuthrich K (2008) NMR structure of the bank vole prion protein at 20 degrees C contains a structured loop of residues 165–171. J Mol Biol 383(2):306–312
467. Pielak RM, Chou JJ (2010) Flu channel drug resistance: a tale of two sites. Protein Cell 1(3):246–258
468. Pielak RM, Chou JJ (2010) Solution NMR structure of the V27A drug resistant mutant of influenza A M2 channel. Biochem Biophys Res Commun 401(1):58–63
469. Pielak RM, Chou JJ (2011) Influenza M2 proton channels. Biochim Biophys Acta 1808(2):522–529
470. Pielak RM, Jason R, Schnell JR, Chou JJ (2009) Mechanism of drug inhibition and drug resistance of influenza A M2 channel. Proc Natl Acad Sci U S A 106(18):7379–7384
471. Pimenta J, Viegas A, Sardinha J, Martins IC, Cabrita EJ, Fontes CM, Prates JA, Pereira RM (2013) NMR solution structure and SRP54M predicted interaction of the N-terminal sequence (1–30) of the ovine Doppel protein. Peptides 49C:32–40. doi:10.1016/j.peptides.2013.08.013

472. Pintér JD (2006) Global optimization – scientific and engineering case studies. Springer, New York, p 153. ISBN:978-0-387-30408-3
473. Polak B, Ribière G (1969) Note sur la convergence des méthodes de directions conjuguées. Rev Fran Informat Rech Opér 16:35–43
474. Polymenidoua M, Trusheimb H, Stallmacha L, Moosa R, Julius JA, Mielea G, Lenzbauerb C, Aguzzia A (2008) Canine MDCK cell lines are refractory to infection with human and mouse prions. Vaccine 26(21):2601–2614
475. Powell MJD (1970) A new algorithm for unconstrained optimization. In: Rosen JB, Mangasarian OL, Ritter K (eds) Nonlinear programming. Academic, New York, pp 31–66
476. Powell MJD (1970) A Fortran subroutine for solving systems of nonlinear algebraic equations. In: Rabinowitz P (ed) Numerical methods for nonlinear algebraic equations. Gordon and Breach, London, Chapter 7, pp 115–161
477. Powell MJD (1984) Nonconvex minimization calculations and the conjugate gradient method. In: Griffiths DF (ed) Numerical analysis. Lecture notes in mathematics, vol 1066. Springer, Berlin, pp 122–141
478. Powell MJD (1986) Convergence properties of algorithms for nonlinear optimization. SIAM Rev 28(4):487–500
479. Pray L (2008) Discovery of DNA structure and function: Watson and Crick. Nat Educ 1(1): 16–18
480. Premzl M, Sangiorgio L, Strumbo B, Marshall Graves JA, Simonic T, Gready JE (2003) Shadoo, a new protein highly conserved from fish to mammals and with similarity to prion protein. Gene 314:89–102
481. Press WH, Flannery BP, Teukolsky SA, Vetterling WT (1992) Numerical recipes in FORTRAN: the art of scientific computing, 2nd edn. Cambridge University Press, Cambridge, p 414
482. Press WH, Teukolsky SA, Vetterling WT, Flannery BP (2007) Section 10.12. Simulated annealing methods. Numerical recipes: the art of scientific computing, 3rd edn. Cambridge University Press, New York
483. Priola SA, Chesebro B (1995) A single hamster PrP amino acid blocks conversion to protease-resistant PrP in scrapie-infected mouse neuroblastoma cells. J Virol 69(12):7754–7758
484. Prusiner SB (1982) Novel proteinaceous infectious particles cause scrapie. Science 216(4542):136–144
485. Prusiner SB (1997) Prion diseases and the BSE crisis. Science 278(5336):245–251
486. Prusiner SB (1998) Prions (Nobel Lecture). Proc Natl Acad Sci U S A 95(23):13363–13383
487. Prusiner SB, Scott MR, DeArmond SJ, Cohen FE (1998) Prion protein biology. Cell 93(3):337–348
488. Qing LL, Zhao H, Liu LL (2014) Progress on low susceptibility mechanisms of transmissible spongiform encephalopathies. Dongwuxue Yanjiu 35(5):436–445
489. Ragg E, Tagliavini F, Malesani P, Monticelli L, Bugiani O, Forloni G, Salmona M (1999) Determination of solution conformations of PrP106–126, a neurotoxic fragment of prion protein, by 1H NMR and restrained molecular dynamics. Eur J Biochem 266(3):1192–1201
490. Reeves C (1995) Modern heuristics for combinatorial optimization. Wiley, New York
491. Reilly CE (2000) Nonpathogenic prion protein (PrP^C) acts as a cell-surface signal transducer. J Neurol 247(10):819–820
492. Rezaei H, Choiset Y, Eghiaian F, Treguer E, Mentre P, Debey P, Grosclaude J, Haertle T (2002) Amyloidogenic unfolding intermediates differentiate sheep prion protein variants. J Mol Biol 322(4):799–814
493. Rhee SH, Chung CP, Chung CP, Park YJ (2009) Method for preparing a prion-free bond grafting substitute. Research Triangle Park NC USA Patent no. 20090304807. http://www.faqs.org/patents/app/20090304807#ixzz38Fci5wsI. Accessed in May 2015
494. Riek R, Hornemann S, Wider G, Billeter M, Glockshuber R, Wuthrich K (1996) NMR structure of the mouse prion protein domain PrP(121–231). Nature 382(6587):180–182

495. Rivillas-Acevedo L, Grande-Aztatzi R, Lomeli I, Garcia JE, Barrios E, Teloxa S, Vela A, Quintanar L (2011) Spectroscopic and electronic structure studies of copper(II) binding to His111 in the human prion protein fragment 106–115: evaluating the role of protons and Methionine residues. Inorg Chem 50(5):1956–1972
496. Robakis NK, Sawh PR, Wolfe GC, Rubenstein R, Carp RI, Innis MA (1986) Isolation of a cDNA clone encoding the leader peptide of prion protein and expression of the homologous gene in various tissues. Proc Natl Acad Sci U S A 83(17):6377–6381
497. Roberts GW, Lofthouse R, Allsop D, Landon M, Kidd M, Prusiner SB, Crow TJ (1988) CNS amyloid proteins in neurodegenerative diseases. Neurology 38(10):1534–1540
498. Romeijn HE, Smith RL (1994) Simulated annealing for constrained global optimization. J Glob Optim 5(2):101–126
499. Romeijn HE, Zabinski ZB, Graesser DL, Neogi S (1999) New reflection generator for simulated annealing in mixed-integer/continuous global optimization. J Optim Theory Appl 101(2):403–427
500. Romero D, Barron C, Gomez S (1999) The optimal geometry of Lennard-Jones clusters: 148–309. Comput Phys Commun 123(1–3):87–96
501. Rosenbrock HH (1960) An automatic method for finding the greatest or least value of a function. Comput J 3(3):175–184
502. Ross SM (2014) Introduction to probability models, 11th edn. Academic Press (Elsevier): San Diego. ISBN:978-0124079489
503. Rossetti G, Cong X, Caliandro R, Legname G, Carloni P (2011) Common structural traits across pathogenic mutants of the human prion protein and their implications for familial prion diseases. J Mol Biol 411(3):700–712
504. Rossetti G, Giachin G, Legname G, Carloni P (2010) Structural facets of disease-linked human prion protein mutants: a molecular dynamic study. Proteins 78(16):3270–3280
505. Rubenstein R, Kascsak RJ, Papini M, Kascsak R, Carp RI, LaFauci G, Meloen R, Langeveld J (1999) Immune surveillance and antigen conformation determines humoral immune response to the prion protein immunogen. J Neurovirol 5(4):401–413
506. Ryan KJ, Ray CG et al (eds) (2004) Sherris medical microbiology, 4th edn. McGraw Hill, New York, pp 624–628. ISBN:0-8385-8529-9
507. Sachsamanoglou M, Paspaltsis I, Petrakis S, Verghese-Nikolakaki S, Panagiotidis CH, Voigtlander T, Budka H, Langeveld JP, Sklaviadis T (2004) Antigenic profile of human recombinant PrP: generation and characterization of a versatile polyclonal antiserum. J Neuroimmunol 146(1–2):22–32
508. Saez-Cirion A, Nieva JL, Gallaher WR (2003) The hydrophobic internal region of bovine prion protein shares structural and functional properties with HIV type 1 fusion peptide. AIDS Res Hum Retroviruses 19(11):969–978
509. Sakudo A, Lee DC, Li S, Nakamura T, Matsumoto Y, Saeki K, Itohara S, Ikuta K, Onodera T (2005) PrP cooperates with STI1 to regulate SOD activity in PrP-deficient neuronal cell line. Biochem Biophys Res Commun 328(1):14–19
510. Sakudo A, Lee DC, Nishimura T, Li S, Tsuji S, Nakamura T, Matsumoto Y, Saeki K, Itohara S, Ikuta K, Onodera T (2005) Octapeptide repeat region and N-terminal half of hydrophobic region of prion protein (PrP) mediate PrP-dependent activation of superoxide dismutase. Biochem Biophys Res Commun 326(3):600–606
511. Sakudo A, Nakamura I, Lee DC, Saeki K, Ikuta K, Onodera T (2007) Neurotoxic prion protein (PrP) fragment 106–126 requires the N-terminal half of the hydrophobic region of PrP in the PrP-deficient neuronal cell line. Protein Pept Lett 14(1):1–6
512. Sakudo A, Nakamura I, Tsuji S, Ikuta K (2008) GPI-anchorless human prion protein is secreted and glycosylated but lacks superoxide dismutase activity. Int J Mol Med 21(2):217–222
513. Salomon-Ferrer R, Gotz AW, Poole D, Grand SL, Walker RC (2013) Routine microsecond molecular dynamics simulations with AMBER on GPUs. 2. Explicit solvent particle mesh Ewald. J Chem Theory Comput 8(5):1542–1555

514. Sasaki K, Gaikwad J, Hashiguchi S, Kubota T, Sugimura K, Kremer W, Kalbitzer HR, Akasaka K (2008) Reversible monomer-oligomer transition in human prion protein. Prion 2(3):118–122
515. Saupe SJ (2011) The [Het-s] prion of Podospora anserina and its role in heterokaryon incompatibility. Semin Cell Dev Biol 22(5):460–468
516. Sauve S, Buijs D, Gingras G, Aubin Y (2012) Interactions between the conserved hydrophobic region of the prion protein and dodecylphosphocholine micelles. J Biol Chem 287(3):1915–1922
517. Sawaya MR, Sambashivan S, Nelson R, Ivanova MI, Sievers SA, Apostol MI, Thompson MJ, Balbirnie M, Wiltzius JJ, McFarlane HT, Madsen A, Riekel C, Eisenberg D (2007) Atomic structures of amyloid cross-β spines reveal varied steric zippers. Nature 447(7143):453–457
518. Schmerr MJ, Goodwin KR, Cutlip RC (1994) Capillary electrophoresis of the scrapie prion protein from sheep brain. J Chromatogr A 680(2):447–453
519. Schnell JR, Chou JJ (2008) Structure and mechanism of the M2 proton channel of influenza A virus. Nature 451(7178):591–595
520. Schuur PC (1997) Classification of acceptance criteria for the simulated annealing algorithm. Math Oper Res 22(2):266–275
521. Seabury CM, Derr JN (2003) Identification of a novel ovine PrP polymorphism and scrapie-resistant genotypes for St. Croix White and a related composite breed. Cytogenet Genome Res 102(1–4):85–88
522. Sechen C (1988) VLSI placement and global routing using simulated annealing. Kluwer Academic, Boston. ISBN:978-1-4612-8957-9
523. Sekijima M, Motono C, Yamasaki S, Kaneko K, Akiyama Y (2003) Molecular dynamics simulation of dimeric and monomeric forms of human prion protein: insight into dynamics and properties. Biophys J 85(2):1176–1185
524. Senator A, Rachidi W, Lehmann S, Favier A, Benboubetra M (2004) Prion protein protects against DNA damage induced by paraquat in cultured cells. Free Radic Biol Med 37(8): 1224–1230
525. Shamsir MS, Dalby AR (2005) One gene, two diseases and three conformations: molecular dynamics simulations of mutants of human prion protein at room temperature and elevated temperatures. PROTEINS: Struct Funct Bioinf 59(2):275–290
526. Shi Q, Dong XP (2011) (Ctm)PrP and ER stress: a neurotoxic mechanism of some special PrP mutants. Prion 5(3):123–125
527. Shin W, Lee B, Hong S, Ryou C, Kwon M (2008) Cloning and expression of a prion protein (PrP) gene from Korean bovine (Bos taurus coreanae) and production of rabbit anti-bovine PrP antibody. Biotechnol Lett 30(10):1705–1711
528. Shinagawa M, Munekata E, Doi S, Takahashi K, Goto H, Sato G (1986) Immunoreactivity of a synthetic pentadecapeptide corresponding to the N-terminal region of the scrapie prion protein. J Gen Virol 67(Pt 8):1745–1750
529. Siarry P, Berthiau G, Durbin F, Haussy J (1997) Enhanced simulated annealing for globally minimizing functions of many-continuous variables. ACM Trans Math Softw 23(2):209–228
530. Sicher M, Mohr S, Goedecker S (2010) Efficient moves for global geometry optimization methods and their application to binary systems. J Chem Phys 134(4):44–106
531. Sigurdson CJ, Joshi-Barr S, Bett C, Winson O, Manco G, Schwarz P, Rülicke T, Nilsson KP, Margalith I, Raeber A, Peretz D, Hornemann S, Wüthrich K, Aguzzi A (2011) Spongiform encephalopathy in transgenic mice expressing a point mutation in the $\beta 2$-$\alpha 2$ loop of the prion protein. J Neurosci 31(39):13840–13847
532. Sigurdson CJ, Manco G, Schwarz P, Liberski P, Hoover EA, Hornemann S, Polymenidou M, Miller MW, Glatzel M, Aguzzi A (2006) Strain fidelity of chronic wasting disease upon murine adaptation. J Virol 80(24):12303–12311
533. Sigurdson CJ, Nilsson KP, Hornemann S, Heikenwalder M, Manco G, Schwarz P, Ott D, Rülicke T, Liberski PP, Julius C, Falsig J, Stitz L, Wüthrich K, Aguzzi A (2009) De novo generation of a transmissible spongiform encephalopathy by mouse transgenesis. Proc Natl Acad Sci U S A 106(1):304–309

534. Sigurdson CJ, Nilsson KP, Hornemann S, Manco G, Fernández-Borges N, Schwarz P, Castilla J, Wüthrich K, Aguzzi A (2010) A molecular switch controls interspecies prion disease transmission in mice. J Clin Invest 120(7):2590–2599
535. Simone AD, Pedone C, Vitagliano L (2008) Structure, dynamics, and stability of assemblies of the human prion fragment SNQNNF. Biochem Biophys Res Commun 366(3):800–806
536. Sinkala Z (2006) Soliton/exciton transport in proteins. J Theory Biol 241(4):919–927
537. Smith CJ, Drake AF, Banfield BA, Bloomberg GB, Palmer MS, Clarke AR, Collinge J (1997) Conformational properties of the prion octa-repeat and hydrophobic sequences. FEBS Lett 405(3):378–384
538. Snyman JA (2005) Practical mathematical optimization: an introduction to basic optimization theory and classical and new gradient-based algorithms. Springer, New York. ISBN:978-0387243481
539. Sorensen DC (1994) Minimization of a large scale quadratic function subject to an ellipsoidal constraint. Department of Computational and Applied Mathematics, Rice University, Technical Report TR94-27
540. Soto C (2011) Prion hypothesis: the end of the controversy? Trends Biochem Sci 36(3):151–158
541. Soto C, Castilla J (2004) The controversial protein-only hypothesis of prion propagation. Nat Med 10 Suppl:S63–S67
542. Souan L, Tal Y, Felling Y, Cohen IR, Taraboulos A, Mor F (2001) Modulation of proteinase-K resistant prion protein by prion peptide immunization. Eur J Immunol 31(8):2338–2346
543. Stahl N, Prusiner SB (1991) Prions and prion proteins. The FASEB J 5(13):2799–2807
544. Stanker LH, Scotcher MC, Lin A, McGarvey J, Prusiner SB, Hnasko R (2012) Novel epitopes identified by anti-PrP monoclonal antibodies produced following immunization of Prnp0/0 Balb/cJ mice with purified scrapie prions. Hybridoma (Larchmt) 31(5):314–324
545. Starck V (1996) Implementation of simulated annealing optimization method for APLAC circuit simulator. Master thesis of Helsinki University of Technology
546. Steihaug T (1983) The conjugate gradient method and trust regions in large scale optimization. SIAM J Numer Anal 20(3):626–637
547. Stewart P, Shen C, Zhao D, Goldmann W (2009) Genetic analysis of the SPRN gene in ruminants reveals polymorphisms in the alanine-rich segment of shadoo protein. J Gen Virol 90(Pt 10):2575–2580
548. Stote R, Dejaegere A, Kuznetsov D, Falquet L (1999) Molecular dynamics simulations CHARMM. http://www.ch.embnet.org/MD_tutorial/
549. Strodel B, Lee JW, Whittleston CS, Wales DJ (2010) Transmembrane structures for Alzheimer's $A\beta$(1-42) oligomers. J Am Chem Soc 132(38):13300–13312
550. Sugita Y, Okamoto Y (1999) Replica-exchange molecular dynamics method for protein folding. Chem Phys Lett 314(1–2):141–151
551. Sun J, Zhang JP (2001) Global Convergence of Conjugate Gradient Methods without Line Search. Ann Oper Res 103(1–4):161–173
552. Sunde M, Serpell LC, Bartlam M, Fraser PE, Pepys MB, Blake CC (1997) Common core structure of amyloid fibrils by synchrotron X-ray diffraction. J Mol Biol 273(3):729–739
553. Supattapone S, Muramoto T, Legname G, Mehlhorn I, Cohen FE, DeArmond SJ, Prusiner SB, Scott MR (2001) Identification of two prion protein regions that modify scrapie incubation time. J Virol 75(3):1408–1413
554. Suzuki T, Kurokawa T, Hashimoto H, Sugiyama M (2002) cDNA sequence and tissue expression of Fugu rubripes prion protein-like: a candidate for the teleost orthologue of tetrapod PrPs. Biochem Biophys Res Commun 294(4):912–917
555. Sweeting B, Brown E, Chakrabartty A, Pai EF (2009) The structure of rabbit PrP^C: clues into species barrier and prion disease. Canadian Light Source 2009 Activity Report 28:72–73. www.lightsource.ca/science/pdf/activity_report_2009/28.pdf
556. Sweeting B, Brown E, Khan MQ, Chakrabartty A, Pai EF (2013) N-terminal helix-cap in α-helix 2 modulates β-state misfolding in rabbit and hamster prion proteins. PLoS ONE 8(5):e63047

557. Sweeting B, Brown E, Chakrabartty A, Pai EF (2009) The structure of rabbit PrP^C: clues into species barrier and prion disease susceptibility. Can Light Source 28:72–73. http://www.lightsource.ca/science/pdf/activity_report_2009/28.pdf
558. Sweeting B, Khan MQ, Chakrabartty A, Pai EF (2010) Structural factors underlying the species barrier and susceptibility to infection in prion disease. Biochem Cell Biol 88(2): 195–202
559. Swendsen RH, Wang JS (1986) Replica Monte Carlo simulation of spin-glasses. Phys Rev Lett 57(21):2607–2609
560. Szu H, Hartley R (1987) Fast simulated annealing. Phys Lett A 122(3–4):157–162
561. Tabrett CA, Harrison CF, Schmidt B, Bellingham SA, Hardy T, Sanejouand YH, Hill AF, Hogg PJ (2010) Changing the solvent accessibility of the prion protein disulfide bond markedly influences its trafficking and effect on cell function. Biochem J 428(2):169–182
562. Takahashi K, Shinagawa M, Doi S, Sasaki S, Goto H, Sato G (1986) Purification of scrapie agent from infected animal brains and raising of antibodies to the purified fraction. Microbiol Immunol 30(2):123–131
563. Takekida K, Kikuchi Y, Yamazaki T, Kakeya T, Takatori K, Tanamoto K, Sawada J, Tanimura A (2002) Study on the detection of prion protein in food products by a competitive enzyme-linked immunosorbent assay. Shokuhin Eiseigaku Zasshi 43(3):173–177
564. Takahashi H, Takahashi RH, Hasegawa H, Horiuchi M, Shinagawa M, Yokoyama T, Kimura K, Haritani M, Kurata T, Nagashima K (1999) Characterization of antibodies raised against bovine-PrP-peptides. J Neurovirol 5(3):300–307
565. Takeuchi H (2006) Clever and efficient method for searching optimal geometries of Lennard-Jones clusters. J Chem Inf Model 46(5):2066–2070
566. Tang H, Fu Y, Cui Y, He Y, Zeng X, Ploplis VA, Castellino FJ, Luo Y (2009) Fibrinogen has chaperone-like activity. Biochem Biophys Res Commun 378(3):662–667
567. Tcherkasskaya O, Sanders W, Chynwat V, Davidson EA, Orser CS (2003) The role of hydrophobic interactions in amyloidogenesis: example of prion-related polypeptides. J Biomol Struct Dyn 21(3):353–365
568. Thachuk C, Shmygelska A, Hoos HH (2007) A replica exchange Monte Carlo algorithm for protein folding in the HP model. BMC Bioinform 8:342–362
569. Timmes AG, Moore RA, Fischer ER, Priola SA (2013) Recombinant prion protein refolded with lipid and RNA has the biochemical hallmarks of a prion but lacks in vivo infectivity. PLoS ONE 8(7):e71081
570. Touati-Ahmed D, Storey C (1990) Efficient hybrid conjugate gradient techniques. J Optim Theory Appl 64(2):379–397
571. Tsai HHG (2005) Understanding the biophysical mechanisms of protein folding, misfolding, and aggregation at molecular level (in Chinese). Chem (The Chinese Chem Soc of Taipei) 63:601–612
572. Tseng CY, Yu CP, Lee HC (2009) Integrity of H1 helix in prion protein revealed by molecular dynamic simulations to be especially vulnerable to changes in the relative orientation of H1 and its S1 flank. Eur Biophys J 38(5):601–611
573. Truant R, Atwal RS, Desmond C, Munsie L, Tran T (2008) Huntington's disease: revisiting the aggregation hypothesis in polyglutamine neurodegenerative diseases. The FEBS J 275(17):4252–4262
574. Tulip WR, Varghese JN, Laver WG, Webster RG, Colman PM (1992) Refined crystal structure of the influenza virus N9 neuraminidase-NC41 Fab complex. J Mol Biol 227(1):122–148
575. Turi I, Kallay C, Szikszai D, Pappalardo G, Di Natale G, De Bona P, Rizzarelli E, Sovago I (2010) Nickel(II) complexes of the multihistidine peptide fragments of human prion protein. J Inorg Biochem 104(8):885–891
576. Uchida L, Heriyanto A, Thongchai C, Hanh TT, Horiuchi M, Ishihara K, Tamura Y, Muramatsu Y (2014) Genetic diversity in the prion protein gene (PRNP) of domestic cattle and water buffaloes in Vietnam, Indonesia and Thailand. J Vet Med Sci 76(7):1001–1008

577. van der Spoel D, Lindahl E, Hess B, van Buuren AR, Apol E, Meulenhoff PJ, Tieleman DP, Sijbers ALTM, Feenstra KA, van Drunen R, Berendsen HJC (2010) Gromacs user manual version 4.5.4. www.gromacs.org
578. Vanderbilt D, Louie SG (1984) A Monte Carlo simulated annealing approach to optimization over continuous variables. J Comput Phys 56(2):259–271
579. Valensin D, Gajda K, Gralka E, Valensin G, Kamysz W, Kozlowski H (2010) Copper binding to chicken and human prion protein amylodogenic regions: differences and similarities revealed by Ni2+ as a diamagnetic probe. J Inorg Biochem 104(1):71–78
580. Varghese JN, Epa VC, Colman PM (1995) Three-dimensional structure of the complex of 4-guanidino-NeuSAc2en and influenza virus neuraminidase. Protein Sci 4(6):1081–1087
581. Verdier JM (2012) Prions and prion diseases: new developments. NOVA Science Publishers, New York. ISBN:978-1-62100-027-3, Preface, pp vii–ix
582. Vidal RVV (1993) Applied simulated annealing. Lecture notes in economics and mathematical systems, vol 396. Springer, Berlin/New York
583. Vidal E, Fernández-Borges N, Pintado B, Ordóñez M, Márquez M, Fondevila D, Torres JM, Pumarola M, Castilla J (2013) Bovine spongiform encephalopathy induces misfolding of alleged prion-resistant species cellular prion protein without altering its pathobiological features. J Neurosci 33(18):7778–7786
584. Villa A, Mark AE, Saracino GAA, Cosentino U, Pitea D, Moro G, Salmona M (2006) Conformational polymorphism of the PrP 106–126 peptide in different environments: a molecular dynamics study. J Phys Chem B 110(3):1423–1428
585. Vilches S, Vergara C, Nicolas O, Sanclimens G, Merino S, Varon S, Acosta GA, Albericio F, Royo M, Del Rio JA, Gavin R (2013) Neurotoxicity of prion peptides mimicking the central domain of the cellular prion protein. PLoS ONE 8(8):e70881
586. Viles JH, Donne D, Kroon G, Prusiner SB, Cohen FE, Dyson HJ, Wright PE (2001) Local structural plasticity of the prion protein. Analysis of NMR relaxation dynamics. Biochemistry 40(9):2743–2753
587. Vilette D, Andreoletti O, Archer F, Madelaine MF, Vilotte JL, Lehmann S, Laude H (2001) Ex vivo propagation of infectious sheep scrapie agent in heterologous epithelial cells expressing ovine prion protein. Proc Natl Acad Sci U S A 98(7):4055–4059
588. Villa A, Mark AE, Saracino GAA, Cosentino U, Pitea D, Moro G, Salmona M (2006) Conformational polymorphism of the PrP 106–126 peptide in different environments: a molecular dynamics study. J Phys Chem B 110(3):1423–1428
589. Vitagliano L, Stanzione F, Simone AD, Esposito L (2009) Dynamics and stability of amyloid-like steric zipper assemblies with hydrophobic dry interfaces. Biopolymers 91(12):1161–1171
590. Vol'pina OM, Zhmak MN, Obozhaia MB, Titova MA, Koroev DO, Volkova TD, Egorov AA, Ivanov VT (2001) Antibodies against synthetic fragments of the prion protein for the diagnosis of bovine spongiform encephalopathy. Bioorg Khim 27(5):352–358
591. Vorberg I, Martin HG, Eberhard P, Suzette AP (2003) Multiple amino acid residues within the rabbit prion protein inhibit formation of its abnormal isoform. J Virol 77(3):2003–2009
592. Wade WF, Dees C, German TL, Marsh RF (1987) Immunochemical characterization of proteins from scrapie-infected hamster brain, using immunoblot analysis. Am J Vet Res 48(7):1077–1081
593. Wadsworth JD, Asante EA, Desbruslais M, Linehan JM, Joiner S, Gowland I, Welch J, Stone L, Lloyd SE, Hill AF, Brandner S, Collinge J (2004) Human prion protein with valine 129 prevents expression of variant CJD phenotype. Science 306(5702):1793–1796
594. Wagoner VA, Cheon M, Chang I, Hall CK (2011) Computer simulation study of amyloid fibril formation by palindromic sequences in prion peptides. Proteins Struct Funct Bioinf 79(7):2132–2145
595. Wagoner VA (2010) Computer simulation studies of self-assembly of fibril forming peptides with an intermediate resolution protein model. Ph.D. thesis, North Carolina State University, Raleigh
596. Wales DJ, Doye JPK (1997) Global optimization by Basin-Hopping and the lowest energy structures of Lennard-Jones clusters containing up to 110 atoms. J Phys Chem A 101(28):5111–5116

597. Walker R (2010) Amber tutorials: TUTORIAL B1: simulating a small fragment of DNA, section 5: Running Minimization and MD (in explicit solvent) http://ambermd.org/tutorials/basic/tutorial1/section5.htm. Access in May 2015
598. Wang L (2001) Intelligent optimization algorithms with applications. Tsinghua University Press & Springer: Beijing
599. Wang PP, Chen DS (1996) Continuous optimization by a variant of simulated annealing. Comput Optim Appl 6(1):59–71
600. Wang JF, Chou KC (2009) Insight into the molecular switch mechanism of human Rab5a from molecular dynamics simulations. Biochem Biophys Res Commun 390(3):608–612
601. Wang JF, Chou KC (2010) Insights from studying the mutation-induced allostery in the M2 proton channel by molecular dynamics. Protein Eng Des Sel 23(8):663–666
602. Wang JF, Gong K, Wei DQ, Li YX, Chou KC (2009) Molecular dynamics studies on the interactions of PTP1B with inhibitors: from the first phosphate binding site to the second one. Protein Eng Des Sel 22(6):349–355
603. Wang X, He L, Zhao C, Du W, Lin J (2013) Gold complexes inhibit the aggregation of prion neuropeptides. J Biol Inorg Chem 18(7):767–778
604. Wang YC, Lv J, Zhu L, Ma YM (2010) Crystal structure prediction via particle-swarm optimization. Phys Rev B 82(9):094116
605. Wang Y, Miao M, Lv J, Zhu L, Yin K, Liu H, Ma Y (2012) An effective structure prediction method for layered materials based on 2D particle swarm optimization algorithm. J Chem Phys 137(22):224108. doi:10.1063/1.4769731
606. Wang J, Pielak RM, McClintock MA, Chou JJ (2009) Solution structure and functional analysis of the influenza B proton channel. Nat Struct Mol Biol 16(12):1267–1271
607. Wang CY, Shi Z (1997) The research advances in facility location problem. Adv in Math 26:113–122
608. Wang JF, Wei DQ, Chen C, Li Y, Chou KC (2008) Molecular modeling of two CYP2C19 SNPs and its implications for personalized drug design. Protein Pept Lett 15(1):27–32
609. Wang JF, Wei DQ, Li L, Chou KC (2008) Review: drug candidates from traditional chinese medicines. Curr Top Med Chem 8(18):1656–1665
610. Wang JF, Wei DQ, Chou KC (2009) Insights from investigating the interactions of adamantane-based drugs with the M2 proton channel from the H1N1 swine virus. Biochem Biophys Res Commun 388(2):413–417
611. Wang JF, Wei DQ, Du HL, Li YX, Chou KC (2008) Molecular modeling studies on NADP-dependence of Candida tropicalis strain xylose reductase. Open Bioinf J 2:72–79
612. Wang JF, Wei DQ, Li L, Zheng SY, Li YX, Chou KC (2007) 3D structure modeling of cytochrome P450 2C19 and its implication for personalized drug design. Biochem Biophys Res Commun 355(2):513–519
613. Wang JF, Wei DQ, Lin Y, Wang YH, Du HL, Li YX, Chou KC (2007) Insights from modeling the 3D structure of NAD(P)H-dependent D-xylose reductase of Pichia stipitis and its binding interactions with NAD and NADP. Biochem Biophys Res Commun 359(2):323–329
614. Wang YJ, Xiu NH (2012) Nonlinear optimization theory and methods. Science Press, Beijing. ISBN:9787030330284, in Chinese
615. Wang JF, Yan JY, Wei DQ, Chou KC (2009) Binding of CYP2C9 with diverse drugs and its implications for metabolic mechanism. Med Chem 5(3):263–270
616. Wang H, Yang L, Kouadir M, Tan R, Wu W, Zou H, Wang J, Khan SH, Li D, Zhou X, Yin X, Wang Y, Zhao D (2013) Expression and distribution of laminin receptor precursor/laminin receptor in rabbit tissues. J Mol Neurosci 51(2):591–601
617. Wang F, Yin S, Wang X, Zha L, Sy MS, Ma J (2010) Role of the highly conserved middle region of prion protein (PrP) in PrP-lipid interaction. Biochemistry 49(37):8169–8176
618. Wang Y, Zhao S, Bai L, Fan J, Liu E (2013) Expression systems and species used for transgenic animal bioreactors. Biomed Res Int 2013:580463. doi:10.1155/2013/580463
619. Wegner C, Romer A, Schmalzbauer R, Lorenz H, Windl O, Kretzschmar HA (2002) Mutant prion protein acquires resistance to protease in mouse neuroblastoma cells. J Gen Virol 83(Pt 5):1237–1245

620. Wei DQ, Sirois S, Du QS, Arias HR, Chou KC (2005) Theoretical studies of Alzheimer's disease drug candidate [(2,4-dimethoxy) benzylidene]-anabaseine dihydrochloride (GTS-21) and its derivatives. Biochem Biophys Res Commun 338(2):1059–1064
621. Wei H, Wang CH, Du QS, Meng J, Chou KC (2009) Investigation into adamantane-based M2 inhibitors with FB-QSAR. Med Chem 5(4):305–317
622. Weinberger E (1990) Correlated and uncorrelated fitness landscapes and how to tell the difference. Biol Cybern 63(5):325–336
623. Weissmann C (1996) The ninth Datta lecture. Molecular biology of transmissible spongiform encephalopathies. FEBS Lett 389(1):3–11
624. Weissmann C (2004) The state of the prion. Nat Rev Microbiol 2(11):861–871
625. Wen Y, Li J, Xiong MQ, Peng Y, Yao WM, Hong J, Lin DH (2010) Solution structure and dynamics of the I214V mutant of the rabbit prion protein. PLoS ONE 5(10):e13273
626. Wen Y, Li J, Yao WM, Xiong MQ, Hong J, Peng Y, Xiao GF, Lin DH (2010) Unique structural characteristics of the rabbit prion protein. J Biol Chem 285(41):31682–31693
627. Weydt P, La Spada AR (2006) Targeting protein aggregation in neurodegeneration–lessons from polyglutamine disorders. Expert Opin Ther Targets 10(4):505–513
628. Wikipedia, the free encyclopedia (en.wikipedia.org/wiki/): Epidemic model, Compartmental models in epidemiology, Mathematical modelling of infectious disease, Markov chain Monte Carlo, Parallel tempering, Metropolis-Hastings algorithm, etc (and references therein)
629. Wiley CA, Burrola PG, Buchmeier MJ, Wooddell MK, Barry RA, Prusiner SB, Lampert PW (1987) Immuno-gold localization of prion filaments in scrapie-infected hamster brains. Lab Invest 57(6):646–656
630. Wille LT (1987) Minimum-energy configurations of atomic clusters: new results obtained by simulated annealing. Chem Phys Lett 133(5):405–410
631. Wiltzius JJ, Landau M, Nelson R, Sawaya MR, Apostol MI, Goldschmidt L, Soriaga AB, Cascio D, Rajashankar K, Eisenberg D (2009) Molecular mechanisms for protein-encoded inheritance. Nat Struct Mol Biol 16(9):973–978
632. Wolf MD, Landman U (1998) Genetic algorithms for structural cluster optimization. J Phys Chem A 102(30):6129–6137
633. Wolfe P (1969) Convergence conditions for ascent methods. SIAM Rev 11(2):226–235
634. Wolfe PH (1975) A method of conjugate subgradients of minimizing nondifferentiable convex functions. Math Program Study 3:145–173
635. Wolfe PH (1976) Finding the nearest point in polytope. Math Program Study 11(1):128–149
636. Wormell RL (1954) New fibres from proteins. Academic, New York, p 106
637. Xi YG, Cardone F, Pocchiari M (1994) Detection of proteinase-resistant protein (PrP) in small brain tissue samples from Creutzfeldt-Jakob disease patients. J Neurol Sci 124(2):171–173
638. Xiang Y, Cheng L, Cai W, Shao X (2004) Structural distribution of Lennard-Jones clusters containing 562 to 1000 atoms. J Phys Chem A 108(44):9516–9520
639. Xiang Y, Jiang H, Cai W, Shao X (2004) An efficient method based on lattice construction and the genetic algorithm for optimization of large Lennard-Jones clusters. J Phys Chem A 108(16):3586–3592
640. Xiao XL, Han J, Zhang L, Chen L, Zhang J, Chen XL, Zhou W, Jiang HY, Zhang BY, Liu Y, Dong XP (2006) Protein expression of human neuron-specific enolase and its antiserum preparation. Nan Fang Yi Ke Da Xue Xue Bao 26(11):1543–1547
641. Xiao X, Chou KC (2007) Digital coding of amino acids based on hydrophobic index. Protein Pept Lett 14(9):871–875
642. Xiao X, Lin WZ, Chou KC (2008) Using grey dynamic modeling and pseudo amino acid composition to predict protein structural classes. J Comput Chem 29(12):2018–2024
643. Xiao X, Shao SH, Chou KC (2006) A probability cellular automaton model for hepatitis B viral infections. Biochem Biophys Res Commun 342(2):605–610
644. Xiao X, Shao S, Ding Y, Huang Z, Chen X, Chou KC (2005) An application of gene comparative image for predicting the effect on replication ratio by HBV virus gene missense mutation. J Theor Biol 235(4):555–565

645. Xiao X, Shao S, Ding Y, Huang Z, Huang Y, Chou KC (2005) Using complexity measure factor to predict protein subcellular location. Amino Acids 28(1):57–61
646. Xiao X, Shao SH, Huang ZD, Chou KC (2006) Using pseudo amino acid composition to predict protein structural classes: approached with complexity measure factor. J Comput Chem 27(4):478–482
647. Xiao X, Wang P, Chou KC (2008) Predicting protein structural classes with pseudo amino acid composition: an approach using geometric moments of cellular automaton image. J Theory Biol 254(3):691–696
648. Xiao X, Wang P, Chou KC (2009) GPCR-CA: a cellular automaton image approach for predicting G-protein-coupled receptor functional classes. J Comput Chem 30(9): 1414–1423
649. Xiao YH, Wei ZX, Wang ZG (2008) A limited memory BFGS-type method for large-scale unconstrained optimization. J Comput Math Appl 56(4):1001–1009
650. Xin B, Chen J, Pan F (2009) Problem difficulty analysis for particle swarm optimization: deception and modality. In: 2009 proceedings of the 1st ACM/SIGEVO summit on genetic and evolutionary computation, Shanghai, pp 623–630. ISBN:978-1-60558-326-6, doi:10.1145/1543834.1543919
651. Xu H, He X, Zheng H, Huang LJ, Hou F, Yu Z, de la Cruz MJ, Borkowski B, Zhang X, Chen ZJ, Jiang QX (2014) Structural basis for the prion-like MAVS filaments in antiviral innate immunity. Elife 3:e01489
652. Xue GL (1993) Parallel two-level simulated annealing. Proceedings of the 7th international conference on Supercomputing, pp 357–366. ISBN:0-89791-600-X, doi:10.1145/165939.166011
653. Xue GL (1994) Improvement on the Northby algorithm for molecular conformation: better solutions. J Glob Optim 4(4):425–440
654. Xue GL (1994) Molecular conformation on the CM-5 by parallel two-level simulated annealing. J Glob Optim 4(2):187–208
655. Xue GL (2002) Optimization in computational molecular biology. In: Pardalos PM, Reende MGC (eds) Handbook of applied optimization. Oxford University Press, Oxford. Chapter 18.13, pp 885–900
656. Xue GL, Maier RS, Rosen JB (1992) Minimizing the Lennard-Jones potential function on a massively parallel computer. Proceedings of the 6th international conference on Supercomputing, pp 409–416. ISBN:0-89791-485-6, doi:10.1145/143369.143443
657. Yamaguchi K, Matsumoto T, Kuwata K (2008) Critical region for amyloid fibril formation of mouse prion protein: unusual amyloidogenic properties of the helix 2 peptide. Biochem 47(50):13242–13251
658. Yan X, Huang JJ, Zhou Z, Chen J, Liang Y (2014) How does domain replacement affect fibril formation of the rabbit/human prion proteins. PLoS ONE 9(11):e113238
659. Yang SC, Levine H, Onuchic JN, Cox DL (2005) Structure of infectious prions: stabilization by domain swapping. FASEB J 19(13):1778–1782
660. Yang YT, Xu CX (2007) A compact limited memory method for large scale unconstrained optimization. Eur J Oper Res 180(1):48–56
661. Ye T, Xu R, Huang W (2011) Global optimization of binary Lennard-Jones clusters using three perturbation operators. J Chem Inf Model 51(3):572–577
662. Yeung L (2011) Theoretical studies of apolipoprotein interactions. Master of Applied Physics thesis, RMIT University
663. Yokoyama T, Itohara S, Yuasa N (1996) Detection of species specific epitopes of mouse and hamster prion proteins (PrPs) by anti-peptide antibodies. Arch Virol 141(3–4):763–769
664. Yokoyama T, Kimura K, Tagawa Y, Yuasa N (1995) Preparation and characterization of antibodies against mouse prion protein (PrP) peptides. Clin Diagn Lab Immunol 2(2): 172–176
665. Young MM, Skillman AG, Kuntz ID (1999) A rapid method for exploring the protein structure universe. Proteins 34(3):317–332

666. Yost CS, Lopez CD, Prusiner SB, Myers RM, Lingappa VR (1990) Non-hydrophobic extracytoplasmic determinant of stop transfer in the prion protein. Nature 343(6259): 669–672
667. Yu HF, Wang DW (2008) Research on food-chain algorithm and its parameters. Front Electr Electron Eng China 3(4):394–398
668. Yuan YX (2000) A review of trust region algorithms for optimization. In: ICIAM 99: proceedings of the fourth international congress on industrial & applied mathematics, Edinburgh. Oxford University Press, Oxford, pp 271–282. doi:10.1.1.45.9964
669. Yuan FF, Biffin S, Brazier MW, Suzrez M, Cappai R, Hill AF, Collins SJ, Sullivan JS, Middleton D, Multhaup G, Geczy AF, Masters CL (2005) Detection of prion epitopes on PrP^C and PrP^{Sc} of transmissible spongiform encephalopathies using specific monoclonal antibodies to PrP. Immunol Cell Biol 83(6):632–637
670. Yuan YX, Sun WY (1997) Optimization theory and methods. Chinese Science Press: Beijing. Chapter 13, pp 559–598
671. Yuan GL, Wei ZX, Wu YL (2010) Modified limited memory BFGS method with nonmonotone line search for unconstrained optimization. J Korean Math Soc 47(4):767–788
672. Zahn R, Guntert P, von Schroetter C, Wüthrich K (2003) NMR structure of a variant human prion protein with two disulfide bridges. J Mol Biol 326(1):225–234
673. Zahn R, Liu AZ, Luhrs T, Riek R, Schroetter CV, Garcia FL, Billeter M, Calzolai L, Wider G, Wuthrich K (2000) NMR solution structure of the human prion protein. Proc Natl Acad Sci U S A 97(1):145–150
674. Zanuy D, Ma B, Nussinov R (2003) Short peptide amyloid organization: stabilities and conformations of the islet amyloid peptide NFGAIL. Biophys J 84(3):1884–1894
675. Zeng YH, Guo YZ, Xiao RQ, Yang L, Yu LZ, Li ML (2009) Using the augmented Chou's pseudo amino acid composition for predicting protein submitochondria locations based on auto covariance approach. J Theory Biol 259(2):366–372
676. Zhang JP (2003) A brief review on results and computational algorithms for minimizing the Lennard-Jones potential. arXiv.org > physics > arXiv:1101.0039v1: arxiv.org/PS_cache/arxiv/pdf/1101/1101.0039v1.pdf or adsabs.harvard.edu/abs/2011arXiv1101.0039Z
677. Zhang JP (2004) Derivative-free hybrid methods in global optimization and their applications to data mining, protein folding, unfolding, misfolding. The University of Ballarat, Doctoral Dissertation, Ballarat
678. Zhang JP (2009) Studies on the structural stability of rabbit prion probed by molecular dynamics simulations. J Biomol Struct Dyn 27(2):159–162
679. Zhang JP (2010) Studies on the structural stability of rabbit prion protein probed by molecular dynamics simulations of its wild-type and mutants. J Theory Biol 264(1):119–122
680. Zhang JP (2011) Optimal molecular structures of prion AGAAAAGA amyloid fibrils formatted by simulated annealing. J Mol Model 17(1):173–179
681. Zhang JP (2011) Comparison studies of the structural stability of rabbit prion protein with human and mouse prion proteins. J Theory Biol 269(1):88–95
682. Zhang JP (2011) The structural stability of wild-type horse prion protein. J Biomol Struct Dyn 29(2):369–377
683. Zhang JP (2011) Molecular dynamics – practical application – mechanism underlying the resistance to prion diseases in rabbits. LAP Lambert Academic Publishing: Saarbrücken. ISBN:978-3-8465-4843-1
684. Zhang JP (2011) Atomic-resolution structures of prion AGAAAAGA amyloid fibrils. In: Halcheck IP, Nancy R (eds) Vernon amyloids: composition, functions, and pathology. Nova Science Publishers, Hauppauge. ISBN:9781621005384 (hardcover), Chapter 10: 177–186
685. Zhang JP (2011) Practical global optimization computing methods in molecular modelling – for atom-resolution structures of amyloid fibrils, 1st Version. LAP Lambert Academic Publishing: Saarbrücken. ISBN:978-3-8465-2139-7

686. Zhang JP (2011) Derivative-free hybrid methods in global optimization and applications – in December 2010, 1st Version. LAP Lambert Academic Publishing: Saarbrücken. ISBN:978-3-8454-3580-0, Chapter 4.2: 92–95
687. Zhang JP (2011) The Lennard-Jones potential minimization problem for prion AGAAAAGA amyloid fibril molecular modeling. arXiv:1106.1584
688. Zhang JP (2011) An effective simulated annealing refined replica exchange Markov chain Monte Carlo algorithm for the infectious disease model of H1N1 influenza pandemic. World J Model Simul 7(1):29–39
689. Zhang JP (2012) Molecular dynamics studies on the structural stability of wild-type rabbit prion protein: surface electrostatic charge distributions. In: Battik C, Belhassine K (eds) Bioinformatics research: new developments. NOVA Science Publishers, New York, Chapter 7, pp 131–8. ISBN:978-1-61942-363-3
690. Zhang JP (2012) Computational potential energy minimization studies on the prion AGAAAAGA amyloid fibril molecular structures. In: Bendict JB (ed) Recent advances in crystallography. InTech Open Access Publisher: Rijeka, 19 Sept 2012, Chapter 12, pp 297–312. ISBN:978-953-51-0754-5, doi:10.5772/47733
691. Zhang JP (2014) Simulated annealing: in mathematical global optimization computation, hybrid with local or global search, and practical applications in crystallography and molecular modelling of prion amyloid fibrils. In: Tsuzuki M, Martins TDC (eds) Simulated annealing: strategies, potential uses and advantages. NOVA Science Publishers, Hauppauge, Chapter 1, pp 1–34. ISBN:978-1-63117-268-7
692. Zhang JP (2015) The hybrid idea of optimization methods applied to the energy minimization of (prion) protein structures. In: Proceeding of the workshop on continuous optimization: theory, methods and applications, Ballarat, Apr 2015, pp 16–17
693. Zhang JP, Hou YT, Wang YJ, Wang CY, Zhang XS (2012) The LBFGS quasi-Newtonian method for molecular modeling prion AGAAAAGA amyloid fibrils. Nat Sci 4(12A):1097–1108
694. Zhang ZQ, Chen H, Bai HJ, Lai LH (2007) Molecular dynamics simulations on the oligomer formation process of the GNNQQNY peptide from yeast prion protein Sup35. Biophys J 93(5):1484–1492
695. Zhang ZQ, Chen H, Lai LH (2007) Identification of amyloid fibril-forming segments based on structure and residue-based statistical potential. Bioinf 23(17):2218–2225
696. Zhang JP, Gao DG, Yearwood J (2011) A novel canonical dual computational approach for prion AGAAAAGA amyloid fibril molecular modeling. J Theory Biol 284(1):149–157
697. Zhang JP, Liu DDW (2011) Molecular dynamics studies on the structural stability of wild-type dog prion protein. J Biomol Struct Dyn 28(6):861–869
698. Zhang JP, Sun J, Wu CZ (2011) Optimal atomic-resolution structures of prion AGAAAAGA amyloid fibrils. J Theory Biol 279(1):17–28
699. Zhang Y, Swietnicki W, Zagorski MG, Surewicz WK, Soennichsen FD (2000) Solution structure of the E200K variant of human prion protein. Implications for the mechanism of pathogenesis in familial prion diseases. J Biol Chem 275(43):33650–33654
700. Zhang JP, Varghese JN, Epa VC (2006) Studies on the conformational stability of the rabbit prion protein. CSIRO Preventative Health National Research Flagship Science Retreat, Aitken Hill, Melbourne, 12–15 Sept 2006 Poster in Excellence
701. Zhang JP, Wang F, Chatterjee S (2015) Molecular dynamics studies on buffalo prion protein. J Biomol Struct Dyn. doi:10.1080/07391102.2015.1052849
702. Zhang JP, Wang F, Zhang YL (2015) Molecular dynamics studies on the NMR structures of rabbit prion protein wild-type and mutants: surface electrostatic charge distributions. J Biomol Struct Dyn 33(6):1326–1335
703. Zhang JP, Zhang YL (2013) Molecular dynamics studies on 3D structures of the hydrophobic region PrP(109-136). Acta Biochimicaet Biophysica Sinica 45(6):509–519
704. Zhang JP, Zhang YL (2014) Molecular dynamics studies on the NMR and X-ray structures of rabbit prion proteins. J Theory Biol 342:70–82

705. Zhang X, Zhou X, Ding T, Gan W, Yang L, Yin X, Zhao D (2012) Polymorphisms of SPRN (shadow of prion protein homology) in three breeds of sheep in China. Virus Genes 44(3):548–550
706. Zhao DM (2012) Transmissible spongiform encephalopathies (in Chinese). China Agricultural University Press: Beijing. ISBN:978-7-5655-0428-0
707. Zhao X, Dong X, Zhou W (2000) Preparation of polyclonal antibody to human prion protein using the expressed GST-PrP fusion protein as antigen. Zhonghua Shi Yan He Lin Chuang Bing Du Xue Za Zhi 14(2):131–133
708. Zhao H, Liu LL, Du SH, Wang SQ, Zhang YP (2012) Comparative analysis of the Shadoo gene between cattle and buffalo reveals significant differences. PLoS ONE 7(10):e46601
709. Zhao ZS, Xu B, Wang LM, Zhou XF, He JL, Liu ZY, Wang HT, Tian YJ (2011) Three dimensional carbon-nanotube polymers. ACS Nano 5(9):7226–7234
710. Zheng J, Ma BY, Tsai CJ, Nussinov R (2006) Structural stability and dynamics of an amyloid-forming peptide GNNQQNY from the yeast prion Sup-35. Biophy J 91(3):824–833
711. Zheng W, Wang L, Hong Y, Sha Y (2009) PrP 106–126 peptide disrupts lipid membranes influence of C-terminal amidation. Biochem Biophys Res Commun 379(2):298–303
712. Zhong LH (2010) Exposure of hydrophobic core in human prion protein pathogenic mutant H187R. J Biomol Struct Dyn 28(3):355–361
713. Zhong L, Xie J (2009) Investigation of the effect of glycosylation on human prion protein by molecular dynamics. J Biomol Struct Dyn 26(5):525–533
714. Zhou XB, Chen C, Li ZC, Zou XY (2007) Using Chou's amphiphilic pseudo-amino acid composition and support vector machine for prediction of enzyme subfamily classes. J Theory Biol 248(3):546–551
715. Zhou GP, Deng MH (1984) An extension of Chou's graphical rules for deriving enzyme kinetic equations to system involving parallel reaction pathways. Biochem J 222(1):169–176
716. Zhou GP, Huang RB (2013) The pH-triggered conversion of the PrP^C to PrP^{Sc}. Curr Top Med Chem 13(10):1152–1163
717. Zhou Z, Yan X, Pan K, Chen J, Xie ZS, Xiao GF, Yang FQ, Liang Y (2011) Fibril formation of the rabbit/human/bovine prion proteins. Biophys J 101(6):1483–1492
718. Zhu C, Byrd RH, Nocedal J (1997) L-BFGS-B: Algorithm 778: L-BFGS-B, FORTRAN routines for large scale bound constrained optimization. ACM Trans Math Softw 23(4):550–560. Morales JL, Nocedal J (2011) Remark on "algorithm 778: L-BFGS-B: Fortran subroutines for large-scale bound constrained optimization". J ACM Trans Math Softw 38(1). doi:10.1145/2049662.2049669
719. Zhu L, Wang H, Wang YC, Lv J, Ma YM, Cui QL, Ma YM, Zou GT (2011) Substitutional alloy of Bi and Te at high pressure. Phys Rev Lett 106(14):145501–145504
720. Zocche SH, Canes SL, Debbas V, Martins LFR (2011) Cellular prion protein (PrP^C) and superoxide dismutase (SOD) in vascular cells under oxidative stress. Exp Toxicol Pathol 63(3):229–236
721. Zou ZH, Bird RH, Schnabel RB (1997) A stochastic/perturbation global optimization algorithm for distance geometry problems. J Glob Optim 11(1):91–105
722. Zoutendijk G (1970) Nonlinear programming, computational methods. In: Abadie J (ed) Integer and nonlinear programming. North-Holland, Amsterdam, pp 37–86
723. Zweckstetter M (2013) Conserved amyloid core structure of stop mutants of the human prion protein. Prion 7(3):193–197

Index

J. Zhang, *Molecular Structures and Structural Dynamics of Prion Proteins and Prions*, Focus on Structural Biology 9, DOI 10.1007/978-94-017-7318-8

MIX
Papier aus verantwortungsvollen Quellen
Paper from responsible sources
FSC® C105338

If you have any concerns about our products,
you can contact us on
ProductSafety@springernature.com

In case Publisher is established outside the EU,
the EU authorized representative is:
Springer Nature Customer Service Center GmbH
Europaplatz 3, 69115 Heidelberg, Germany

Printed by Libri Plureos GmbH
in Hamburg, Germany